MUM.
July 7/8

IGNEOUS PETROLOGY

Series

Developments in Petrology

Developments in Petrology 7

IGNEOUS PETROLOGY

CHARLES J. HUGHES

Department of Geology, Memorial University of Newfoundland, St. John's, Newfoundland, Canada

ELSEVIER SCIENTIFIC PUBLISHING COMPANY
AMSTERDAM — OXFORD — NEW YORK   1982

ELSEVIER SCIENTIFIC PUBLISHING COMPANY
Molenwerf 1
P.O. Box 211, 1000 AE Amsterdam, The Netherlands

Distributors for the United States and Canada:

ELSEVIER SCIENCE PUBLISHING COMPANY INC.
52, Vanderbilt Avenue
New York, N.Y. 10017

(xvi + 551 pp., with 79 figs.; 56 tables; 752 references)

ISBN 0-444-42011-8 (Vol. 7)
ISBN 0-444-41562-9 (Series)

Printed in The Netherlands

PREFACE

For many years there has been a noticeable lack of senior undergraduate texts in igneous petrology that might bridge a gap between, on the one hand, several excellent introductory petrography-orientated texts and, on the other hand, more advanced texts heavily predicated on a thermodynamic and/or chemical approach to the subject. Furthermore, in the last decade and a half, say since the appearance of the prescient synthesis of Arthur Holmes' *Principles of Physical Geology*, second edition, in 1965, we have witnessed a profound re-orientation of igneous petrology in relation to global tectonics. A giant step has been taken in clothing R.A. Daly's historic comment,

"A final philosophy of earth history must be largely founded upon the unshakable facts known about igneous rocks"

with evidence and conclusions from both young and old igneous rocks. Ophiolites, Proterozoic igneous rocks, and a few other conspicuous growth areas apart, we now seem to be entering a relatively quiet period of amassing and interpreting a considerable amount of increasingly sophisticated analytical data about igneous rocks, coupled however with a slackening in the rate of new major advances in synthesis and understanding. Thus at this particular time it seems appropriate to attempt to integrate the structure of recent conceptual advances with the firm foundation provided by classical petrology.

The approach in this book is hence predicated initially on field and other direct observations and on an historical background. This approach leads naturally from observational facts about extrusive and intrusive rocks and a classical mineralogical classification, via differentiation processes married to experimental work, to the evolution of ideas on igneous rock series and the recent explosion of the relationship of these to plate tectonics and inferred mantle processes, and to an indication of some current problematical areas. Students do seem to benefit from an appreciation of the historical path of discovery and evolution of ideas about igneous rocks in order to make intelligible the present body of knowledge and thrust of contemporary research.

I have assumed that a reader of this book will have had some elementary training in the physical sciences and some first courses in geology including introductory petrology, and ideally has access to a good thin-section collection of igneous rock suites and the opportunity to read up or attend parallel courses in relevant aspects of geochemistry and isotope geology.

This approach is well illustrated by the choice of references — a perennial and important problem in texts. Many classic and benchmark papers are included, together with a large number of up-to-date review articles and books, and a selection of more specific papers that the writer personally has found interesting and informative, particularly those which may express contrary or modifying opinions to contemporary conventional wisdom. In so far as this book may fail to achieve a compromise between the desirable attributes of readability, brevity, and information source, the reader may be provoked into reading some of these references. As Gilbert Highet in his *The Art of Teaching* so appositely recommends, there is no comparison between the benefit to be derived from looking up work in the original and that from tamely accepting someone else's commentary.

My advice therefore to readers of this book is to use it as I will — with marginal notes and additions, queries, expressions of dissent, etc., as not only is any book imperfect but also our subject is continually evolving. Any suggestions and rectifications will be gratefully received and acknowledged.

I should like to take this opportunity to record a great debt to teachers and comrades over the years. Foremost among these is my former teacher and supervisor, Professor L.R. Wager. Fashions may change in igneous petrology as in other things, but a man's stature does not. Lawrence Wager, mountaineer and field geologist, perceptive petrographer and innovative thinker, possessed the true scientific knack of being able to apply himself assiduously to important problems capable of a solution at the time. A major part of his legacy to igneous petrology is a now dispersed group of igneous petrologists that received their training under him at Oxford and in the Hebrides and Greenland, and elsewhere. I also recall with great pleasure the camaraderie of geologists over the years in Zambia, Zimbabwe, Union of South Africa, Ghana, Egypt, Morocco, Spain, France, Federal Republic of Germany, Switzerland, Scandinavia, Eire, United Kingdom, U.S.A., Canada, Iceland, India, Nepal, Malaysia, Australia, New Zealand and Japan, and especially the stimulus afforded by undergraduate and graduate students and colleagues at Memorial University of Newfoundland over the last fourteen years.

Among colleagues at MUN I should especially like to thank in the context of this work Glenys Woodland for the major part of the typing with able assistance from Lillian Murphy, Cynthia Neary and Betty Andrews; Clifford Wood, Gary McManus and colleagues for drafting services and advice; and Jeanne Mills, Adele Poynter and George Einarson for help with references. I owe a particular debt to my friend and colleague, Dr. John G. Malpas, for his pains in cheerfully reading the text and making numerous suggestions for improvement despite other heavy claims on his time.

CHARLES J. HUGHES
St. John's, Newfoundland
January 1981

GLOSSARY

Units used in this book and their conversions to SI units:

Concentration	ppb	(parts per billion)		
	ppm	(parts per million)		
	vol.%	(per cent by volume)		
	wt.%	(per cent by weight)		
Energy	cal.	(calorie)	4.184 J	(joule)
Length	Å	(ångström)	10^{-10} m	(meter)
	fathom	(nautical fathom)	1.8288 m	
Mass	t	(metric ton, tonne)	10^6 g	(gram)
Pressure	atm.	(atmosphere)	$1.013 \cdot 10^5$ Pa	(pascal)
	bar		10^5 Pa	
Temperature	°C	(degree Celsius)	K	(kelvin)
Time	day		86,400 s	(second)
	a	(year)	$3.16 \cdot 10^7$ s	
Viscosity	P	(poise)	10^{-5} N m^{-2} s^{-1}	

S.I. unit prefixes:

μ (micro)	10^{-6}		k (kilo)	10^3
m (milli)	10^{-3}		M (mega)	10^6
c (centi)	10^{-2}		G (giga)	10^9

Abbreviations, acronyms, and symbols used in text:

AFM variation diagram	alkalis: iron oxide: magnesia
An	proportion by weight of anorthite, the higher-temperature component, in the solid-solution series anorthite—albite; for example, An$_{81}$ signifies a composition of 81 wt.% anorthite
BABI	basaltic achondrite best initial
CFSE	crystal field stabilization energy
CI	crystallization index
C.I.	colour index
CMAS system	$CaO-MgO-Al_2O_3-SiO_2$
DI	differentiation index
DLVL	depleted low-velocity layer
E	explosion index

En	proportion by weight of enstatite in the solid-solution series enstatite—ferrosilite (see note above on An)
FAMOUS	French American Mid-Ocean Undersea Study
FI	fractionation index

Abbreviations, acronyms, and symbols used in text (continued):

FMA variation diagram	*see AFM* variation diagram
Fo	proportion by weight of forsterite in the solid-solution series forsterite—fayalite (see note above on An)
h	vertical distance in kilometres to Benioff seismic zone
HREE	heavy rare-earth elements
IAT	island-arc tholeiite
K_{55}-value	wt.% K_2O in a series at a silica content of 55 wt.%
LIL	large-ion lithospheric
LKT	low-potassium tholeiite
LREE	light rare-earth elements
LVL	low-velocity layer
M-value	$100Mg/(Mg + total Fe)$
Mg ratio	$100Mg/(Mg + Fe^{2+})$
MORB	mid-ocean ridge basalt
OIT	oceanic-island tholeiite
PHMP	primordial hot mantle plume
Q	dimensionless quantity expressing seismic measure of anelasticity within the mantle
REE	rare-earth elements
SBZ	supra-Benioff seismic zone
SI	solidification index
$(^{87}Sr/^{86}Sr)^0$	initial $^{87}Sr/^{86}Sr$ ratio

See also general index for references to further explanation and usage in context of the above.

Frequently used abbreviations of normative mineral molecules for CIPW classification:

ab = albite	fo = forsterite	ne = nepheline
ac = acmite	fs = ferrosilite	ns = sodium metasilicate
an = anorthite	hm = hematite	ol = olivine
C = corundum	hy = hypersthene	or = orthoclase
di = diopside	il = ilmenite	Q = quartz
en = enstatite	lc = leucite	wo = wollastonite
fa = fayalite	mt = magnetite	

Chemical symbols and elements:

Ac	actinium	B	boron	Cd	cadmium
Ag	silver	Ba	barium	Ce	cerium
Al	aluminum	Be	beryllium	Cl	chlorine
Ar	argon	Bi	bismuth	Co	cobalt
As	arsenic	Br	bromine	Cr	chromium
At	astatine	C	carbon	Cs	caesium
Au	gold	Ca	calcium	Cu	copper

Chemical symbols and elements (continued):

Dy	dysprosium	Mo	molybdenum	Sb	antimony
Er	erbium	N	nitrogen	Sc	scandium
Eu	europium	Na	sodium	Se	selenium
F	fluorine	Nb	niobium	Si	silicon
Fe	iron	Nd	neodymium	Sm	samarium
Fr	francium	Ne	neon	Sn	tin
Ga	gallium	Ni	nickel	Sr	strontium
Gd	gadolinium	O	oxygen	Ta	tantalum
Ge	germanium	Os	osmium	Tb	terbium
H	hydrogen	P	phosphorus	Tc	technetium
He	helium	Pa	protactinium	Te	tellurium
Hf	hafnium	Pb	lead	Th	thorium
Hg	mercury	Pd	palladium	Ti	titanium
Ho	holmium	Pm	promethium	Tl	thallium
I	iodine	Po	polonium	Tm	thulium
In	indium	Pr	praseodymium	U	uranium
Ir	iridium	Pt	platinum	V	vanadium
K	potassium	Ra	radium	W	tungsten
Kr	krypton	Rb	rubidium	Xe	xenon
La	lanthanum	Re	rhenium	Y	yttrium
Li	lithium	Rh	rhodium	Yb	ytterbium
Lu	lutetium	Rn	radon	Zn	zinc
Mg	magnesium	Ru	ruthenium	Zr	zirconium
Mn	manganese	S	sulphur		

CONTENTS

MINERALOGY OF IGNEOUS ROCKS

1.1. INTRODUCTION

Igneous rocks, apart from those few which are wholly or partly glassy, are composed of minerals and a knowledge of these minerals therefore is one logical starting point for study. Indeed the microscopic determination in thin sections of the content, abundance, and textural relations of component minerals remains, as it was for classical petrographers, the most widely used tool in the investigation of igneous rocks.

A **mineral** is a naturally-formed chemical compound having a definite chemical composition and crystalline structure reflecting an ordered arrangement of constituent atoms. The words "definite chemical composition" need qualification: a certain amount of **crystalline solution** (alternatively referred to as **solid solution** or **diadochy**) is possibly whereby atoms, essentially of similar sizes, may substitute to a varying extent for each other within certain crystal lattices. This may at first sight appear to the student to be an unwelcome complication. The possibilities, however, for diadochy between major elements are few in number, and the resultant mineral compositions vary in a systematic manner that can reveal a great deal about magma compositions, temperatures of crystallization, fractional crystallization processes and the like (see Chapter 6), and thus prove to be a most useful tool in our understanding of the genesis of igneous rocks. The characteristic crystalline structures of the different minerals (with some degree of variation resulting from this crystalline solution) are reflected in differing optical properties which readily facilitate identification in thin section.

The actual number of distinct mineral species found in any one igneous rock is small, commonly no more than about half a dozen. This in part reflects the small number of chemical elements that are at all abundant in igneous rock compositions, the possibility of substitution between some of these elements within crystal lattices, and the possibility of further substitution such that most remaining elements present in small concentrations in a magma accommodate themselves in the crystal lattices of the common minerals as dispersed elements rather than form distinct mineral species. This observed small number of minerals in any igneous rock also reflects more fundamentally the operation of the phase rule and some close approach to equilibrium conditions of crystallization, thus explaining the systematic way in which magmas of similar compositions have crystallized to give similar proportions of the same small set of minerals.

In addition to this restraint, the range of magma compositions has well-defined limits, and the *total* number of mineral species found in all but rare and compositionally exceptional igneous rocks such as carbonatite is thus also very limited. The olivines, pyroxenes, amphiboles, micas, feldspars, feldspathoids and quartz, plus a few oxide species in fact make up ~99% of the overall content of igneous rocks. The feldspars alone comprise ~50%, and pyroxenes and quartz a further 25%, of igneous rocks in general (although not necessarily, of course, of each individual igneous rock).

1.2. CHEMICAL CONSIDERATIONS

1.2.1. The chemical elements

Just as igneous rocks are composed of minerals, so minerals are composed of chemical elements. The periodic table of elements will be familiar to most readers of this book; it is reproduced here (Table 1.1) for reference purposes, and includes the atomic numbers and atomic weights of the first 92 elements including all 90 known to exist naturally on Earth*. The horizontal rows are the seven periods corresponding to the number of electron shells surrounding an atomic nucleus. The major vertical columns contain groups of elements that are related by a similar number and configuration of electrons in an outer shell of up to eight electrons, and hence have comparable chemical properties — properties that commonly show progressive shifts within a group related to the differing atomic weights of its members. Well-recognized groups among the commoner elements include the alkali elements (Li, Na, K, Rb, Cs), the alkaline earth elements (Ca, Sr, Ba), the halogens (F, Cl, Br, I), and the inert or noble gases (He, Ne, Ar, Kr, Xe).

Note that a systematic progression when tabulated in order of increasing atomic number is interrupted in periods 4 to 7 by sets of elements occurring immediately after group IIA. Following the simple Bohr model of the atom, with progressively higher atomic numbers increasing by integral steps, electrons have to be added one by one to the envelope of electron shells to balance the increase in the number of protons in the nucleus. These electrons are added to the outer shell (the one whose electron configuration has a decisive influence on chemical properties) in atoms up to calcium, atomic number 20. After calcium and succeeding group-IIA elements, however, a certain number of electrons are added progressively to inner electron shells,

---

*Technetium, atomic number 43, has no stable nuclides and does not occur naturally on Earth; its presence in certain stars has, however, been inferred from examination of their spectra indicating that thermonuclear synthesis was proceeding there at or just before the time that their light now reaching us was emitted. Promethium, atomic number 61, similarly has not been detected on Earth.

TABLE 1.1

The periodic table — comprising the 90 elements known to occur naturally on Earth

Group / Period	IA	IIA	Transition elements	IIIA	IVA	VA	VIA	VIIIA	Inert gases
1	1 Hydrogen 1.008								2 He Helium 4.003
2	3 Li Lithium 6.939	4 Be Beryllium 9.012		5 B Boron 10.81	6 C Carbon 12.01	7 N Nitrogen 14.01	8 O Oxygen 16.000	9 F Fluorine 19.00	10 Ne Neon 20.18
3	11 Na Sodium 22.99	12 Mg Magnesium 24.31		13 Al Aluminum 26.98	14 Si Silicon 28.09	15 P Phosphorus 30.97	16 S Sulphur 32.06	17 Cl Chlorine 35.45	18 Ar Argon 39.95
4	19 K Potassium 39.10	20 Ca Calcium 40.08	21–30	31 Ga Gallium 69.72	32 Ge Germanium 72.59	33 As Arsenic 74.92	34 Se Selenium 78.96	35 Br Bromine 79.91	36 Kr Krypton 83.30
5	37 Rb Rubidium 85.47	38 Sr Strontium 87.62	39–48	49 In Indium 114.8	50 Sn Tin 118.7	51 Sb Antimony 121.8	52 Te Tellurium 127.6	53 I Iodine 126.9	54 Xe Xenon 131.3
6	55 Cs Caesium 132.9	56 Ba Barium 137.3	57–80	81 Tl Thallium 204.4	82 Pb Lead 207.2	83 Bi Bismuth 209.0	84 Po Polonium (210)	85 At Astatine (210)	86 Rn Radon (222)
7	87 Fr Francium (223)	88 Ra Radium (226)	89–103						

TABLE 1.1 (*continued*)

Transition elements:

Group / Period	IIIB	IVB	VB	VIB	VIIB	VIII			IB	IIB
4	21 Sc Scandium 44.96	22 Ti Titanium 47.90	23 V Vanadium 50.94	24 Cr Chromium 52.00	25 Mn Manganese 54.94	26 Fe Iron 55.85	27 Co Cobalt 58.93	28 Ni Nickel 58.71	29 Cu Copper 63.54	30 Zn Zinc 65.37
5	39 Y Yttrium 88.91	40 Zr Zirconium 91.22	41 Nb Niobium 92.91	42 Mo Molybdenum 95.94	43 [Tc] Technetium (99)	44 Ru Ruthenium 101.1	45 Rh Rhodium 102.9	46 Pd Palladium 106.4	47 Ag Silver 107.9	48 Cd Cadmium 112.4
6	57—71 Lanthanides	72 Hf Hafnium 178.5	73 Ta Tantalum 180.9	74 W Tungsten 183.9	75 Re Rhenium 186.2	76 Os Osmium 190.2	77 Ir Iridium 192.2	78 Pt Platinum 195.1	79 Au Gold 197.0	80 Hg Mercury 200.6
7	89—103 Actinides									

Rare-earth elements (REE) (syn. "inner transition elements"):

Lanthanides
(syn. first series of inner transition elements):

57	La	Lanthanum	138.9
58	Ce	Cerium	140.1
59	Pr	Praseodymium	140.9
60	Nd	Neodymium	144.2
61	[Pm]	Promethium	(147)
62	Sm	Samarium	150.4
63	Eu	Europium	152.0
64	Gd	Gadolinium	157.3
65	Tb	Terbium	158.9
66	Dy	Dysprosium	162.5
67	Ho	Holmium	164.9
68	Er	Erbium	167.3
69	Tm	Thulium	168.9
70	Yb	Ytterbium	173.0
71	Lu	Lutetium	175.0

Actinides
(syn. second series of inner transitional elements):

89	Ac	Actinium	(227)
90	Th	Thorium	232.0
91	Pa	Protactinium	(231)
92	U	Uranium	238.0

Element symbols in square brackets indicate elements not known on Earth; atomic weights in parentheses indicate the atomic weight of the most stable isotope of a radioactive element.

before a resumption of the pattern of addition of electrons to the outer shell.

In elements *21* to *30* in period *4*, and again in elements *39* to *48* in period *5*, electrons are progressively added essentially to the electron shell just inside the outermost thus increasing its content of electrons from eight to eighteen, resulting in two series of ten transition elements. These electrons in the next to outermost shell interact in a complex manner with the two electrons (or, in some elements, one electron) in the outer shell, resulting in a less systematic progression of valency states (and often multiple valencies) within these transition elements. The presence of any of these transition elements either as an essential component or as a dispersed element in a crystal lattice results in the phenomena of colour and pleochroism, explicable in terms of crystal field theory.

A third series of 24 transition elements, *57* to *80*, occurs in period *6*. This larger number of transition elements is caused by the progressive infilling in elements *58* to *71* of electrons into an inner shell, in this case the fourth shell, and *not* the next outermost shell which would be the fifth shell, thus further increasing the content of electrons in the fourth shell from eighteen to thirty-two. Changes in the electron configuration of a relatively deeply buried shell such as this have very little effect on chemical properties. This group of 14 elements plus its forerunner lanthanum, that thus differ only in the number of extra electrons in the fourth shell are known as the **rare-earth elements** (REE), or alternatively as the lanthanides, or more rarely as the first series of inner transition elements. The lanthanides have closely similar chemical properties; however, one feature of significance is a steady contraction in ionic radius (the lanthanide contraction) from the light (large) REE of low atomic number to the heavy (small) REE of high atomic number within the lanthanide series of elements, a phenomenon that can in fact be predicted by computation of the attractive effect of the positive charge of the nucleus on their electron configurations. The remaining transition elements, *72* to *81*, of period *6* are characterized by the completion of the progressive infilling of the next outermost shell of electrons to its full complement of 18 electrons in an analogous manner to that of the transition elements in periods *4* and *5*.

The **actinides**, the second series of inner transition elements, have atomic numbers *89* to *103* inclusive, with electron configurations similarly produced by the infilling of an inner shell, in this case the fifth and again not the next outermost. Although all of the actinides have been produced synthetically, all of these large actinide nuclides are radioactive. Only two, uranium and thorium, possess isotopes with sufficiently long half-lives to be found in nature in significant amounts; their transient daughter-products include the other light actinides, actinium and protactinium, along with other elements of lower atomic number, *84* to *88*, that occur naturally only as a result of the radioactive decay of uranium and thorium. The transition elements and rare-earth elements are shown in detail in separate tabulations in Table 1.1.

Elements that are closely comparable structurally and in atomic radius (by reason of the effect of the lanthanide contraction) and hence in their chemical behaviour, include the pairs Zr—Hf and Nb—Ta, and the group of platinum metals (Ru, Rh, Pd, Os, Ir, Pt).

If used without qualification the term **transition elements** is taken to refer only to the transition elements of period *4*, atomic numbers *21* to *30* (Sc, Ti, V, Cr, Mn, Fe, Co, Ni, Cu, Zn), as these are commoner than elements in other transition series. Similarly the term rare-earth elements (REE) is universally used by petrologists synonymously with the term lanthanides, although to be precise the rare-earth elements include both lanthanides and actinides.

1.2.2. Clarkes

The abundance of the chemical elements in the Earth reflects: (1) a period of thermonuclear synthesis of unknown duration in our solar system producing a theoretically consistent, if not entirely predictable, pattern of relative element and isotope abundances; and, (2) aggregation of a sample of the products of this into our planet ~4.6 Ga ago (along with the formation of other planets and parental bodies of meteorites). There is isotopic evidence (J. H. Reynolds, 1960) from meteorites indicating that event (2) followed the end of event (1) closely, within a maximum time span of 120—290 Ma, depending on assumptions made as to the duration of thermonuclear synthesis. Shortly after event (2) the heat produced by aggregation and a high rate of radioactive decay resulted in substantial melting of planet Earth and fractionation on a massive scale into an inner core of presumed predominant nickel—iron composition (which is still partly molten) and an outer mantle of silicate material. Although there is indisputable evidence of inhomogeneity within the present-day upper mantle, several convergent lines of evidence suggest that much of it may be composed of garnet lherzolite containing some two-thirds of forsteritic olivine, and variable proportions of harzburgite and eclogite.

Partial melting of the upper mantle and ensuing fractionation and crystallization processes coupled with other superficial processes have resulted in the formation of a crust over the whole Earth of different composition to that of the mantle. At present ~65% of the Earth's surface is underlain by a thin oceanic crust some 6.5 km in thickness composed mainly of young igneous rocks of basaltic composition, none known to be older than Mesozoic in age. The remaining 35% of the Earth's surface is underlain by continental crust, on the average ~35 km thick, composed of more variable rocks. In probable order of abundance these comprise granulites, overlain by granitic gneisses and granites (sensu lato), overlain in turn by various sediments and their metamorphic derivatives. Known ages of formation of crustal rocks range back to ~3.8 Ga.

We have become aware comparatively recently that philosophically the crust of the Earth must be regarded as a kinetic phenomenon. Oceanic crust is being continuously produced by igneous activity at ocean ridges and is concurrently destroyed at inferred subduction zones (much of it ultimately to form eclogite in the mantle?) on a massive scale. The amount of continental crust, on the other hand, has slowly been increased, the rate of increase, especially in proto-Archean and Archean times (see Chapter 12), being a highly debatable point. Thus perceived, it is obvious that the formation of the crust of the Earth — our habitat and the only part amenable to direct observation — is an ongoing igneous phenomenon of the first magnitude. This was, of course, memorably stated by Daly (1933):

> "A final philosophy of the Earth's history must be largely founded upon the unshakable facts known about igneous rocks."

Nevertheless, this accessible crust, important as it may be to us, the end-product of complex and long-continued igneous and other processes, represents only just over 0.4% of the bulk of the Earth or ~0.6% of the bulk of the mantle.

Clearly in the foregoing there is a vast field for speculation, hypothesis and debate which is not our present purpose. The compositions of volcanic rocks and magmas — the ultimate source of crustal rocks — are limited to a well-defined spectrum in composition, in terms of silica content, for example, between roughly 40% and 75% with corresponding variations in the contents and proportions of other major oxides (see Table 4.1 and accompanying discussion). As a kind of yardstick for igneous rock compositions it is useful to have some idea of the average abundance of each element by weight in crustal rocks, a figure known as the **clarke** of an element. Estimates are in part oblique and necessarily crude and it seems that older presentations of clarkes have traditionally been based on estimations of abundances in *continental* crust solely. Those that employ a 1:1 weighting of average granite and average basalt in arriving at estimates of continental crust composition (see Krauskopf, 1967, appendix 3, with references) as here, for example, probably still overestimate the overall proportion of granitic material (see also B. Mason, 1966, pp. 41—50 for discussion). With these uncertainties in mind, Tables 1.2 and 1.3 nevertheless present a consensus on data of clarkes for all 90 elements occurring naturally on Earth.

About 98.5% of both average crustal material and of all common igneous rocks is composed of only eight **major elements** (see Table 1.2), O, Si, Al, Fe, Mg, Ca, Na and K, that each have a clarke of over 1%. These are presented not quite in order of abundance, but in the conventional order of oxides in analyses of igneous rocks. The weighting given to continental crust as noted above probably inflates the quoted clarkes of Na and K at the expense of that of Mg in overall crustal material. Next in abundance come three elements Ti, H, and P, with clarkes between 0.1% and 1%. They

TABLE 1.2

Abundance of major elements in continental crust (after B. Mason, 1966, p. 48, table 3.4)

Element	Weight per cent — wt.%	Atom per cent* — at.%	Volume per cent* — vol.%
O	46.6	62.6	93.8
Si	27.7	21.2	0.9
Al	8.1	6.5	0.5
Fe	5.0	1.9	0.4
Mg	2.1	1.8	0.3
Ca	3.6	1.9	1.0
Na	2.8	2.6	1.3
K	2.6	1.4	1.8
	98.5		

*Recalculated to 100%.

TABLE 1.3

Clarkes of minor and trace elements (after compilation by Krauskopf, 1967, appendix III, pp. 637—640) (with references to data sources)

Element	Clarke (ppm)
Ti	5,700
H	1,400
P	1,050
Mn	950
F	625
Ba	425
Sr	375
S	260
C	200
Zr	165
V, Cl, Cr	100—150
Rb, Ni, Zn, Ce, Cu	50—90
Y, Nd, La, Co, Sc, Li, N, Nb	20—35
Ga, Pb, B	10—15
Th, Sm, Gd, Pr, Dy	5—10
Yb, Hf, Cs, Be, Er, U, Br, Ta, Sn	2—3
As, Ge, Mo, W, Ho, Eu, Tb	1—2
Lu, Tm, I, Tl, Cd, Sb, Bi, In	0.1—1
Hg, Ag, Se	0.05—0.1
Au, the six platinum metals, Te, Re	0.05
The five inert gases and seven elements of high atomic numbers ($84—89$ and 91) present as radioactive nuclides in the decay schemes of U and Th	very low

are sometimes known as the **minor elements,** or alternatively included with the major elements as opposed to the trace elements (see below). In igneous rocks Ti and P typically occur in the common accessory minerals ilmenite, sphene and apatite. H of course is an essential component of double-chain inosilicates, phyllosilicates, and some other mineral species. It is the above major and minor elements, generally reported as weight per cent (wt.%) oxides, that are the essential constituents of an igneous rock analysis.

The remainder of the elements all with clarkes under 0.1% are known collectively as **trace elements.** In igneous rock analyses they are generally reported as parts per million (ppm) of the element by weight, a figure which is numerically equal to grammes per metric ton (g/t). In igneous rocks these trace elements generally occur dispersed in small concentrations within the lattices of major silicate mineral species according to certain principles of crystalline solution discussed below, in which event they are known as **dispersed elements.** Mn, for example, although nearly as abundant as P, does not form pyrogenetic minerals in which it is an essential component (as P is in apatite) but proxies for ferrous iron in crystalline solution in mafic silicate minerals*. Some trace elements, however, notably Zr and Cr, may form separate mineral species in igneous rocks, e.g., zircon and chromite, although Cr also occurs in significant concentrations dispersed in augite. Many of the remaining elements, although generally totally dispersed in the commoner igneous rocks, may become concentrated by processes associated with pegmatite formation, carbonatites, hydrothermal vein deposits, evaporites, etc., to form occurrences of discrete, albeit often rare, minerals in which they may be essential rather than dispersed constituents (e.g., Be in beryl, Nb in pyrochlore, Sn in cassiterite, Sr in strontianite). Some elements may be so dispersed in habit that neither in igneous rocks nor by any natural process do they ever occur in minerals of which they are an essential component; rather they only occur in small concentrations proxying for other specific elements, e.g., Ga in Al minerals, Rb in K minerals, and Hf in Zr minerals.

1.2.3. Valency

Although the nature of inter-atomic bonding in silicate minerals is only in part ionic and in part covalent, the concept of valency is useful as it can be strictly applied to the formulae of all silicate minerals, i.e. their formulae must always "balance" in terms of the valency of the (potential) cations and anions of the elements that they contain. Valency, always a small integral number, refers to the number of electrons that must be subtracted from (*cations*) or added to (*anions*) the outermost electron shell of an atom in order to convert it into an ion in which the number of electrons in that

---

*The content of manganese is, however, traditionally reported as MnO and included in major-element analyses of igneous rocks.

TABLE 1.4

Valency states of the major elements and some of the commoner trace elements of igneous rocks

	Major elements	Trace elements
Cations:		
Monovalent	H^+, Na^+, K^+	Li^+, Rb^+, Cs^+
Divalent	Ca^{2+}, Mg^{2+}, Fe^{2+} (ferrous iron)	Be^{2+}, Sr^{2+}, Ba^{2+}, Mn^{2+}, Co^{2+}, Ni^{2+}, Eu^{2+}
Trivalent	Al^{3+}, Fe^{3+} (ferric iron)	B^{3+}, Ga^{3+}, Cr^{3+}, V^{3+}, Sc^{3+}, Y^{3+}, REE (including Eu^{3+})
Tetravalent	Si^{4+}, Ti^{4+}	Ge^{4+}, Zr^{4+}, Hf^{4+}, Sn^{4+}
Pentavalent	P^{5+}	V^{5+}, Nb^{5+}, Ta^{5+}
Anions:		
Monovalent	OH^-	F^-, Cl^-
Divalent	O^{2-}	S^{2-}

outer shell comes to equal zero or eight (noble-gas configuration). For all but the transition elements valency is very simply and obviously related to the position of an element within the groups of the periodic table. Some of the transition elements, however, have more than one possible valency state; under the oxidation conditions normally pertaining to igneous rock formation one major element, iron, can enter minerals in two valency states, namely ferrous iron, Fe^{2+}, and ferric iron, Fe^{3+}. Table 1.4 shows the valencies of major and minor elements on the left, and of some of the commoner trace elements on the right.

1.3. FACTORS GOVERNING STRUCTURES OF SILICATE MINERALS

1.3.1. Relative abundance of oxygen to cations in igneous and crustal rocks

As igneous rock compositions can be represented in terms of oxides, it might be asked why there appears to be just enough oxygen to go round, allowing for some small latitude in ferrous/ferric ratios. This is actually a most interesting question to which an answer (here necessarily abbreviated) would have to include reference to the following:

(1) Primordial Earth composition was deficient in oxygen to the extent that some siderophile elements existed in the metallic state, by analogy with the compositions of meteorites including irons and stony irons of which the orbits and ages indicate that they too represent planetary matter of our own solar system.

(2) Early heating up of the Earth resulted in melting and thorough fractionation of an iron—nickel melt that sank to form the core of the Earth

in a manner reminiscent of blast furnace operation, leaving an outer-mantle rock of silicate composition relatively enriched in lithophile elements, analogous to blast furnace slag.

(3) The compositions of all rock-forming silicate minerals can in fact be expressed in terms of oxides in which valency requirements are obviously met. For example forsterite, generally written Mg_2SiO_4, can also be written $2MgO \cdot SiO_2$; orthoclase, usually written $KAlSi_3O_8$, can be written $K_2O \cdot Al_2O_3 \cdot 6SiO_2$ and so on. The notation $KAlSi_3O_8$ better illustrates the essential structure of orthoclase, namely a tectosilicate (basic formula SiO_2, see below) in which one atom in four of tetravalent silicon is diadochically replaced by trivalent aluminium with the addition of one atom of monovalent potassium per unit formula to restore charge equilibrium. For this reason mineral formulae are not written out in oxide form except in calculations relating to normative computations which are directly based on relative amounts of component oxides (see Chapter 4).

(4) The processes of igneous rock formation, viz. partial melting of silicate material of mantle or crust to form wholly or partly molten magma and its subsequent emplacement and solidification does not further alter oxygen to cation proportions*. Magmas have also been shown to contain very small overall amounts of immiscible droplets of molten sulphide, including for example $CuFeS_2$, chalcopyrite, so that very small amounts of some chalcophile elements may not in fact be combined with oxygen but with sulphur. This quantitatively, however, is a very minor consideration.

(5) All magmas contain some **volatile constituents** that would be gaseous at atmospheric pressure and temperature. Predominant among them is water (see Chapter 6), which is partly dissociated, $2H_2O \rightleftharpoons 2H_2 + O_2$. Equilibrium is always well to the left of this equation under all upper-mantle, crustal and atmospheric conditions but is relatively displaced to the right at higher temperatures. Hydrogen, because of its small size, might diffuse selectively from a magma chamber, thus leaving the magma relatively slightly richer in oxygen, but this is likely to be a very small effect only.

(6) More importantly, the Earth's hydrosphere—atmosphere system (an end-product of long-continued igneous activity!), although predominantly water, contains excess oxygen. This is due to photosynthesis and the fact that hydrogen can escape the Earth's gravitational field from an extremely tenuous upper atmosphere although only at a minutely small rate. There is good evidence that the oxygen and ozone content of the atmosphere has in fact slowly increased during the history of the Earth, and indeed the evolution of the higher forms of life has been critically dependent on this process.

---

*Some diabases have been reported to contain metallic iron, but are found only in the exceptional circumstance where the diabase has intruded strata containing coal seams, and a local reaction involving the reduction of the magma by carbon apparently occurred.

1.3.2. The role of oxygen in the structure of silicate minerals

It so happens that oxygen as well as being the most abundant element by weight in mantle and crustal rocks has a large atomic radius and is consequently far and away the most abundant constituent by volume. Oxygen constitutes some 94% by volume of average crustal rock! (see Table 1.2). In silicate crystal structures the cations of other elements have perforce to fit between large oxygen anions, as in general the cations are much smaller than the oxygen anions. Picturing ions to have the shape to a first approximation of spheres, this fitting together becomes a geometrical problem and given the range of cation/oxygen ion ratios, the following structures are possible in order of increasing cation size: *four-fold coordination*, in which each small cation lies at the centre of four oxygen anions that are arranged symmetrically at the corners of a tetrahedron; *six-fold coordination*, in which each cation lies at the centre of six oxygen anions arranged symmetrically at the corners of an octahedron; *eight-fold coordination* in which a cation lies at the centre of eight oxygen anions arranged symmetrically at the corners of a cube; *ten-* and *twelve-fold coordinations*, which are geometrically more complicated arrangements in which cation size approaches and equals anion size, respectively.

1.3.3. Ionic radii

At this stage it will be apparent that a knowledge of ionic radii would be very useful. Although easy to conceptualise, the actual calculation or measurement of ionic radii is, in fact, imprecise: not only may bonding be in part covalent, but also in general ions become distorted from a spherical shape due to differential bonding forces acting on an ion in different three-dimensional crystal structures. A careful reconsideration of ionic radii, with reference to previous work and specifically designed for use in silicate geochemistry, has been undertaken by Whittaker and Muntus (1970). It would seem wise therefore to follow these authors even although their figures will appear a little unfamiliar to readers of older texts. Note also the complicating factors discussed by Shannon and Prewitt (1969). Table 1.5 gives these recalculated ionic radii for the same selected set of elements as in Table 1.4 arranged broadly in order of increasing size with minor departures from this order to reflect natural groupings and thus to facilitate comparisons. The radii are quoted in Ångström units ($1Å = 10^{-7}$ mm). This is not in accordance with units of the SI system, but the familiar Ångström units are retained here as it is convenient, as in all measuring systems, to have units of the same order as that of the thing being measured.

A consideration of cation/anion ratios derived from these figures shows that of the major-element ions Si^{4+} would be expected to enter readily into four-fold coordination with O^{2-}. Similarly, Fe^{2+} and Mg^{2+} should enter into

TABLE 1.5

Ionic radii in Ångström units (Å) according to coordination number (after Whittaker and Muntus, 1970)

Coordination number	III	IV	VI	VIII	X	XII
Cations:						
B^{3+}	0.10	0.20				
P^{5+}		0.25				
Be^{2+}	0.25	0.35				
Si^{4+}		0.34	0.48			
Ge^{4+}		0.48	0.62			
Al^{3+}		0.47	0.61			
Ga^{3+}		0.55	0.70			
V^{5+}		0.44	0.62			
Ti^{4+}			0.69			
Cr^{3+}			0.70			
V^{3+}			0.72			
Fe^{3+}		0.57	0.73			
Ni^{2+}			0.77			
Co^{2+}			0.83			
Fe^{2+}		0.71	0.86			
Mg^{2+}		0.66	0.80			
Li^{+}		0.68	0.82			
Nb^{5+}			0.72			
Ta^{5+}			0.72	0.77		
Sn^{4+}			0.77			
Hf^{4+}			0.79	0.91		
Zr^{4+}			0.80	0.92		
Sc^{3+}			0.83	0.95		
Y^{3+}			0.98	1.10		
La^{3+}			1.13	1.26		
Ce^{3+}			1.09	1.22		
Eu^{2+}			1.25	1.33		
Eu^{3+}			1.03	1.15		
Lu^{3+}			0.94	1.05		
Ca^{2+}			1.08	1.20	1.36	1.43
Na^{+}			1.10	1.24		
Sr^{2+}			1.21	1.33	1.40	1.48
Ba^{2+}			1.44	1.50	1.60	1.68
K^{+}			1.46	1.59	1.67	1.68
Rb^{+}			1.57	1.68	1.74	1.81
Cs^{+}			1.78	1.82	1.89	1.96
Anions:						
O^{2-}	1.28	1.30	1.32	1.34		
F^{-}	1.22	1.23	1.25			
Cl^{-}		1.67	1.72			
S^{2-}		1.56	1.72	1.78		

six-fold coordination with O^{2-}, with possible diadochy between them. Ca^{2+} and Na^+ should enter into eight-fold coordination with O^{2-}, and could also possibly replace each other diadochically if the charge difference between them could be otherwise balanced in the crystal structure. K^+, easily the largest of the major cations, should occur in twelve-fold coordination with O^{2-}. The ionic radius of Al^{3+} places it between the ideal values of four- and six-fold coordinations with O^{2-}, so that either or both could perhaps be anticipated; similar considerations apply to Fe^{3+} although, being significantly larger than Al^{3+}, Fe^{3+} should be relatively more prone to enter into six-fold coordination with O^{2-}.

1.3.4. The SiO₄ tetrahedron

Oxygen and silicon together account for nearly 75% by weight and approximately 95% by volume of igneous rocks. It just so happens that the ratio of the size of the silicon ion to that of the oxygen ion is close to the optimum for four-fold coordination. Not surprisingly therefore SiO_4 tetrahedra prove to be the basic building blocks of silicate minerals*.

1.3.5. Resultant silicate structures

In the lattices of silicate minerals SiO_4 tetrahedra are linked together in a small number of easily conceptualized ways that are enumerated below (with classical derivations of names in parentheses). Remaining cations fit between the relatively large oxygen atoms of the resultant structures in six- or eight-fold or higher coordinations. A few silicate minerals have slightly more complex structures that involve the interlocking of SiO_4 tetrahedra in combinations of these fundamental styles.

(1) **Nesosilicates** (Gr. *nesos*, an island). the SiO_4 tetrahedra are not linked directly to one another by sharing oxygen atoms but are entirely connected to intervening cations, so that each SiO_4 tetrahedron is isolated (syn. *orthosilicates*).

(2) **Sorosilicates** (Gr. *soros*, a group; L. *soror*, a sister): the SiO_4 tetrahedra are linked in pairs by sharing one oxygen atom; each pair is separated from other pairs by intervening cations.

(3) **Cyclosilicates** (Gr. *kyklos*, a circle): the SiO_4 tetrahedra are linked in essentially planar rings, generally of six, with columnar stacking of the rings on top of another (syn. *ring silicates*).

(4) **Inosilicates** (Gr. *inos*, a thread): the SiO_4 tetrahedra are arranged in long parallel chains, either a single chain, or a double chain in which alternate tetrahedra in each chain are cross-linked (syn. *chain silicates*).

(5) **Phyllosilicates** (Gr. *phyllon*, a leaf): the SiO_4 tetrahedra are arranged

---

*Only in stishovite, a very high-pressure polymorph of silica, is there known to be a different coordination of silicon with oxygen, namely eight-fold.

TABLE 1.6

Structural classes of rock-forming silicate minerals

Name	Description in terms of SiO_4 tetrahedra	Resultant SiO ratio	Major silicate groups	Examples
Nesosilicates	isolated tetrahedra	$-SiO_4$	olivines garnets	Mg_2SiO_4, forsterite $Fe_3Al_2Si_3O_{12}$, almandine
Sorosilicates	pairs of tetrahedra sharing 1 oxygen atom	$-Si_2O_7$	melilites	$Ca_2MgSi_2O_7$, åkermanite
Cyclosilicates	rings of 6 tetrahedra each sharing 2 oxygen atoms	$-Si_6O_{18}$	—	$Be_3Al_2Si_6O_{18}$, beryl
Inosilicates	(a) single chain of tetrahedra each showing 2 oxygen atoms	$-SiO_3$	pyroxenes	$MgSiO_3$, enstatite $CaFeSi_2O_6$, hedenbergite
	(b) double chain of tetrahedra with alternate tetrahedra cross-linked so that tetrahedra share 2 and 3 oxygen atoms alternately	$-Si_4O_{11}$	amphiboles	$Ca_2Mg_5Si_8O_{22}(OH)_2$, tremolite $Na_2Fe_2^{3+}Fe_3^{2+}Si_8O_{22}(OH)_2$, riebeckite
Phyllosilicates	sheets of tetrahedra each sharing 3 oxygen atoms	$-Si_4O_{10}$	micas chlorites serpentine minerals	$KAl_2AlSi_3O_{10}(OH)_2$, muscovite $Mg_6Si_4O_{10}(OH)_8$, serpentine
Tectosilicates	three-dimensional framework of linked tetrahedra each sharing all 4 oxygen atoms	SiO_2	silica minerals feldspars feldspathoids	SiO_2, quartz $NaAlSi_3O_8$, albite $KAlSi_2O_6$, leucite

in continuous parallel planes, each tetrahedron sharing three oxygen atoms with its neighbours (syn. *sheet silicates*).

(6) Tectosilicates (Gr. *tekton*, a builder): the SiO_4 tetrahedra form an interconnected three-dimensional lattice in which all oxygen atoms are shared (syn. *framework silicates*).

These arrangements lead inevitably to certain silicon/oxygen ratios in the formulae of each group so constructed as shown in Table 1.6. The serious student should satisfy himself that he can derive these basic units of formulae, using appropriate diagrammatic sketches (see, e.g., Hatch et al., 1972, part 1), as this is a most useful peg on which to hang a working knowledge of silicate mineral compositions. For the most part these are obvious, the hardest to see possibly being the Si_4O_{10} "repetition unit" of phyllosilicates, which in plan view of the sheet has a confusing hexagonal pattern suggesting six units of silicon in the basic formulae (biotite, of course, does have a platy pseudo-hexagonal habit).

Some obvious relationships of crystalline structure to crystal form should be noted, for example the hexagonal symmetry of cyclosilicates, the two prominent prismatic cleavages of inosilicates parallel to the chains of SiO_4 tetrahedra, and the pronounced basal cleavage of phyllosilicates.

Owing to a progressively increasing silicon/oxygen ratio from nesosilicates to tectosilicates, there is a reciprocal decreasing proportion of other cations, decreasing to nil in silica, SiO_2. Nesosilicates are thus the most basic in the sense of containing relatively the most cations other than silica, and basic rocks hence contain greater amounts of nesosilicates and inosilicates than do acid rocks. Formulae tend to be more complex and unit-cell size tends to increase going from nesosilicates to tectosilicates (silica polymorphs constituting an exception). Densities, melting points and refractive indices also tend to decrease from nesosilicates to tectosilicates, but all three properties are much influenced in addition by the nature of other cations present in crystalline solution (silica again is anomalous in having a high melting point).

1.4. PRINCIPLES GOVERNING CRYSTALLINE SOLUTION

Historically this subject has been progressively treated at increasing levels of complexity and is more fully reviewed in texts on mineralogy and geochemistry (e.g., B. Mason, 1966; Frye, 1974). Some milestones along the way are as follows.

Goldschmidt (1954, chapters 3 and 6, with reference to his earlier classic work) formulated empirical rules governing diadochy, based in the main on ion size and valency. The most important consideration is ion size (see Table 1.5). As would be expected the substitution of one ion for another in a crystal lattice is apparently only possible if their ionic radii are approximately the same — say within 15%. Thus diadochy is commonly found in silicate minerals within the following groups or pairs of ions: Mg^{2+}, Fe^{2+}, Mn^{2+}, and Ti^{4+}; Zr^{4+} and Hf^{4+}; Ca^{2+} and Na^+; K^+ and Ba^{2+}; O^{2-}, OH^-, and

F^-. Within possible substitution ranges the smaller cations commonly tend to be more readily incorporated into crystals during the progress of crystallization; for example, first-formed olivines and pyroxenes invariably have a higher Mg/Fe ratio than the crystallizing magma within which conversely the Mg/Fe ratio necessarily falls as crystallization proceeds. Analogously, the K/Rb ratio of crystallizing feldspars is always a little higher than that of the crystallizing magma in which K/Rb ratio similarly falls as crystallization proceeds. Ions of the same size, other factors being equal, would enter a crystal lattice in the same proportion as their relative concentrations in the crystallizing magma; for example hafnium is "camouflaged" in this way in zirconium-bearing minerals. In general the consideration of ionic size outweighs that of charge; however, it is found that a proxying ion of higher charge may be preferentially "captured" by early-forming crystals and that the ion with the lower charge may this only be "admitted" as its relative concentration increases in the crystallizing magma; for example, calcium enters before sodium in the plagioclase crystalline solution series. In the case of diadochy such as this by ions of differing valency, the overall charge requirements in the crystal structure must be met in order to preserve electrostatic neutrality, for example Na^+Si^{4+} for $Ca^{2+}Al^{3+}$ in plagioclase. This consideration explains why the diadochic substitution of ions with valencies differing by more than one is less commonly observed, even although ionic size requirements may be met, as there may be difficulty in accomodating the charge difference.

A further important consideration governing the possible extent of diadochy is temperature of formation. At higher temperatures a higher degree of crystalline solution is possible than at lower temperatures. Thus, despite the widely differing ionic radii of K^+ and Na^+, more or less complete solid solution between the compositions of orthoclase and sodic plagioclase is possible at (water-poor) magma liquidus temperatures to produce the familiar sanidine (and rarer anorthoclase) phenocrysts of certain lavas. At lower temperatures, below a boundary curve known as the solvus (see Chapter 6), a complete range of crystalline solution in alkali feldspars is no longer possible (under equilibrium conditions). Thus for alkali feldspars formed in this lower temperature range, and for sanidines that have re-equilibrated at lower temperatures, two feldspars are found, a potassium-rich one and a sodium-rich one. Where this re-equilibration has occurred the two feldspars are found intimately intergrown on various scales ranging down to the submicroscopic, resulting in the range of perthites (syn. perthitic intergrowths). The process of formation of these from a homogeneous parent-crystal is known as **exsolution**. Although the term perthite is specific to intergrowths of alkali feldspar, the operation of the process of exsolution is more generally observed. For example, in the pyroxenes there is found to be only a limited degree of crystalline solution between a calcium-rich monoclinic augite and a calcium-poor orthorhombic hypersthene even at magmatic temperatures. On very slow cooling nevertheless exsolution may cause the formation of exsolved lamellae of augite in host hypersthene (Henry lamellae) and the reciprocal formation of exsolved lamellae of hypersthene in augite (typically along 001).

Differences in temperature and in pressure too can also affect coordination and hence the lattice site where a given element may preferentially enter a given crystal structure. For example, relatively high pressures favour a six-fold coordination for aluminium as shown by the minerals jadeite ($NaAlSi_2O_6$) and omphacite (intermediate in composition between jadeite and diopside). Relatively high temperatures, on the other hand, favour a four-fold co-ordination for aluminum, as for example in the igneous amphiboles where the content of aluminum in four-fold coordination approaches one in four of the silicon atoms. The amphiboles of metamorphic rocks reveal a pro-gressively increasing proportion of Al proxying in this way for Si with increases in temperature of formation, viz. actinolite of low-greenschist facies via sub-aluminous hornblende to hornblende of amphibolite facies.

Because the nature of the bonding in silicates is partly ionic and partly covalent in character another complexity is added to considerations based solely on ionic properties. Ringwood (1955a) has drawn attention to the fact that, other things being equal, a cation of low electronegativity (i.e. more prone to form ionic bonds) will tend to be preferentially incorporated into a crystal lattice. Some Li for example, in spite of the negative feature of a low charge, enters the lattice of olivine crystals. Ringwood (1955b) has also drawn attention to the possible role of complex formation in governing trace-element behaviour in crystallizing, volatile-rich magmas. A more complete set of principles governing the structure of and crystalline solution in silicate minerals is developed by Pauling (1960, with reference to his earlier work).

More recently, an important consideration governing diadochy, at least as far as the transition elements are concerned, is that of crystal field stabi-lization energy (CFSE) (see Burns, 1970). Crystal field theory considers the effect of the electrostatic field of adjacent ions on the unfilled orbitals of transition-element ions. Certain transition-element ions are relatively more stable than others in the electrostatic field of a particular lattice site in a crystal relative to the less regular coordination of a melt. Based on calcu-lated CFSE-values, it can be predicted that transition elements will pref-erentially enter into a spinel lattice, for example, in the following orders: Ni, Cu, Co, Fe, Mn, Zn for divalent ions, and Cr, Co, V, Ti, Fe, Sc, for trivalent ions [Burns and Fyfe (1964); see also Campbell and Borley (1974) for an interesting application of crystal field theory to the observed sequence of pyroxene compositions in the Jimberlana intrusion, Australia].

The qualitative statements made above as to preference can be quantified by measuring partition coefficients, i.e. the ratio of concentrations of a proxying trace element to major element in both liquid and crystal, ideally groundmass and phenocryst compositions of a porphyritic fine-grained rock [see, however, Allègre and Minster (1978, pp. 13—15) for a discussion of difficulties inherent in this superficially straightforward procedure].

1.5. NOMENCLATURE, COMPOSITION, AND PARAGENESIS OF IGNEOUS ROCK-FORMING MINERALS

For easy reference a list of compositions of igneous rock-forming minerals (including all but very uncommon species and some species found in relatively very uncommon rocks such as carbonatites) is presented in Table 1.7. The silicate species are arranged according to their silicate structure pattern as discussed above, and are followed by oxides and other miscellaneous chemical groupings. The symbol (s) indicates that a particular silicate mineral has a more complex crystal structure than simple representatives of its class. Common major-element diadochy is indicated by an appropriate notation in parentheses in the formula and/or an indication (e) of theoretical end-members in a crystalline solution series, but trace-element and other minor diadochy is not generally indicated.

It is important to realize that many mineral names are in common use in igneous petrology with multiple connotations arising from no less than three differing patterns of usage: (1) an end-member in a crystalline solution series; (2) a defined compositional range in a crystalline solution series; and (3) a member of a crystalline solution series with an imprecisely-defined compositional range, sometimes with other mineral or compositional names used as a modifier. Thus, the term forsterite may denote the pure magnesian end-member, Mg_2SiO_4, of the olivine series, *or* olivine in the compositional range Fo_{90}—Fo_{100}; again the term "forsteritic olivine", commonly applied, for example, to phenocrysts in basaltic rocks, refers to crystals with a range of compositions that actually straddle the forsterite—chrysolite boundary (chrysolite lies within the compositional range Fo_{70}—Fo_{90}). Similarly, the "calcic plagioclase" of mafic rocks generally indicates a bytownite or labradorite composition. Again the term hypersthene has a precise compositional connotation but is sometimes loosely used synonymously with orthopyroxene, or more commonly for pleochroic orthopyroxenes. Context and familiarity generally make clear what is meant, and confusion seldom arises in practice from this admittedly lax usage of terminology.

A difficult exercise in condensation is presented in Table 1.7, and the reader is referred to Deer et al. (1966) and to Hatch et al. (1972) for further detailed information on crystal structure, composition and paragenesis, all of which topics are treated in these two references from the point of view of a petrologist, an important consideration as we are not here concerned with mineralogy as an end in itself. An attempt at a very broad paragenetic grouping and relative abundances is indicated by the letters which are used only in the context of this tabulation, explained in the notes to the table.

TABLE 1.7

Nomenclature, composition, and paragenesis of igneous rock-forming minerals

Name	Composition	Paragenesis	Abundanc
Nesosilicates:			
Olivine	$(Mg,Fe)_2SiO_4$		
forsterite (e)	Mg_2SiO_4		
fayalite (e)	Fe_2SiO_4		
forsterite	$Fo_{90}-Fo_{100}$	M	c
chrysolite	$Fo_{70}-Fo_{90}$	A, UA, B, UB	c
hyalosiderite	$Fo_{50}-Fo_{70}$	A	f
hortonolite	$Fo_{30}-Fo_{50}$	A	f
ferrohortonolite	$Fo_{10}-Fo_{30}$	A, B	f
fayalite	Fo_0-Fo_{10}	F	f
Monticellite	$CaMgSiO_4$	UA	r
Garnet	generally $R_3^{2+}R_2^{3+}Si_3O_{12}$		
pyrope (e)	$Mg_3Al_2Si_3O_{12}$		
almandine (e)	$Fe_3Al_2Si_3O_{12}$		
spessartite (e)	$Mn_3Al_2Si_3O_{12}$		
pyralspite	$(Mg,Fe,Mn)_3Al_2Si_3O_{12}$		
pyrope-rich pyralspite		M	c
almandine-rich pyralspite		F, G	f
spessartite-rich pyralspite		P	f
andradite (syn. melanite)	$Ca_3Fe_2^{3+}Si_3O_{12}$	N (per)	r
Zircon	$ZrSiO_4$	F, G, N	a, c
Sphene (s)	$CaTiSiO_5$	F, G, N	a, c
Topaz	$Al_2SiO_4(OH,F)_2$	P	r
Epidote group (s)	$Ca_2R_3^{3+}(SiO_4)_3(OH)$		
zoisite and clinozoisite	$Ca_2Al_3(SiO_4)_3(OH)$	D in B, F, G	c
epidote	$Ca_2Al_{3-2}Fe_{0-1}^{3+}(SiO_4)_3(OH)$	D in B, F, G	c
piemonite	$Ca_2(Mn,Fe,Al)_3(SiO_4)_3(OH)$	D in F	r
allanite (syn. orthite)	$(Ca,Ce)_2(Fe^{2+},Fe^{3+},Al)_3(SiO_4)_3$ (OH)	F, G	a, f
Sorosilicates:			
Melilite		UA	f
åkermanite (e)	$Ca_2MgSi_2O_7$		
gehlenite (e)	$Ca_2Al_2SiO_7$		
Cyclosilicates:			
Beryl (e)	$Be_3Al_2Si_6O_{18}$	P	f
Tourmaline	complex and variable Al-borosilicate with many other cations	P	c
Cordierite	$Al_3(Mg,Fe)_2AlSi_5O_{18}$	G	r

TABLE 1.7 (*continued*)

Name	Composition	Paragenesis	Abundan
Inosilicates:			
Pyroxenes			
wollastonite (e)	$CaSiO_3$		
diopside (e)	$CaMgSi_2O_6$		
hedenbergite (e)	$CaFeSi_2O_6$		
enstatite (e)	$MgSiO_3$		
ferrosilite (e)	$FeSiO_3$		
jadeite (e)	$NaAlSi_2O_6$		
aegirine (e)	$NaFe^{3+}Si_2O_6$		
Tschermak's molecule (e)	$CaAl_2SiO_6$		
Clinopyroxenes	generally $(Ca, Mg, Fe)SiO_3$		
diopside		M	c
endiopside		M	c
salite	(see Fig. 1.1 for compositional range)	A	c
augite		B	c
ferroaugite		B	f
subcalcic augite		B	f
pigeonite		B	f
hedenbergite		F	f

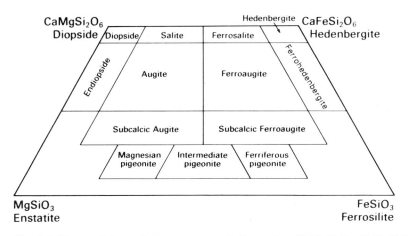

$CaMgSi_2O_6$ — Diopside | Hedenbergite $CaFeSi_2O_6$ — Hedenbergite

Fig. 1.1. Nomenclature of clinopyroxenes in the system $CaMgSi_2O_6$—$CaFeSi_2O_6$—$Mg_2Si_2O_6$—$Fe_2Si_2O_6$ (after Poldervaart and Hess, 1951, p.474, fig. 1).

titanaugite (and titansalite)	significant Ti content in augite or salite	A	c
omphacite	intermediate between diopside and jadeite	M	c
aegirine—augite	intermediate between aegirine and augite or ferroaugite	N (per)	c

TABLE 1.7 (*continued*)

Name	Composition	Paragenesis	Abunda
aegirine	$NaFe^{3+}Si_2O_6$	N (per)	c
spodumene	$LiAlSi_2O_6$	P	f
wollastonite (s)	$CaSiO_3$	UA	r
Orthopyroxenes	$(Mg,Fe)SiO_3$		
enstatite	$En_{90}-En_{100}$	M	c
bronzite	$En_{70}-En_{90}$	UB, B	c
hypersthene	$En_{50}-En_{70}$	F	c
ferrohypersthene	$En_{30}-En_{50}$	F	r
eulite	$En_{10}-En_{30}$		
ferrosilite	En_0-En_{10}		
Amphiboles	generally $X_{2-3}Y_5Z_8O_{22}(OH)_2$ where X includes Ca, Na, K; Y includes Mg, Fe^{2+}, Fe^{3+}, Al, Ti; and Z includes Si, Al (up to 1 in 4 of Si)		
tremolite	$Ca_2Mg_5Si_8O_{22}(OH)_2$	D in B, UB	c
actinolite	$Ca_2(Mg,Fe)_5Si_8O_{22}(OH)_2$	D in B	c
common hornblende	$(Na,K)_{0-1}Ca_2(Mg,Fe^{2+},Fe^{3+}, Al)_5Al_{2-1}Si_{6-7}O_{22}(OH,F)_2$	F	c
barkevikite	high Fe content; low Fe^{3+}/Fe^{2+} ratio	A, N	f
basaltic hornblende (syn. lamprobolite)	high Fe^{3+}/Fe^{2+} ratio; O takes place of OH	F	c
kaersutite	high Ti content	A, N	f
kataphorite	high Na and Fe; aluminous	A	r
arfedsonite	high Na and Fe^{2+}; slightly aluminous	N (per)	c
riebeckite	$Na_2Fe_3^{2+}Fe_2^{3+}Si_8O_{22}(OH)_2$	F (per)	c
aenigmatite (syn. cossyrite)	$Na_2Fe_5^{2+}TiSi_6O_{20}$	F (per), N (per)	f
cummingtonite	$(Mg,Fe)_7Si_8O_{22}(OH)_2$	F	r
Phyllosilicates:			
muscovite	$KAl_2AlSi_3O_{10}(OH)_2$	G, P	c
muscovite, var. sericite	$KAl_2AlSi_3O_{10}(OH)_2$	D in G, F	c
phlogopite (e)	$KMg_3AlSi_3O_{10}(OH)_2$		
annite (e)	$KFe_3AlSi_3O_{10}(OH)_2$		
phlogopitic mica	$KMg_{3-2}Fe_{0-1}AlSi_3O_{10}(OH)_2$	UA	c
biotite	$KMg_{2-0}Fe_{1-3}AlSi_3O_{10}(OH)_2$	A, F, G	c
lepidolite (e)	$KLi_2AlSi_4O_{10}(OH)_2$	P	f
prehnite	$Ca_2AlAlSi_3O_{10}(OH)_2$	D in B	f
talc	$Mg_3Si_4O_{10}(OH)_2$	D in UB	c
pyrophyllite	$Al_2Si_4O_{10}(OH)_2$	D in F	r
chlorite	$(Mg,Fe,Al)_6(Si,Al)_4O_{10}(OH)_8$	D in B, F, G	c
serpentine minerals	$Mg_6Si_4O_{10}(OH)_8$	D in UA, UB	c
kaolinite	$Al_4Si_4O_{10}(OH)_8$	D in F	f

TABLE 1.7 (*continued*)

Name	Composition	Paragenesis	Abundan
Tectosilicates:			
Silica polymorphs			
low-temperature or α-quartz	SiO_2	P	c
high-temperature or β-quartz*	SiO_2	F, G	c
tridymite*	SiO_2	B, F	f
cristobalite*	SiO_2	F	r
Plagioclase feldspar			
anorthite (e)	$CaAl_2Si_2O_8$		
albite (e)	$NaAlSi_3O_8$		
anorthite	$An_{90}-An_{100}$		
bytownite	$An_{70}-An_{90}$	A, B	c
labradorite	$An_{50}-An_{70}$	A, B	c
andesine	$An_{30}-An_{50}$	A, B, F	c
oligoclase	$An_{10}-An_{30}$	F, G	c
albite	An_0-An_{10}	P; D in B, F	c
Alkali feldspar			
orthoclase (e)	$KAlSi_3O_8$		
albite (e)	$NaAlSi_3O_8$		
sanidine, mesoperthite, cryptoperthite	$(K,Na)AlSi_3O_8$	F	c
perthitic orthoclase	$KAlSi_3O_8$ + exsolved albite	G	c
microcline	$KAlSi_3O_8$	G, P	c
anorthoclase	$(Na,K)AlSi_3O_8$	N	f
Feldspathoids			
nepheline	$NaAlSiO_4$	N	c
cancrinite	$3nepheline \cdot CaCo_3$	D in N	c
sodalite	$3nepheline \cdot NaCl$	N	f
nosean	$6nepheline \cdot Na_2SO_4$	N	r
analcite	$NaAlSi_2O_6 \cdot H_2O$	A	f
leucite	$KAlSi_2O_6$	A	f
kalsilite	$KAlSiO_4$	A	r
Petalite	$LiAlSi_4O_{10}$	P	r
Zeolites	hydrated aluminosilicates of alkali and alkaline earth elements	D in A, B	f
Oxides:			
Ilmenite	$FeTiO_3$	A, B	c
Leucoxene	TiO_2	D in A, B	c
Picroilmenite (syn. magnesioilmenite)	Mg-bearing ilmenite	UA	f
Spinel (e)	$MgAl_2O_4$		
Hercynite (e)	$Fe^{2+}Al_2O_4$		

TABLE 1.7 (*continued*)

Name	Composition	Paragenesis	Abunda
Magnetite	$Fe^{2+}Fe_2^{3+}O_4$	A, B, F, G	a, c
Ulvöspinel	$Fe^{2+}TiO_4$	B	c
Chromite	$(Fe,Mg)(Cr,Al)_2O_4$	UB	c
Gahnite	$ZnAl_2O_4$	P	r
Cassiterite	SnO_2	P	f
Rutile	TiO_2	P	r
Perovskite	$CaTiO_3$	UA	f
Corundum	Al_2O_3	P	r
Pyrochlore	Nb_2O_5 (+ other elements)	C	c
Columbo-tantalite	$(Fe,Mn)(Nb,Ta)_2O_6$	P	f
Miscellaneous:			
Apatite	$3Ca_3(PO_4)_2 \cdot Ca(F,Cl,OH)_2$	A, B, C, F, G, N, P	a, c
Monazite	$(Ce,La,Th)PO_4$	G, P	a, f
Fluorite	CaF_2	P	f
Calcite	$CaCO_3$	C; D in A, B, F, G, N	c
Magnesite	$MgCO_3$	D in B	f
Pyrite	FeS_2	B; D in B, F	c
Pyrrhotite	$Fe_{(n-1)}S_n$	B	c

Paragenesis: A, UA = alkalic basic (A) and ultrabasic (UA) rocks of low to very low silica activity; B, UB = basic (B) and ultrabasic (UB) rocks of subalkaline affinity (i.e. excluding A and UA); C = carbonatites; D = deuterically altered, degraded or metamorphosed (low grade) igneous rocks of specified group; F = oversaturated felsic rocks including both intermediate and acid compositions; G = deep-seated granitic rocks; M = mantle-derived nodules; N = undersaturated felsic rocks; P = (granitic) pegmatites and various other late-stage granitic differentiates; (per) = peralkaline varieties of F and N.
Abundance: a = accessory mineral; c = very common mineral (or widespread if accessory); f = fairly common mineral; r = rare mineral.
N.B. The designations c, f, and r are not to be taken as an indication of absolute abundance, but are estimates with respect to the paragenetic group indicated by the capital letter designation.
*Always paramorphed by α-quartz.

Although the amphiboles offer many parallels to pyroxenes (notably, orthorhombic and monoclinic members, a Ca-rich group, a Ca-poor group, and varieties rich in Na and Fe^{3+}) they exhibit more complex diadochy and are in consequence even less amenable to outline presentation. Furthermore, optical parameters commonly do not uniquely define amphibole compositions so that analysis is frequently required. A proposed new overall nomenclature of amphiboles (Leake, 1978) includes a chemical codification of accepted and widely used names and an accompanying recommendation for extinction of no less than about 300 varietal names of this difficult group. However, some of the more commonly met with, petrographically recognized, varieties are included in Table 1.7.

1.6. RECOGNITION OF MINERALS IN THIN SECTION

Familiarity with igneous rocks in hand-specimen is very desirable and much can be learnt this way. Hand-specimen examination, however, has obvious limitations and a more detailed examination is essential for many rocks. The examination by petrographic microscope of a thin section of the rock is the universally used method, supplemented in more sophisiticated work by electron probe analysis, especially useful for small or zoned or partly altered crystals. Several texts dealing with the optical mineralogy of the rock-forming minerals include Kerr (1959), Heinrich (1965), and Deer et al. (1966); at least one of these will prove indispensable.

What the petrologist wants to know and what the thin section can reveal is the mineralogy and composition of the igneous rock, intergrain textures indicating crystallization history, intragrain exsolution textures indicating cooling history, optical parameters indicative of variation in composition within crystalline solution series, and the nature and extent of alteration and deformation processes, and the like.

With practice most igneous rock minerals can be recognized on sight from such immediately obvious optical and crystallographic features as relief, colour, pleochroism, anisotropy and degree of birefringence, twinning and cleavage. One has to learn how to mentally integrate the differing appearances of the generally random orientations and sizes presented by different grains of the same mineral species in a rock depending on the accidental direction and location of sectioning with respect to each grain. This mental integration, essential for a mastery of thin-section petrography, involves for example the simultaneous identification of a number of grains of a mineral species within the field of view showing a range of interference colours up to the maximum value defined by slide thickness and mineral birefringence.

Additional optical properties take a little time to determine precisely but add to the number of clues for confirmation or identification if necessary, or determination of variation in composition of crystalline solutions. These additional properties include extinction angle, whether the mineral is fast or slow relative to crystal form, cleavage, or twin composition plane, and the optic sign, 2V, dispersion, and orientation of optic plane.

Optical parameters, fascinating as they may be, are of course not an end in themselves but rather a means to an end in petrographic studies. Techniques and short cuts based on an appreciation of optical properties are, however, very useful to the petrographer, and a few miscellaneous tips follow.

It is preferable to work with the diaphragm setting some way closed so as to enhance relief. When working with the high-power objective lens it is often useful to insert the condenser; this reduces relief but reveals colour in small grains such as, for example, the tinted augites of microcyrstalline alkaline mafic rocks or translucent chromite edges. Pleochroism is best tested by carefully rotating to and fro 90° from one extinction position to another, previously obtained under crossed nicols. This technique, rather than random rotation of the stage, is particularly effective in detecting the often subtle pleochroism of orthopyroxene for example. Testing for weak birefringence is best done with the sensitive tint inserted; this will reveal, for example, the aggregate birefingence of cryptocrystalline groundmass material or the characteristic twinning of large perovskite grains. The sensitive tint will, of course, quickly distinguish between first-order white and high-order white interference colours. Orientations showing the least birefringence are desirable for the interpretation of all figures, particularly of uniaxial figures where the centre of the cross must be seen to remain in view during rotation of the stage, as off-centre figures with low 2V resemble off-centre uniaxial figures. Note the confusingly wide and diffuse isogyres of minerals of low birefringence such as nepheline and apatite. A useful orientation in inosilicates is the one showing maximum birefringence; this yields the extinction angle to c (if the mineral is monoclinic) and X and Z absorption colours (if the mineral is pleochroic). A basal section of an inosilicate, recognized by two sets of sharp characteristic cleavage traces, will of course yield the third absorption colour corresponding to Y. In all orientations of perthite the albite component has a higher birefringence than the alkali feldspar component, and will thus appear brighter under crossed nicols. The estimation of plagioclase compositions in thin section using extinction angles is notoriously unreliable, unless done on a universal stage. The least accurate procedure is random searching for maximum symmetrical extinction; better for intermediate plagioclase compositions is observation of symmetrical extinction angle in the orientation perpendicular to the 001 cleavage; better still is the utilization of Carlsbad albite twins if present. A distinctive pattern of alteration, even if incipient, can often be used, for example, to differentiate pigeonite or orthopyroxene grains from augite grains, or to differentiate between the plagioclase and alkali feldspar of an intermediate to acid plutonic rock, particularly where twinning is not well developed or masked by the alteration. Degrees of undersaturation in aphanitic mafic rocks can be recognized by the progressive appearance of olivine, nepheline, and perovskite in the groundmass with increasing undersaturation. The less commonly encountered peralkaline and peraluminous mineral parageneses can quickly be recognized. In oversaturated rocks the presence and habit of small amounts of quartz can readily be detected because its birefringence is higher than that of both alkali feldspar and plagioclase of andesine and oligoclase compositions, so that *some* of the quartz will appear the brightest or most yellowish of all light minerals in the thin section.

In a similar manner to learning to read in sentences rather than in individual words or letters, with experience a petrographer will often come to recognize at a glance one of the many distinctive sets of igneous mineralogies and textures; the *rock* thus properly becomes the unit of recognition and description rather than its individual mineral components.

Chapter 2

VOLCANIC ACTIVITY

2.1. INTRODUCTION

It is impossible to compress into one chapter the range of material that occupies whole books on the fascinating subject of volcanology, as for example, those by Bullard (1962), Rittmann (1962), Ollier (1969), G. A. Macdonald (1972), and the very readable and informative book by P. Francis (1976). However, our present interest lies in how, as petrologists, a study of active and recent volcanic activity may help us to interpret the geological record of volcanic rocks. With this theme in mind, the way to condense is clearer. A corollary — what we may legitimately infer about magma from a study of volcanic activity — is deferred until Chapter 6.

To anticipate some of the results of this, we can state that magma is a melt, the composition of which generally can be expressed mainly as silicates, containing dissolved gases, and with, or without, included crystals. When magma erupts to the surface, the resultant kind of volcanic activity is largely a function of its viscosity and gas content. Magma viscosity is itself largely a function of temperature and of chemical composition, especially gas content (see Chapter 6). These parameters appear to vary in a broadly consistent manner with composition, a consistency which justified our considering volcanic activity quite simply under the broad headings of basalt, andesite, rhyolite, and kimberlite. Our study will conclude with an overview of the difficult and variable group of fragmental volcanic rocks.

2.2. VOLCANIC ACTIVITY DESCRIBED BY NATURAL GROUPINGS

2.2.1. Basalts

2.2.1.1. Subaerial basaltic eruptions
Basalt eruption commonly characterizes areas of crustal tension and many major active basalt fields exhibit parallel tensional gashes extending as open fissures from the surface to accessible depths of 30 m or more over lengths up to a kilometre. Basalt extrusion characteristically occurs as fissure eruptions where basalt magma ascends to the surface within a narrow fissure, often initiating a new fissure with successive eruptions. Since basalt lava flows readily, basalt flows are relatively thin and may cover large areas. The well-known Laki fissure eruption of 1783, the largest historically

recorded, produced ~12—15 km³ of lava covering an area of 565 km² from a fissure system ~30 km in length. The fissure is now marked by a line of prominent cinder cones and smaller spatter cones formed in the waning stages of the eruption that lasted for seven months. The even larger prehistoric Thjórsá flow, also in Iceland, covers 800 km², maximum length of flow at least 130 km, and is thought to be the largest postglacial flow on Earth.

Smaller recent fissure eruptions of Kilauea have been observed to begin with considerable explosive ejection of fragmentary material at points along a newly developed small fissure, changing rapidly to fountains of incandescent lava up to some hundreds of metres in height. At this stage a nearly continuous curtain of fountaining lava may overlie the upwelling and vesiculating magma in the fissure. Many phases of fountaining may occur within a few days before the eruption is over and the line of the fissure comes to be represented by numerous scattered small steep spatter cones and the characteristically somewhat broader cinder cones (see G. A. Macdonald, 1972, pp. 9—20).

Notwithstanding the explosive products formed at the beginning and end of basaltic fissure eruptions, and the fragmentation of lava in the fountains caused by vigorous near-surface vesiculation (most of which coalesces), the overall explosion index of a basaltic fissure eruption is very low, commonly around 3.

The **explosion index**, E, generally expressed as a percentage ratio to the nearest integer, is defined as the ratio of amount of tephra/total amount of tephra plus lava. Tephra is considered to include all the nett resultant fragmentary products of an eruption, i.e. both cold country rock and fragmented hot lava. Estimation is obviously imprecise, but nevertheless the explosion index is an informative and widely used parameter.

A graphic account of a rifting episode accompanying fissure eruption in northern Iceland has been given by Björnsson et al. (1977). In 1975—1976 a 80-km segment of the plate boundary, characterized by a fault-and-fissure swarm and two central volcanoes, was the locale of numerous shallow earthquakes followed by minor basalt extrusion. Similar rifting has apparently occurred at intervals of 100—150 years, i.e. episodic on the time scale of human history, but virtually continuous on a geological time scale.

Repeated eruption is indeed the rule in basaltic provinces. The consequent superposition of numerous flows, commonly only a few metres or less in thickness, gives rise to monotonous sequences, and on erosion to characteristic terraced topography.

Theoretically there should be a dyke accompanying each flow, and dykes should become more numerous downwards within the basalt pile. This geometrical relationship is abundantly realised in the rather special case of sheeted dyke complexes associated with ophiolites, but frequently there is an embarrassing lack of dykes in subaerial basalt terrain even where it is well dissected and exposed. One amusing example of a dyke apparently feeding a recent lava flow, photographed by the writer in the Snake River field, proved on closer examination to be a flow that had flowed back into an open fissure! Several separate centres of extrusion, recognized by dyke swarms, higher proportions of pyroclastics, presence of pahoehoe rather than more distal aa-type flows, have recently been demonstrated in the Columbia River basalt field, and eruptions from these centres must presumably have fed those relatively large surrounding areas where dykes are sparse or lacking (Swanson et al., 1975).

Proximal basalt flows are relatively free running — they are apparently the least viscous of major lave types — and vesiculate readily forming numerous small spherical vesicles. Surfaces tend to be smooth, often in recent flows showing an iridescent blue tarnish, an effect possibly due to the combustion of released gases. Although broadly smooth, "ropy lava" formed by the buckling of a thin congealed crust by fast-running lava below, and shallow elliptical domes some 10—20 m in diameter often breached by a tension gash along their long axis, are common on the surface of these **pahoehoe flows**. Recent pahoehoe flows are commonly seen to consist of an aggregate of elongate sacs, somewhat resembling pillow lavas in cross-section, and formed in a comparable fashion by the continual pouring of highly mobile lava into sacs during flow. These sacs, in constrast to pillow lavas, are markedly elongated in the overall flow direction. Large lava tunnels, elongated in the direction of flow, have been observed to underlie some thick pahoehoe flows.

More distal portions of basalt flows have been observed to move at a barely perceptible pace in a steeply moving front, accompanied by auto-brecciation and a noise comparable to that of breaking crockery, i.e. in a highly viscous fashion. The colour of freshly exposed surfaces on moving lava flows of this type indicates a considerably lower temperature than that of flowing pahoehoe-type lava. Vesicles, if present, are scattered, large and angular, gas release being near complete by this stage. Surfaces of these **aa flows** are unpleasant to traverse being extremely rough and composed of chaotic angular blocks with a surface relief typically up to ~1 m. Cross-sections generally show a conspicuous amount of blocky scoriaceous tops and autobreccia.

The frequently illustrated columnar jointing of basalt is generally restricted to thick flows. Regular large-scale, vertical columnar jointing is characteristic of the lower parts of ponded lava flows which came to rest before crystallizing. This may be overlain by a zone of spectacular smaller-scale curved joint patterns, overlain, in turn, by a non-jointed scoriaceous top. Under static conditions the jointing results in vertical hexagonal columns, but as a result of the stresses in still slowly moving lava this pattern may give way to conjugate shear joints.

An observer cannot help being impressed by the apparent contrast between

the variable structures seen in a recent basalt field and the rather monotonous and structureless sequences of basalt flows in the older geological
record. However, this could perhaps be a superficial impression in view,
for example, of the detailed work on the structures in the Columbia River
basalt field referred to above.

The products of fissure eruptions may be superimposed to form a basaltic
shield volcano with low-angle slopes commonly no steeper than 8 to 10°,
often even less in the upper portions. In plan view a shield volcano may be
roughly elliptical with its long axis following the dominant trend of fissure
eruption. A somewhat equivocal example is Hekla, Iceland. The relatively
steep flanks of Hekla, however, reflect the presence of a proportion of more
fractionated lava products of higher viscosity in addition to basalt. Mature
shield volcanoes commonly tend to assume a more radially symmetrical
circular plan. These range in size from the world's largest ones in Hawaii to
quite small ones. Mauna Loa, for example, has a sub-elliptical plan, subaerial
diameters roughly 60 by 100 km, and vertical height above sea level 4170 m.
It should be noted, however, that Mauna Loa is but one of a cluster of five
coalescing major shield volcanoes that have built up the island of Hawaii,
diameter at sea level nearly 150 km, from ocean-floor depths. Three of
them, older than Mauna Loa, are either extinct or inactive; the smallest and
youngest, Kilauea, is very actively growing and extending the land area of
Hawaii southeastward. Skjaldbreidur in Iceland, situated some 70 km northeast of Hekla, is circular in plan, diameter 10 km with a constructional
height of 600 m.

In some large shield volcanoes, the continued accession of hot magma
within geologically short periods of time raises the country rock temperature
resulting in locally high geothermal gradients. Also, because of the constructional height of the volcanic edifice, fresh magma may not always have
sufficient hydrostatic head to reach the surface as fissure eruptions. Under
these conditions high-level magma chambers develop within which considerable fractional crystallization may take place. One visible topographic
effect is the formation of large collapse calderas inferred to overlie magma
chambers from which periodic withdrawals of magma may occur and at
times exceed replenishment. These collapse calderas have diameters of the
order of 2—5 km; they commonly contain smaller circular steep-sided
"pit craters", diameters of the order of 0.5 km, that may reveal lava pools*.

2.2.1.2. Subglacial basaltic eruptions

Living by chance as we do, in an interglacial period gives geologists the
opportunity to see, superbly exposed in Iceland, the very young products

*It is amusing to watch the thin congealing crust of a basalt lava lake developing triple
junctions, spreading ridges with associated transform fault systems, subduction zones,
etc., and to speculate whether the relatively higher density of a congealed solid crust or
convection within the underlying liquid material is the prime cause of these phenomena.

of subglacial basaltic eruption, which incidentally offer some analogies with and insight into submarine eruption (see Sigvaldason, 1968). In several places in Iceland the products of a series of eruptions from one fissure system have been contained by an ice sheet so as to form rather steep elongate ridges. The ice has so recently melted that little erosion has occurred since. A good example is the Laugarvatn region some 80 km east of Reykjavik (J.G. Jones, 1970), situated at the southern end of one such ridge, the Kalfstindar ridge, some 25 km long and some 1—2 km in width. Vertical sequences up to 500 m thick reveal pillow lavas akin to submarine pillow lavas underlying some 200 m of pillow breccias and massive palagonite tuffs formed from glassy basaltic shards. These are strikingly yellow fragmental rocks forming the characteristic **moberg** topography of parts of central Iceland. The observed upward transition testifies to a marked change from quiet extrusion of pillow lavas to violent explosive disintegration both of pillows to form pillow breccias, and of basaltic magma to form shards of basaltic glass. Vesiculation at low confining pressures has been proposed as a possible mechanism for this explosivity, but this may not be an adequate explanation (see next section). In places the buildup of volcanic products exceeded the ice sheet in height, so that eventually flat-lying subaerially extruded flows came to overlie the palagonite tuffs*. In this event, because the lava flows have much greater resistance to erosion than the underlying fragmental rocks, steep-sided, flat-topped hills and ridges result, the **tuyas** of Mathews (1947). On the flanks of these ridges thick, steeply cross-bedded, poorly graded units are common, indicating extensive reworking of tuffaceous material in water under turbulent conditions. Sometimes these rocks incorporate spherical basalt pillows ~50 cm in diameter, distinct from the larger pillows of the underlying subaqueously erupted pillow lava sequence. These spherical pillows represent products of subaerial eruption that rolled into meltwater down the flanks. The overall envelope shape of the cross-bedded material flanking a moberg ridge commonly consists of one or more flat terraces presumably filling subglacial meltwater chambers of large dimensions adjacent to the ridge.

Basaltic eruption periodically recurs under parts of the present-day Icelandic ice-caps; one manifestation of these is the sudden escape of meltwater in catastrophic amounts — the **jokulhlaups** — as from the presently active subglacial volcano Grimsvötn (Thorarinsson et al., 1960, p. 148).

2.2.1.3. Submarine basaltic eruptions

Submarine basaltic eruptions involving **pillow lava** formation have been observed and filmed at shallow depths off southeastern Hawaii. Little or no explosive activity accompanied this, and divers were able to approach

---

*Palagonite is a yellowish hydrogel, a hydration product of basaltic glass, that also cements the glass and basalt fragments (see G. P. L. Walker, 1965, p. 35.)

within arm's length of, and even collect from, newly forming pillows. Incandescent magma was seen to erupt at the lava—water interface in a bulbous mass around which, in a matter of seconds, a thin dark congealed, glassy crust formed continuously as it swelled. The sac or pillow was seen to distend with the inflow of lava rather like blowing up a balloon; rupture then resulted in the formation anew of another pillow. Recent pillow lavas have been photographed and recovered at depths continuously down to 5000 m in this same area (J. G. Moore, 1965).

Fresh-looking pillow lavas with no sediment accumulation or manganese coating have been photographed at depth along the ocean ridges — mute testimony to igneous activity on a scale now inferred to be an order greater than all subaerial volcanic activity combined, unwitnessed by man and unsuspected by most geologists until the early 1960's.

In the geological record basaltic pillow lavas are frequently associated both with massive flows and with tuffaceous rocks also of basaltic composition. An idealized sequence might be massive lavas (not always present), followed upwards by pillow lavas, overlain in turn by **pillow breccias** and **bedded tuffs**. This kind of sequence, in whole or in part, may be repeated in a cyclical manner (Carlisle, 1963). One such cycle as we have seen would be typical of the Icelandic subglacial sequences.

The observation that pillows can form in shallow water with no explosive activity suggests that explosive vesiculation at depths shallower than that equivalent to the critical containing pressure is not the reason for the formation of these pyroclastic rocks (Sigvaldason, 1968). It has been shown (Colgate and Sigurgeirsson, 1973) that extremely violent autocatalytic explosions can result from the intimate mixing of hot lava and water with large intersurface areas. This probably occurred at Surtsey when rapidly spaced phreatic explosions of great violence resulted from the intimate penetration of seawater through a mass of hot cinders towards an active vent. Limiting depths below which this kind of explosive process would likely not occur are given as 130 and 700 m, depending on the precise mechanism (Peckover et al., 1973). It would appear, therefore, that pillows can and do form at any water depth, but that accompanying violent explosive activity producing submarine pyroclastic material may only (or may not) occur at depths of a few hundred metres or less, i.e. not, for example, at ocean ridges. Given appropriately shallow-water conditions perhaps the presence or lack of pillow breccias and hyalotuff material reflects the rate of eruption — a high rate favouring conditions for brecciation. It is worth noting the genetic and textural distinctions drawn by Honnorez and Kirst (1975) between **hyaloclastites** — "volcaniclastic rocks generated by non-explosive granulation of volcanic glass which takes place when basaltic magmas are quenched by contact with water" — and **hyalotuffs** — "the actual pyroclastic rocks which are generated by phreato-magmatic and phreatic explosions taking place when basaltic volcanoes erupt into shallow waters".

2.2.2. Andesites

2.2.2.1. Subaerial andesitic eruptions

Typical of subaerial andesitic volcanism are **stratovolcanoes**, alternatively known as **composite volcanoes**, those magnificent transients of the geologic scene, with upward steepening concave flanks, their beauty commonly enhanced even at moderate latitudes at the present day by a capping of permanent snow and ice. Their aloof grace belies a struggle of rapid and violent growth against continuous erosion, the latter inevitably triumphant. Fujiyama, Mt. Mayon, and Mt. Egmont, situated in Japan, the Philippines, and New Zealand respectively, are among the best known of large present-day symmetrical examples, and receive their due homage from pilgrims, photographers, stamp designers and the like. The two highest are Guallatiri, Chile, 6060 m, and Cotopaxi, Ecuador, 5897 m. Despite differences in altitude, the constructive elevation of stratovolcanoes is commonly up to 2000 m, exceptionally 3000 m, above base level. The volume of products of a stratovolcano is commonly of the order of 100 km^3, erupted within a geologically brief period of time.

The names stratovolcano and composite volcano derive from their being composed of interbedded tephra and flows, the latter present in subordinate amounts; E is thus high, usually greater than 90, commonly around 95, sometimes even 99. Their shape derives from the angle of rest of the predominating tephra coupled with erosional processes. A high proportion of large ejected blocks near the summit vent area results in slopes up to 35° in inclination. The somewhat shallower lower slopes underlain by high proportions of lapilli and ash are prone to mass wastage by radial gullying, rapid stream erosion and mud flows, thus further reducing the angle of slope until it merges into a surrounding extensive ring-plain underlain by laharic and alluvial outwash material.

Mudflows are exceedingly common, and are initiated, for example, by (i) ash fall and heavy rains often associated with atmospheric convection during eruption and accompanied by frightening thunder, lightning and darkness, or (ii) catastrophic breach of a summit crater lake, itself possibly augmented by melting of snow and ice during an eruption. These volcanic mudflows termed lahars contain a variable and usually high proportion of muddy ash and exhibit a pronounced lack of sorting. They usually consist entirely of fragments of volcanic material derived from the flanks of the stratovolcano, although the distal portions of some huge lahars may incorporate non-volcanic material from over-run drainage systems. Much of the ring plain immediately surrounding the stratovolcano is underlain by laharic material, which in proximal zones may contain irregularly scattered block mounds of laharic debris of large size. Reworking of laharic material results in an outer surrounding area underlain by immature alluvial volcanic sediments. Laharic material bears a gross physical resemblance to till. Striated cobbles have been recorded in lahars but abrasion is never so intense as to produce "flat-iron" fragments of hard rocks as in some glacial moraine. The distinguishing of lithified laharic material from tillite in the geological record is not unequivocal, however, and indeed a diamictite in stratovolcanic terrain could have had a dual origin: in the North Island of New Zealand, for example, it has been demonstrated that some recent tills have been subsequently subjected to mass movement as mudflows. Dimensions of individual lahars vary greatly. The large well-documented Osceola mudflow at Mount Rainier (Crandell and Mullineaux, 1967) travelled ~60 km along a radial valley, submerging parts of it during its passage to depths of over 100 m, and spread out in a lobate mass over 160 km² of surrounding lowland leaving a deposit up to 20 m thick. Numerous lahars around the comparatively modest strativolcano of Vesuvius have been recorded in historic times; one of them buried Herculaneum.

The summit crater area of a stratovolcano is often complex and variable in detail. Where a young stratovolcano is actively erupting at short intervals, the orifice or vent is a small feature of the overall constructional shape of the volcano. In older, less active volcanoes, there may be time between eruptions for the crater area to be enlarged by inward collapse of the walls to form a typical **crater caldera** with a diameter of the order of a kilometre. In the intervals between eruption this can be seen to be an area of intense fumarole activity, and is commonly also the site of a crater lake. Small eruptions may result in small subsidiary ash cones or tholoid domes entirely within a crater caldera. Lavas may erupt not only from within the central summit crater area but also from the upper flanks of the volcano. Large explosive eruptions may destroy the radial symmetry of the summit area and result in land forms like the somma of Vesuvius after the eruption of A.D. 79, and that of Mount St. Helens after the eruption of 1980.

Catastrophic explosive eruptions which seem to characterize the late stage of some but not all stratovolcanoes, may result in the partial collapse of the upper structure of a stratovolcano to form an **explosion caldera**, typically of the order 4—10 km in diameter. Crater Lake, Oregon, on the site of the ancestral Mount Mazama is a familiar example. Caldera formation in this manner may be repeated more than once on the site of the same volcano as for example at Tongariro, New Zealand. A more detailed discussion of this phenomenon is given in the succeeding section as it characterizes a form of rhyolitic extrusive activity.

The rates of growth and decay of stratovolcanoes are surprisingly rapid. In the central part of the North Island of New Zealand, for example, (Grindley, 1965) there are three well-documented andesitic stratovolcanoes, Tongariro (1968 m), Ruapehu (2796 m), and Ngauruhoe (2290 m). Andesitic volcanism apparently began not much before early Pleistocene time as indicated by the influx of clastic andesitic material into nearby sedimentary sequences. Tongariro, the oldest of the trio, is now an extinct basal wreck, reduced by a combination of collapse and explosion at many different centres, followed by deep erosion. It is surrounded by the remnants of a laharic ring plain, itself deeply dissected and preserved on spurs and planezes, with dips up to 30° radially outward. It is at times difficult, even in these recent deposits with landforms to go on as well, to distinguish with confidence between laharic material and glacial till. Ruapehu, situated some 20 km to the south of Tongariro, is a mature high stratovolcano with a historical record of activity limited to ash deposits and tholoids in the summit crater lake region. Ruapehu is suffering ungoing erosion and is surrounded by a laharic ring plain. In 1953 a lahar flow, resulting from the partial breach of the summit crater lake, overwhelmed and destroyed a railway bridge in the path of an oncoming express train with the resulting loss of 151 lives. The third volcano, the very youthful cone of Ngauruhoe, has formed a constructional relief of ~1000 m above the surrounding plain in only 10,000 years. Ash eruptions are continual, occurring every year or two, and the pyroclastic rocks on the steep unconsolidated slopes are interbedded with minor lava flows of which there have been several during the century or so of European observation.

Mount Rainier in the state of Washington, one of the High Cascades peaks, is another example of a large stratovolcano, fully described and illustrated by Fiske et al. (1963). Constructed since the early Pleistocene of parental pyroxene andesite, from a base level of ~2400 m, the summit of this imposing volcano at the present day reaches 4392 m and is inferred to have had a maximum height ~300 m higher than this. The last significant eruption produced a pumice layer ~600 years ago, and the volcano is presently undergoing spectacular denudation by glacial erosion, avalanches, rock falls, mudflows and floods. The present-day volcano is constructed on a basement of young rocks which represent the products of a complicated history of volcanism over a 20-Ma time span from late Eocene to late Miocene. These include the upper Eocene Ohanapecosh Formation, 3000 m in thickness, composed predominantly of sediments, mudflows and subaqueous pyroclastic flows. These were folded, regionally metamorphosed to zeolite facies, uplifted, and deeply eroded before being overlain by up to 2400 m of ash flows, basalts and basaltic andesites of around early Miocene age. These, in turn, were folded, faulted, uplifted and eroded before the emplacement of the late Miocene Tatoosh pluton, one of several comparable intrusions in the area with outcrop diameters of the order of 10—20 km. The predominant rock type of the Tatoosh pluton is hornblende-biotite granodiorite, but quartz monzonite is also abundant, and lesser amounts of pyroxene quartz-diorite and granophyre occur. The pluton is a high-level intrusion with the form of multiple sills. It has breached its roof at several places as shown by plugs associated with lava flows, welded tuffs and pumice lapilli tuffs, overlying vesiculated and autobrecciated varieties of the pluton. The foregoing emphasizes the enormous variety of rocks emplaced within a geologically short interval of time in igneous activity of this type. Their relative youth enables relative ages and stratigraphy to be worked out with confidence at Mount Rainier. However, similar complexities in the older geological record would prove very difficult to decipher, and geochronological investigation would not have sufficient resolution to assign relative ages.

Geological mapping both in recently active and in older stratovolcanic terrains is commonly bedeviled by details such as numerous local uncon-formities of little time significance and great lateral facies variations*. Broad facies mapping is at times practical, a concept formulated by Parsons (1968). For example, a proximal "vent facies" would include coarse tephra inter-bedded with lavas, minor intrusions and diatremes, and is commonly a zone of more or less intense propylitization. An "alluvial facies" would encompass the considerable volumes of volcanogenic epiclastic deposits accumulated in a more distal environment interbedded with tuffs and ashflow deposits. A "plutonic facies", applicable only to older dissected terrain, would include any large, associated sub-volcanic intrusions typically of epizonal ring dike/cauldron subsidence habit. A "marine facies" applicable to island or coastal stratovolcanoes would include both epiclastic and pyroclastic material laid down in seawater (C.J. Hughes and Brückner, 1971). The amount of volcanic material represented in penecontemporaneous sedimentary rocks is indeed large in volcanicity of this type and commonly, one suspects, still underestimated.

2.2.2.2. Submarine andesitic volcanism

Many present-day islands characterized by andesitic volcanism may be underlain by a pedestal composed mainly of rocks of basaltic compositions rather than andesitic (see Chapter 8), and thus submarine andesitic volcanics may not be abundant in the volcanic record. Lacking contemporary obser-vation, we must turn to the geological record for evidence. An informative account of an Ordovician volcanic marine sequence, 6 km in thickness, passing upwards from predominantly basaltic to some more andesitic rock types, and interpreted as a fossil island arc, is given by Kean and Strong (1975). A thick sequence of basaltic pillow lavas, in part brecciated, passes upwards into pyroclastic and volcanoclastic materials of andesitic derivation, interbedded with pillowed andesite flows, overlain in turn by shallow-water limestones, andesitic agglomerates and some subaerial tuffs. The account includes petrographic descriptions and chemical analyses of the now some-what altered rocks of the sequence.

---

*Study, for example, the complex association of lavas, mudflows, ash and nuée ardente deposits resulting from but one recent eruption of Mayon (P. Francis, 1976, pp. 197—201, especially fig. 48), and the comparable deposits resulting from eruptions of Mount St. Helens, the most continually active of the High Cascades volcanoes in Holocene times (see Crandell and Mullineaux, 1975).

In a manner reminiscent of meteorite "finds" and "falls", one notes that andesite is the most abundant lava type seen to erupt at the present day, but that for reasons of high explosion index and proneness to denudation andesite flows are rare in the geological record.

2.2.3. Rhyolites

2.2.3.1. Subaerial rhyolitic lava flows

Rhyolitic magma typically intrudes uppermost crustal levels as domes, of the order of up to hundreds of metres in diameter, which may or may not breach the surface, i.e. they may technically be intrusive rather than extrusive. This may be an academic distinction as for example with Syowa Sinzan, a dome of hypersthene dacite which erupted in 1944 in Hokkaido, the northern island of the Japanese archipelago. This "intruded" within a carapace of a few metres of superficial soil, etc., and reached a constructional height of 150 m in ten months (Yagi, 1953). Erosion has already removed most of the soil cover but isolated pieces remain. Extrusive domes (synonym tholoids) are commonly surrounded by a mantle of talus, the pieces of which were often originally formed by marginal autobrecciation of the rhyolite itself, and may sometimes be difficult to identify as talus in the geological record. Where the tholoid is large or extruded on a slope, as for example the upper slope of a stratovolcano, a short, thick viscous flow or coulée may develop from it but seldom attains a length of more than 1 km. See, for example, the photographs, drawings and description of the (trachytic) tholoid and coulée erupted on Tristan da Cunha in 1961 (P. E. Baker et al., 1964). Flow banding, commonly accentuated by alternating bands of glassy and spherulitic material, is common, and is frequently preserved even in old, completely devitrified rhyolites as alternating bands of different colour. Autobrecciation is common in addition to flow banding, and in fact these two structures are rarely absent from rhyolitic lava flows or from near-surface rhyolitic intrusions. Unusually large rhyolitic lava flows, as for example the large flow west of Yellowstone, may become autobrecciated throughout in their distal portions and thus come to resemble a pile of bouldery rubble. Pieces in a well-mixed autobreccia such as this can appear remarkably different and include juxtaposed glassy, flow-banded, and lithic varieties of different colours; close examination, however, will reveal that they are all of the same rhyolitic composition with similar phenocryst content (monolithologic), a feature, of course, of all autobreccias and necessarily so because they belong to the same extrusion.

2.2.3.2. Ash flows and recent ignimbrites

Let us consider now the far more common form of eruption of material of rhyolitic compositions, namely that produced by an inferred ash-flow mechanism. It has been estimated that at least 95% of subaerial rhyolites erupted not as lava flows but as ash flows. Ash flows are presumed to be comparable to observed **nuées ardentes**; if so, some must have been vastly greater in scale than any nuées of which we have historical record. The rock type produced by an inferred ash flow can be called an **ash-flow tuff** or alternatively **ignimbrite** (which may, or may not, be wholly or partially welded, and hence is more generally applicable than the term "welded tuff"). The term **sillar** has been applied to unwelded ignimbrites that are however indurated and coherent by reason of recrystallization during cooling (see Fenner 1948). The overwhelming majority of ignimbrites are rhyolitic (sensu lato) in composition and possess characteristic structures distinct from those of lava flows of comparable compositions.

In Indonesia, a part of the world where nuées ardentes have frequently been witnessed, they have been subdivided (see van Bemmelen, 1949, p. 193) into two. The first, **nuées d'avalanche**, typically originate where parts of a still hot but extremely viscous and essentially solid rhyolite coulée on the steep flank of a stratovolcano may break away. The released blocks roll down, spall and break up, liberating much hot gas from within the rhyolite. This process of mechanical and explosive disintegration results in a small avalanching hot, dusty cloud, tending to roll downhill because of its impetus and bulk density. The residual deposits of a nuée d'avalanche are negligible, but the hot cloud is capable of searing the lungs and killing. A village in its path may be left structurally intact, with no fire damage even to dry grass roofs, but the life of humans and livestock may be extinguished. Indonesians call this the "black death". The second much larger and more catastrophic type, a **nuée d'explosion**, comparable to Pelée and Katmai, is the one referred to above as parental to ignimbrites.

At Mont Pelée, Martinique, Lesser Antilles, West Indies, a viscous plug of rhyolite was pushed out of a dome in a small volcanic complex over a period of nine months. The curious, continually crumbling "spine" so produced, ~200 m in diameter, slickensided along parts of its outer contacts, had a total constructive height of 850 m, and was evidently a reflection of great underlying pressure. From places around the foot of the spine, several nuées ardentes were erupted over a period of months. Sudden intermittent explosive release of hot vesiculating magma, gas, dust and rock fragments produced not only voluminous clouds of ash and debris, but also in each eruption a dense billowing nuée, dull-red in colour, hugging the ground and travelling downslope at a velocity of ~60 km/hr. Several of the nuées conveniently became channelized into radial valleys away from habitation. One, however, on the morning of May 8, 1902, followed a different course and devastated the town of St. Pierre, killing almost instantaneously all 30,000 inhabitants except, so the story goes, the fortunate although badly burnt inmate of a subterranean dungeon that had only one tiny aperture that faced away from the advancing nuée. The sole geological evidence of the passage of the nuée was a layer of 30 cm of unconsolidated vitric ash, insignificant one might add when compared with the thickness and the areal extent of some ignimbrites in the geological record.

The much larger eruption of Katmai, 1912, was associated with the synchronous formation of a collapse caldera ~5 by 3 km across. The nuée, witnessed from a boat at sea, was channelized in a deep glacial valley. Products of the eruption reached a maximum of some hundreds of metres in thickness with a generally flat top surface. Rising steam, generated probably in large part from the underlying water-logged ground, escaped in numerous fumaroles that lasted up to several decades, resulting in the name of the "Valley of the 10,000 Smokes". Subsequent rapid gully erosion has revealed a little of the structure of the upper part of this ignimbrite or ash-flow deposit. A detailed account of the stratigraphy of the ejectamenta resulting from this eruption and its complexity in detail is given by Curtis (1968).

Indian legend, handed down by word of mouth over more than 200 generations, records the catastrophic eruption of the ancestral stratovolcano, Mount Mazama, ~8000 years before present, to produce the present Crater Lake (a lake-filled explosion caldera ~6 km in diameter with a small younger volcanic cone inside it), together with thick ignimbrite near the volcano, and an extensive ash deposit over much of the northwestern U.S.A. (Howel Williams, 1942). The fragmental products that have been identified and mapped are estimated to aggregate to 40 km³, a figure that corresponds well enough with the estimated 60 km³ representing the vanished upper part of the ancestral Mount Mazama.

The Bishop Tuff is a very large composite Pleistocene ignimbrite, erupted ~0.7 Ma ago, that outcrops over ~1000 km² in eastern California, just S.E. of Mono Lake. With an estimated average aggregate thickness of 150 m, the several component ignimbrite member flows constitute a total volume of 150 km³. To this figure should be added a further inferred volume of 350 km³ of intracaldera ignimbrite almost wholly buried within the huge parental Long Valley caldera nearby, ~17 × 32 km in size. Allowance should also be made for an estimated 300 km³ of accompanying air-fall ash material dispersed over much of the western U.S.A. thus making a grand total, allowing for the porosity of the pyroclastic rocks, of ~600 km³ of magma (R. A. Bailey et al., 1976). As a result of this young age, the form and structure of the Bishop Tuff outside the caldera area are easily comprehensible, and examination is facilitated by deep gorges incised into its remarkably flat upper surface.

The account of the Bishop Tuff by Gilbert (1938) is a model of lucid petrographic description and interpretation, which, although just postdating Marshall's (1935) classic account of New Zealand ignimbrites (Marshall coined the term ignimbrite), anticipates a great deal of the concepts in the classic syntheses on ash flows of R. L. Smith (1960) and Ross and R. L. Smith (1961)*.

---

*See the excellent historical introduction by C. E. Chapin and W. E. Elston and other specialised articles in the *Special Paper of the Geological Society of America* entitled "Ash-flow tuffs" (Chapin and Elston, 1979).

The geologic setting of the area was one of continual rhyolitic and andesitic volcanism in Miocene times followed by the extrusion in early Pliocene times of olivine basalts and a large rhyolite mass. Numerous normal faults have been active up to the present time and movement on some is younger than the Bishop Tuff which itself overlies a Pleistocene moraine. Since the eruption of the Bishop Tuff, there have been basalt flows, andesite domes, rhyolite domes and explosive rhyolite eruptions depositing rhyolitic pumice. Some of the ash-flow eruptions were so closely spaced in time that their resultant stacked ignimbrites behave as one cooling unit. Other more widely-spaced ash-flow eruptions evidently permitted some alluvial reworking of soft, ashy ignimbrite tops between successive eruptions.

A comparison of these geologically very young ash-flow deposits with many older ignimbrites in the geological record including the Precambrian leaves little room to doubt their overall similarity and a presumed similar mode of formation.

2.2.3.3. Structure and origin of ignimbrites

The four components of ignimbrites are *vitric shards*, variable proportions of *pumice fragments* and *crystals* (often partly broken), all representing the parental magma of the eruption, and a further variable proportion of angular *lithic fragments* reamed from the walls of the vent during the extrusion process. Although the relative proportions of these four components often appear to show little or no variation within individual outcrops of ignimbrite, detailed studies of sections of young ignimbrites show that this first impression of homogeneity may be misleading.

In addition to being sandwiched between a thin, underlying ground surge deposit and a thin, overlying ash deposit, all resulting from the one eruptive episode, individual ignimbrite units may reveal a variation in the size and frequency of clasts. The tendency is for the larger lithic clasts to be concentrated towards the base and for pumice fragments to become concentrated towards the top. The very lowest part of the ignimbrite may be poor in lithic fragments and show a reverse grading in clast size (Sparks et al., 1973). Roobol (1976) has demonstrated a considerable amount of fractionation between lower welded ignimbrite zones relatively enriched in lithic and crystal material and upper soft zones relatively richer in vitric material. These fine-grained, soft, ashy tops of ignimbrites are particularly prone to erosion and are thus commonly not preserved in situ in the geological record, so that older ignimbrites often give the impression of being crystal-rich. As well as this possible *vertical* variation reflecting the behaviour of clasts in the flow regime of the nuée and accompanying ground surge and ash clouds, regional investigation often reveals a *lateral* variation. Kuno et al. (1964), for example, studied some 200 outcrops over an area of some 10^4 km^2 of a large Pleistocene ignimbrite originating from the Towada caldera, Japan, and documented a significant lateral variation in the abundance and maximum size of lithic clasts, the larger lithic clasts as would be expected proving to be more abundant near the vent area. See also Briggs (1976a, b) for documentation of lateral variation within a New Zealand ignimbrite unit.

Superimposed on these mechanical effects are the post-accumulation flattening, welding, and devitrification processes that account for much of the distinctive and varied appearances of ignimbrites. R. L. Smith (1960) and Ross and R. L. Smith (1961) in landmark review papers suggest accumulation of ignimbrites at temperatures around 600°C and above. The very marked vertical variation in appearance within most ignimbrites reflects: (i) a progressively higher degree of flattening and welding of the still hot shards and pumice fragments (**fiamme**) towards the base (but often excluding a lowest zone of less welded material in contact with cold bedrock), resulting in the **eutaxitic structure** that is specific to ignimbrites; and (ii) a variable degree of **primary devitrification** of glassy material during cooling, generally more complete towards the top of the ignimbrite where devitrification was catalysed by the upward passage of hot gases. A young ignimbrite devoid of lithic fragments (which on incorporation are cold and thus reduce the bulk temperature of the ignimbrite) could, therefore, show an upward variation in texture as follows: (a) a relatively unwelded base; (b) vitrophyre of dense, black, obsidian-like appearance; (c) vitrophyre with scattered spherulites produced during high-temperature, primary devitrification; (d) lithoidal (i.e. devitrified and crystalline as opposed to glassy) matrix with prominent, flattened, black pumice fragments; (e) overall lithoidal rock; (f) similar but increasingly porous on a small scale upwards; and (g) unconsolidated ashy top resembling air-fall material. In the geological record, lithification and a later **secondary devitrification** (commonly complete in all originally glassy acid igneous rocks older than the Tertiary) can be anticipated. Many ignimbrites show vertical columnar jointing, developed on a more massive scale than that of some basalts and diabases but for the same reason — cooling of a static mass.

Once the nature of these complications is appreciated, even individual ignimbrites, both young and old, can often be identified in outcrop in areas exposing many ignimbrites, not only on the basis of a distinctive proportion of crystals and lithic fragments but also on the basis of their structure and appearance.

Several closely spaced nuées may result in superimposed ignimbrites that accumulate over such a short period of time that they effectively become one **cooling unit**, with an overall upward pattern of variation in degree of welding, devitrification, etc., superimposed on the stack of individual flows.

Some apparently cogenetic ignimbrites of one cooling unit show distinct chemical variation in the sense of a more basic composition upwards (usually over a rather small compositional range). The ignimbrites resulting from the explosion caldera of Mount Mazama, for example, show a transition from pale-grey dacitic varieties below to darker-grey andesitic varieties above. One of the clearest descriptions and documentation of this phenomenon is that by Katsui (1963, pp. 641—646), who records increases in density, darkness in colour, and phenocryst content upwards resulting in a change from rhyolitic pumice to dacitic ignimbrite in erupted products associated with caldera formation of Shikotsu and Mashu volcanoes in late Pleistocene time in Hokkaido, Japan.

This phenomenon is interpreted as reflecting the progressive evacuation of a zoned magma chamber from the top downwards. There is independent isotopic evidence for the existence of a long-lived (0.14 Ma) differentiating magma chamber below the andesite volcano of Irazú, Costa Rica, (Allègre and Condomines, 1976), such longevity would favour processes of crystal fractionation and upward diffusion of water within the magma chamber (see Chapter 7). The overall effect of these inferred processes results in the production of compositionally-zoned magma chambers with increasing water contents upwards and developing in their upper parts relatively acid magmas under pressures potentially greater than confining pressure. The weakest link in the covering rocks between this potentially explosive magma chamber and the surface, such as a previously formed volcanic conduit, a fault, a joint system, or a newly formed fissure, may thus become the site of explosive eruption of nuée ardente type. Catastrophic evacuation of the zoned magma chamber from the top downwards could, therefore, result in the observed reverse zoning in the accumulated ignimbrite product of the eruption.

An instructive parallel has been drawn (Healy, 1962) between the "flashing" of water into steam in periodic geyser eruptions and the paroxysmic eruption of vesiculating ignimbrite that causes large near-surface chambers of acidic magma to lose magma, heat, and gas, and much of their eruptive potential in catastrophic evacuations, leaving behind only relatively small amounts of degassed magma that may later erupt as domes and tholoids.

It should be noted that the larger ignimbrites plus their accompanying air-fall ash deposits represent, by up to two orders of magnitude, the largest individual volcanic eruptions on Earth. In some instances their volume (equivalent to that of the catastrophically evacuated magma chamber) exceeds that of the entire products of an andesitic volcano, a factor of petrogenetic significance.

2.2.3.4. Geologic setting of rhyolite eruption

Rhyolite eruptions occur in diverse geologic settings. The most obvious from present-day land forms are those associated with the formation of explosion calderas centred on mature andesitic stratovolcanoes, Crater Lake being an outstanding example.

Within the San Juan volcanic field of Colorado the detailed compilation of Steven and Lipman (1976) shows that eighteen major ignimbrites were deposited and as many related calderas developed in late Oligocene time over a short time span (30—26 Ma ago). The ignimbrites overlie and overlap the products of a coalescing assemblage of early Oligocene (35—30 Ma ago) stratovolcanoes that erupted voluminous andesitic material over an area of $2.5 \cdot 10^4$ km^2. The development of the calderas is believed to chronicle the emplacement of successive segments of an underlying shallow batholith, strongly differentiated in places to produce large volumes of silicic rock, its position indicated by a major gravity low with sharp marginal gradients. The ignimbrites have an estimated total volume in excess of 9000 km^3; the largest ignimbrite unit, the Fish Canyon Tuff, has an estimated volume in excess of 3000 km^3 and is associated with the enormous La Garita caldera ~30—40 km in diameter (in part obliterated by later caldera eruptions). These ignimbrites, therefore, appear to belong to a definable later stage of igneous activity than the preceding andesites.

Not all calderas, however, are associated, even tenuously, with andesitic volcanoes. Some calderas associated with voluminous rhyolitic pumice and ignimbrite vent in Mexico (J. V. Wright and G. P. L. Walker, 1977) is covered example the Taupo volcanic zone of the North Island of New Zealand. The Taupo zone forms a southern extension of the back-arc spreading Lau Basin where the latter intersects continental crust in New Zealand. Overall, the Taupo zone is a large complex graben structure of Pleistocene age ~200 km in length and ~50 km in width. Numerous active normal faults, both bounding and within the graben, have resulted in a maximum aggregate downthrow of 3.5 km. The graben is filled with young volcanic rocks; extrapolated estimates based on surface exposures indicate that these comprise at least ~40 km^3 of high alumina basalt, ~800 km^3 of andesite, and 16,000 km^3 of greatly predominating rhyolitic material (Healy, 1962). The products of rhyolitic eruption overlap the edges of this great volcano-tectonic depression making exact delineation of its shape difficult without recourse to interpretation of geophysical measurements. Four large shallow calderas, ranging from 15 to 30 km in diameter, now mainly occupied by lakes, have apparently acted as the centres for large ignimbrite eruptions. These were succeeded by the eruption of rhyolite domes emplaced both around the edges and in the central parts of some of the calderas, followed in turn by violent ash-fall eruptions producing pumice deposits which blanket the area (Ewart and Stipp, 1968).

Voluminous ignimbrites of the classic Basin and Range province extend over $1.5 \cdot 10^5$ km^2 in central and eastern Nevada and southwest Utah. The central part of the province has been lucidly described by E. F. Cook (1965), who records ignimbrite sequences up to 1 km thick. Radiometric dating indicates that although there is some overlap, the ignimbrites, from 34 to 22 Ma old are mainly younger than high-K andesites, 37—33 Ma old, present in the same general area (McKee and Silberman, 1970). Crustal extension, involving the characteristic half-graben faulting, and associated with bimodal basalt/rhyolite volcanism, began suddenly at a significantly later date. The most sharply defined dates in the above generalized sequence are the onset of andesitic volcanism (37 ± 1 Ma) and the commencement of crustal extension (16.5 ± 0.5 Ma), both apparently occurring almost simultaneously over this part of the Great Basin*. The voluminous early ignimbrites are of calc-alkaline affinity and may represent the later stages of the preceding andesitic volcanism. Younger ignimbrites, however, tend to be highly silicic and are petrographically and chemically similar to rhyolites in areas of bimodal basalt—rhyolite volcanism (Noble, 1972). Several calderas have been recorded (see, e.g. Noble et al., 1968) but many of the ignimbrites are considered to have been erupted from linear fissures. Evidence, partly circumstantial, for a linear vent zone associated with younger (15 Ma) rhyolitic ash and lava flows in northwest Nevada is given by Korringa (1973). Ignimbrite plugs and dykes fill vents from which some Oligocene ignimbrite was erupted in Nye County, southern Nevada (H. E. Cook, 1968). A massive fissure vent, 1 km long, 60 m wide in its deepest part exposed and flaring rapidly upwards to over 400 m wide over a vertical distance of 400 m, merges with an ash-flow unit, 25 Ma in age, in west-central Nevada, and has been interpreted as a feeder to it (Ekren and Byers, 1976).

The question of source vents for large ignimbrites in apparently calderaless terrain remains somewhat problematical (see, e.g., the discussion in G. A. Macdonald, 1972, pp. 268—271). They would have to be of a fair size to feed the quantities involved in large ash-flow eruptions, and notwithstanding the above discoveries, they appear to be embarrassingly rare in the geological record, even in appropriately dissected terrain. One inferred igimbrite vent in Mexico (J. V. Wright and G. P. L. Walker, 1977) is covered by a "co-ignimbrite lag-fall deposit" consisting mainly of large lithic fragments, believed to have been formed by the collapse of an eruptive column of nuée ardente material.

In the older geological record, volcanic rocks including rhyolitic lavas and ash flows can be seen to be associated in space and time with shallow epizone intrusions, commonly in subsided central blocks (see ring complexes in next chapter). Examples include the classic Devonian complexes of Glen Coe and Ben Nevis, Scotland, (E. B. Bailey, 1960) and the Jurassic "Younger Granites" of northern Nigeria (Jacobson et al., 1958). In the latter area there are numerous ring dykes and cauldron subsidence intrusions emplaced at high crustal levels, and in places the magmas breached the surface resulting in rhyolitic flows, some of them distinctively peralkaline, matching closely the compositions of the intrusive rocks. Surface landforms of the volcanic activity are not preserved, but large volcanoes do not appear to have been formed.

---

*Note again, in passing, how the relative ages are easily obtained in these youthful rocks, but in contrast how radiometric data alone would scarcely resolve these events in say Palaeozoic or older rocks.

Some large circular or sub-circular structures in rhyolite volcanic fields could, therefore, reflect the position of underlying epizonal magma chambers. Vents might be aligned around the circumference of the structure or perhaps within it. Some kind of connection in this way of surface volcanism with underlying magma chambers is suggested by the phenomenon of resurgent cauldrons, of which the Valles Caldera, New Mexico, is an outstanding example.

The Valles caldera is apparently just old enough to encompass a complete sequence of events relating to the development of the caldera and young enough for the evidence to be well preserved. The history of the caldera began with the eruption of the Bandelier tuff — actually very large masses of ignimbrite forming two cooling units, each ~ 200 km^3 in volume, erupted 1.4 and 1.0 Ma ago. A 20-km-diameter caldera formed during the later eruption and largely obliterated evidence of a caldera of comparable size associated with the earlier eruption. The younger caldera became the site of further intermittent eruption of rhyolite domes and pyroclastics over a time span of a million years to the present (Doell et al., 1968). R. L. Smith and R. A. Bailey (1968), in a comparison of the Valles caldera with similar structures elsewhere, drew attention to: (1) early doming of an area some-what greater than that of the area of the calderas and the generation of ring fracturing prior to any large ignimbrite extrusion; and (2) resurgent doming early in the post-collapse history of the caldera. A resurgent dome, some 10 km in diameter, occupies a central position within the caldera and displays an intricate pattern of normal faulting producing structural relief up to 1 km. Post-caldera volcanism occurs both within the central resurgent dome and around the original caldera rim zone, comprising domes, plugs, lava flows and pyroclastic cones; the eruptive material is silicic rhyolite throughout.

Smith and Bailey demonstrate that these phenomena are paralleled by several other examples; they are thus not fortuitous but related presumably to a common cause that most likely is magmatic pressure. Resurgent calderas then are held to reflect the accession of large bodies of underlying acid magma by large-scale stoping along ring fractures to shallow levels in the Earth's crust and thus distinct from andesitic stratovolcanoes. At Valles and related examples we witness the surface manifestations of this; in older central complexes we see only the complementary intrusive rocks (see Chapter 3).

The contrast between the readily observed continual relatively quiet upwelling of magma in areas of active basaltic volcanism and the legitimately inferred rare catastrophic nature of the larger rhyolitic eruptions is indeed striking. Catastrophism is not a favoured word with geologists who are conventionally trained to think in terms of "uniformitarianism", or perhaps better "actualism" (see discussion in Holmes, 1965, p. 44). However, we should not forget the puny time scale of recorded human history, some 0.005 Ma at best, compared with that of geological history. Some geological phenomena could be so infrequent as to be indeed statistically unlikely to have been recorded, and events leading to the production of the large ignimbrites would appear to belong to this category, perhaps fortunately so in view of their catastrophic nature. Ironically in this connection, one civilization, the Minoan, may not have had the opportunity to record an event of this dimension because it was apparently extinguished by the eruption of Santorini about 1470 B.C. Ruins have been excavated from a layer of ~30 m of pumice lapilli at Santorini, a picturesque innocent-looking group of islands that form an incomplete rim to a caldera~10 km in diameter (Bond and Sparks, 1976)*.

One further actualistic comment is that one cannot fail to be impressed with the frequency of young ignimbrites in many recently glaciated areas of the northern hemisphere. Their abundance in the Holocene of Hokkaido, the northernmost island of Japan for example, just does not square with their lesser abundance in the preceding geological record there. One must reckon, of course, just as with andesite stratovolcanoes, with rapid erosion and the eventual preservation of much igneous material, particularly the unwelded ashy tops of ignimbrites, as epiclastic deposits. Nevertheless, it may be that a marginal reduction in confining pressure on high-level magma chambers, caused by the removal of ice by melting in an interglacial period, could have triggered the explosive release of material to form ash flows.

2.2.3.5. Subaqueous rhyolitic volcanism

Fragmental products are the rule in subaqueous rhyolitic volcanism. A rare example of a "pseudo-pillow" structure developed in dacite which extruded into water has been described from the Aso caldera, Kyushu, Japan (Watanabe and Katsui, 1976):

> ..."The mechanism of formation of the pseudo-pillow lavas can be interpreted as follows: The viscous lava is cracked with curved or spheroidal fissures during movement and cooling in water. Water penetrates into the fissures, and cooling contraction produces columnar joints perpendicular to the fissure. Then, pillow-like blocks split away, and the outer rim of the blocks cuts the flow layers obliquely. This is an important feature that distinguishes the acidic pseudo-pillows from basaltic pillows."

The formation of these pseudo-pillows thus occurred when the lava was essentially already solidified as opposed to true pillows which form a bulbous sac of liquid magma. These pseudo-pillows are somewhat of a curiosity.

On a much larger scale, Fiske (1963) has given a classic description of voluminous **subaqueous pyroclastic flows** that constitute almost half of the 3000 m thick Ohanapecosh Formation in the eastern part of Mount Rainier National Park, Washington.

---

*See, however, Pichler and Schiering (1977) for evidence that a devastating earthquake may have been the actual cause of the demise of the Minoan civilization.

... "The Ohanapecosh subaqueous pyroclastic flows are extensive, nonwelded deposits of lapilli—tuff or fine tuff—breccia ranging in thickness from 10 to more than 200 feet. They are interbedded with thinner and generally finer turbidity-current and ash-fall deposits formed by smaller and more water-rich slumps of pyroclastic debris from the underwater volcanoes and by ash falls that rained into the water.

The three main types of flows are thought to be related to three different kinds of volcanic activity. The most common flows — those containing a variety of lithic fragments and variable amounts of pumice — were probably produced by underwater phreatic eruptions. The flows rich in pumice and glass shards were probably caused by underwater eruptions of rapidly vesiculating magma which, on land, would have produced hot ash flows or ash falls. The least common flows — those containing only one or two kinds of lithic fragments — were probably derived from fairly homogeneous bodies, such as domes, spines, and lava flows that were erupted into water and fragmented by steam-blast explosions."

The parental magmatic material ranges in composition from rhyolite through dacite to andesite.

The petrography and structure of a subaqueous pyroclastic flow unit, the Caradocian Frondderw Ash of the Bala district, North Wales, has been described and illustrated in considerable detail by Schiener (1970) and usefully contrasted with a subaerial vitric tuff and a water-lain vitric tuff occurring in the same area.

Niem (1977) describes two thick tuff units of Mississippian age from the deep-marine Ouachita flysch basin, and infers that:

... "These tuffs were probably formed by highly explosive eruptions of vesiculating acidic magma from a vent or fissure that produced incandescent avalanches of pyroclastic debris and accompanying ash clouds. The hot turbulent suspensions were rapidly quenched by sea water to form steam-inflated density slurries that flowed into the Ouachita basin. Pyroclastic flows created thick, density-graded, pumiceous vitric-crystal tuff. Numerous smaller density slurries following the main flow in rapid succession deposited the overlying bedded pumiceous tuff. Toward the end of each volcanic eruption, continuous settling of fine ash formed thick, fine-grained upper vitric tuff."

Niem also comments that:

"the interaction of volcanic and sedimentary processes in the formation of submarine pyroclastic deposits generally has been overlooked by volcanologists and sedimentologists".

Considerable work on acid submarine volcanism has been accomplished in the "Green Tuff" regions of Japan in connection with the well-known Kuroko stratabound sulphide deposits which are of considerable economic importance. The following comments are taken from the well-referenced reviews of T. Sato (1974) and Lambert and T. Sato (1974). Subsidence in these regions below sea level in latest Oligocene time was followed by violent volcanism building up volcanic and clastic sequences up to several thousands of metres in thickness. Composition of the erupting volcanic material changed with time from andesitic to more felsic. It is evident that the major Kuroko deposits occur at a similar stratigraphic horizon, ~13 Ma in age, in close association with rhyodacitic volcanic rocks comprising lava domes and closely associated breccias. These deposits each appear to be the products of a single volcanic cycle of a submarine volcano, the mineralization belonging to its waning stages. In a typical Kuroko deposit, Kosaka, nine separate domes were emplaced at or close to the sea floor during part of a short eruptive cycle. Tuff breccias underlying the ore deposits are considered to have been produced by magmatic explosions followed by repeated steam explosions as the hot felsic lavas came into contact with seawater (in a manner reminiscent of the Moberg formations of Iceland, see p. 33). The lavas continued to rise forming domes, of which the flanks were brecciated by further steam explosions. The Kuroko deposits formed adjacent to the brecciated domes in the waning stages of the eruptive cycle. Some later explosive activity is indicated by brecciated and transported ore, and by breccia dykes which contain fragments of basement rocks, rhyolitic lavas, and ore.

2.2.4. Kimberlites

The last type of volcanic activity to be described is that which has been called the **diatremic association**. It is the characteristic mode of eruption of kimberlite, of most highly alkalic basic rocks such as melilites, ankaratrites and nephelinites, and of some alkali basalts. Some hypabyssal rocks show evidence of diatremic emplacement, notably lamprophyres, carbonatites, and some members of the appinitic association. Furthermore, diatremic activity is not confined to these classes of rocks, but characterizes in addition near-surface igneous activity of intermediate and acid rocks. Diatremes develop at crustal depths shallower than the critical vesiculation depth of the magma [the "saturation level" of McBirney (1974)]. Under these circumstances and given adequate amounts of gas, the generation and release of magmatic gas produces a fluidized system generated at depth in which gas is the predominant constituent by volume. The system then undergoes enormous volumetric expansion as confining pressure is reduced during its upward ascent to the surface. The typically upward-flaring kimberlite "pipes" are the outstanding example of diatremes. The mechanism of kimberlitic diatreme emplacement indicates that some of the ejected material may have achieved velocities in excess of Mach 3, with the temperature of the gas phase below 0°C due to adiabatic expansion!

Where diatremes become enlarged beyond a certain size considerable spalling off and down-sliding of large blocks occurs around the steep flared sides of the vent, and often an aggregate circular fault develops at the surface of the order of up to 2 km in diameter with a downward displacement within.

The beds preserved within the upper part of the vent may consist largely of bedded pyroclastics formed in the earlier stages of the eruption; these beds often become tilted due to slumping with appreciable inward dips up to 25° or even 50°. These may be intruded in a complex manner by tuffisite veins and are commonly overlain by washed-in and reworked pyroclastics of the original outer surrounding constructive cone (Hawthorne, 1975; Lorenz, 1975). Shallow, often water-filled depressions result, termed maars in their type area of Eifel in Germany, surrounded by a shallow rim of ejecta. Recent examples are commonly either very poorly exposed or not exposed at all. Well-exposed examples of such diatremes showing the above features, associated with alkali basalt of Carboniferous age, occur in the Elie district of Fifeshire, Scotland, where they have been well documented both from excellent coastline exposures and from intersections in coal mine workings (E. H. Francis, 1970). A detailed discussion of examples of "maar-type" volcanoes and their eroded remnants , diatremes, is provided by McGetchin and Ullrich (1973).

It should be noted that superficially similar landforms can result from phreatic explosive eruptions where rising magma of more usual type and behaviour comes into contact with water-logged ground. In the Auckland district of New Zealand, for example, numerous Holocene eruptions of common alkali basalt generally form low-angle flows around small shield volcanoes with little associated explosive activity. Where, however, vents happen to have occurred in low-lying ground with unconsolidated, water-logged sediments, inferred phreatic eruptions have resulted in shallow craters ~1 km across.

Comparable landforms to the surface manifestation of diatremes on Earth occur on the surface of the Moon and may be distinguished with some difficulty from impact craters of comparable size by slightly lower profiles, but more strikingly by their localized occurrence in some instances in strings within rilles in a more organized manner than that produced by chance meteorite impacts (Baldwin, 1965).

Some lapilli tuff cones as for example in Oahu, Hawaiian islands, contain a large proportion of rounded ultramafic cobbles, typical of the nodular inclusions common in highly alkaline basalts, associated with spatter of nephelinite composition. Presumably, a high rate of magmatic elutriation followed at high crustal levels by a high rate of diatremic gas flow would be necessary to transport these dense nodules to the surface.

The diatremic association described above would seem to be equatable with the "Plinian" type of eruption occurring in some stratovolcanoes, in eroded examples of which shallow diatremes of subterranean tuffisite are not uncommon. In these instances the role of heated meteoric water relatively to that of juvenile gas may become increasingly important. The heating of initially cold water to magmatic temperatures within a confined space could lead to extremely high pressures with consequent explosivity and high projectile velocity of tephra from vents (McBirney, 1974).

TABLE 2.1

Simplified classification of fragmental rocks resulting from volcanic activity

AUTOCLASTIC ROCKS (produced by relative movement within an igneous body)	extrusive:	*flow breccias*
	intrusive:	*intrusion breccias*
PYROCLASTIC ROCKS (produced by explosive activity)	subterranean:	*explosion breccias* *intrusive breccias*
	surface:	*air-fall material* *ignimbrites* *base surge deposits*
	submarine:	*pillow breccias and hyalotuffs* *subaqueous pyroclastic flows*
EPICLASTIC ROCKS volcanogenic material (reworked by sedimentary processes)	subaerial and subaqueous *volcanic sediments* and *lahars*	

2.3. FRAGMENTAL VOLCANIC ROCKS

So far the subject of volcanism has been presented in terms of volcanic activity. The resultant products are recognizable readily enough when they are lava or ash flows. The distinguishing of the many kinds of fragmental rocks resulting from volcanic activity (and near-surface igneous activity) remains, however a constant source of practical difficulty in the field. Two indispensable reviews dealing with these difficult groups of rocks are those by Fisher (1966, with references to his earlier papers) and by Parsons (1969). As a basis for a brief outline discussion here, a simplified classification, in large part composed of rock types already incidentally referred to in this chapter, is presented in Table 2.1. This classification is not without imperfections and overlaps: recent mappable ignimbrite units, for example, comprise an extensive upper part of air-fall material; subaqueous pyroclastic flows show depositional structures; epiclastic rocks include varieties showing very different degrees of proximal volcanic derivation; hyalotuffs in particular could be further subdivided, etc. (see Silvestri, 1965). However, increasingly complete classifications designed to take account of all possible genetically distinct rock types rapidly become unwieldly. The synopsis of Table 2.1, therefore, is a pragmatic one in which genetic considerations in part give precedence to those based essentially on readily distinguishable rock types.

2.3.1. Autoclastic rocks

Autoclastic flow breccias result from the continued flow of highly viscous lavas, resulting in stresses set up between adjacent bodies of lava moving with differing velocities that in the time available can only be accommodated by rupture. Common examples are autobrecciated rhyolite and andesite flows and the blocky tops of some distal basalt flows. Fragments are angular, tending to become subrounded with movement, and show a range in size up to several centimetres or even decimetres. An account of the mechanism of the process of flow folding accompanied by autobrecciation, illustrated by colour photographs of rhyolites from the very large Lebombo flows, is given by Wachendorf (1973).

Autoclastic intrusion breccias originate in a similar manner subterraneously and characterize the margins of rhyolitic plugs and domes emplaced close to the surface as degassed highly viscous magmas. Intrusive rocks of other compositions only rarely show autobrecciation.

A characteristic of all autoclastic breccias is that they are necessarily *monolithologic*, i.e. they are composed of fragments of the same parental igneous rock. Mixing in a large, moving, rubbly pile of autobrecciated rhyolite may bring together superficially different fragments of variable colour, crystallinity, and development of flow banding, but closer examination will reveal their monolithologic character.

2.3.2. Pyroclastic rocks

Subterranean pyroclastic rocks are formed by the explosive escape and passage of gas which may be juvenile and/or heated meteoric water. They occupy diatremes and smaller vents and veins. A convincing analogy has been made with the industrial process of fluidization (D. L. Reynolds, 1954, p. 577):

...''Fluidization is an industrial process in which gas is passed through a bed of fine-grained solid particles in order to facilitate mixing and chemical reaction. At a particular rate of gas flow the bed expands and the individual particles become free to move. With increase in the rate of gas flow a bubble phase forms and travels upwards through the expanded bed in which the particles are violently agitated; the bed is now said to be fluidized. With continued increase in the rate of gas flow more and more of the gas travels as bubbles containing suspended solids, until ultimately the solid particles become entirely entrained and transported by the gas.''

Where the rate of gas flow is relatively low, only sand-sized and smaller particles may be transported rapidly upwards with the escaping gases, and the larger fragments will remain in the vent. The latter become exceedingly well-rounded due to this sand-blasting action, and come to resemble in lithology a coarse, closely-packed conglomerate. Notwithstanding their rounded appearance, they have come to be called **explosion breccias** (Richey, 1932). The large rounded fragments, commonly the size of a football, are generally composed of the adjacent country rock; next to the vent the country rock is often cracked and large angular pieces may occur in the wall zone of the vent. Some movement of fragments within the vent may occur, however, as during the passage of gas there is a violent jigging action and small relatively heavy fragments in particular tend to work downwards as a consequence of this.

With a higher rate of gas flow fragmental material of all sizes may become physically transported with the exception of large spalled and slumped masses of country rock in the wall zone. The characteristic rock type produced is an **intrusive breccia** (also known as **intrusive tuff breccia** or **tuffisite**), in which the larger fragments are typically less well rounded and tend to be relatively more widely spaced than in the explosion breccias. Tuffisites generally occur in smaller bodies than the vent explosion breccias, ranging downwards in size to narrow dyke-like bodies a few centimetres wide. The tuffisites display intrusive contacts with country rock or with other vent pyroclastics. Given their inferred mode of origin, it is not surprising that tuffisites often contain exotic fragments carried from lower crustal levels. The extreme example of this, of course, is the range of mantle and crustal samples incorporated in kimberlite diatremes. Also characteristic are fragments of epigranite transported upwards in tuffisite into overlying volcanic terrain in vents of much more modest dimensions and vertical extent associated with intermediate and acid volcanism.

The two types of subterranean pyroclastic breccia occur in close spatial and time association in the Isle of Rum, Scotland (C. J. Hughes, 1960), where the more deeper-seated tuffisites contain fragments of vesiculating magma in addition to the usual mixture of country rock fragments. It is important to bear in mind that in nature there will be a range in scale, rates of gas flow, and perhaps a complicated sequential history of subterranean explosive events within one vent complex, so that considerable variability in lithologies is to be anticipated. Note also the description by Bowes and A. E. Wright (1967) of apparently deeper-seated explosion-breccia pipes, some of them blind, near Kentallen, Scotland associated with the intrusion of gas-charged basic magmas of kentallenite affinity.

Air-fall material forms perhaps the most easily appreciated group of fragmental volcanic rocks. It lends itself to description based on such readily observable parameters as:

(1) Grain size: ash, **lapilli**, **blocks** or **bombs** (the upper and lower size limits of lapilli vary according to different writers; a reasonable consensus is 2 and 64 mm). Lapilli and coarser material will contain some proportion of finer-grained material interstitially.

(2) Type of ejected material: **blocks** (of country rock); **bombs** (of hot lava); **vitric** (small fragments of glassy igneous material); **crystal** (originating from magma phenocrysts); **lithic** (country rock, usually volcanic).

(3) Chemical composition of parental magma.

(4) Lithification: **tuff** from ash; **lapilli tuff** from lapilli; **agglomerate** from bombs; **volcanic breccia** from blocks.

Composite names result such as "basaltic agglomerate", "andesitic crystal tuff", "rhyolitic vitric ash", etc. Further subdivisions can be made based on the relative proportions of shards, crystals and lithic fragments.

Isopleths of tephra thickness and maximum included grain size have been constructed, for example, by Thorarinsson (1968), for the score or so of successive eruptions of Hekla in historically recorded time, and by Self (1976) for several recent pumice fall deposits in Terceira, Azores. These isopleths show, as might be expected, concentration of coarse tephra near the vent, and elliptical shapes that are greatly elongated along the direction of prevailing wind with the vent lying close to one end. Interesting attempts to estimate distances which tephra of varying grain sizes may undergo air-borne transport are summarized by Fisher (1964). The conclusion is reached that one factor governing this is the individual "explosiveness" of an eruption. In this connection, Fundali and Melson (1971) and McBirney (1974) record the incongruously high "muzzle velocity" of some large ejected blocks, more likely attributable to the localized violent expansion on heating of meteoric water within some vents than to very high magmatic overpressures. McGetchin and Ullrich (1973) have shown that even for the most violently explosive diatremic eruption the maximum horizontal *ballistic* range of large blocks (that travel farthest under these conditions) is ~8 km, assuming a projection angle that gives maximum range; see also L. Wilson (1972) for a parallel calculation with comparable results. In general, the larger fragments are not accelerated to the same extent as small ones, but the latter undergo the braking effect of atmospheric friction to a much greater degree, to the extent that fragments in the size range 1—30 cm, for example, are commonly restricted to within 1 km of crater rims. By this same token, yet smaller particles are prone to be entrained in winds at different levels and thus have a considerably extended *wind-borne* range.

In the geological record, as opposed to historically observed eruptions, air-fall tuffs are relatively rarely preserved in their exact attitude of accumulation. This is readily understandable as a thin, relatively soft blanket of air-fall material deposited over hill and dale alike would be extremely prone to reworking (see epiclastic rocks below).

Ignimbrites are a special case of pyroclastic rocks resulting from the explosive disintegration of vesiculating magma in a vent formed during the catastrophic evacuation of a shallow magma chamber and the further disintegration of pumice fragments in an ensuing nuée ardente. Their inferred eruptive history and association in the geological record with ground surge deposits below and air-fall ashes above (where preserved) is illustrated by Sparks et al. (1973).

Base surge, also known as "ground surge" or an "ash hurricane", is a phenomenon that has been observed in subaerial nuclear explosions, and in powerful, often phreatic eruptions such as those of Surtsey, Iceland, and Taal, Philippines. A laterally spreading cloud from the base of the eruptive column, often laden with water, ash, lapilli and blocks, travels outwardly hugging the ground with initial velocities as high as 50 m/s. The effect of its passage is akin to sandblasting, commonly intense at 1 km from the eruption centre, and still perceptible at 5 km. There is little evidence of charring of trees, etc., so that temperatures are not as high as in a nuée. The deposits resulting from witnessed eruptions are cross-bedded and dune-bedded ash and lapilli with some contained blocks, wavelength 4—15 m, thickness up to 1 m. As with air-fall material the deposits are prone to erosion and are thus rare in the geological record (see J. G. Moore, 1967; Fisher and Waters, 1970; Sparks and G. P. L. Walker, 1973).

The submarine formation of pyroclastic rocks such as pillow breccias and hyalotuffs, and subaqueous pyroclastic flows has been discussed above in connection with submarine basalt and rhyolite eruption, respectively.

2.3.3. Epiclastic rocks

These are more customarily the domain of the sedimentary petrologist. Nevertheless, much igneous material finds its way very quickly into the stratigraphical record as sedimentary rock.

Two environments where this is manifestly true are: (i) andesitic strato-volcanoes, with explosion index typically in the high nineties, and subject moreover to extremely rapid denudation; and (ii) rhyolitic eruptions, wherein much fine-grained vitreous material associated with nuées becomes relatively enriched in accompanying widespread air-fall deposits and in the upper non-welded parts of ignimbrites which are very prone to rapid erosion. It is curious how igneous petrologists tend to neglect this component of volcanically-produced material in the sedimentary record when studying a volcanic province (see C. J. Hughes, 1976; Roobol, 1976). The volcanic derivation of much subaerial and marine sediment in volcanic terrain is revealed by such characters as the outlines of preserved shards, and abundance of euhedral plagioclase and paramorphed high-pressure quartz crystals ("volcanic" quartz), and other phenocryst constituents under the microscope. Angular, often poorly sorted, pieces of volcanic rock that clearly could not have withstood much transportation are also cogent evidence of proximal derivation and quick processes of accumulation and burial. Given a constructional volcanic relief and rapid denudation, it is not surprising that the rate of accumulation of volcanic sediments can be very high.

In conclusion, it must be re-emphasized that for the successful field interpretation of fragmental volcanic rocks in the geological record there is no substitute for seeing recent and active volcanic provinces, coupled with examination and discussion of outcrops of older rocks.

Chapter 3

FORMS AND STRUCTURES OF INTRUSIVE ROCKS

3.1. INTRODUCTION

Whereas volcanic activity can be witnessed, intrusive activity can not, and is therefore a matter of legitimate geological inference. An excellent account of the structures of igneous rocks is given by Hills (1963, ch. 12). A more genetic approach from the point of view of an igneous petrologist is attempted here, but the structural details and controls so well illustrated and described by Hills should not be lost sight of. See also the various papers in *Mechanism of Igneous Intrusion* (Newall and Rast, 1970).

Intrusive rocks are conventionally subdivided into hypabyssal and plutonic. It is generally held that the hypabyssal rocks have been emplaced as small dykes, sills, and plugs near the surface, have cooled quickly, and are fine-grained; and conversely that plutonic rocks were emplaced as larger masses at deeper crustal levels, have cooled slowly, and are coarse-grained. These two sets of concepts although satisfyingly logical are not always in practice consistently distinct. For example, some lamprophyric and other dykes were emplaced at depths comparable to those of penecontemporaneous plutonic rocks, and some granitic ring-dyke intrusions emplaced at shallow crustal levels may be as coarse-grained as much deeper-seated plutonic granites.

Following Daly's (1933, p. 41) classic observation on the predominance of extrusive basalt and intrusive granite amongst exposed rocks a two-fold grouping of igneous activity was proposed by W. Q. Kennedy and E. M. Anderson (1938). A **volcanic association** comprised

"...not only the superficial lava flows and vent intrusions but, in addition, all intrusive masses which are genetically related to a cycle of volcanic activity and originate in the same magmatic source"

and a **plutonic association** comprised

"...the great subjacent stocks and batholiths together with the diverse minor intrusions of such abyssal masses".

In addition to these a **diatremic association** was subsequently proposed (P. G. Harris et al., 1970, p. 197) to comprise rocks injected by gas fluidization.

This tripartite scheme of visualizing igneous activity has the attraction of distinguishing between rocks that were emplaced mainly as a result either of liquid or of solid or of gas movement. That is to say that rocks of the volcanic association (and including the large basic intrusions) were emplaced in a liquid condition (although quickly cooled representatives indicate in fact that intermediate and acid members are only rarely completely aphyric); that rocks of the plutonic association (and of broadly granitic compositions) were mainly solid at the time of their final emplacement, containing only minor proportions of silicate melt; and that rocks of the diatremic association show evidence of the passage of large volumes of gas.

However, for purposes both of description and comprehension, it seems to the writer that a pragmatic subdivision of the intrusive rocks would be a fourfold one as follows:

(1) Basic intrusions into continental crust.
(2) Ophiolite association.
(3) Sub-volcanic and central complexes.
(4) Deep-seated granitic rocks.

A rationale for this treatment which involves some minor overlap is incorporated within the following sections. Note also that, for the sake of continuity and completeness in describing volcanic activity, an account of the diatremic association, strictly, of course, a subterranean phenomenon, has already been given in Chapter 2 in the sections dealing with kimberlite and pyroclastic rocks. The chapter concludes with some guidelines on features to map in intrusive igneous rocks, predicated on their specific forms and structures.

3.2. BASIC INTRUSIONS INTO CONTINENTAL CRUST

3.2.1. Dykes

Basic dykes seldom come singly but commonly occur in **swarms** usually with a recognizable structural control on their attitude. Individual basic dykes are generally quite thin, up to several metres wide, discordant, and vertical at time of intrusion. Intuitively conceived of by geologists as a feature of crustal "tension", E. M. Anderson (1951) has shown how they may indeed be emplaced along planes perpendicular to minimum horizontal stress, whether this be in fact compressive or tensile. This relationship is well known by the numerous basic dykes exposed in those areas of Iceland where erosion has bitten deep into the Tertiary lava pile particularly in the region of the eastern fjords (G. P. L. Walker, 1958). There the dykes are locally very numerous and tend to occur in parallel swarms striking NNE, that is to say parallel to the strike of the ocean ridge and present-day fissures and fissure eruptions.

Swarms of dykes may occur in areas where any associated volcanic rocks, if they were once present, have now been completely eroded away. For example, Clifford (1968) has drawn attention to the existence of several distinct periods of basic dyke injection in the Canadian Shield, each with a differing strike direction, and presumably therefore signaling periods of different stress regimes in the craton.

Dykes are frequently very abundant locally and may indicate volcanic centres. For example, in eastern Iceland a local abundance of dykes coupled with subtle quaquaversal dips in the lava pile, zones of hydrothermal alteration, and a relatively greater abundance of pyroclastics and fractionated rock types signals the site of Tertiary shield volcanoes such as Breiddalur (G. P. L. Walker, 1963), exhumed from the lava pile. Similarly localized swarms of parallel dykes in areas of flows of pahoehoe type associated with relatively abundant pyroclastics were probably feeders to the Columbia River basalt plateau (Swanson et al., 1975).

Rarely is a relationship between basic dykes and volcanic centres better displayed than in the early Tertiary igneous province of the Inner Hebrides, western Scotland. A regional vertical dyke swarm with a slightly curving strike extends for some 400 km along a line of central complexes that are believed to represent eroded volcanoes and on into northern England. Around two of the central complexes, Mull and Arran, basic dykes are very abundant indeed. On the southeast coast of Mull, for example, 375 dykes with an aggregate thickness of 763 m occur across a distance of 20 km; similarly in Arran, there are 525 dykes, aggregate thickness 1844 m within 24 km. These figures imply crustal extension of the order of 4% and 7% over these distances, respectively (Richey et al., 1961, p. 111).

A radial pattern of dyke intrusion (of basic and other compositions) is sometimes conspicuous around some central complexes and intrusions, presumably reflecting a localized stress pattern; Rum, Inner Hebrides, and Spanish Peaks, Colorado, are commonly figured examples.

An interesting example of the inferred relationship between crustal stress and a regional dyke swarm is apparent in a crustal flexure developed along some several hundred kilometres of the coast of east Greenland (Wager and Deer, 1939). Early Tertiary basalt flows constitute a thin flat-lying sequence in the interior Watkins Mountains but assume a much greater thickness and a gradually increasing seaward dip towards the coast, reaching maximum observed dips of $\sim 30°$ to the southeast along the coast itself. Emplaced at intervals in this zone are several well-known intrusions and intrusive complexes, such as Skaergaard, Kangerdlugssuaq and Lille. A swarm of basic dykes with dips inclined more or less normal to the lavas is most prolific near the coast where flows and dykes thus dip $\sim 30°$ SE and $\sim 60°$ NW, respectively. Where densest, the dyke material equals or even exceeds the country rock in amount. The abundance pattern does suggest a genetic relationship between the flexuring and dyke intrusion, i.e. the dykes were possibly emplaced in planes other than the vertical in response to a local stress pattern rather than necessarily having been tilted to their present attitude by later earth movements. Analogues of this major crustal structure occur in southeast Africa (the Lebombo monocline), the north coast of the Gulf of Aden, and the west coast of India, near Bombay. Their position and age in all instances correlate closely with the inception of inferred crustal separation due to plate movements.

Basic dyke swarms sometimes show examples of **multiple intrusion**, i.e. successive intrusions of the same composition, at its most spectacular in sheeted dyke complexes (see Section 3.3).

Basic dykes, unlike most minor intrusions of other compositions, are commonly aphyric, i.e. intruded at or even possibly above their liquidus temperatures at the pressure obtaining at the crustal level of emplacement. However, where phenocrysts are present, these are often seen to be concentrated, sometimes gradationally, sometimes quite sharply, towards the centre of the dyke by **flow differentiation** (Bhattacharji, 1967). This is a physical phenomenon related to flow in a liquid carrying solid particles; its efficiency depends on viscosity, flow rate, and particle size (see Chapter 7). Such a dyke therefore does not represent the "composite intrusion" of a porphyritic magma closely succeeding an aphyric one; nor are the phenocrysts present in greater abundance towards the centre simply because of a longer cooling time permitting their growth, as there still remains too great a disparity in grain size between the phenocrysts and matrix that may contain the same mineral species as the phenocrysts for this simple supposition to be true.

Basic dykes must have been emplaced rapidly for silicate melts around liquidus temperatures (probably near 2.65 g cm^{-3}). Under these conditions and blocking. It has been shown that for a planar body of basic magma emplaced in the upper crust conduction rates are such that basic magma will cool from liquidus temperatures to solidus temperatures roughly according to the equation:

$$t = 0.01d^2 \qquad\qquad\qquad (3.1)$$

where t = time in years and d = thickness in metres, i.e. a dyke 1 m wide would be expected to become completely solid in ~3 days*.

Basic dykes, and indeed other minor intrusions such as cone sheets (see below), should not be overlooked as a source of useful compositional evidence of the magmatic affinity of an igneous rock series (see Chapter 8).

3.2.2. Sills

The development to any large extent of lateral sill intrusion from basic dyke feeders is restricted to areas of flat or gently-dipping sedimentary

---

*The subject of cooling rates has been thoroughly covered by F. G. Smith (1963) and by Jaeger (1968).

strata. Sill formation occurs where the average density of a column of rock above the level of sill intrusion is less than the density of basic magma at liquidus temperatures (probably near 2.65 g/cm^3). Under these conditions less work is done when a body of basic magma intrudes laterally along some convenient plane thus raising the overlying strata than when it raises itself to the surface. Once the magma has solidified and cooled to produce a sill of diabase with a density of ~3.0 g cm^{-3}, the average density of the column of rock is increased and later sill intrusions will occur at higher levels until eventually lava eruption at the surface may occur instead of sill intrusion. Thick sequences of sediments that are relatively young and incompletely lithified at the time of eruption would thus be expected to be a favorable venue for large-scale sill injection because of their low density. Bradley (1965) has discussed the mechanics of lateral emplacement of sills probably prefaced by the passage of steam of juvenile and/or meteoric origin. The base of a competent cap-rock in a sedimentary sequence of varied lithologies is often a preferred plane of intrusion.

There is an element of forceful intrusion involved in sill intrusion and physical investigations have shown that a preferred ideal three-dimensional shape for a sill-like body resembles that of a shallow saucer concave-upwards, i.e. lopolithic. Although intrusion is frequently constrained by preferred bedding planes resulting in **concordant intrusions** in horizontal or near-horizontal strata, there is thus also a tendency for low-angle cross-cutting relationships to develop resulting in **discordant intrusion.**

Where basic magma erupts into poorly consolidated and waterlogged sediments disruption of the magma commonly occurs producing numerous blind ovoid bodies within the envelope of the zone of intrusion, apparently disrupted feeders, and smaller, less regular bodies of diabase on scales varying from tens of metres or more down to centimetres. These rocks are technically intrusive rocks, but the conditions of eruption could presumably grade from those of submarine intrusion to extrusion. See O. T. Jones and Pugh (1948) for a description and discussion of the phenomenon.

Sills are generally considerably thicker than dykes are wide and thus take much longer to cool according to eq. 3.1. There is a chance therefore, in the time interval necessary for cooling, of multiple intrusion resulting perhaps in magma mixing rather than visibly separate intrusions, and also there is the possibility of limited fractionation. Well-known examples illustrating these phenomena include the following.

The compositionally homogeneous Whin sill of northern England concordantly intrudes cyclically-bedded Carboniferous shales, sandstones, and limestones. Over an area of at least 3500 km^2 as proved by outcrops and numerous borings it maintains stratigraphical horizon for long distances with only minor jumps. Thickness varies up to 70 m, average ~25 m, thus indicating the intrusion of ~100 km^3 of magma. The lateral extent to thickness ratio is of the order of 1000:1.

Sill intrusion on a gigantic scale has occurred in the Karroo area of South Africa where basic sills and lavas of early Triassic age intrude and overlie thick sedimentary series of the Karroo system of Permo-Triassic age (F. Walker and Poldervaart, 1949). Some of the thicker sills approach 500 m in thickness, and the total volume of basic rocks in the province has been estimated at near $5 \cdot 10^5$ km^3, a lower limit in view of extensive erosion of the associated extrusive rocks and sediments. Some of the thicker sills approach the theoretical saucer shape and form small lopoliths some several tens of kilometres in diameter.

The well-known Palisades sill outcrops on the right bank of the Hudson River and forms a prominent escarpment over most of its exposed strike length of 80 km. It is the lowermost unit of the several tholeiitic sills and flows intruding into and extruded upon the now gently-dipping Triassic and early Jurassic sediments of New Jersey and New York State. This sill is a multiple intrusion ~300 m in maximum thickness in which internal contacts can be discerned in the northern part and apparent magma mixing occurred towards the south (K. R. Walker, 1969). The famous "olivine layer" reflects this event, i.e. it is not apparently a function of static fractionation but originated in a kinetic system of moving and mixing magmas. Apart from the olivine layer and chilled contacts, the sill rocks are mostly monotonous looking diabase; this apparent monotony, however, is belied by their mineralogical interest and evidence of a pattern of coherent internal differentiation within the sill mainly by congelation crystallization (see Chapter 7).

The Shiant Isles sill of western Scotland (Johnston, 1953), although only 150 m thick, shows much more pronounced mineral differentiation. Appreciable amounts of early crystallizing olivine sank towards the base and were thus concentrated there to form picritic varieties whereas complementary leucocratic analcite-bearing rocks are found towards the top, all within an envelope of marginally chilled rocks. The parental magma was an alkali basalt that was less viscous than the tholeiitic parental magmas of the Whin and Palisades sills and therefore permitted more crystal settling within the lesser cooling time available.

The Shonkin Sag "laccolith" of "mafic phonolite" in the Highwood Mountains of Montana is actually a flat discus-shaped sill, 3 km in diameter and up to 80 m thick, composed of fractionated olivine leucitite. Within an envelope of marginally chilled rocks containing phenocrysts of augite, (pseudo-)leucite, and olivine, gravitational differentiation is very apparent despite the relatively small size of the intrusion. This has been attributed to a very low viscosity characteristic of alkaline mafic magmas and to the presence of primocrysts on intrusion. Thick upper and lower layers of shonkinite, an accumulative rock with varying amounts of euhedral cumulus crystals of calcium-rich augite, each show increasing amounts of the cumulus augite upwards, a curious feature attributed to congealing outpacing a more complete fallout of abundant augite primocrysts. Between the shonkinite layers are sandwiched more fractionated rock types. These sill rocks have yielded extraordinarily precise and consistent estimates of crystallization temperatures, pressures, oxygen fugacity and silica activity, based upon the mineral assemblages present (Nash and Wilkinson, 1970).

3.2.3. Large basic intrusions

In some respects these show the phenomena exhibited by differentiated basic sills but to a greater degree. Repeated multiple injections create a large magma chamber, the cooling of which can be a lengthy process (estimated cooling time for the Skaergaard Intrusion, 0.02 Ma, and the Bushveld Intrusion, the largest igneous body exposed in the world, 1 Ma). Layered cumulate sequences within a chilled margin and adjoining congelation cumulates are typical. The description and interpretation of structures of layered basic intrusions, precise analytical work on the cumulus phases, the remarkable tie-in with experimental work at low pressures, elaboration of a cumulus mechanism and fractionation, has been a massive centrepiece of igneous petrology in the mid-twentieth century, fittingly summarized by Wager and G. M. Brown (1968). One commonly thinks of a lopolith as being the usual form of large basic intrusions, but examination of some classic examples shows that viewed in section they either are, or equally well could be, **funnel-shaped intrusions.** Mapping of the large Michikamau anorthosite pluton of Labrador (Emslie, 1970) reveals that it too has a funnel shape; a comparable interpretation is indicated by mining exploration at Sudbury and geophysical work on the Bushveld. In plan the funnel is frequently elliptical, sometimes a very elongated shape as for example the Muskox Intrusion. Also of problematic origin but showing strong affinities with each other are the late Archaean canoe-shaped intrusions known as the Great Dyke of Rhodesia, and the Jimberlana Norite and Binneringie Dyke of Western Australia (the term dyke is a misnomer for these intrusions).

One interesting question with respect to the larger intrusions is how so much magma came to be generated in one place within a geologically short period of time, i.e. no longer than that of the cooling time of the intrusion. In the case of the Sudbury Intrusion, it has been established beyond reasonable doubt that intrusion postdated at no great time interval a large meteorite impact (French, 1967; Dence, 1972). Presumably therefore such factors as residual rise in temperature following shock metamorphism, lower confining pressure due to cratering with upward mantle transport due to rebound and isostatic recovery resulted in the production of basic magma on a very large scale. Such an origin now seems possible for the Bushveld Intrusion too (Rhodes, 1975). However, areas of voluminous eruption of flood basalts are another favoured venue of large basic intrusions, which may have acted as staging-points for magma during its ascent to surface. For example, the

Duluth gabbro is spatially associated with the voluminous Keweenawan lavas, and the enormous, newly discovered Dufek Intrusion (Behrendt et al., 1980) is of comparable age and spatially associated with the Ferrar Group diabases and lavas, Antarctica. The Muskox Intrusion represents crystallization products from a large subterranean magma chamber which was replenished at intervals but was not a closed system, as it apparently also acted as the proximate feeder for Coppermine basalt lavas (Irvine and C. H. Smith, 1967). The Skaergaard Intrusion by contrast appears to have behaved very much as a closed system (Wager and Deer, 1939) emplaced within and penecontemporaneously with the eruption of flood basalts on the coast of east Greenland during early Tertiary plate separation.

3.3. OPHIOLITE ASSOCIATION

In marked contrast to continental crust, where basic intrusions are "accidental" additions, oceanic crust is *essentially* composed of basic intrusive and extrusive rocks. The evidence of seismic data, of direct sampling particularly from the much faulted segment of the ocean—ridge system in the North Atlantic south of Iceland, of detailed examination of ophiolite complexes, and of their tectonic setting leaves little room for doubt as to the validity of the correlation between ophiolite and oceanic crustal material. Nevertheless it should be stressed that the precise mechanism by which all the components of an ophiolite suite are produced is still a matter of debate, as is the nature of the exceptional process, or more likely processes, (obduction) by which an ophiolite suite is emplaced on continental crust. Less than 0.001% of oceanic crust escapes subduction and becomes incorporated as ophiolite in continental crust (Coleman, 1977, p. 16). Certainly, armed only with a conventional background in intrusive rock forms, a geologist would have as much difficulty in comprehending an ophiolite suite today as did the early workers on the "alpine periodotites" or the Troodos Massif of Cyprus, for example. Hence the justification for considering ophiolite as a distinct association, quite apart from the consideration that if its correlation with oceanic crust is correct, rocks of the ophiolite association underlie ~65% of the Earth's crust!

Historically, the term ophiolite was first used a long time ago by A. Brongniart in 1821 to describe a recurring four-fold association in the Alps of cherts, diabases and spilites, gabbros, and ultrabasic rocks, and the term was employed by B. Lotti in 1886 in the same sense. In the early twentieth century, however, it was a three-fold association of radiolarian cherts, pillow lavas, and serpentinized ultramafic rocks (the latter commonly known as "alpine peridotites") that came to be emphasized, notably by G. Steinmann, and thus these rocks came to be known colloquially by geologists as "the Steinmann trinity".

For a contemporary overview of ophiolite we may take as a convenient starting point a statement produced by the Penrose Conference in 1972, and reported by conference participants in the December 1972 issue of *Geotimes*:

> "'Ophiolite' refers to a distinctive assemblage of mafic to ultramafic rocks. It should not be used as a rock name or as a lithologic unit in mapping. In a completely developed ophiolite the rock types occur in the following sequence, starting from the bottom and working up:
>
> Ultramafic complex, consisting of variable proportions of harzburgite, lherzolite and dunite, usually with a metamorphic tectonic fabric (more or less serpentinized).
>
> Gabbroic complex, ordinarily with cumulus textures commonly containing cumulus peridotites and pyroxenites and usually less deformed than the ultramafic complex.
>
> Mafic sheeted dike complex.
>
> Mafic volcanic complex, commonly pillowed.
>
> Associated rock types include (1) an overlying sedimentary section typically including ribbon cherts, thin shale interbeds, and minor limestones; (2) podiform bodies of chromite generally associated with dunite; (3) sodic felsic intrusive and extrusive rocks.
>
> Faulted contacts between mappable units are common. Whole sections may be missing. An ophiolite may be incomplete, dismembered, or metamorphosed, in which case it should be called a partial, dismembered, or metamorphosed ophiolite. Although ophiolite generally is interpreted to be oceanic crust and upper mantle the use of the term should be independent of its supposed origin."*

Ophiolites can be either autochthonous or allochthonous. Autochthonous ophiolites, such as those in California and New Guinea, are bounded by steep reversed faults against blue schists which contain mineralogical evidence of upward transport of the order of 30 km (Ernst, 1971). Allochthonous ophiolites, such as those of western Newfoundland and Oman, occur in klippen with low-angle thrust contacts at their base. The classic west Newfoundland ophiolites (Malpas, 1977, with references to earlier work) provide a thick, well-exposed, and complete section of an ophiolite complex taken here as a model as indicated in Fig. 3.1.

Note the presence of a thin **aureole** of underlying amphibolite which shows a sharp decrease in metamorphic grade downwards. These are complicated rocks, of uncertain provenance, apparently deformed during the movement of the ophiolite slice. They have been recrystallized in part by the conduction of heat from the overlying ophiolite slice and in part by frictional heat developed during emplacement. It is unusual in ophiolite suites and of considerable interest to find rocks of the composition of **spinel lherzolites** interpreted by Malpas as reflecting the diapiric ascent of relatively undepleted mantle material from depths of around 60 km. The

---

*Note also that many reported occurrences of ophiolite in the world rely solely on the occurrence together of pillow lavas and cherts, i.e. only where fully preserved and exposed can the ophiolite association be considered mainly intrusive.

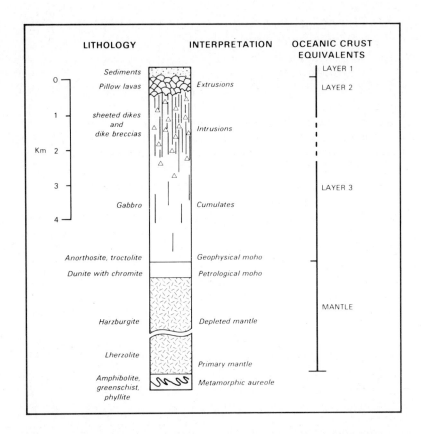

Fig. 3.1. Diagrammatic section of an allochthonous ophiolite sheet based mainly on the work of J. G. Malpas in western Newfoundland (slightly modified after Malpas and Strong, 1975, p. 1050, fig. 4).

mantle tectonites are composed mainly of variably serpentinized harzburgites of which the primary mineralogy is essentially olivine + orthopyroxene + brown chrome spinel. These harzburgites, in contrast to the lherzolites, contain little or no diopsidic augite and are considered to represent mantle material depleted by partial-melting processes. They show considerable flow folding and are cut by dunite and enstatolite veins. This is the unit of an ophiolite suite equatable with **alpine peridotite** in the older literature. Above this comes a succession of layered **cumulate rocks** over 1 km thick; cumulate assemblages in upward sequence give rise to dunite, feldspathic dunite, troctolite and anorthosite rock types. The boundary of the lowest olivine-rich cumulate rocks (containing small amounts of small, black, cumulus chromite crystals) with the underlying harzburgites (containing relatively large grains of a brown chrome spinel) can be narrowed down in the field to within a few metres in west Newfoundland, and is seen to be a marked nonconformity in some Oman exposures. A comparable fundamental distinction between these two types of ultramafic peridotites in the remarkably fresh and unserpentinized ophiolite suite of New Guinea was commented on by Davies (1971). The colloquial but apt name **petrological moho** has been applied to this junction by Malpas. Note how the (seismically defined) Mohorovičić discontinuity occurs within a narrow zone a few hundred metres higher in the ophiolite cumulate sequence where olivine cumulates are overlain by cumulates containing plagioclase as an additional cumulus phase; these upper cumulates thus approximate to basic rock in their elastic constants, as opposed to the various types of ultramafic rock below. Above this well-defined cumulate sequence comes ~2000 m of **basic rocks.** These show good layering in many places but either no layering or only faint layering in some places; they vary in grain size and texture from rocks describable as gabbroic (cumulitic?) to diabasic. Some pegmatitic patches, veins, and intrusion breccias of hornblende gabbro and relatively leucocratic rocks of dioritic and trondheimitic affinity, "plagiogranites", are also present. Extrusive equivalents of these more evolved rocks are lacking, a fact attributed by G. P. L. Walker (1971) to the effect of the confining pressure of seawater inhibiting a submarine nuée form of extrusion. Careful elucidation of these gabbroic rocks may well prove the key to testing models of ophiolite genesis (see p. 279). The intrusive basic rocks are overlain by **submarine basalt flows,** massive in the lower part, pillowed above. These flows are presumably fed by the conspicuous **vertical basic dykes** which increase in abundance downward through the lava sequence and become less abundant again further downward in the gabbros. At maximum abundance near the base of the lavas they form a **sheeted dyke complex** in which the abundance of dykes approaches or attains 100%, an example par excellence of multiple intrusion. This unit of an ophiolite suite was early recognized and described in the Troodos Massif in Cyprus, but the interpretation of these puzzling rocks led to controversy that was not finally resolved until the period of acceptance of sea-floor spreading. A remarkable feature of a sheeted-dyke complex is that close inspection shows that many of the dykes exhibit only one chilled margin; those that do so must presumably each have been split and intruded in turn by a younger dyke. A statistical analysis of this phenomenon in one locality in Cyprus has revealed a polarity and hence an inferred sea-floor spreading vector (Kidd and Cann, 1974).

Superimposed on this "ophiolite stratigraphy" are several complicating factors originating either at the time of formation of oceanic crust or during later tectonic emplacement:

(1) Present-day ocean ridges are offset by transform fault systems and in sections of relatively low spreading rate are affected by normal faulting striking parallel to the spreading axis and producing the observed high relief of scarplands adjacent to the ridge, for example in the North Atlantic and

also in the back-arc Lau Basin. This faulting would, of course, be retained in an ophiolite complex.

(2) In many ophiolite suites considerable brecciation has affected the level occupied by the dykes. In the Bay of Islands ophiolite complex (Hank Williams and Malpas, 1972), for example:

> "the brecciation is chiefly localized in the dike horizon but it locally affects under-lying gabbros and overlying volcanic rocks. There is nothing to suggest that the brecciation is the result of brittle deformation. Rather the textures are comparable to those developed in the igneous rocks through gas action or fluidization. The brecciation largely predates metamorphism in the dike rocks and it is therefore also interpreted as a feature related to early Paleozoic plate accretion."

(3) Very high thermal gradients at ocean ridges have resulted in static burial metamorphism of oceanic crust. This commonly ranges through zeolite and greenschist facies to amphibolite facies within the basic members of an ophiolite suite, indicative of very high thermal gradients, of the order of $150°/km$.

(4) This metamorphism is accompanied by pervasive metasomatism likely reflecting the accession of seawater to the upper parts of the ophiolite sequence.

(5) Under these conditions hydrothermal fluid circulation is an "inevitable consequence" (Norton and Knight, 1977). The return of heated solutions to the rock—seawater interface in localized sites results in the formation of stratabound cupriferous pyrite deposits that characteristically occur in ophiolite suites (see Andrews and Fyfe, 1976).

(6) During the process of "obduction", thrusting, slicing and breaking of the oceanic crust must occur, still further complicating the sequence of rocks displayed in accessible ophiolite complexes. It will be appreciated from the above that the deciphering of a particular outcrop or area in ophiolite terrain is not easy, and that some conceptual overview is essential in comprehending the geological relationships.

The actual mechanism of formation of new oceanic crust still has some mystifying features. Illustrative of this is an article by Brock (1974) drawing attention to the apparent physical impossibility of spreading in lavas and gabbros that overlie and underlie respectively the intruded sheeted dyke layer in which crustal extension must have occurred. This essentially logical but paradoxical observation provoked two models of ophiolite genesis to explain the apparent anomaly, one involving a continuous magma chamber underlying the ridge (Church and Riccio, 1974), the other involving numer-ous small overlapping and cross-cutting tholeiitic magma chambers (Strong and Malpas, 1975), both at the level of the gabbros. In this respect, Flower et al. (1977) deduce, from the variation in content of major and trace elements:

"distinct cycles of low pressure fractionation operating independently within a complex network of magma storage reservoirs beneath the crustal spreading axis."

At the time of writing, a definitive review of the challenging and rapidly developing subject of ophiolites is that of Coleman (1977). See also the *Proceedings of the International Ophiolite Symposium, Cyprus, 1979* (Panayiotou, 1980), particularly the introductory review article by Gass (1980).

3.4. SUBVOLCANIC AND CENTRAL COMPLEXES

Many central complexes are associated with volcanic rocks of the same age and comparable chemical compositions, the extrusive rocks often being preserved in down-faulted blocks that are structurally associated with the intrusive rocks of the complex. Some complexes without associated extrusive rocks occur in more deeply dissected terrain where lavas if originally present could easily have been eroded away. Some central complexes indeed are exposed at relatively shallow erosional levels at which the fine details of clearly associated subvolcanic intrusion structures such as diatremes plus complementary intrusive bodies of crystallized degassed magma and cone sheets are preserved (an outstanding example of this is Ardnamurchan, see p. 72). Hence the commonly used term **subvolcanic** to describe such complexes. However, some structurally comparable intrusion complexes could be "blind" in the sense that surface eruption need not necessarily have occurred, and lacking direct evidence of such surface eruption, the non-genetic term **central complex** is thus more appropriately retained for them, and could well be used for the entire group.

In central complexes we are dealing with mainly permissive intrusion of essentially liquid magmas of intermediate to acid compositions into the upper continental crust. This may be associated in time with varying proportions of basic magma and an equivalent range of lava compositions, the whole forming part of the volcanic association of W. Q. Kennedy and E. M. Anderson (1938). It should be noted, however, that apart from dykes, sills (if appropriate conditions for sill formation exist), and local magma chambers within large basaltic shield volcanoes and other areas of basalt eruption, a large proportion of basic magma that reaches upper crustal levels apparently goes on to erupt at the surface. In contrast, bodies of the intermediate and particularly the acid magmas seem to work their way up through the crust by gradual stoping processes on various scales, and only some of them come so close to the surface by this mechanism that subaerial eruption by ash-flow mechanism or less commonly by lava flow ensues. The apparent paradox that relatively dense basic magmas commonly reach the surface whereas the relatively lighter acidic magmas commonly do not can therefore be explained partly by the much lower viscosities of basic magmas, but probably more fundamentally by the fact that bodies of acid magma are not in continuous connection with magma in lower crust and upper-mantle levels, whereas basic magma commonly is and is thus subject to a considerable hydrostatic head. Furthermore, the escape of dissolved water at shallow depths from ascending granitic magmas (inherently more water-rich than basic magmas) results in an increase in melting point and therefore crystallization before the magmas actually reach the surface, particularly in compositions approaching that of ideal granite where there is little or no difference between liquidus and solidus temperatures (P. G. Harris et al., 1970).

Epizone ring dykes and associated cauldron-subsidence intrusions characterize high-level granites, but are rare among basic intrusions. It would seem that the relative densities of the respective magmas and country rock are responsible for this difference. In a rare example of a basic ring dyke in the island of Rum, Inner Hebrides, Scotland, the subsided block of country rock is in fact composed of a layered ultrabasic cumulate sequence of high average density. At least some of the so-called basic ring dykes of the neighbouring complex of Ardnamurchan may likely have the form of funnel-shaped intrusions combined with sill-like apophyses (Wells, 1953) rather than ring dykes.

The phenomenon of large-scale stoping of country rock permitting the concomitant uprise of magma has been very widely documented and must be regarded as the normal way in which essentially liquid acid magmas are emplaced in the upper crust [see Daly (1933, p. 267 ff.), with references to early work by Daly and others]. A classic demonstration is provided by the several Devonian complexes of Lorne, western Scotland, (E. B. Bailey, 1960; Johnstone, 1973). Large cylindrically shaped masses of country rock, ~5—20 km in diameter, overlain in places by Devonian lavas penecontemporaneous with the intrusive rocks, have demonstrably sunk distances up to the order of at least 1000 m. The surface of displacement is in places a steep arcuate fault containing flinty crush rock; more commonly the sunken blocks are margined by ring-dyke intrusions ranging up to some hundreds of metres in width usually with small outward hade. The ring dykes and arcuate faults together completely encircle the cylindrical subsided block. Where the erosion level permits, the subsided block can be seen to be overlain by a thick tabular intrusion known by the somewhat awkward name of cauldron-subsidence intrusion which is in continuity with the underlying ring dyke. Ring-dyke and cauldron-subsidence intrusion thus refer merely to geometrically distinct parts of one and the same permissive intrusion that envelops the subsided block of country rock. These field relationships are convincingly demonstrated by excellent exposures in the region of Ben Nevis, Glen Coe, and Loch Etive, where dissection commonly exceeds 1 km.

Multiple and composite intrusions are common in central complexes. Successive intrusions about any one centre generally trend towards a more ideal granite composition with time, although it should be noted that in many such complexes the compositional range is often quite narrow, from granodiorite to adamellite, for example.

Structurally, successive ring dykes and their cauldron-subsidence intrusion cappings may transect earlier intrusions in a complex manner. The determination of age relationships is often complicated by faulting along contacts and the presence of screens of country rock, which are often the hornfelsed selvedges adjacent to earlier intrusions, separating successive intrusions. In the Etive Complex in the Lorne area referred to above, a parallel swarm of porphyry dykes intervened between successive ring-dyke intrusions, thus making the determination of relative age easy in this particular instance.

Although successive intrusions may be disposed around approximately the same geometrical centre as that of earlier intrusions, many complexes display a marked shift in centre over distances of up to a few kilometres with time, thus justifying the descriptive terms "centre 1", "centre 2", etc. Nevertheless, the intrusions are generally so closely grouped in time and space as to warrant the use of the term "central complex" for the whole. The term ring complex is also used in the same context, and derives from the roughly circular outcrop pattern of ring dykes.

Petrographically, ring-dyke and cauldron-subsidence intrusions, together, of course, forming single intrusive units, are generally homophanous, i.e. lacking primary igneous foliation or other directional properties except joint patterns. Within one intrusive unit the texture may vary, for example, from a chilled marginal facies which may be porphyritic to more coarsely crystalline in the interior parts where the rocks are commonly equigranular but may be coarsely porphyritic in places. Some of these textural varieties may well overlap in appearance those from successive intrusions, but during careful field mapping and allowing for this kind of gradational variation, the different intrusive units can often be recognized. Richey (1928) in mapping the Mourne Mountains used a subscript notation to indicate successive intrusive units, e.g., G_1, G_2, G_3, etc., a procedure that has been widely followed.

Numerous ring-dyke intrusions of early Tertiary age are disposed around three closely spaced centres in the island of Mull, Inner Hebrides, Scotland, and have been described in the classic "Mull Memoir" (E. B. Bailey et al., 1924). The accompanying "one-inch" sheet remains probably the most detailed map of an igneous complex ever produced on that scale. The case of Mull demonstrates the association of the intrusive rocks with an overlying volcano caldera and is also characterized by another type of subvolcanic intrusion — cone sheets.

Cone sheets were first recognized by Harker (1904) in the nearby Hebridean island of Skye and were originally termed by him "inclined sheets". Cone sheet, however, is an appropriate term to use as these intrusions are arcuate in outcrop plan dipping inwards with the form of a partial cone, apex downwards. They invariably form part of a cone-sheet complex composed of numerous cone sheets, in which the outer ones have shallow dips, as low as 20—30°, whereas the inner cone sheets are much more steeply inclined, with dips as high as 70—80°. Very roughly speaking, a cone-sheet complex has a common focus zone, inferred to be situated ~2—3 km below surface level at their time of formation, and not far from the general level of the top of ring-dyke and cauldron-subsidence intrusions belonging to the same complex. Individual cone sheets are usually a few metres in thickness and are thus much smaller than the usually more massive ring dykes. Compositionally, cone sheets differ from ring dykes in comprising basic types in addition to intermediate and acid varieties. Whereas ring dykes are permissively emplaced, analysis of cone-sheet geometry convincingly indicates (E. M. Anderson, 1951) that they are forceful intrusions produced by magmatic overpressures at focus depth.

A very shallow depth of subvolcanic activity is represented in part of the neighbouring island of Rum. Here dykes and diatremes of intrusive tuff breccias and explosion breccias occur around a subsided block ~3 km in diameter and are, in turn, cut by felsitic bodies crystallized from degassed acid magma in a manner reminiscent of resurgent cauldrons (C. J. Hughes, 1960). It would seem that in Rum acid magmas had worked upwards by processes of cauldron subsidence to very shallow depths of crust, certainly to depths shallower than their critical vesiculation depth. The resultant kind of diatremic activity thus provides a link between intrusive and extrusive phenomena.

An extraordinarily complete three-dimensional picture of sub-volcanic igneous activity containing all the above intrusive forms can be obtained by studying the geology of the compact area of the peninsula of Ardnamurchan, yet another centre of early Tertiary volcanism in western Scotland, situated between Mull and Rum. Intrusive activity occurred around three successive, closely spaced centres within a short time span, possibly as short as 1 Ma. Ring dykes, cone sheets, etc., were emplaced at appropriate levels below surface but, with the concomitant buildup of a volcanic pile, necessarily at successively higher absolute levels. Subsequent erosion and Pleistocene glaciation has resulted in a subdued relief of no more than ~500 m but excellent exposures, thus resulting in three essentially horizontal sections of the three centres representing three different depths in a sub-volcanic complex. Centre 3, the youngest, is dissected at the deepest structural level and is now recognized by the disposition of concentric, composite, ring-dyke intrusions; centre 2 by both ring dykes and cone sheets; and centre 1 by cone sheets, diatremes and plugs. An outline geological map and a striking air photograph mosaic of the complex is included in an account of Tertiary igneous activity in Scotland by Stewart (1965).

Another source of information on these classic Hebridean igneous complexes is the *Tertiary Volcanic Districts* by Richey et al. (1961), a publication in the excellent series of regional geology handbooks produced by the British Geological Survey. A considerable emphasis on these Hebridean occurrences has been placed because happy accidents of structural levels exposed, excellent exposures, relatively young age and hence resolution of radiometric ages, extent of study and good accounts make them classic. They may be a little untypical in that a definite association of the central complexes with volcanic activity is displayed, whereas as stated above this is not necessarily always the case. In any event this section cannot conclude without at least passing reference to other areas where comparable phenomena are displayed and have been lucidly described.

These would include the classic New Hampshire complexes (Billings, 1945); the numerous ring complexes of the Younger Granite Province of northern Nigeria (Jacobson et al., 1958) together with similar complexes to the north in Aïr (R. Black and Girod, 1970) comprising such items of petrological interest as a perfectly circular gabbro ring dyke, radius 33 km, and a complete ring of peralkaline granite of which the diameter does not exceed 2 km; cauldron subsidence, granitic rocks and crustal fracturing in southeast Australia (Hills, 1959); comparable structures in the carbonatite complexes of Alnö, Sweden, (von Eckermann, 1966) and of the Chilwa district, Malawi, (Garson, 1966). Probably the most majestic example in the world of cauldron-subsidence intrusions and associated igneous activity must be that discussed and figured by Myers (1975) from the Coastal Batholith of Peru, magnificently exposed in mountainous rock desert with a relief up to 4000 m that readily permits a three-dimensional overview to be obtained.

3.5. DEEP-SEATED GRANITIC ROCKS

3.5.1. Introduction

Consideration of deep-seated granitic rocks demands some introductory historical and genetic perspective, although here space requires that this be much condensed and simplified, and a somewhat more philosophical approach to their description as compared with foregoing sections in this chapter.

The most abundant igneous-looking rocks exposed in continental crust are various granites (s.l.) of a very different appearance to that of the much less abundant rocks of central complexes. The latter occur in anorogenic terrain, are generally homogeneous and homophanous in outcrop appearance, have a generally low colour index, and a quartz content approaching a ternary minimum proportion in some granitic varieties, sometimes with iron-rich and/or peralkaline members, and some reveal patterns of trace-element content consistent with prolonged fractionation of liquid magma.

The granitic rocks to be considered have been emplaced in regionally metamorphosed rocks typically during or soon after deformation. They are commonly inhomogeneous, often foliated and very rarely homophanous, and have relatively high colour indices and low quartz contents; they are often peraluminous and never peralkaline, and do not for the most part have trace-element patterns ascribable to strong fractionation. They typically occur in large batholiths, originally defined as large intrusions with no visible floor or roots*, the subjacent intrusions of Daly (1933).

It would be a considerable task to assemble a complete dossier of "case-histories" of different granite intrusions and complexes. It would certainly reveal enormous ranges in structural styles, compositions, contact relationships, etc. No one geologist has seen it all, and not surprisingly therefore attempts at generalizations concerning the origin of granitic rocks have generated controversy and sharp differences of opinion even among eminent geologists who have had much field experience. For illustrative purposes, we can commence by considering the opposing viewpoints of N. L. Bowen and H. H. Read.

3.5.2. Contrasted approaches of Bowen and Read

Based primarily on experimental work, Bowen (1928) argued plausibly and at length for *fractionation* processes to produce the full range of intermediate and acid igneous rocks from parental basic magma. Certainly this process is possible and it has certainly been operative beyond doubt in some complexes (typically the anorogenic ones) in producing relatively small amounts of felsic differentiates. However, to produce the vast observed volumes of deep-seated granitic rocks would require prodigious amounts of parental basic magma, of which there is frequently embarrasingly little sign in the type of granitic terrain considered here. This, of course, is not to deny that the very existence of continental crust reflects massive fractionation processes in earlier stages of the Earth's history, or that this is to some degree a continuing process. However, what we are here concerned with is the proximate origin of these deep-seated granites typically occurring in terrain of regionally metamorphosed rocks.

Based mainly on extensive fieldwork in high-grade metamorphic terrain, an origin by *granitization* was argued with matching fervour by Read (1957) in a series of papers conveniently edited into one book, *The Granite Controversy*. A continuum was claimed from deep-seated syn-orogenic granitized material via late-orogenic plutons to high-level post-orogenic intrusions.

---

*Nowadays the term batholith is often applied, perhaps less happily, to any large granitic mass over 100 km² in outcrop area, in which usage it would encompass many of the large floored cauldron-subsidence intrusions. An alternative term for a recognizably distinct granitic body that does not carry any connotation of size or genesis is pluton.

3.5.3. Granitization defined

Before proceeding further we had better define what we mean by the emotive term granitization: **granitization** includes those mineralogical, physical, chemical, and kinetic processes by which an essentially solid rock is transformed into one of granitic appearance. Migmatites, inextricably linked with the subject of granitization, faithfully record the varied mineralogical, physical, chemical, and kinetic response of continental crustal rocks to elevated temperatures. *Mineralogically*, the prograde reaction of mica to feldspar and coarsening of grain size is very evident. *Physically*, migmatites reveal a loss of rigidity by their commonly developed agmatitic structures and flow folding. The proportion of melt phase produced at any one stage of the development of a migmatite complex remains a controversial question: to the simplistic concept that the neosome component of a migmatite can be equated in its entirety with a former melt phase, several opposing lines of compositional and textural evidence can be put (see C. J. Hughes, 1970b, and references therein). *Chemically*, the case for metasomatism in migmatite terrain both on a local and regional scale is strong, and it has been demonstrated that diffusion of the less dense ionic or molecular units such as Na, K, Rb, H_2O, etc., to areas of lower chemical potential (in general, upwards within the crust to cooler regions at lower pressure) could take place readily within the medium of, and in part by the movement of, a water-saturated silicate pore fluid (see Marmo, 1967). *Kinetically*, migmatite complexes, because they are relatively light, tend to rise diapirically (see, e.g., Hutchison, 1970; Haller, 1971, pp. 157—200), and in so doing effect a degree of internal homogenization, and also the possible segregation of homogeneous melt phases of more acid compositions than that of the overall composition of the migmatite complex itself.

Note that crustal anatexis can commence at temperatures much below the "dry" solidus and produce water-saturated melt of a minimum composition related to rock composition (von Platen, 1965). Note also, however, that the amount of melt generated at this "wet" solidus temperature or even higher temperature is very strongly buffered by the amount of available water. Kinetic upward movement to regions of lower pressure exerts a strong freezing influence on such water-saturated melt (P. G. Harris et al., 1970), so that it is physically impossible for partially liquid magmas (migmas) produced by anatectic processes within the crust in this way to be emplaced in their totality at high crustal levels. They may, however, be able to generate and segregate quantitatively minor bodies of magma of lower temperature, more ideal composition that may penetrate to higher crustal levels, thus in part vindicating Read's proposed continuum.

The reality of granitization has been convincingly demonstrated by work in the Coast Range, British Columbia. A summary (Hutchison, 1970) of the large-scale investigations completed by the Geological Survey of Canada in

this well-exposed terrain of considerable relief could well serve as a model
for the development of some of the granitic rocks considered here.

> "The metamorphic framework in Prince Rupert—Skeena region of the Coast
> Mountains of British Columbia comprises schist, gneiss, and migmatite displaying
> progressive regional metamorphism that overlaps the Barrovian and Idahoan Facies
> Series. ... Plutonic rocks, which were probably an integral part of the early meta-
> morphic framework, have apparently been mobilized during metamorphism and
> continued to move out of their original environment while metamorphism waned,
> some even deforming the pre-existing fabric.

> Within the framework, four main plutonic styles have been recognized:

> (1) Autochthonous, migmatitic, plutonic complexes.
> (2) Para-autochthonous, steep-walled (tadpole) plutons.
> (3) Para-autochthonous, tongue-shaped recumbent plutons.
> (4) Allochthonous, intrusive plutons.

> Quartz-diorite and granodiorite are the most common plutonic rocks. Diorite and
> quartz-monzonite are less common: gabbro and especially granite are rare.

> In the course of moving from the sites of generation to the zones of emplacement,
> the plutonic rock became:

> (1) more homogeneous.
> (2) less migmatitic, and impoverished in inclusions.
> (3) less foliated.
> (4) more acidic, more biotite-rich.
> (5) a rock containing plagioclase of lower average anorthite content and more
> complex oscillatory zoned crystals."

We seem to be left then with a minority of "liquid" granites (and other
rock types) mainly contained to anorogenic central complexes, and a majority
of "orogenic" granites typically found in metamorphic terrain. It is this
apparent two-fold division which has been followed for descriptive purposes
in the overall subdivision of this chapter. However, it is appropriate at this
stage to refer to a notable landmark in granite investigation and classification
— the synthesis of A. F. Buddington.

3.5.4. Approach of Buddington

Disregarding in the first instance thorny questions of genesis, Buddington
(1959) compiled case histories of North American granites. The evidence
from this true geological approach to the problem leads to a natural three-
fold division of granitic bodies based on their style of emplacement, in turn
related to host-rock temperature and rheidity and thus roughly to crustal
depth at time of emplacement. The ensuing subdivision into **katazone-**,
metazone- and **epizone** granites is widely followed, and is adopted here as
a framework for description in Tables 3.1—3.3, which are largely self-
explanatory. Admittedly some parameters added and tabulated by the writer

may be open to criticism on the grounds of overgeneralization. In general, the epizone granites are those found in central complexes discussed above, whereas the deeper-seated mesozone and katazone granites are those found in abundance in metamorphic terrain.

The granite "space problem" is one that has long exercised petrologists in view of the large outcrop areas underlain by granite — what happened to the pre-existing rock? More than one mechanism can be seen to be operative in answering this apparent problem which is interrelated with critical evidence from contact relationships and internal structures.

3.5.5. Katazonal granitic rocks

In katazonal conditions we are in the domain of migmatites, granitization phenomena, and the possible site of generation of some granitic rocks. All that may be apparent in the way of contact relationships of an autochthonous migmatite complex is a fairly gradual increase in the proportion of neosome material often over distances of several kilometres or more in outcrop. Also under conditions which must still be regarded as katazonal it is possible for migmatite complexes to begin to move upwards in a rheid manner forming incipient diapirs, mushroom- and tadpole-shaped bodies. Intrusive contacts will not be displayed however, rather a somewhat narrower zone of "contact migmatites" between country rock and the now para-autochthonous migmatite complex.

On the subject of migmatites, there is little doubt among petrologists that partial anatexis has played a significant role in their formation. At upper amphibolite facies conditions, under which the majority of migmatites are found, temperatures are such that water-saturated silicate melts of granitic compositions are just possible. Nevertheless, considerable differences in emphasis has been placed on anatexis by different investigators working on migmatites. A somewhat rigid approach is that followed in the definitive book by Mehnert (1968), similar to an approach based on experimental work (von Platen, 1965). Both these writers categorically equate the neosome component of migmatite with partial melt. However, closer field examination reveals that much neosome material is not only variable in composition but commonly nowhere near the minimum compositions that would be expected if they represented partial melts. Workers with wide experience in migmatite terrain have therefore questioned the above simplistic assumption of Mehnert and von Platen both on this ground and also on textural and physical considerations. See particularly in this regard a review of potassic migmatites in southern Finland by Härme (1965). A. J. R. White (1966) considers that metamorphic differentiation can adequately account for the mineralogy and chemistry of migmatites from the Palmer region of South Australia. P. E. Brown (1967) and Misch (1968) have both given strong reasons for believing that regional metasomatism has

TABLE 3.1

Overall structural features of granitic bodies

Nomen-clature of Budding-ton (1959)	Approximate depth of emplacement		Intrusion style	Overall structure	Internal structure
Epizone granite	(exact limits impossible as emplacement style depends on the temperature and rheidity of country rock which varies with depth according to the geothermal gradient which itself varies from place to place)	0—9 km	"permissive"	ring-dyke intrusions connected to overlying cauldron-subsidence intrusions	generally homophan (structurel
Mesozone granite		7—16 km	"forceful"	diapirs	igneous foliæ common; domain of nite tecto ± superim regional de mation
Katazone granite		> 12 km	rheid adjustment	incipient diapirs of migmatite	tendency to chaotic; lc any pre-ex planar str
			"granitization"	migmatite complexes in situ	

occurred in the migmatite terrains of Sutherland, Scotland, and Washington, U.S.A., respectively. King (1965, p. 233) maintained that selective melting is not even the main factor in migmatite genesis. A cautionary note is sounded by Härme (1966) who gently reminds us that:

"a process proved possible in the laboratory is not always the sole possibility in nature"

ntact relationship		Temperature difference between granite and country rock	Probable physical state	Resolution of "space problem"
lled contact against ornfels		extreme	liquid ± phenocrysts	large-scale magmatic stoping
ally variable; illing generally not ovious	phyllonite along fault contact	variable — increasing →	crystal mush, probably largely solid	forceful pushing aside of flanks of diapir; incorporation of "pendants" or smaller "septa" in roof zone
	schist aureole			
	narrow zone of contact migmatite			
id zone of regional/contact igmatite		very little	solid with incipient silicate melt	large-scale plastic deformation of migmatite complex and country-rock envelope
ation in proportion of neosome aterial in regional migmatite omplex		nil		replacement and recrystallization

a piece of geological philosophy that should never be forgotten. C. J. Hughes (1970b) considered that those migmatites with prominent dark biotite-rich selvedges may have undergone incipient melting only. The above articles with their illustrations and references provide an overview of the spectrum of geological opinion on these fascinating rocks.

Internal structures of migmatites are very variable; see Mehnert (1968) for excellent illustrations and a classificatory descriptive scheme. In many,

TABLE 3.2

Petrological features of granitic bodies

Type (as in Table 3.1)	Nature of inclusions	Common rock types
Epizone	inclusions rare; some angular "accidental" inclusions of country rock near contacts; cognate inclusions uncommon	diorite to granite, including sy and quartz-syenite; ideal gra compositions include felsitic granophyric textural varietie high quartz contents commo overall; seldom peraluminou granitic and syenitic varietie commonly rich in Fe and/or peralkaline; pegmatites uncommon
Mesozone	inclusions common; rounded "cognate" inclusions of debatable origin, plus inclusions of country rock near upper contacts	diorite to granite, excluding syenite and quartz-syenite; quartz content varieties con overall; ideal granite compo rare; silicic varieties often p uminous; pegmatites comm localized autometasomatism resulting in greisening, tour-malinization, and kaoliniza-tion
Katazone	inclusions numerous, representing the metamorphic component of migmatite	generally inhomogeneous, bu locally homogeneous; bulk composition variable and di to determine; commonly p aluminous

veinlets and patches of neosome are seen to occur both concordantly and discordantly with respect to pre-existing structural planes in a coherent metamorphic country-rock framework. In some, a more pervasive type of crystallization of large feldspars (in response to metasomatism?) imparts a "nebulitic" appearance.

There is commonly a variable proportion of neosome to palaeosome not only from one part to another of a migmatitic complex but also within different palaeosome rock types. Mica schists, for example, commonly host relatively abundant neosome. Quartzites, on the other hand, commonly have little neosome presumably because they are compositionally further removed from granitic composition and are thus relatively slow and difficult to recrystallize and transform to material of granitic aspect. Amphibolite layers in migmatite very commonly appear as neosome-poor boudins sepa-

Dark minerals	Nature of alkali feldspar		Relationship to solvus at time of emplacement
fayalite, heden-bergite, aegirine and riebeckite often found in granitic and syenitic com-positions	mesoperthite		hypersolvus
muscovite com-mon in granitic compositions; also pinite after cordierite, both reflecting a peraluminous condition	perthitic orthoclase or perthitic microcline	determination complicated by subsolidus exsolution and recrystallization	subsolvus
"metamorphic" minerals such as garnet, cor-dierite, and sillimanite common	microcline		relatively low-tem-perature subsolvus equilibrium attained

(left margin, spanning all rows: biotite and hornblende common throughout)

rated by granitic, even pegmatitic, patches [see, e.g., the illustrations in Ramberg (1956)]. Migmatized mafic rocks often exhibit a reticulate network of quantitatively minor trondheimitic neosome, prior to more extensive assimilation processes involving metasomatism (see, e.g., Lobjoit, 1969). In autochthonous migmatite terrain such resister beds may permit the elucidation of a ghost stratigraphy.

In para-autochthonous migmatites particularly, the palaeosome component loses its coherency, banding of neosome and palaeosome becomes swirly and intricately flow-folded, and agmatitic structures with subangular disorientated blocks of palaeosome are found. In para-autochthonous migmatites too, neosome becomes relatively more abundant, commonly even predominant, and volumes of more or less homogeneous granitic rock may occur.

TABLE 3.3

Geological relationships, economic potential, and origin of granitic bodies

Type (as in Tables 3.1 and 3.2)	Relationship, if any, to comagmatic volcanic rocks	Relationship to regional metamorphism	Economic potential		Origin (still controversial)
Epizone	indubitable in a majority of cases, and inferred in many of remainder; intrusive rocks linked to volcanism by occurrence of such distinctive sub-volcanic features as cone sheets, diatremes, and near-surface crystallization of intrusive bodies of degassed magma, e.g., felsite	"post-orogenic"; generally more appropriately termed "anorogenic" (i.e. no genetic relationship whatsoever to regional metamorphism)	porphyry copper type deposits; mineralized breccia pipes; pyrophyllite		fractionation of mantle-derived melts ± some possible assimilation of crustal material; also some crust-derived melts
Mesozone	very doubtful (but direct evidence usually lacking as the amount of erosion necessary to expose mesazone intrusions would have removed consanguineous volcanic rocks if ever present)	"late-orogenic"	kaolin, china clay; vein mineralization (Sn, W, Cu, etc.)	simple and complex pegmatites	diapiric ascent of lower crustal material re-mobilized under katazonal conditions
Katazone	none	"syn-orogenic"	barren		crustal "granitization" processes

Assuming that migmatites signal some degree of partial melting, an autochthonous migmatite complex therefore reflects regional attainment of the necessary temperature conditions, and contact migmatites merely the local attainment of this temperature adjacent to a para-autochthonous or allochthonous migmatite body transferring heat to the upper crust. However, the temperature difference in the latter case between "igneous" and "country" rocks is small, plastic deformation is the rule, and contacts are vaguely defined and concordant. The development of contact migmatites around an allochthonous body of migmatitic granite may thus present the illusion of granitization in situ.

3.5.6. Mesozonal granitic rocks

Mesozone granite contacts embrace a great variety of structural styles related mainly to different country rock temperatures at the time of emplacement of the igneous body, and present a continuum from katazonal conditions upwards to relatively cool country rocks. Usually, a distinct metamorphic aureole, varying mainly with size of the intrusion, reflects the prolonged conduction of heat away from the igneous rock into cooler country rocks.

At relatively greatest depths contact migmatites may be found, usually in a rather narrow zone measurable in tens or hundreds of metres. Still at considerable depths where country rock temperature is elevated a marked penetrative deformation and ensuing new schistosity accompanied by a distinct episode of recrystallization may be superimposed on the country rocks related to the stress and heat produced by the forceful intrusion of the granitic body. At shallower levels and cooler country rock conditions, however, continued movement reflecting continued internal "push" within the pluton is taken up along a relatively cool contact, so that retrogression to phyllonites and even brittle faulting may occur.

Whereas some compressive stress must be set up around the flanks of a forcefully intruded mesozone granite body often resulting in a marked development of flattening in the plane of a new schistosity as noted above, a degree of extension may occur over the top of a diapir. This is most evident where the country rock is cool or has vertical planes of weakness and thus tends to deform not by flattening and extension but by breaking; under these latter conditions, **roof pendants** of all sizes down to narrow septa* a few metres across may be enveloped by the granite. Sometimes a dilated "stratigraphy" of metamorphic cover rock is preserved in this fashion, as for example in the main Donegal Granite (Pitcher and Read, 1959), not

---

*The term "septum" (pl. septa) has also been applied by American petrologists to the screens of country rock that commonly separate successive ring-dyke intrusions in epizone terrain.

to be confused with the in situ ghost stratigraphy of an autochthonous migmatite complex.

In the above discussion it must be borne in mind that differences in geothermal gradient will result in differing actual depths of cover rocks characterizing the different contact relationships.

The internal structures of mesozone granites have provided much material for classic discussions of granite tectonics associated with the name of E. Cloos and summarized in Balk (1937). Indeed, a primary igneous foliation is seldom lacking from mesozone granites, sometimes obvious as shown by aligned feldspar tablets and/or subparallel biotite crystals, sometimes so subtle as to be discernible only to a quarryman's practiced eye and touch. This foliation is taken as evidence of final emplacement as a crystal mush under the influence of directed pressure attributable to the forceful intrusion, as this igneous foliation will parallel the walls of the intrusion and is thus distinct from any regional foliation that the igneous rock might later acquire.

3.5.7. Discussion

Our first analysis led to the supposition that there might be two kinds of granite: (1) fractionated silicate melts in anorogenic central complexes, and (2) granitized crustal rocks forming katazonal and mesozonal bodies in orogenic terrain. How valid is this assumption?

Buddington has demonstrated that there appear to be continua, on the one hand between granites of mesozonal and epizonal affinities, and on the other hand between those of the katazone and mesozone. The argument that there is no continuum in any one granite terrain between all three is unfortunately a circular one as no terrain exposes all three levels. Perhaps the mesozone—epizone continuum could be interpreted as the essentially liquid type of granitic working its way up through intermediate to high crustal levels by processes of both stoping and diapiric uprise. Anorogenic epizone intrusions must presumably have penetrated the mesozone somehow, the katazone too for that matter. It should be borne in mind that temperature gradients in anorogenic terrain, particularly older basement rocks, are much lower than in orogenic terrain. The katazone—mesozone continuum is indeed to be expected, so that in sum it would appear that Buddington's natural three-fold division does not necessarily invalidate a genetic two-fold division.

Nevertheless it must be admitted that the petrogenesis of many mesozone granites continues to be a source of great controversy, particularly in the situation in which they seem to be abundant in Phanerozoic time, namely a supra-Benioff zone one. Are they to be conceived of as crustal remelts or mantle fractionates? An interesting example of this, the Sierra Nevada "batholith" (actually composed of numerous groups of composite intrusions), is discussed in Chapter 7, and it appears that some compromise is possible between these two contrasted ideas, i.e. the intrusions could represent the products of mantle melts undergoing both fractionation and assimilation processes. (If you don't happen to include the right question among your multiple working hypotheses, you won't come up with the right answer, so be flexible!)

3.6. FEATURES TO MAP AND SAMPLE IN INTRUSIVE ROCKS

Generalities do not have much value and we must consider the distinctive features shown by each of the groups of intrusive igneous rocks considered in this chapter.

Aphyric dykes and small sills of basic composition are usually chemically homogeneous varying   only in a slightly coarser grain size inwards and are thus easy to sample. Porphyritic rocks of this category may likely show flow differentiation (see Chapter 7), a crude form of fractionation affecting only the phenocrysts present on intrusion. Sampling may thus yield rocks either relatively enriched or depleted in early formed primocrysts, a not uncommon situation with porphyritic mafic rocks.

In larger basic sills, layered basic intrusions, and anorthosite bodies the attitude of any layered structures can be mapped just as in a sedimentary sequence. What is desirable is a regular sampling of a layered igneous se-quence with a view to the later detailed investigation of the succession of cumulus phases and of any cryptic variation that they may reveal. Phase layering can usually be detected in the field and provides a basis for separating a layered sequence into mappable units. The Bushveld Intrusion, for example, can be subdivided in this way into units of ultramafic rocks, norites, gabbros, and ferrogabbros. Careful fieldwork can lead to separation into quite detailed map units based upon phase layering [see the coloured map of the Skaergaard Intrusion included in Wager and G. M. Brown (1968, facing p. 244)]. In addition, major rhythmic layers if present can often be mapped. Rhythmic layering connotes repetition of phase-layered cumulus sequences commonly on a scale of ~10—100 m and is usually attributed to fresh inflows of magma (see, e.g., G. M. Brown, 1956; C. J. Hughes, 1970a).

Conventional wisdom has it that sampling and analysis of the chilled margin material yields the composition of the parental magma of the intrusion, obviously an important factor. However, repeated intrusions successively replenishing a magma chamber need not necessarily have had compositions identical with each other or with the chilled margin material. The Cap Edward Holm Intrusion in east Greenland, for example, is believed to have been emplaced within a fracture formed by splitting of a cone-sheet intrusion, the material of which thus simulates chilled margin material but need not have been the same as the parental magma of the subsequent major intrusion (Elsdon, 1969). Also, it is not improbable that some metasomatism across the igneous contact may have accompanied the development of the autometamorphic textural recrystallization shown by many chilled margin rocks of large intrusions.

In ophiolite complexes one would record such varied features as (1) fabrics in the tectonite unit; (2) enstatolite, dunite and other veins, and other noticeable departures from a harzburgitic mineralogy; (3) attitude of layering and phases present in the overlying cumulate sequence; (4) existence of any layering in, mineralogy of, and variations within the overlying gabbroic rocks (with the knowledge that there may stand much to be revealed from more detailed studies of this component of the ophiolite suite); (5) strike and abundance of sheeted dykes and any apparent unusual petrographic varieties such as, for example, the dyke members of high-magnesium basalt composition of a sheeted dyke complex recorded by Gale (1973); and (6) massive or pillowed varieties, and any distinctive stratigraphical horizons or indications of mineralization in the lava sequence.

The intrusive rocks of central complexes may well show textural and even compositional variation (see description of zoned plutons in Chapter 7) within mappable intrusive units, requiring careful fieldwork to distinguish successive intrusions that may be multiple or composite, conveniently designated G_1, G_2, G_3, etc. (see p. 71).

In rocks of the diatremic association one is mapping bodies of possibly different kinds of breccias (see Chapter 2); an examination of the lithologies of breccia fragments for provenance is often important in evaluating the degree of relative movement of fragments in the diatreme. A magmatic component in the diatreme may be represented by varying proportions of shards, phenocrysts, vesicular masses, or may be completely lacking.

Deep-seated granites generally show such mappable features as joint patterns, igneous foliation (to be distinguished from any possible metamorphic foliation), aplite vein direction, pegmatites, mineralized veins, and zones of alteration. Granite contacts always merit careful examination for structural style and conditions of emplacement.

Criteria for establishing the relative ages of igneous rocks include the following: (1) normal stratigraphic principles of superposition in lava flows; (2) absolute age determinations, which may reveal a surprising amount of fine detail in relatively young and undeformed igneous rocks, as for example the accounts of the San Juan volcanic field by Steven and Lipman (1976) and of the Boulder batholith region by Robinson et al. (1968), but cannot be relied upon for relative ages in older provinces or igneous rock complexes; (3) structural truncation of relatively earlier igneous bodies by relatively later ones, as for example in many ring complexes and calderas; (4) unequivocal chilled contact of an intrusive igneous rock against older; (5) veining of one igneous rock by another [but beware of the phenomenon of "back-veining"; see definition in Chapter 7, and C.J. Hughes (1960, p. 114; plate X, fig. 2)]; (6) inclusions of one rock type (if sufficiently distinctive) within another; and (7) relative dating of two intrusive igneous bodies even if not in contact by reference to an intervening period of deformation or of intrusion such as a dyke swarm.

Chapter 4

CLASSIFICATION OF IGNEOUS ROCKS

4.1. INTRODUCTION

Historically, some igneous rock terms have been in field use for a considerable time to describe such obviously distinctive rock types as trap, felsite, granite, porphyry, etc. With the application of the petrological microscope to petrographical studies beginning in the latter third of the nineteenth century, the precise mineral composition of all but glassy and very fine-grained rocks could be established. A great many new igneous rock names were formulated based on the content, proportions, and to some extent the textural relationships of their constituent minerals. The classic work of Johannsen (1931—1939) and the more concise book by Holmes (1920) are excellent etymological sources for this heroic age of petrography. There was an undeniable tendency to trim a newly collected igneous rock to a convenient size and shape, bestow upon it a varietal name, and put it in a museum showcase. This treatment may be all very well for fossils and minerals which belong to distinct species, but is an obvious source of potential difficulties in dealing with the continuous and broad spectrum of compositions and textures of igneous rocks. For example, communication may suffer between petrologists who may individually be inclined to be either "splitters" or "lumpers" and thus work with differently sized pigeonholes in a classification scheme. More significantly, legitimate genetic interrelationships may be lost sight of in a wealth of mineralogical and textural complexity.

The mineralogy of an igneous rock is a function of a more fundamental attribute, namely its chemical composition. For example, magma of essentially the same chemical composition could yield such distinct textural varieties as granite, or microgranite, granophyre, quartz-porphyry, aplite, felsite, rhyolite, or obsidian, depending on the accident of intrusion or extrusion, size of the igneous body, and cooling history. Also, genetic relationships between igneous rocks of different compositions within the one province, i.e. those emplaced in the same localized area and at approximately the same time, may likely be the result of physico-chemical processes such as fractionation, assimilation, and hybridization that result in definable variation trends in magma compositions. For these reasons, therefore, classifications based on chemical composition or compositional parameters have been proposed and are widely used by research petrologists at the present time.

The purpose, however, of any classification scheme is to communicate accurately, conveniently, and appropriately in context. A field classification appropriate to a company report without any microscope or chemical investigation could be refined by microscope work to a terminology appropriate say to a definitive geological survey report on an area. Further research work on the petrology and petrogenesis of the same rocks may well utilize a different terminology based on chemical analyses and derived parameters.

For our present purposes we will first review the presentation of chemical data, then see how chemistry and mineralogy are interrelated, and finally build up a reasonable petrographic classification scheme based upon mineralogy alone, and independent of the kind of petrogenetic inferences that are left until later chapters. It is, however, important to be aware from the outset of the essential but often uneasy alliance between chemistry and mineralogy in the classification of igneous rocks.

4.2. PRESENTATION OF COMPOSITIONAL DATA

4.2.1. Igneous rock analyses

Table 4.1 presents chemical analyses both of some common rock types such as tholeiite basalt, alkali basalt, andesite, dacite, rhyolite, trachyte, and granite, and also of some rarer more extreme varieties in order to portray the range of chemical variation found in the major-element contents of igneous rocks*. Even a cursory study of the data shows that there is by no means a regular variation from those rocks low in silica content to ones high in silica, but rather a wide spectrum of compositions exhibiting considerable differences between the relative proportions of other oxides, notably in the relative contents of the alkali oxides, Na_2O and K_2O.

The analyses have been tabulated in order of increasing silica content — a widely followed procedure. One reason for this is because, as we shall see later, fractionation, believed to be a dominant factor in the differentiation of igneous rocks, leads to an overall increase in silica content of an igneous rock series. Although the rocks in Table 4.1 have been selected for their diversity and do *not* belong to any one genetically interrelated series of igneous rocks which would be expected to reveal a more regular variation, it is convenient to tabulate the analyses in this manner.

In common petrological usage some of the major-element oxides have acquired names that are not the same as their chemical names and should

---

*The order in which the major oxides are presented in igneous rock analyses is in the main that of decreasing cation valency, conveniently so in that the two generally most abundant oxides, SiO_2 and Al_2O_3, appear first. Some analyses quote the much less abundant oxide TiO_2 immediately after SiO_2, presumably to preserve the valency order, though no one goes so far as to put P_2O_5 at the head of the list!

not, of course, be confused with the names of elements as Table 4.2 makes clear.

H_2O^- in a rock analysis refers to that water which is lost from a sample by heating (and reheating until a constant weight is attained) to 105°C. It is thus merely a measure of the degree of dampness of the powdered sample as received by the analyst and has no relevance to igneous rock composition. For comparative purposes it could well be omitted. H_2O^+, on the other hand, refers to that water which is an essential component of such minerals as biotite, hornblende, epidote, analcite, etc., and thus is an essential constituent of the rock. Sometimes all water is deleted from an analysis and the remaining oxides recast to 100% for comparative purposes. This arbitrary procedure, followed, for example, in calculation of the CIPW norm (see p. 97), parallels that in nature where on eruption of extrusive rocks, water together with other volatile constituents is largely lost from a magma system from which an equivalent plutonic rock would retain magmatic water. In analysis 1 the very high content of H_2O^+ reflects not a water-rich magma but the fact that the igneous rock has been metamorphosed and contains secondary hydrous minerals such as serpentine, chlorite and amphibole. In view of this high water content some doubts may be entertained whether the amounts of other major oxides have remained relatively fixed during metamorphic recrystallization. This is indeed one of the areas of current debate in igneous petrology where conclusions on magmatic affinity based on chemical compositions of older, altered, and metamorphosed igneous rocks are concerned (see Chapter 14).

Common mineral constituents of igneous rocks can be expressed stoichiometrically in terms of oxides, and thus a chemical analysis of an igneous rock presented in terms of oxides should be expected to sum to 100% to a first approximation. However, there may be some sulphide present and halide may substitute for hydroxyl in minerals such as apatite and white micas. In analysis 12, for example, there is appreciable halide and the total would come to 100.05 wt.% were not an appropriate subtraction made to allow for the fact that some cation must be combined with the halide and not with oxygen. After correcting the analysis in this manner (and similarly if sulphide were present) and allowing for small unavoidable analytical errors the total of a superior analysis should come to between 99.5 and 100.25 wt.%.

In classical silicate analysis results are conventionally calculated and presented to two decimal places, i.e. three significant figures for most major oxides and four significant figures for silica and alumina contents. This implied degree of accuracy is just not attainable even with the most careful work whether by classical or rapid silicate analysis*. The style of analysis 10 where oxide contents are given to the number of significant figures justified by the analytical method used, even if less tidy in appearance, is scientifically a more correct procedure.

---

*For footnote, see p. 92.

TABLE 4.1

Chemical compositions (in wt.%) of selected igneous rocks

	1	2	3	4	5	6
SiO_2	41.61	42.55	43.49	45.4	46.82	49.3
Al_2O_3	2.70	16.66	12.66	14.7	14.50	17.6
Fe_2O_3	5.63	7.05	2.06	4.1	5.26	1.9
FeO	4.35	3.42	9.45	9.2	6.78	6.8
MnO	0.17	0.18	0.23	0.2	0.20	0.1
NiO	0.18	—	—	—	—	—
MgO	30.58	4.08	12.68	7.8	7.61	7.1
CaO	4.29	9.35	12.44	10.5	11.07	11.7
Na_2O	0.15	6.25	3.08	3.0	2.56	2.7
K_2O	0.03	4.20	1.32	1.0	0.36	0.1
TiO_2	0.31	2.84	2.13	3.0	1.95	1.4
P_2O_5	0.02	0.73	0.48	0.4	0.22	0.
H_2O^+	8.81	1.53	0.30	—	1.74	0.4
H_2O^-	0.22	0.73	—	—	0.75	0.5
CO_2	—	0.43	—	—	—	—
Cr_2O_3	0.32	—	—	—	—	0.4
ZrO_2	—	—	—	—	—	—
F, Cl	—	—	—	—	—	—
Total	99.37	100.00	100.32	99.3	99.82	100.

1 = average of five analyses of *peridotitic komatiite*, Komati Formation, Barberton greenstone belt (M.J. Viljoen and R.P. Viljoen, 1969, p.72, table 1, analysis 5).

2 = *nepheline leucitite*, Nyiragongo volcano, Western Rift Valley (Sahama, 1962, p.11, table 2, analysis 2).

3 = *basanite*, La Grille volcano, Grande Comore (Strong, 1972, p.192, table 6, analysis 9).

4 = average of 35 analyses of *alkali olivine basalt*, Hawaii (G.A. Macdonald, 1968, p.502, table 8).

5 = *olivine tholeiite*, Thingmuli, Iceland (Carmichael, 1964, p.439, table 2, analysis 2).

6 = average composition of *oceanic tholeiitic basalt* (Engel et al., 1965, p.721, table 2).

---

*Classical silicate analysis for major oxides, in use since the earliest days of chemical petrology, is an exacting gravimetric procedure, best executed by a trained chemist. It involves dissolving the rock powder, then precipitating, drying, and weighing oxides and other compounds of major elements in turn, under carefully controlled conditions of pH, etc. The procedure is in part sequential, and thus any mistake generally entails time-consuming reduplication. A batch of six analyses will take a full week to complete. Fairbairn (1953) sent the same synthetic glass of granitic composition to eleven leading laboratories around the world for chemical analysis. Results showed that the precision of classical silicate analysis was by no means as good as had been complacently assumed. Rapid methods of silicate analysis, widely adopted in the sixties, include colour and flame photometry and titration methods. In general, redeterminations of individual elements

7	8	9	10	11	12	
54.70	54.91	62.90	65.0	71.98	76.25	SiO_2
18.07	12.73	16.15	15.4	13.13	10.86	Al_2O_3
3.90	1.02	1.37	2.46	1.33	1.23	Fe_2O_3
3.75	6.41	3.09	2.40	1.64	0.76	FeO
0.16	0.17	0.14	0.06	0.14	0.03	MnO
—	—	—	—	—	—	NiO
3.32	11.29	0.61	2.6	0.56	0.18	MgO
8.17	8.34	1.89	6.4	1.15	0.37	CaO
3.69	1.88	6.17	3.88	2.98	4.68	Na_2O
1.88	0.67	5.58	0.95	4.93	4.65	K_2O
0.85	0.48	0.41	0.74	0.37	0.11	TiO_2
—	0.10	0.17	0.18	0.19	0.01	P_2O_5
—	1.02	0.13	0.1	1.38	0.50	H_2O^+
—	0.72	0.26	—	0.39	—	H_2O^-
—	—	0.84	—	—	—	CO_2
—	—	—	—	—	—	Cr_2O_3
—	—	0.13	—	—	0.10	ZrO_2
—	—	—	—	—	0.32	F, Cl
8.49	99.74	99.84	100.17	100.17	100.05	Total
					0.10	less O for F, Cl
					99.95	Total

7 = average composition of *andesite* from Mt. Egmont, New Zealand (Gow, 1968, p.182, table 4, analysis A).

8 = *pyroxene andesite*, Ohakune, Tongariro National Park, New Zealand (Ewart, 1965a, p.90, table B-B, analysis 9).

9 = *trachyte*, Nandewar volcano, New South Wales (Abbot, 1969, p.127, table 3, analysis 12).

10 = *dacite*, Mt. Hood, Cascades (Wise, 1969, p.992, table 12, analysis 100).

11 = *granite*, Skye, Inner Hebrides, Scotland (Tuttle and Bowen, 1958, p.114, table 20, analysis F2-239).

12 = *riebeckite-aegirine granite*, Kudaru, northern Nigeria (Jacobson et al., 1958, p.17, table 7, analysis 1).

can be readily repeated. A batch of six analyses can be completed in two days by a technician. Accuracy closely approaches that of analysis by classical methods and is adequate for petrological purposes.

More recently, integrated schemes use "AA" (atomic absorption) and "XRF" (X-ray fluorescence), the latter capable of mechanised input of samples and computerised output of data. Some major oxides (H_2O, CO_2, Fe_2O_3, Na_2O, P_2O_5) are not amenable to either of these methods and are determined separately by rapid methods. These schemes give acceptable accuracy and are capable of even more streamlined operation than "classical" rapid methods.

Maxwell (1968) gives outlines of classical silicate analysis (pp. 536—539), rapid methods (pp. 540 and 541), and a combined XRF and rapid methods analytical scheme used by

TABLE 4.2

Petrological nomenclature of major-element oxides

Element symbol	Element name	Cation	Chemical formula of oxide	Commonly used oxide name
Si	silicon	Si^{4+}	SiO_2	silica
Al	aluminum	Al^{3+}	Al_2O_3	alumina
Fe	iron	Fe^{3+}	Fe_2O_3	ferric oxide
		Fe^{2+}	FeO	ferrous oxide
Mn	manganese	Mn^{2+}	MnO	manganese oxide
Mg	magnesium	Mg^{2+}	MgO	magnesia
Ca	calcium	Ca^{2+}	CaO	lime
Na	sodium	Na^+	Na_2O	soda
K	potassium	K^+	K_2O	potash
Ti	titanium	Ti^{4+}	TiO_2	titania or TiO_2
P	phosphorus	P^{5+}	P_2O_5	phosphorus pentoxide or P_2O_5
H	hydrogen	H^+	H_2O	water

Other less common elements known as trace elements are, of course, present in igneous rocks in addition to the major elements. They generally occur dispersed in the crystal lattices of major minerals and are thus not present as distinct mineral species; exceptions to this would include, for example, the minerals chromite and zircon containing essential chromium and zirconium. Although varying in amount from rock to rock, the relatively more abundant trace elements include MnO and NiO (usually present in solid solution in minerals containing ferrous iron and therefore combined with the FeO figure when not reported separately) and BaO and SrO (which are similarly combined with CaO). Contents of Cr_2O_3 and ZrO_2 may be appreciable enough to warrant being reported separately, particularly in some ultramafic and peralkaline rocks respectively. Other trace elements where determined, and also Mn, Ni, Ba, Sr, Cr and Zr, where not reported as oxides, are usually quoted in ppm (parts per million), equal to g/t (grams per tonne). Simple calculations using the appropriate atomic weights convert

---

the Geological Survey of Canada (pp. 542 and 543), together with comprehensive details, in the text of his book *Rock and Mineral Analysis*.

For trace elements a choice of emission spectrography, AA, XRF, and other methods are available. The choice depends on the accuracy and sensitivity of each method for a specific element.

Note in passing the meanings of the following terms commonly employed in referring to analytical data: *precision* (the reproducibility of an analytically determined value by one method); *accuracy* (the approach of an analytically determined value to the true value); *sensitivity* (the lower limit of concentration that can be measured by a specified method for a specified element). At low concentrations near the sensitivity the accuracy deteriorates to the point where an element can merely be detected rather than measured.

weight per cent oxide to ppm and vice versa: e.g., 0.32 wt.% Cr_2O_3 is equivalent to $104/152 \times 3200 = 2189$ ppm Cr.

It might be wondered whether the aggregate total of trace elements may become significant enough to affect the total of weight per cent major oxides in a rock analysis. In classical silicate analysis most trace elements are precipitated and weighed with the chemically similar major oxides and are thus already included in the total; in theory a subtraction should therefore be made from the appropriate major-oxide content for trace elements determined separately by other methods. In the case of some rapid methods which may be specific for some major elements, any subsequently determined trace-element content should be added. However, for practical purposes these considerations are academic, as the sum of trace elements (excluding Mn which is usually reported with the major oxides as MnO) is generally so small (of the order of 0.2 wt.%) that it does not materially affect the total rock analysis within its limits of accuracy.

CO_2 is an important and significant primary constituent of carbonatite, kimberlite, some lamprophyres, nepheline syenites, etc. However, some CO_2 may be introduced by groundwater or hydrothermal solutions into igneous rocks including those of the more common compositions that do not usually contain CO_2 and its presence therefore may indicate alteration.

4.2.2. Norms

Chemical analyses of igneous rocks are available with increasing facility by modern analytical techniques and in ever increasing numbers. Comparison of numerical chemical data is tedious and does not always reveal at first sight some of the more subtle but fundamental differences between rocks, so the idea was evolved of recasting a chemical analysis of an igneous rock into component minerals, something more tangible and familiar for petrologists to cope with. This idea has been universally adopted and a calculated normative mineralogy or norm commonly accompanies chemical analyses in research papers. The norm will not necessarily be precisely the same as the mode for several reasons. One is that the most commonly used norm is calculated as *weight* per cent (wt.%) of minerals from the weight per cent of oxides obtained by analysis, whereas a measured or estimated *mode* is, of course, the *volume* per cent (vol.%) of the actual minerals present (theoretically it is possible to convert a measured volumetric mode to a quantitative mode by adjusting the volume per cent of each mineral by a factor related to its density, but this refinement is seldom followed by petrologists). Another reason is that relatively complex hydroxyl-bearing mineral species such as hornblende and biotite, that are common in granitic rocks for example, are generally excluded from normative calculations and anhydrous pyroxenes used instead. A third reason is that in calculating the norm the various oxides are assigned to appropriate mineral species in a fixed sequence and the results expressed as the simplified end-members of solid-solution series. This procedure, although related to observed mineral parageneses as far as possible, does not exactly match the more subtle possibilities of solid solution in naturally occurring mineral species, for example, Fe_2O_3 in natural orthoclase, K_2O in natural plagioclase, Na_2O and TiO_2 in natural augite, etc. However, when all is said and done, the fact remains that normative mineralogy matches modal mineralogy for basic rocks rather closely and conveniently; the equivalence is a little less close for granitic rocks, where besides the omission of hornblende and biotite from the norm, a problem arises in assigning normative albite between total normative alkali feldspar and plagioclase unless an assumption on plagioclase composition is made (see Le Maitre, 1976). However, the advantage for comparative purposes of everyone following the same procedure for all rocks generally outweighs minor inconveniences.

The above remarks refer to the **CIPW norm**, first proposed by the American petrologists W. Cross, J.P. Iddings, L.V. Pirsson, and H.S. Washington in 1902 and accompanied by a classificatory nomenclature based upon possible normative mineral parageneses. Although the nomenclature, which introduced at one fell swoop rather too many new and "artificial" terms for digestion by petrologists, has not caught on, their normative scheme is in universal and everyday use, particularly in the English-speaking world.

In the comparable and perhaps even more elegant Niggli norm, first proposed by the Swiss petrologist Paul Niggli in 1920, and widely used on the continent of Europe, molecular amounts of four groups of oxides: al (Al_2O_3 + Cr_2O_3 + rare-earth element oxides), fm (FeO + $\frac{1}{2}$ Fe_2O_3 + MgO + MnO), c (CaO + SrO + BaO) and alk (Na_2O + K_2O) are recast to 100%. A computed "quartz value", qz, can be either positive for oversaturated rocks or negative for undersaturated rocks. Other parameters are k, the ratio of K_2O to sum of alkalis in alk, and mg, the ratio of MgO to sum of the divalent oxides in fm. All parameters are retained in molecular proportions as opposed to weight per cent minerals in the CIPW norm. A lucid account of the Niggli norm with a worked example is given by Barth (1952, pp. 74—77), and a spirited defence of the molecular norm as opposed to the CIPW norm is made by Niggli (1952, pp. 388 and 389) himself.

From time to time other normative schemes have been proposed, notably the "metamorphic" norms of Barth (1959) utilizing minerals appropriate to different metamorphic facies groupings, of which the *mesonorm* containing hornblende and biotite would for example relate more closely to the modal composition of mesozone granitic rocks than the CIPW norm. Subsequent discussion (O. Larsen and Sørensen, 1960), however, has highlighted some of the difficulties and ramifications involved wherever these more complicated molecules are introduced.

Whatever may be the intrinsic merits of other schemes, it is fair to say that the CIPW norm has achieved a popularity in common use that is not at all likely to be displaced.

4.2.3. Procedure for calculating the CIPW norm

Full details, together with tables facilitating calculation, are given in Holmes (1930) and Johannsen (vol. 1, 1931). Whereas norm calculation lends itself to computer programming, it is nonetheless very instructive for students to work out a few norms, and at times it is convenient to be able to work out one or two, or obtain normative parameters without running to a computer. If a pocket calculator is available the procedure takes only a few minutes. A simplified procedure, embodying all common normative minerals, which serves for all but a very few compositionally unusual and uncommon igneous rocks, is as follows. (Fig. 4.1 shows a worked example, analysis 2 of Table 4.1.)

(1) (a) Add amounts of MnO, NiO (if reported separately) to FeO.
    (b) Add amounts of BaO, SrO (if reported separately) to CaO.
(2) Adjust Fe_2O_3 if desired*.
(3) Recast adjusted analysis to 100 wt.% omitting H_2O and any CO_2.
(4) Express as molecular amounts, obtained by dividing the weights per cent of oxides by their respective molecular weights, multiplying by 1000 and entering to nearest integer for convenience in calculation.

Now begin to construct normative minerals in the following order, using the molecular proportions as indicated by the formulas in the tabulation below.

(5) Use all P_2O_5 to make apatite.
(6) Use all TiO_2 to make ilmenite.
(7) Use all K_2O to make orthoclase [provisional, see (14), (d)].
(8) (a) Use all Na_2O to make albite [provisional, see (14), (c)].
    (b) Use any excess Na_2O to make acmite [i.e. if not enough Al_2O_3 to satisfy (8), (a)].
    (c) Use any remaining excess of Na_2O to make sodium metasilicate [i.e. if not enough Fe_2O_3 to satisfy (8), (b)].
(9) (a) Use remaining Al_2O_3 to make anorthite.

---

*An anomalously high Fe_2O_3 content, for example in mafic volcanic rocks that may have undergone post-crystallization oxidation, is often arbitrarily reduced to 2 wt.%, or to some proportion of total iron oxides [e.g., to 20 wt.% of total iron oxides (C.J. Hughes and Hussey, 1976, p. 485)], or to some comparable appropriate figure [e.g., to 1.5 wt.% + TiO_2 value (Irvine and Baragar, 1971, p. 526)]. The excess Fe_2O_3 is then added to FeO content with an appropriate adjustment for the different weight per cent of combined oxygen. $0.9 \times$ "surplus" $Fe_2O_3 =$ "new" FeO to be added to original FeO of analysis. The reason for doing this is that anomalously high Fe_2O_3 contents in weathered rocks result in high normative magnetite and consequently high values of available silica content, resulting in turn in misleading indications of silica activity and hence magmatic affinity from a raw norm not using corrected values. The differing corrections used by various workers (or the lack of any correction for obviously oxidized rocks) makes comparative work on published norms exceedingly tiresome and the reader is warned that there is not consensus among petrologists on dealing with this matter.

ROCK NAME _Nepheline leucitite. Heart of dyke, Nyiragongo upper craters_

Source of analysis _Sahama, 1962, Trans. geol. Soc. Edinburgh. V.19, p.11, Table 2, analysis._

Oxide	RAW ANALYSIS Weight percent	ME only. MnO etc. added. Fe₂O₃ adjusted	Recast to 100%	Divide by molecular weights to give molecular proportions Multiply by 1000 and report to nearest integer for ease in calculation	ap	il	or provisional	ab provisional	ac peralkaline rocks only	an peraluminous rocks only	mt C	calculate :: of MgO & FeO	di	hy	Q provisional	UNDERSATURATED ROCKS ONLY hy revised	ol revised	ab	ne revised	or	lc
SiO₂	42·55	42·55	43·95	÷ 60 732			276	624	38				278	33	-517	17		208	24	168
Al₂O₃	16·66	16·66	17·21	÷ 102 169			46	104	19									104	4	42
Fe₂O₃	7·05	2·03	2·10	÷ 160 13							13									
FeO	3·42	8·12	8·39	÷ 72 117	37						13		54	13		13				
MgO	4·08	4·08	4·21	÷ 40 105									85	20		20				
CaO	9·35	9·35	9·66	÷ 56 173	15					19			139							
Na₂O	6·25	6·25	6·46	÷ 62 104				104										104		
K₂O	4·20	4·20	4·34	÷ 94 46		46													4	42
TiO₂	2·84	2·84	2·93	÷ 80 37	37															
P₂O₅	0·73	0·73	0·75	÷ 142 5	5															
MnO	0.18			Calculate wt % normative minerals by multiplying by appropriate mol. wt and dividing by 1000 1·6	5·6	-	-	-	5·3	-	3·0		wo 16·1	en		en	fo 1·4	-	295	2·2	18-
NiO													en 8·5	fs		fs	fa 1·3				
BaO													fs 7·1								
SrO																					
H₂O	2·26																				
CO₂	0·43																				
others																					
TOTAL	100·00	96·81	100·0																		

Column annotation (vertical text): calculate :: of MgO & FeO = 61 : 39 ; remaing FeO = 105 MgO and 67 FeO = 105 ; (i) part or→ ne saves 84 leaves 84 ; (ii) all hy → ol saves 17. (iii) part or→ ne saves 116. ; (i) all hy → ol saves 17. (iii) all ab → ne saves 116.

REPORT NORM

Q	-		wo 16·1	hm -
C	-	di {	en 8·5	il 5·6
or 2·2			fs 7·1	ap 1·6
ab	-		en -	TOTAL 99·9
an 5·3		hy {	fs -	
ne 29·5		ol {	fo 1·4	
lc 18·3			fa 1·3	
ac	-		mt 3·0	

Fig. 4.1 Worked example of a CIPW norm calculation. Note examples of parameters obtained from the norms: normative colour index = Σ(di + hy + ol + mt + hm + il) = 43; differentiation index, D.I. = Σ(Q + or + ab + ne + lc) = 50.

    (b) Use any excess Al_2O_3 to make corundum [i.e. if not enough CaO to satisfy (9), (a)].

(10) (a) Use all Fe_2O_3 to make magnetite.

    (b) Use any excess Fe_2O_3 to make hematite [i.e. if not enough FeO to satisfy (10), (a)].

(11) Note molecular ratio of MgO and remaining FeO at this stage.

(12) Use any remaining CaO to make diopside, composed of equal molecular amounts of wollastonite and hypersthene, the latter preserving the molecular ratio of MgO and FeO noted in (11).

(13) Use remaining MgO and FeO to make hypersthene [provisional, see (14), (b)].

(14) (a) Use any remaining SiO_2 to make quartz. However, the rock may be undersaturated with respect to silica, and on summation of molecular amounts of SiO_2 already used in the formation of the above minerals it may thus become apparent that there is not enough SiO_2 present in the rock analysis to make the above (saturated) minerals. In this eventuality proceed as follows:

(b) Convert hypersthene to olivine, to extent necessary to rectify deficiency, again preserving the MgO/FeO ratio of (11) in both hypersthene and olivine, $2hy = 1ol + 1Q$.

If there is still not enough SiO_2, after converting all hypersthene to olivine:

(c) Convert albite to nepheline, to the extent necessary, $1ab = 1ne + 4Q$.

If there is still not enough SiO_2, after converting all albite to nepheline:

(d) Convert orthoclase to leucite, to the extent necessary, $1or = 1lc + 2Q$.

(15) Now convert the molecular amounts of "normative minerals" to weight per cent by multiplying by their respective molecular weights and dividing by 1000 (Table 4.3). Results should be quoted to one decimal place only as original analytical errors plus inevitable "rounding-off" procedures involved in the calculation certainly do not justify quoting some

TABLE 4.3

Common normative minerals

	Normative mineral	Symbol	Formula	Molecular weight
	quartz	Q	SiO_2	60
	corundum	C	Al_2O_3	102
	orthoclase	or	$K_2O \cdot Al_2O_3 \cdot 6SiO_2$	556
	albite	ab	$Na_2O \cdot Al_2O_3 \cdot 6SiO_2$	524
	anorthite	an	$CaO \cdot Al_2O_3 \cdot 2SiO_2$	278
	nepheline	ne	$Na_2O \cdot Al_2O_3 \cdot 2SiO_2$	284
	leucite	lc	$K_2O \cdot Al_2O_3 \cdot 4SiO_2$	436
	acmite	ac	$Na_2O \cdot Fe_2O_3 \cdot 4SiO_2$	462
	sodium metasilicate	ns	$Na_2O \cdot SiO_2$	122
Diopside, di	wollastonite	wo	$CaO \cdot SiO_2$	116
	enstatite	en	$MgO \cdot SiO_2$	100
	ferrosilite	fs	$FeO \cdot SiO_2$	132
Hypersthene, hy	enstatite	en	$MgO \cdot SiO_2$	100
	ferrosilite	fs	$FeO \cdot SiO_2$	132
Olivine, ol	forsterite	fo	$2MgO \cdot SiO_2$	140
	fayalite	fa	$2FeO \cdot SiO_2$	204
	magnetite	mt	$FeO \cdot Fe_2O_3$	232
	hematite	hm	Fe_2O_3	160
	ilmenite	il	$FeO \cdot TiO_2$	152
	apatite	ap	$3CaO \cdot P_2O_5$	310

more significant figures. (Take care with the unit formula of olivine, the source of a common mistake.) The total, of course, should be 100 wt.%, in practice often between 99.8 and 100.2 wt.%, again due to inevitable and perhaps cumulative "rounding-off" errors.

4.3. REFLECTION OF CHEMISTRY IN MINERALOGY

The relative proportions of minerals in an igneous rock necessarily reflect variations in the relative proportions of major oxides which not only are few in number but also allowing for common diadochies can be conveniently reduced to five groups: SiO_2; Al_2O_3; ($FeO + Fe_2O_3 + MgO$); CaO; and ($Na_2O + K_2O$). The generally smaller amounts of TiO_2 and P_2O_5 are reflected by the presence of ilmenite or sphene and apatite, common examples of minor accessory minerals that are generally present but in small amounts only and do not affect the position of an igneous rock in a classification scheme. Let us now see how the relative abundances of these five groups of major oxides are indeed reflected in the mineralogy.

4.3.1. Silica content

Silica is the most abundant oxide and shows the greatest *absolute* variation in weight per cent in igneous rocks (although it should be noted that several other oxides show a greater *relative* variation). Early recognition of this fact led to the broad subdivision of igneous rocks given in Table 4.4, based upon their silica content as determined by analysis. The terms are still widely used and understood, and indeed are frequently employed for descriptive purposes for rocks that have not actually been analysed but merely correlated on the basis of their general appearance and mineralogy with one of the above groupings. The terminology derives from the fact that silica (alone with P_2O_5 of the major oxides) forms an acid solution with water, alumina is amphoteric, and the remaining major oxides are bases*.

TABLE 4.4

Subdivision of igneous rocks by silica content

Name	Silica content (wt.%)	Common examples
Acid rocks	>66	rhyolite, granite
Intermediate rocks	52—66	andesite, diorite, trachyte, syenite
Basic rocks	45—52	basalt, gabbro
Ultrabasic rocks	<45	peridotitic komatiite, peridotite

*For footnote, see p.101.

4.3.2. Principle of silica saturation

The more elegant concept of silica saturation formulated by Shand (1950) in 1927 relates chemistry more closely to the actual mineralogy. It is observed, for example, that forsteritic olivine is not found in igneous rocks in equilibrium with free quartz whereas a magnesian orthopyroxene may coexist with free quartz. Forsteritic olivine can thus be termed "undersaturated" with respect to quartz whereas orthopyroxene would be "saturated". Again kalsilite, $KAlSiO_4$, and leucite, $KAlSi_2O_6$, are never found with free quartz, whereas of course orthoclase, $KAlSi_3O_8$, often occurs with free quartz. The two feldspathoids are thus undersaturated and orthoclase saturated. Similarly, nepheline and analcite are undersaturated and albite is saturated. Within the range of plagioclase compositions, albite, oligoclase and andesine are often found with free quartz, but more calcic compositions rarely so. It is thus a relatively simple matter to determine whether free quartz is present in a crystalline igneous rock, and conversely if quartz is absent, to suspect the presence of and search for any appropriate undersaturated minerals. Thus in place of a determined silica content necessitating analysis, igneous rocks can quite simply be grouped on the basis of their mineralogy into **oversaturated** rocks and **undersaturated** rocks, and note made of the amounts of quartz or undersaturated minerals which characterize the two groups respectively. A further refinement, following the same principle, is the concept of the **silica activity** of a melt calculated from the proportions of different oxides present at specified temperature, and the reflection in igneous rocks of different silica activity levels, essentially increasing degrees of undersaturation, by certain diagnostic mineral assemblages (see Chapter 8).

4.3.3. Alumina saturation|

End-member components of the feldspars, in which most of the Al_2O_3 in igneous rocks occurs, are orthoclase, $K_2O \cdot Al_2O_3 \cdot 6SiO_2$, albite, $Na_2O \cdot Al_2O_3 \cdot 6SiO_2$, and anorthite, $CaO \cdot Al_2O_3 \cdot 2SiO_2$. Written thus in oxide form, it is apparent that in molecular proportions alumina preserves a 1:1 ratio to oxides of the alkali or alkaline-earth elements in all feldspars and feldspathoids. In a similar manner to the idea of silica saturation and again following Shand (1950) how might any variations in this ratio be reflected mineralogically in igneous rocks? **Peraluminous** rocks, i.e. those oversaturated with respect to alumina, where the molecular proportion of Al_2O_3 exceeds that of $Na_2O + K_2O + CaO$ combined, might contain free alumina (corundum, Al_2O_3, is a rare constituent of some syenite pegmatites), or alumina combined with silica if the rock were oversaturated with respect to silica in addition

---

*In applying the descriptive term acid to igneous rocks there is thus no connotation of hydrogen ion concentration — indeed petrologists may speak, for example, of an "alkali" or "peralkaline" acid rock, somewhat to the consternation of chemists.

to being oversaturated with respect to alumina (andalusite is known from some granites and granite pegmatites). The usual indicator of a peraluminous rock, however, is the much more common alumina-rich mineral muscovite, $K_2O \cdot 3Al_2O_3 \cdot 6SiO_2 \cdot 2H_2O$ found in some granitic rocks and pegmatites. Almandine garnet, the most aluminous of the dark minerals, and cordierite, or more commonly its chloritic alteration product, pinite, are also found in some peraluminous granites. **Peralkaline** rocks, on the other hand, in which the molecular amount of the alkali oxides ($Na_2O + K_2O$) exceeds that of alumina, perforce contain one or more minerals of the aegirine, riebeckite, arfvedsonite, or aenigmatite classes in which Na_2O is an essential component of the inosilicate. The two terms, peraluminous and peralkaline, are in common use with this precise compositional and mineralogical connotation. Two intermediate conditions of alumina saturation are known as **subaluminous** and **metaluminous** in which the molecular amount of Al_2O_3 approximately equals or exceeds respectively that of ($Na_2O + K_2O$) but does not in the latter case exceed that of $Na_2O + K_2O + CaO$. These latter two terms, which reflect the more usual condition in basic and acid igneous rocks respectively, are not encountered nearly so frequently as the terms peraluminous and peralkaline which refer to more extreme and distinctive compositions.

4.3.4. Colour index

Feldspars, feldspathoids, quartz, and muscovite and typically colourless or light in colour except where turbid or tinted due to strong alteration or exceptionally dark due to perthitic or antiperthitic exsolution (as in some anorthosites). They are known collectively as **light** or **felsic** minerals. The colour of igneous pyroxenes, amphiboles, olivine, biotite, and opaque minerals is by contrast dark (although olivine when quite fresh is a rather pale green), and these minerals are known as **dark** or **mafic** minerals. They are also known as **ferromagnesian** minerals as they all contain essential iron and/or magnesium oxides. The parallel term, **cafemic** mineral, recognizes the essential lime in amphiboles and clinopyroxenes, and ferromagnesian is at times employed in a more restricted sense to include only the olivine and lime-poor pyroxene series, in which the only essential major-element cations are indeed iron and magnesium.

One obvious observable parameter of a crystalline rock therefore is the relative amount of light and dark minerals as defined above, and this is conventionally recorded as the **colour index**, defined as the volume per cent of dark minerals quoted to the nearest integer. The overall density of the dark silicate minerals is $\sim 20\%$ greater than that of the light minerals, so that the actual weight per cent of dark minerals is significantly greater than their observed volume per cent in the rock. This becomes of some importance when comparison is made between the *mode* and the *norm*, particularly when considering the relatively very dense opaque minerals.

Some texts give precise definitions of the terms leucocratic (C.I. <30), mesocratic (C.I. = 30—60), melanocratic (C.I. = 60—90), and hypermelanic (C.I. >90). However, most petrologists obstinately do not adhere to this usage and instead tend to use the terms leucocratic and melanocratic as qualifiers with a purely relative meaning. For example, a syenite with a colour index of 30 might well be called a melanocratic syenite because it possesses a high colour index *for a syenite*. Similarly, a gabbro with a colour index of 30 might well be called a leucocratic gabbro, or leucogabbro, because it possesses a low colour index *for a gabbro*. For igneous rocks with a colour index greater than 90, the term hypermelanic has now been replaced by **ultramafic**.

4.3.5. Feldspar proportions

Feldspar species are the commonest minerals in igneous rocks, in which they constitute, on the average, over 50 vol.%. The feldspars of igneous rocks fall into two distinct mineralogical groupings, plagioclase and alkali feldspar. Pyrogenetic plagioclase is generally optically homogeneous (although commonly antiperthitic in many anorthosites) and very rarely, if ever, more sodic than oligoclase. Alkali feldspar is much more variable in appearance and comprises optically homogeneous sanidine and anorthoclase in young volcanic rocks, cryptoperthites in older volcanic rocks, and orthoclase and microline that are frequently perthitic in plutonic rocks. In many granitic rocks, pronounced exsolution and/or late-stage recrystallization and alteration have resulted in the formation of discrete plagioclase that is very close to pure albite in composition, and is thus compositionally as well as texturally distinct from the pyrogenetic plagioclase. This albite together with the albite more intimately mixed with potassium feldspar in perthites is conventionally included with the potassium feldspar as alkali feldspar. In most crystalline rocks an estimate of the proportion of alkali feldspar and plagioclase as thus defined can readily be made, and is another parameter commonly employed in a mineralogical classification of igneous rocks, reflecting the relative amounts of alkali oxides and lime.

A classification utilizing a few mineralogical parameters can thus reflect rather closely the range of fundamental chemical variation in the major oxides that constitute igneous rocks and should thus provide us with a convenient and appropriate scheme of mineralogical classification.

4.4. CLASSIFICATION

4.4.1. Occurrence and grain size

Igneous rocks can be either extrusive or intrusive, the latter comprising both hypabyssal and plutonic rocks. Thus three names could be applied to rocks of the same chemical composition, depending on their mode of eruption. A good example of this would be the very common terms basalt, diabase*, and gabbro. Many hypabyssal rocks, however, are aphanitic and petrologists frequently therefore apply the same name to them as to the volcanic rock of equivalent composition. Terms applicable to grain size are more specifically defined as follows:

(1) **Phaneritic** (visibly crystalline, with aid of hand lens if necessary):

    (a) very coarse-grained   >30 mm
    (b) coarse-grained   5—30 mm
    (c) medium-grained   1—5 mm
    (d) fine-grained   <1 mm, but crystallinity discernible using hand lens if necessary

(2) **Aphanitic** (crystallinity not discernible even with aid of hand lens):

    (a) microcrystalline   crystals visible under microscope
    (b) cryptocrystalline   crystals too small to be capable of being resolved optically, although an aggregate birefringence may be evident
    (c) glassy or hyaline   material is glass, in practice seldom completely free of crystallites and/or microlites

Grain size refers to the average diameter of crystals in a more or less equigranular rock (such as most gabbros and granitic rocks), or to the average diameter of groundmass crystals in a porphyritic rock in which the dimensions of the phenocrysts are specified separately, e.g., "a microcrystalline andesite with plagioclase phenocrysts up to 5 mm". The prefix micro usefully connotes a fine-grained phaneritic rock, e.g., microdiorite, microgranite, etc. Note, however, that the terms "fine-grained" and "coarse-grained" are very often employed loosely in a relative sense. For example, a petrologist with customary perversity could well refer to a granite of 2 mm grain size as "fine-grained" because that would be fine-grained *for a granite*, and so on. To avoid confusion one can refer to grain sizes quantitatively in millimetres.

4.4.2. Rationale of classification adopted in this chapter

Following a two-fold division into aphanitic and phaneritic, most simplified classification schemes rely on a tabulation of mineralogical features such as presence and amount of quartz or of undersaturated minerals, colour index, and feldspar proportions as discussed above. It is undeniably difficult, however, to encompass a desirable igneous rock nomenclature into one convenient tabular form as different classificatory criteria are in fact used

---

*Diabase is synonymous with dolerite in European usage where diabase has the connotation of a pre-Cenozoic somewhat altered dolerite.

for different groups of rocks. The trick is first to determine to what group of igneous rocks an unknown rock belongs, and then proceed to an appropriate nomenclature. For this practical purpose of classification therefore, and for this purpose alone, it is useful to subdivide igneous rocks into eight groups as follows and to consider their nomenclature separately:

(1) Aphanitic mafic rocks
(2) Aphanitic intermediate and acid rocks
(3) Cumulates
(4) Other mafic and ultramafic plutonic rocks
(5) Granitic plutonic rocks
(6) Lamprophyres and kimberlites
(7) Carbonatites and associated rocks
(8) Spilites and keratophyres

What is presented here is a practical classification based on hand-specimen and thin-section mineralogical examination only. Trying to follow some of the more rigid schemes of classification based on petrography alone will likely lead only to frustration, because of the uncertainty about the exact composition of partly glassy rocks, etc., that only an analysis could resolve (see, e.g., Sabine, 1978). A modern igneous rock series nomenclature is developed in Chapter 8. Further interesting comments on a philosophy of igneous rock classification will be found in Streckeisen (1967) and Middlemost (1971). Streckeisen (1973) sets out the I.U.G.S. Commission proposals on plutonic rock nomenclature which are incorporated here in simplified form.

4.4.3. Aphanitic mafic rocks

Aphanitic mafic rocks are usually dark in appearance, often vesicular when extrusive, commonly aphyric or containing a small proportion of phenocrysts, and thus form a distinctive group unlikely to be confused with other rocks. They are by far the most abundant volcanic rock group present in the Earth's crust.

Some oversimplified classifications subdivide them with respect to the presence or absence of quartz, feldspathoid, or alkali feldspar, leading to the use of such terms as quartz basalt, olivine basalt, etc. This kind of treatment is hopelessly out of line with modern ideas. For example, the term olivine basalt (now falling into disuse) could refer equally to two common and distinct mafic rock types, namely olivine tholeiite and alkali basalt. The distinction between quartz basalt (in modern terms, a quartz tholeiite) and olivine tholeiite, on the other hand, is not all that important and indeed both may, and characteristically do, contain an intersertal mesostasis of oversaturated composition (see Chapter 5).

The fact is that mafic magmas (particularly as represented by their ex-

trusive products which have been quickly cooled and presumably are equivalent to magma compositions having lost only volatile constituents) assume
very great importance indeed in igneous rock petrogenetic theory. They
show distinct compositional differences related in part to their conditions of
formation in the mantle, and in turn they are considered to be parental to
distinct igneous rock series that are recognisable in exposed crustal rocks.

What is of great importance in major-element differences between fresh
aphanitic mafic rocks is a rather subtle variation in alkali content, which
has a great influence on the silica activity of the magma, and which in turn
is reflected mineralogically. So, although the subject of igneous rock series
receives much fuller treatment in Chapter 8, we can hardly leave the subject
of aphanitic mafic rocks without mentioning some modern terms and
concepts, particularly as for the most part the rocks can be distinguished
on the basis of their observable mineral content.

Why should small differences in alkali content have such a profound
effect on silica activity and hence mineralogy? The reason is because of the
high silica content of alkali feldspars. If Na_2O and K_2O were to be incorporated in feldspars as the components of silica-saturated albite and orthoclase, it can be easily calculated that every 1 wt.% Na_2O would require
~6 wt.% silica, and similarly every 1 wt.% K_2O ~4 wt.% silica. So, of two
basalts, chemically comparable say in content of bases save that one contained say 1.5 wt.% more Na_2O and 1 wt.% more K_2O than the other, the
former would require some 13 wt.% more SiO_2 to crystallize saturated
minerals. This order of variation in content of alkalis is indeed characteristic
of the compositional range of aphanitic mafic rocks. However, greater
amounts of alkalis in aphanitic mafic rocks correlate generally not with
high silica contents but with low ones. Those mafic rocks with high alkali
contents therefore are relatively highly undersaturated and must crystallize
large amounts of undersaturated minerals. This concept, long familiar to
students of petrology and petrogenesis, is now finding quantitative expression
as the level of silica activity.

Basalts with the lowest content of alkalis (or, alternatively expressed, the
highest silica activity) are known as **tholeiites,** including both quartz tholeiite
and olivine tholeiite. Mineralogically, the tholeiite series is characterized by
early olivine reacting with the liquid magma to form a ferromagnesian
pyroxene, although direct textural evidence of this reaction may be lacking
in a particular tholeiite; in any event, the groundmass of tholeiite is devoid
of olivine and contains either two pyroxenes, a calcium-poor and a calcium-
rich one, or a subcalcic augite.

A more alkaline basalt family is known simply as **alkali basalt.** Both in the
mode and the norm alkali basalts do not contain ferromagnesian pyroxene,
but consist essentially of olivine, calcic augite, and plagioclase. Nepheline
generally occurs in the norm but not in the mode. One reason for this is
that significant soda is always present in the augite of alkali basalts, express-

ible as a jadeite component, $NaAlSi_2O_6$. Chemically, this represents an undersaturated equivalent to albite, just as a component of "Tschermak's molecule", $CaAl_2SiO_6$, represents an undersaturated equivalent to anorthite. The norm thus faithfully reveals a degree of undersaturation that remains occult in the modal mineralogy due to solid solution in the augite.

More alkaline mafic rocks contain modal nepheline and plagioclase and comprise **basanites** and **tephrites**. Yet more alkaline rocks, the **nephelinites** and **olivine-nephelinites**, contain no plagioclase at all. The extreme of silica undersaturation is reflected by the appearance of **melilite**, which can be regarded compositionally as an undersaturated equivalent of augite, together with **perovskite**, $CaTiO_3$, which crystallizes in place of ilmenite from melts of very low silica activity.

A modern simplified classification of aphanitic mafic* rocks would then appear as in Table 4.5, using alkalinity and MgO as ordinates, and recognizing the mineralogical reflection of variations in these two parameters.

In general terms it can be said that: (1) increasing degrees of alkalinity (along with consistently related and even more marked patterns of concentration of TiO_2, P_2O_5, rare-earth elements and other incompatible trace elements) represent a genetic fingerprint imposed by processes of partial melting and zone refining during magma generation; and (2) decreasing MgO content reflects fractionation, principally of magnesian olivine but perhaps also of magnesian pyroxenes, during the ascent of magma through the upper mantle and crust. The chosen chemical parameters coupled with mineralogy therefore not only provide a reasonable coverage of different rock types as exposed in crustal rocks (our present objective) but are also apparently not without some petrogenetic significance. In passing, one should, however, remark on the dangers of any scheme of classification that relies solely on petrogenetic assumptions as these often prove to become outdated. What may well prove of great petrogenetic significance in the near future is the possible extent of plume generation and parental mantle depletion as expressed in subtle chemical variations (particularly in trace-element content) of mafic magmas, superimposed on grosser mineralogical differences.

One obvious omission from Table 4.5 is transalkaline basalt; another would be high-alumina basalt. In so far as both of these, though occurring in different tectonic environments, have a chemistry somewhere between typical tholeiites and alkali basalts, but are not easily distinguishable on the basis of their optically observable mineralogy, further consideration of them and other varieties is deferred until Chapters 10 and 11.

It is possible that high-magnesium basalt may not come to be held synonymous with komatiites, which incidentally have been subdivided into

---

*Basalt strictly defined contains essential augite and plagioclase and no feldspathoid, and the wider term *mafic* must be used to include tephrites, nephelinites, etc. Similarly, the term *basic* does not include all rocks considered in this section as some of the more alkaline ones are in fact *ultrabasic* (SiO_2 <45 wt.%).

TABLE 4.5

Nomenclature of aphanitic crystalline mafic rocks

	No olivine in groundmass	Groundmass olivine usually present	Feldspathoid present	No plagioclase present	Melilite present
		Increasing content of alkalis coupled with lower silica activity $\longrightarrow$			
	quartz tholeiite (O) A H P m	alkali basalt (O) A P	tephrite A P N (leucite tephrite)	nephelinite A N (leucitite)	melilitite A N M (leucite melilitite)
Higher MgO content $\longrightarrow$	olivine tholeiite O A H P m	alkali olivine basalt O A P (shoshonite)	basanite O A P N (leucite basanite)	olivine nephelinite O A N (olivine leucitite)	olivine melilitite O A N M (olivine-leucite melilitite)
	komatiite; high-Mg basalt O A H P				
Porphyritic accumulative rocks, also high in MgO	oceanite O A H P	ankaramite O A P			

(1) *Abbreviations used for essential silicate phases present*: O = forsteritic olivine; A = augite; H = hypersthene, more commonly represented by pigeonite with augite, or sub-calcic augite in the groundmass of tholeiites; P = plagioclase; N = nepheline; M = melilite; m = mesostasis of acid composition typically present in intersertal texture in tholeiite.

(2) Brackets indicate that olivine may or may not be present.

(3) Underlining indicates phases that are commonly, but not necessarily invariably, present as phenocrysts as well as in groundmass.

(4) Double underlining indicates olivine that is characteristically present as phenocrysts but does not occur in the groundmass (the characteristic condition of tholeiitic rocks).

(5) Rock names in brackets refer to equivalent relatively potassium-rich varieties characterized by the additional presence of a potassium-bearing phase, e.g., sanidine in shoshonite, leucite in all other named varieties, and the very rare mineral kalsilite ($KAlSiO_4$) in some very undersaturated (here unnamed) varieties. Leucite may occur to the exclusion of nepheline in some extremely potassium-rich varieties (see section on highly potassic rocks in Chapter 8).

(6) This classification uses common petrographical names in their normal mineralogical context. Different criteria, often based on the norm, must be employed to categorize analyses of glassy or partly glassy rocks of the group. For example, basanite can be

"basaltic komatiites" and "peridoditic komatiites" with less or more than 20 wt.% MgO, respectively.

The distinction between alkali olivine basalt and alkali basalt is falling into disuse and most rocks termed alkali basalt today contain olivine.

Another "law" of petrography is revealed by this nomenclature — the rarer the rock type, the greater the number of varietal names! To the right of Table 4.5 mafic rocks are increasingly rare; nevertheless, varietal names there become increasingly common and olivine-nephelinite, for instance, has the confusing synonyms ankaratrite and nepheline basalt, all three being used about equally frequently*. Limburgite is an extrusive mafic rock of alkaline affinity containing olivine and augite phenocrysts, more rarely hornblende or biotite phenocrysts in addition, in a glassy matrix. Petrographically, limburgite has some affinity with lamprophyre. Analysis is obviously necessary to place rocks such as this in a classification scheme.

Another possible parameter of chemical variation is not only total alkali content but also the K_2O/Na_2O ratio. In many aphanitic mafic rocks K_2O is present entirely within the plagioclase. Where, however, the content of K_2O is high enough to lead to the crystallization of an additional phase containing essential potassium, varietal names such as shoshonite and leucitite are encountered. For some very rare, ultra-potassic, undersaturated rocks in which a potassic phenocryst phase is present to the exclusion of nepheline, names such as **ugandite** (olivine, augite and leucite phenocrysts) and **mafurite** (olivine, augite and kalsilite phenocrysts) apply (see also discussion of highly potassic rocks in Chapter 10).

Accumulative rocks, such as oceanite and ankaramite (not to be confused with ankaratrite), are those which are conspicuously porphyritic and which can be shown to have been mechanically enriched in the appropriate primo-cryst constituent.

Among hypabyssal rocks, one which has generally acquired and retained a distinct name is **diabase**, which has a connotation of texture as well as indicating a hypabyssal rock. Grain size in diabase is not precisely defined and can range from that comparable to basalts up to that of gabbro. Diabase can be distinguished from gabbro, however, by the common occurrence of ophitic (or diabasic) texture as opposed to gabbroic texture (see Chapter 5). One might speak of a basalt dyke but not of a diabase flow even if the latter had the appropriate grain size and texture. Some diabases may be porphyritic, e.g., "feldsparphyric diabase" (phenocrysts of plagioclase understood); the majority are not. Some diabase sill bodies are thick enough to have undergone fractionation during their cooling history and reveal in thin section unsuspected variation particularly in ferromagnesian mineral content. Some tholeiitic diabases are conspicuously speckled with interstitial

---

*The term nepheline basalt is particularly atrocious as it suggests a tephritic composition; furthermore, basalt by definition contains plagioclase whereas "nepheline basalt" does not.

patches of white material which proves on thin-section examination to be composed of **micropegmatite**, an intergrowth of quartz and alkali feldspar, the crystalline equivalent to the glassy mesostasis in intersertal texture. These are quartz-diabases and this is the typical habit of the quartz in them. In general, however, diabase is a convenient term used to embrace a range in compositions, and varietal names are not common. **Theralite** and **teschenite** are respectively nepheline- and analcite-bearing diabases, i.e. of an alkaline affinity equivalent to tephrite.

4.4.4. *Aphanitic intermediate and acid rocks*

In this group one generally encounters feldspar phenocrysts, small amounts of dark-mineral phenocrysts, and possibly quartz or feldspathoid phenocrysts or neither, in an aphanitic groundmass. Although the actual colour index of the groundmass is low, it may have deceptively deep colour tints, the intensity of body colour tending to increase as grain size diminishes. Microscopic investigation of the groundmass often reveals variable proportions of tabular plagioclase microlites in a glassy, cryptocrystalline, felsitic, or very fine-grained mesostasis that is often brownish in colour. On chemical analysis, this mesostasis generally proves to have a composition expressible as a mixture of quartz and alkali feldspar in ternary minimum proportions. This particular composition (see Chapter 6) is so viscous it crystallizes with great difficulty, which is why it is generally reasonable to equate the glassy or felsitic mesostasis component of the groundmass with this composition. Note that the quartz component of most andesites, latites, and of oversaturated trachytes is occult for this reason. Note also that abundant microlites of sanidine are typical of trachyte.

The main parameters of variation are quartz content (visible or inferred) and feldspar proportions (again an element of inference is involved) as indicated in Table 4.6. One theoretical objection to this classification is that the actual existence or not of quartz phenocrysts in a rock of say dacitic composition will depend partly on the volume per cent of phenocrysts — an accidental attribute; e.g., a dacite with 30 vol.% phenocrysts is more likely to have porphyritic quartz than a rock of the same chemical composition (the more fundamental attribute) but only containing say 10 vol.% phenocrysts. However, as most of the rocks of this category are porphyritic and also tend to contain phenocrysts in roughly comparable amounts, we can make do with the presence or absence of quartz phenocrysts as one of our criteria for a practical field classification, the other criterion being the relative amount of phenocrysts of plagioclase and alkali feldspar. The colour index is not a further significant parameter of variation as it is low throughout the group, decreasing fairly consistently from ~30 in some pyroxene andesites to very low values of 5 or even less in rhyolites. The name of the identifiable dark mineral or minerals can conveniently be used as a qualifier, e.g., biotite dacite, two-pyroxene andesite, etc.

TABLE 4.6

Nomenclature of aphanitic intermediate and acid rocks

← Increasing content of alkalis

Increasing proportion of SiO_2, to other oxides ←

	Alkali feldspar phenocrysts ± plagioclase phenocrysts; no plagioclase in groundmass	Plagioclase phenocrysts ± alkali feldspar phenocrysts; mesostasis predominates over plagioclase microlites in groundmass	Plagioclase phenocrysts; no alkali feldspar phenocrysts; plagioclase microlites conspicuous in groundmass
quartz phenocrysts	rhyolite (quartz porphyry)	rhyodacite	dacite
no quartz phenocrysts, but quartz present in the groundmass	trachyte	latite	andesite
no free quartz; feldspathiod present	phonolite (tinguaite)		

Quartz-porphyry and tinguaite are names given to distinctive equivalent hypabyssal rocks. See text for further explanatory notes.

The term **porphyry**, prefixed by a compositional name, is in common use for aphanitic *hypabyssal* rocks of the appropriate texture and composition, e.g., biotite dacite porphyry. The term porphyry is conventionally *not* applied to an extrusive equivalent that may in fact be equally porphyritic.

Note that "porphyry" prefixed by a plutonic term, e.g., "monzonite porphyry", signifies a porphyritic rock (usually belonging to a small stock as opposed to a smaller hypabyssal intrusion) with a fine-grained, *phaneritic* matrix. "Latite porphyry", for example, has exactly the same compositional range as a monzonite porphyry but has an *aphanitic* matrix (typical of a small hypabyssal intrusion).

The term **quartz-porphyry** is in common use for a hypabyssal rock containing conspicuous porphyritic quartz and feldspar (predominantly alkali feldspar), and is thus equivalent to rhyolite porphyry.

Tinguaite is a hypabyssal peralkaline phonolite in which a significantly longer cooling time than that of the corresponding volcanic rock has permitted the growth of conspicuous acicular aegirine microlites in a random criss-cross manner.

Pantellerite and **comendite** are peralkaline rhyolites; pantellerites, often green in colour, contain significant amounts of iron oxide reflected in abundant sodium—iron inosilicates, whereas comendites, often white in colour, have a low colour index.

Wholly or partly glassy rocks, **obsidians** and **vitrophyres**, pose a problem; as stated above, the glassy component of the rock is very likely to be near the ternary minimum in composition, i.e. rhyolitic, but this is not invariably true. **Pitchstone** is also a glassy rock of rhyolitic composition but, in contrast to the glassy lustre of obsidian, it has a dull resinous appearance. On analysis pitchstone proves to contain significant quantities of water that was presumably absorbed during burial.

Lacking from the classification are some of the varietal names such as hawaiite, mugearite, tristanite, icelandite, banakite, etc., that are being increasingly used for varieties of intermediate rocks in connection with modern series nomenclature. Terms of this type that are employed in this book are indexed, and are defined and described in appropriate chapters.

4.4.5. Cumulates

An important group of igneous rocks are those typically found in layered basic intrusions, subject of a classic text by Wager and G.M. Brown (1968). These are relatively large intrusions in which there has been time for "primocrysts" crystallizing from the magma to apparently settle down under the influences of gravity and convection currents to the floor of the magma chamber to form a sequence of conspicuously layered rocks with distinctive structures. The accumulated primocrysts, which are essentially similar to phenocrysts in those igneous rocks in which crystal settling has not happened, are known as **cumulus** crystals, and the eventual solidification of the trapped liquid between them has produced a later generation of **intercumulus** crystalline material. Considerable interest attaches to the nature and crystallization history of this intercumulus product (see Chapter 7), but for purposes of description cumulate rocks are named quite simply by stating the cumulus phases present in order of abundance, the most abundant first; e.g. an olivine-chromite cumulate, a plagioclase-orthopyroxene cumulate, a plagioclase-hedenbergite-magnetite-apatite cumulate, and so on. This is not unreasonable as cumulus material exceeds intercumulus material in amount. Where, however, intercumulus material is prominent, for example when one or more of the mineral phases it contains differs from the cumulus mineralogy, this too can be easily specified, e.g., "an olivine cumulate with intercumulus orthopyroxene". Indeed, attempting to give different rock names to specimens collected from adjacent layers in a layered cumulate sequence that may well contain subtly or even widely differing proportions of cumulus phases would be absurd, and the cumulate nomenclature is one of the most logical and easy to apply in all igneous nomenclature. Whereas cumulates are the rule in the larger basic intrusions, they are only rarely found in acid intrusions whatever their size, because a much greater magma viscosity inhibits relative movement of crystals and liquid [but see Emeleus (1963), for an account of cumulates developed in some fluorite-bearing granites of southwest Greenland].

4.4.6. Other plutonic mafic and ultramafic rocks

Not all plutonic mafic and ultramafic rocks are necessarily cumulates. In referring, for example, to (1) nodules in mafic flows, (2) inclusions in kimberlites and some lamprophyres, (3) some of the plutonic rock types from an ophiolite suite, (4) the overall composition of a cumulate sequence, or (5) a non-cumulative plutonic body, it is appropriate to employ a selection of mineralogically defined names from the older pre-cumulate terminology. In the examples given above one might well speak for example of a wehrlite nodule in an alkali basalt, or a garnet lherzolite inclusion in a kimberlite, or the harzburgite unit of an ophiolite, or refer to the overall troctolitic nature of an anorthosite cumulate sequence, or again subdivide the Bushveld cumulate sequence into ultramafic, norite, and gabbro zones, or finally quite simply refer to a hypersthene-gabbro intrusion (a homogeneous, non-cumulate body implied).

A reasonable selection of terms is given in Table 4.7, in reference to which note should be taken of the following points:

(1) Hypersthene-gabbros contain more augite than orthopyroxene; norites contain more orthopyroxene than augite.

(2) Anorthosite and other terms such as gabbroic anorthosite and anorthositic gabbro have been variously defined; a reasonable consensus is to call a gabbroic rock with more than 80 vol.% plagioclase an **anorthosite**.

TABLE 4.7

Essential mineralogy of phaneritic mafic and ultramafic rocks

	Quartz	Plagioclase	Augite	Orthopyroxene	Olivine
Quartz-gabbro	×	×	×	—	—
Gabbro	—	×	×	—	—
Olivine-gabbro	—	×	×	—	×
Quartz-hypersthene-gabbro	×	×	×	×	—
Hypersthene-gabbro[1]	—	×	×	×	—
Olivine-hypersthene-gabbro	—	×	×	×	×
Quartz-norite	×	×	×	×	—
Norite[1]	—	×	×	×	—
Olivine-norite	—	×	×	×	×
Troctolite	—	×	+	+	×
Noritic troctolite	—	×	+	×	×
Anorthosite[2]	—	×	+	+	—
Picrite	—	×	×	—	×
Harzburgite	—	+	+	×	×
Wehrlite	—	+	×	+	×
Dunite	—	+	+	+	×
Lherzolite	—	+	×	×	×
Olivine-websterite[4]	—	+	×	×	×
Websterite	—	+	×	×	+
Bronzitite	—	+	+	×	+

Quartz-gabbro through Anorthosite[2]: Basic rocks.

Picrite through Bronzitite: Ultrabasic and ultramafic rocks — ultrabasic rocks[3], ultramafic rocks[3], peridotite[3].

× = essential mineral; + = mineral may be present in small amounts; — = mineral absent. See text for explanation of the above reference marks (1)—(4) and additional comments.

(3) It is worthwhile noting the precise definitions of **ultrabasic** (silica content of rock is less than 45 wt.%), **ultramafic** (colour index is greater than 90), and **peridotite** (an ultramafic rock with more than 40 vol.% olivine), as these widely used and overlapping terms are often confused.

(4) In the case of ultramafic rocks, a "small amount" of other minerals is taken at 10 vol.%; i.e. with more than 10 vol.% augite an olivine-orthopyroxene-augite rock would be termed either lherzolite (if olivine content greater than 40 vol.%), or olivine-websterite (if olivine content between 10 and 40 vol.%).

(5) The additional qualifiers "leucocratic" and "melanocratic" may be used with discretion in a relative sense and not necessarily with a strict numerical definition. A fairly common term is **leucogabbro** which by usage has come to be defined as a gabbroic rock with colour index between 40 and 20, i.e. spanning the range between typical gabbros and anorthosite.

(6) The existence of additional mineral phases can be conveniently indicated by a mineral name qualifier, e.g., spinel lherzolite. If more than a small amount of the extra phase or phases were present, this would require additional mention.

(7) **Hornblende gabbro** contains more augite than hornblende, and a **bojite** is a rock of gabbroic appearance consisting essentially of hornblende and a plagioclase generally of labradorite composition. The significance of hornblende—augite textural relationships in these rocks often proves difficult to establish. In some, hornblende may appear to replace augite, whereas in others hornblende may be a primary crystallization product of water-rich basic magma, or alternatively hornblende may have originated in both ways. Hornblende is the predominant dark mineral in the **appinite** association that ranges in composition from gabbros through diorites and monzonites to syenites. In the more felsic members of this association prominent large euhedral hornblende crystals are unquestionably of early magmatic crystallization.

(8) **Hornblendite** is a hornblende-rich ultramafic rock occurring in an igneous context rather than the more common metamorphic one where the term amphibolite is employed.

(9) **Eucrite** is a gabbro or olivine gabbro characterized by unusually calcic plagioclase of bytownite composition; eucrites are generally cumulate rocks.

(10) Rather than rely on little-used varietal names, it seems reasonable to refer to rarer parageneses by their mineral content, e.g., a "magnetite-hornblende pyroxenite" from an Alaskan-pipe type occurrence.

The terms in Table 4.7 generally refer to phaneritic rocks of *tholeiitic* affinity. The term picrite, however, may embrace rocks of an alkaline affinity. Not only, as we have already noted, are alkaline mafic magmas much less abundant than are tholeiitic magmas, but among intrusive equivalents alkaline rocks are relatively rarer still, the more alkaline mafic magmas in particular often reaching the surface explosively by an inferred rapid ascent through the crust. Some specific names of coarse-grained alkalic mafic rocks that are, however, relatively frequently encountered include: **essexite,** containing plagioclase, titanaugite, generally with small amounts of alkali feldspar and nepheline, and commonly accompanied by biotite and/or hornblende, a plutonic equivalent of alkali basalt and tephrite; **olivine essexite,** containing olivine in addition to the above mineral assemblage, a plutonic equivalent of alkali olivine basalt and basanite; and **kentallenite,** containing approximately equal amounts of olivine, augite, plagioclase and orthoclase, commonly with biotite in addition, and thus a plutonic equivalent of shoshonite. Plutonic equivalents of yet more alkaline mafic rock types such as nephelinites are rare but are represented by the **urtite—jacupirangite** association mentioned below in connection with carbonatites.

4.4.7. Granitic plutonic rocks

The term granitic in the *very* broadest sense would connote a plutonic rock with *granitic* texture (equigranular, hypidiomorphic) and low colour index, and could include all the rocks in Table 4.8. Granite (sensu lato) conventionally means a granitic rock as defined above with more than 10 vol.% quartz. Granite (sensu stricto) is more closely defined, most recently as a granitic rock containing alkali feldspar in at least twice the abundance of plagioclase and more than 20 vol.% free quartz (although it should be noted that different precise definitions of granite have been proposed from time to time).

Leaving aside for the present the thorny question of a genetic classification of granitic rocks, this is a relatively easy group to classify on mineralogical criteria, as the minerals can always be easily ascertained, generally by hand-specimen examination alone. Of the feldspars, plagioclase is often white, creamy, or greenish in colour due to prevalent incipient saussuritic alteration to sericite and epidote/zoisite. Alkali feldspar, on the other hand, is often a characteristic neutral to pink or light-red colour due to the release of some contained Fe_2O_3 as hematite on alteration. These colour guides are not infallible, however, and alkali feldspar for example may be white or grey in colour. Quartz is readily apparent when present (with the aid of a hand lens when in small amounts), and of the two common dark minerals, prismatic hornblende is readily distinguishable from micaceous biotite.

For classification purposes, all agree on a tabular format using quartz content and feldspar proportions as ordinates, and it would seem prudent to follow current I.U.G.S. recommendations. Following Streckeisen (1973) therefore we take 5 and 20 vol.% of free quartz as dividing lines in Table 4.8, though it should be noted that many older texts make a two-fold division of the oversaturated granitic rocks based on a content of 10 vol.% quartz. Note that the term **quartz-monzonite,** as redefined by Streckeisen and in

TABLE 4.8

Classification of granitic plutonic rocks

Proportions of alkali feldspar to plagioclase

	Alkali feldspar: 100 / Plagioclase: 0	90 / 10	65 / 35	35 / 65	10 / 90	0 / 100
>20 vol.% free quartz	alkali feldspar granite	two-feldspar granite	adamellite	granodiorite	tonalite	
5–20 vol.% free quartz	alkali feldspar quartz-syenite	quartz-syenite	quartz-monzonite	quartz-monzodiorite	quartz-diorite	
0–5 vol.% free quartz	alkali feldspar syenite	syenite	monzonite	monzodiorite	diorite	
No free quartz; undersaturated minerals present	nepheline syenite		nepheline monzonite			

← silica saturation increasing

line with the parallel established usage of the term quartz—syenite, has unfortunately often been used synonymously with adamellite in the past; similarly quartz-diorite in many older writings would correlate with tonalite here. See Lyons (1976) for criticism of the I.U.G.S. proposals on these and other points, with yet further discussion by Bateman (1977) and Lyons (1977). However, when all is said and done, there are advantages to be gained in subduing one's personal preferences and agreeing on a common nomenclature.

Nepheline is the commonest undersaturated mineral in felsic plutonic rocks, hence the term "nepheline-syenite". Leucite does not occur in plutonic rocks, although some undersaturated hypabyssal rocks may contain **pseudo-leucite**, a fine-grained mixture of orthoclase and nepheline pseudomorphing leucite. Nepheline-monzonite is known but is not a common rock.

Colour index tends to decrease fairly consistently from ~25—30 in diorite to ~10 in granite. Many plutonic granites have a higher colour index than a corresponding fine-grained rock of equivalent composition in terms of quartz and feldspar proportions, although many high-level granites (**epigranites**) are comparable with the fine-grained rocks in this respect, a fact of some significance (see Chapter 7). The terms leucocratic and melanocratic are often employed in a relative sense with the average colour index value of the rock type in question in mind; use of the colour index itself is, of course, more precise, e.g., "biotite-hornblende granodiorite, colour index 20" conveys a remarkably thorough picture of the mineralogy and implied granitic texture in a few words.

The names of the dark minerals present can be conveniently inserted as qualifying prefixes; the convention is to put the most abundant dark mineral last, nearest the rock name (the opposite order to cumulate nomenclature), so that a biotite-hornblende granodiorite, for example, would contain more hornblende than biotite*.

Other less common minerals that may be present in addition to the ubiquitous feldspars, quartz or nepheline, and the more common dark minerals include the following:

(1) muscovite and/or almandine garnet, more rarely cordierite, usually together with biotite in peraluminous granites.

(2) riebeckite, arfvedsonite, or other soda-amphiboles and/or aegirine or aegirine-augite in peralkaline granites, syenites, and nepheline-syenites (or amphibole of hastingsite type in granite less rich in alkalis).

(3) fayalite and/or hedenbergite in some iron-rich epizonal granites and

---

*Note that in compound geological names the convention is to insert a hyphen between like terms only, so that in the above example a hyphen appears between biotite and hornblende as both are names of minerals, but no hyphen comes between hornblende and granodiorite. However, if a mineral has an essential defining role in naming a rock, as for example, in "quartz-diorite", then a hyphen is employed.

syenites, that may in addition be peralkaline as well. Minerals of the first two groupings above are never found together. Saturation with respect to silica may vary independently of saturation with respect to alumina thus giving rise to the following four possibilities tabulated below and ensuing terminology for three of them, the commonest not having any special name:

Peralkaline	Not peralkaline
Oversaturated with respect to silica:	
"Ekeritic"	
(Q,N)	(Q,b)
Undersaturated with respect to silica:	
"Agpaitic"	"Miascitic"
(F,N)	(F,b)

In this tabulation Q denotes essential quartz, F essential feldspathoid, N essential Na—Fe-inosilicate, and b biotite as the common dark mineral. Considerable confusion reigns over the exact definitions of actual rock types within this classification, owing to use of the overlapping criteria of silica saturation, alumina saturation, ratio of alkali feldspar to plagioclase, and species of dark minerals (the rarer the rock type, the more prolific and less well-defined the nomenclature). Among the peralkaline varieties, **ekerite** is a peralkaline quartz-syenite, **nordmarkite** is a peralkaline syenite, **pulaskite** is a just undersaturated (<5% nepheline) peralkaline syenite, and **foyaite** contains appreciable nepheline (commonly ~20 vol.%); some definitions specify that foyaite is peralkaline, others do not. Dark minerals can, of course, be specified, e.g., "riebeckite granite" or "arfvedsonite-aegirine syenite", and this aids petrographic description and communication considerably.

An **agpaitic index** [molecular ratio $(Na_2O + K_2O)/Al_2O_3$] greater than unity defines a peralkaline rock chemically; this *index* can be employed equally for rocks that may be oversaturated or undersaturated with respect to silica or alumina, even though the *term* "agpaitic" is restricted to rocks that are both peralkaline and undersaturated as shown. Granitic rocks other than epizonal or those with compositions other than granite (sensu stricto) and syenite are far less likely to contain any of the peralkaline or iron-rich minerals of the latter two groupings.

Other varietal names fairly commonly encountered include the following. **Alaskite** is an alkali feldspar granite with very low colour index. **Trondheimite** is a leucocratic tonalite in which the plagioclase is oligoclase rather than andesine. **Charnockite** is a hypersthene granite, with characteristic dull-green hypersthene and often blue, finely rutilated quartz. The hypersthene is conspicuously pleochroic in thin section, like the hypersthene of granulite facies metamorphic rocks. Controversy centres on whether charnockites are granites *metamorphosed* under granulite-facies conditions, or *emplaced and crystallized* under granulite-facies conditions. The charnockitic series of rocks include a range of compositions comparable not only to those of granite (sensu lato) but also to those of more basic compositions, thus giving rise to a plethora of varietal names, e.g., **enderbite** is a charnockitic granodiorite.

4.4.8. Lamprophyres and kimberlites

These are relatively uncommon rocks typically occurring in thin dykes and diatremes. They are distinctive in appearance as they are abundantly porphyritic in dark minerals set in an indeterminate groundmass that is usually grey in colour. A common genetic factor appears to have been an unusually high P_{H_2O} and, in some, a high P_{CO_2} during crystallization. High concentrations of these components are suggested by the frequent occurrence of hydrous minerals such as biotite or phlogopite, hornblende, and analcite, highly serpentinized olivine, prevalent groundmass alteration often involving the formation of carbonate, and the common occurrence of leucocratic ocelli attributed to liquid immiscibility between a silicate melt and a melt richer in H_2O and CO_2. A considerable number of varietal names exists, particularly, as is so often the case in petrology, of rare and unusual varieties. Some of these are by no means well defined and the reader will find it instructive for instance to compare the definition of alnöite given by the American Geological Institute *Glossary of Geology* (Gary et al., 1972), with definitions given by reliable petrographic texts such as Moorhouse (1959) and Hatch et al. (1972).

A field terminology must perforce just state the porphyritic dark minerals present. Thin-section examination will further reveal the groundmass mineralogy (where not glassy or too highly altered), and a convenient simplified classification therefore may be based on these two parameters as shown in Table 4.9.

In this table there is a comparison with that used for aphanitic mafic rocks insofar as increasing richness in MgO is plotted against relatively more undersaturated mineral assemblages. Certain groups of lamprophyres tend to be commonly associated with, and to show mineralogical affinities with, certain rock types, viz. vogesite and minette with granites, spessartite and kersantite with diorites, camptonite with alkali basalts, monchiquite with feldspathoidal mafic rocks, and alnöite with carbonatite. Despite this wide range in composition of associated rocks, lamprophyres themselves lie generally within the range of basi compositions ranging downward in silica content to decisively ultrabasic and very alkaline (monchiqites, alnöites and kimberlites). It is worth noting that biotite, most commonly seen in granitic rocks where it coexists with quartz and is thus technically a saturated

TABLE 4.9

Outline classification of lamprophyres

PHENOCRYST PHASES PRESENT — Overall increase in MgO content →	FEATURES OF GROUNDMASS MINERALOGY — Overall increase in alkalinity →			
	Orthoclase predominant	Plagioclase predominant	Analcite-bearing	Perovskite- and/or melilite-bearing
hornblende	vogesite	spessartite		—
biotite	minette	kersantite (biotite ± augite)	ouachitite (augite + biotite ± hornblende)	
augite	—	camptonite (augite and/or barkevikite)	monchiquite (augite + olivine + biotite)	alnöite (augite + olivine + biotite)
olivine phlogopite	— —	— —	—	kimberlite (olivine + phlogopite)

Mineral names in parentheses refer to phenocryst phases where more than one is commonly present.

mineral, is stoichiometrically equivalent to the following grossly unde
saturated components: 1 kalsilite + 1 leucite + 3 olivine + 2 water. Th
abundance of biotite in many lamprophyres, from which quartz is con
spicuously lacking, thus correlates with their highly basic and generall
undersaturated compositions.

Lamprophyres, often giving the impression of an uncomfortable after
thought in igneous petrology texts, may actually come closer to intratelluri
alkaline magma compositions than aphanitic mafic extrusive equivalents
the more fashionable yardstick of parental magmas. The lack of plagioclase
phenocrysts, the abundance of mafic phenocrysts, and the hydroxyl-bearing
compositions of some of the mafic phases, all reflect crystallization within
a water-rich melt (high P_{H_2O} suppresses the field of plagioclase stability
during the crystallization of mafic melts). The presence of ocelli and deuteric
alteration reflect a retention, at least in part, of magmatic volatile constituents. Thus, the mineralogy of lamprophyres, messy as it may be, gives us
a strong hint that crystallized igneous rocks, particularly those formed at
atmospheric or low pressures, are not necessarily the compositional equivalent of magma. This is not to deny that some lamprophyres may well
represent the crystallization products of unusually water-rich fractions of
magma in the first place.

4.4.9. Carbonatites and associated rocks

Although there are only some 300 known occurrences of carbonatite in
the world with a total surface area of ~500 km², they and their associated
rocks include very varied mineralogical compositions and have a considerable
interest to petrologists out of all proportion to their abundance. Inevitably
therefore they have acquired in some writings a host of rock names, many
of them apotheosising individual localities and mineral parageneses which
may or may not be found worthy of more general application, e.g., sövite,
fenite, jacupirangite, okaite, etc. In general, there are three distinctive rock
types present in most carbonatite complexes: (1) carbonatite intrusions;
(2) altered country rock; and (3) a very distinctive suite of nepheline-rich,
plagioclase-free, alkaline plutonic rocks grading into some of very high
colour indices, together with lamprophyres.

The carbonate in most carbonatites is calcite, more rarely dolomite,
ankerite, or other less common species. The only rock name in common
use is sövite for a calcite carbonatite. Minerals other than carbonate are
ubiquitous and extremely varied, but seldom present in more than small
amounts. Rather than coin varietal names, it would seem preferable to
describe carbonatites in the following manner: for example, "sövite with
5% apatite and 1% pyrochlore", or "dolomite carbonatite with 5% phlogopite
and diopside", etc.

An aureole of highly altered country rock around carbonatite is characteristic, and has acquired the name **fenite** produced by a process of **fenitization**. Fenite strictly defined is a metasomatically altered quartzo-feldspathic rock consisting mainly of alkali feldspar and aegirine. The inferred process of fenitization involves the accession of heat, and the transfer of alkalis, particularly soda, together with variable amounts of H_2O and CO_2 into the country rock. Depending on original country rock composition (the Fen carbonatite is emplaced in basement gneisses), proximity to the carbonatite intrusions, and structural level as exposed by subsequent erosion, the actual rock type can obviously be very variable. The term fenite then may not be universally applicable to carbonatite aureole rocks which may often require detailed description.

Of the accompanying igneous rocks, compositionally somewhat akin to nephelinite but with a wide range in colour index, the following feldspar-free rocks characteristically occur in carbonatite complexes: **urtite** (C.I. = 0—30); **ijolite** (C.I. = 30—60); **melteigite** (C.I. = 60—90); and the ultramafic rock **jacupirangite** (C.I. >90). Nepheline is the predominant light mineral throughout this group. The dark mineral is mainly a clinopyroxene ranging from aegirine in urtite, and aegirine-augite in ijolite, to titanaugite in jacupirangite. Sphene and apatite are common accessory minerals, tending to increase in amount as the colour index rises. In jacupirangite, perovskite may accompany the sphene and melilite may occur.

4.4.10. Spilite and keratophyre

Strictly speaking, these are the names of metamorphic rocks, but because of their abundance and their interest to the igneous petrologist they are included here. In a mineralogical classification of igneous rocks we are considering the pyrogenetic mineralogy and textures arising from the time of emplacement of an igneous body. In its simplest terms this is that produced by crystallization from a liquid. However, this concept would have to be modified to include sub-solidus recrystallization and deuteric alteration in plutonic rocks, particularly in many granites some of which in addition may never have been wholly liquid.

Later metamorphism independent of eruption and cooling may, however, have a considerable effect on igneous rocks. If this is accompanied by a penetrative deformation and recrystallization a metamorphic name may be applicable, e.g., "foliated granite" (in this case a distinction between igneous foliation and metamorphic foliation is necessary), "quartz-sericite schist", "amphibolite", etc. Where the igneous precursor of the metamorphic rock can reasonably be established, terms like "meta-diabase" and "meta-volcanic" can be used.

In very low grades of metamorphism of volcanic rocks in the zeolite and low-greenschist facies a penetrative deformation may be lacking as, for example, in burial metamorphism. Nevertheless it is frequently found that the pyrogenetic mineralogy has been partially and commonly even completely replaced, to a large degree mimetically, by a mineral assemblage appropriate to the lower-temperature conditions. Rocks of which the field structures and mimetically preserved textures show them to have been originally basaltic may thus, for example, come to consist of albite, chlorite, and leucoxene forming the bulk of the rock together with variable amounts of a selection of such minerals as calcite, epidote, prehnite, pumpellyite, and zeolites, often patchily distributed in amygdales and veinlets. Similarly, intermediate and acid rocks will come to contain albite and chlorite pseudomorphing plagioclase and dark mineral phenocrysts respectively in an indeterminate fine-grained matrix.

These rocks are known as **degraded** (syn. **retrograded**) as their present mineralogy reflects equilibrium at a much lower temperature than a pyrogenetic mineralogy. Degradation can often be demonstrated not to have been isochemical and the problems posed by these rocks are more fully discussed in Chapter 14. Anticipating this, we can dismiss an old controversy as to whether these rocks could be primary igneous products, and can now define terms as follows: **spilite,** following Cann (1969),

"is a rock of basaltic texture in which a greenschist or sub-greenschist mineralogy is completely or almost completely developed,"

and **keratophyre,** following C.J. Hughes (1975),

"is a retrograded fine-grained igneous rock of intermediate to acid composition, showing little or no penetrative deformation, in which a pyrogenetic mineralogy is mimetically replaced by an assemblage appropriate to greenschist or lower facies mineralogy."

It is remarkable how commonly these names are applicable, often used as qualifying terms, e.g., "spilitized (or spilitic) basalt, etc. The proper use of these terms immediately alerts the reader to the now well-recognized problems of chemistry and magmatic affinity that these difficult rocks pose for the igneous petrologist.

Chapter 5

PETROGRAPHY OF IGNEOUS ROCKS

5.1. INTRODUCTION

Petrography is the art of rock description. Its purpose is to convey to another geologist an accurate and precise picture of the rock in question in whatever detail is demanded by the appropriate context. The petrography of igneous rocks is largely concerned with the observable features of mineralogy (a function of chemistry) and texture (a function of cooling history). It leads naturally to classification along these lines, a topic we have already discussed in Chapter 4. Increasing use is being made of chemical data in igneous petrology for purposes of both classification and petrogenetic interpretation, but reliance on chemistry alone as a parameter, a temptation easy to succumb to, does not reveal the whole story, and in fact can be misleading. For example, chemical analysis of porphyritic volcanic rocks, particularly where the amounts of phenocryst species are not quoted, may reflect the influence of complex accumulative processes affecting the liquid line of descent (see Chapters 6 and 7). Analyses of cumulate rocks will reflect fortuitous variations in the proportions of cumulus minerals (of which the precise composition is the sole important index of fractionation) plus unknown proportions of intercumulus crystallization of varying type (see Chapter 7). Analyses of degraded igneous rocks may reflect the operation of meta-somatic processes (see Chapter 14). In these instances and many others petrographic detail can supply essential information. For communication in the field and at a more pragmatic level than that of advanced research, for example company and survey reports, petrography is the essential tool. To master petrography, therefore, is a sine qua non for anyone dealing with igneous rocks at whatever level. This is best accomplished by guided study, using one or more petrographical texts* with reference to a good collection of igneous rocks and accompanying thin sections. It is unfortunately com-monplace to see geologists of all levels fail to make their meaning clear or even make unnecessary mistakes for lack of petrographic experience and expertise.

---

*Good introductory petrographical texts, many with an appropriate regional bias in their examples of figured rocks, include the following: *Petrology of the Igneous Rocks*, by Hatch et al. (1972); *Petrology for Students*, by Nockolds et al. (1978); [a sequel to *Petrology for Students*, by Harker (1954)]; *The Study of Rocks in Thin Section*, by Moorhouse (1959) (with numerous Canadian examples); *Petrography*, by Howel Williams et al. (1954); and *A Petrography of Australian Igneous Rocks*, by Joplin (1971).

5.2. PROCEDURE

Rock description begins in the field with observations carefully recorded in a stiff-backed notebook. Traverses and locality descriptions should be precisely described, so that another geologist could retrace the way to the outcrops, using the field notes and the appropriate quoted locality map or air photograph. Adjectives such as large, small, thin, wide, many, should *not* appear in the notebook, however much they may appear to be self-evident to the observer at the time, but should be reduced to measurements, preferably in the metric system.

Volcanic rocks generally show interesting structures, viz. jointing, vesicles, pillows, flow-banding, autobrecciation, etc., but often all that can be determined mineralogically about an aphanitic volcanic rock is its overall colour, appearance and phenocryst content. Distinctive flow units can often be mapped on these bases.

Usually, a fair estimate of the mineralogy of plutonic rocks can be made in the field. Whereas some plutonic igneous bodies are homogeneous others show differentiation in grain size and composition. This poses a problem as to just what to map. Basic, ultrabasic and ultramafic rocks are often layered. This layering can be recorded together with cumulus phases. Mesozone granitic rocks commonly show a foliation, varying from pronounced to very subtle. Sometimes distinct intrusions can be recognized particularly among epizone felsic plutons (G_1, G_2, G_3, etc.), each perhaps varying internally from fine- to coarse-grained facies, perhaps even porphyritic in some places and not in others, but constituting one intrusion. Zones of alteration, perhaps visible as colour changes, can be mapped and may be of considerable economic significance. Pegmatite bodies may reveal zones of different grain size and mineral assemblages. The size and shape of igneous intrusions should, of course, be recorded. Field observations on age relations in intrusive rocks are based on chilled margins, truncation of one intrusion by another, inclusions of one in another, veining of one by another, or an intervening episode of alteration, deformation, or minor intrusions. These, together with contact relationships with country rocks, are often critical observations.

Often precise nomenclature in the field is impractical and the field book will contain names like "fine-grained red granite", "grey dyke rock", "lava type A", pending thin-section investigation of collected samples.

Turning now to written descriptions of igneous rocks, following thin-section investigation, naturally the amount of detailed description will vary greatly according to the context. Nevertheless, it is often useful in working notes to summarize essential details systematically under the following ten headings:

(1) Structure, colour, and general appearance.
(2) Grain size.
(3) Quartz content (if none, look for undersaturated minerals).

(4) Feldspar content and composition (alkali feldspar/plagioclase ratio; An content of plagioclase).

(5) Dark-mineral content and colour index.

(6) Accessory minerals.

(7) Texture.

(8) Alteration.

(9) Name.

(10) Comments.

These are largely self-explanatory and include the bases of classification already mentioned in Chapter 4. The distinction between **structure** and **texture** is not always easy to define but is generally clear in context. Texture strictly refers to the interrelationship of minerals in a rock, but is generally taken to include also structural features visible on a microscopic scale or on the same scale as the grain size of the rock. For example, welded shards of an ignimbrite or aligned crystallites in an obsidian would be considered textural features. On the other hand, the presence of flattened pumice fragments in the same ignimbrite would be designated eutaxitic structure, and similarly flow-banding visible in outcrop of the same obsidian would be considered a structural feature. In general, structural features thus comprise non-textural features visible in outcrop. Note also that in the context of these working notes on petrographical description texture should include the size and shape of the different mineral species in the rock as well as their geometrical interrelationships.

Note that a "straight" sequential description rigidly along the lines of the above headings may become needlessly repetitive and boring. Often what is required most, following some degree of comment on the *classificatory* features of a rock, is allusion to any *unusual* features that it may display. For example, in many contexts "fine-grained diabase" would hardly need amplification along the lines given above as this is such a common and well-known rock. However, if it were, say, vesicular, or contained large phenocrysts of plagioclase, these would be unusual features and worthy of comment. Again the name "biotite-hornblende adamellite, colour index 15", tells a reader a lot about the rock, i.e. quartz content over 20 wt.%, approximately equal amounts of plagioclase and alkali feldspar, nature of and absolute and relative abundance of dark minerals and (implied) medium-grained granitic texture. If the rock say were foliated, or were unusually fine-grained, or contained phenocrysts of alkali-feldspar, or contained accessory tourmaline, *these* kind of features would deserve to be emphasized. One has to be aware of what the various igneous rocks commonly look like before one can really write a description that will be concise and meaningful when read by another geologist, and this is all part of the art of petrography.

Note that whereas the choice of a name for the rock comes *last* in one's working notes, in written descriptions it helps a reader to be given the name *first*.

Often the initial sentence of a petrographical description can usefully combine the rock name with structural features, colour, grain size and general appearance, i.e. encompassing those features that can be appreciated in outcrop. Then further mineralogical and textural details ascertainable from thin-section examination can be added as follows. For aphanitic rocks, many of which are porphyritic, a description of any phenocrysts before groundmass is logical as this follows the sequence of crystallization. In aphanitic mafic rocks it is usual to mention first olivine phenocrysts (if present) followed by other mafic phenocrysts, then plagioclase phenocrysts, and groundmass last. In andesitic rocks, because of the common overall abundance of plagioclase phenocrysts, it is more common to describe them first, then mafic phenocrysts, then groundmass microlites, and mesostasis last. In rhyolitic rocks the descriptive sequence is generally feldspar-, quartz- and mafic phenocrysts, followed by groundmass. Ignimbrites are best described sequentially in terms of the four components that most contain in various proportions: crystals, shard matrix, and pumice and lithic fragments. Mafic cumulate rocks present little problem in description: cumulus minerals first in the order of their abundance in the rock, followed by the mineralogy and texture of intercumulus material. In granitic (sensu lato) rocks it is customary to describe the feldspars first, plagioclase before alkali feldspar, then quartz or nepheline, then dark and accessory minerals, and finally any deuteric alteration. Note that deuteric products do not rank for inclusion as accessory minerals which by definition are part of the pyrogenetic assemblage. Photomicrographs (if they really show the desired feature clearly) and line drawings often prove a serviceable adjunct to verbal descriptions — note, for example, the illustrative quality of many of the drawings in Harker (1954) and in Hatch et al. (1972).

5.3. A WORKING VOCABULARY

Just as a vocabulary arises naturally in a language to fulfill a need and facilitate communication, so a working vocabulary exists in petrography. Much of it is simple and self-explanatory; some relatively specialized terms may save many words of description of specific commonly encountered textures, etc., and are thus useful. If a term fits, use it! Clearly, some petrologists by nature would go further than others in developing a larger and more highly specialized vocabulary, and indeed certain contexts may require it. What is collected below, however, is a practical minimum of useful terms, subdivided by function and coupled in some cases with a discussion as to significance or a reference to another place in this book where a term is more fully discussed in context. Definition of other terms will have to be sought for in a glossary and petrographical texts. Note that the paraphernalia of mineral identification from optical properties ·is dealt with in outline in

Chapter 1; for a petrologist, mineral identification is a means to an end, not an end in itself.

5.3.1. Terms used in description of igneous processes

Lava is a well-understood term; note that it can refer either to an essentially molten extrusive rock or to its solidified product.

Magma, on the other hand, is really an hypothesis concerning the existence of subterranean (and unwitnessed) bodies of molten rock containing variable inferred amounts of crystalline material and of gases in solution. Note that although one tends intuitively to assume a mainly liquid state, such is not specified or demanded by the definition of magma. For deep-seated granites, however, where available evidence suggests at least emplacement and quite likely also evolution in largely crystalline conditions, the term **migma** has been used by some petrologists in place of magma. **Intratelluric crystallization** refers to the process of partial or complete crystallization of magma. It can readily be demonstrated in the case of porphyritic hypabyssal and extrusive rocks that their phenocrysts must have grown at a much slower rate, presumably in intratelluric magma chambers, than their matrices which solidified as a result of rapid chilling (or degassing) during the final stages of extrusion or emplacement. All fine-grained porphyritic rocks, therefore, have had at least two stages of crystallization, the earlier possibly accompanied by fractionation processes (see Chapters 6 and 7).

Pyrogenetic minerals* are those formed by crystallization from a magma or lava, or developed in equilibrium with a silicate melt phase, in the case of migma.

In clear distinction to the above, **deuteric alteration** and the formation of **deuteric minerals** occur when the pyrogenetic mineralogy is affected by reaction with juvenile hydrothermal solutions within an igneous rock body *during its cooling period*, i.e. following the period of pyrogenetic crystallization as broadly defined here. Naturally, lavas show little or no deuteric alteration because water and other volatile constituents are lost by vesicu-

---

*The term *pyrogenetic* was more closely defined by some petrologists in the past to include only anhydrous minerals formed during an assumed early "orthomagmatic" stage of crystallization of magma, and excluding later-stage hydroxyl-bearing minerals such as hornblende and biotite. The assumption, however, that hydroxyl-bearing minerals are always of late formation is untenable for some rocks, as for example, in the case of hornblende phenocrysts in andesite which are demonstrably of early intratelluric magmatic crystallization. Note that in some plutonic rocks a buildup of water concentration occurs in the liquid portion of the magma remaining as crystallization of anhydrous phases proceeds; this may well result in the appearance of new phases in equilibrium with relatively water-rich silicate melt, hornblende in place of augite in the later stages of crystallization of many diorites, for example. It is this "hydatomagmatic" (syn. hydatogenetic) stage that the strict definition (not followed here) of pyrogenetic and orthmagmatic was intended to exclude.

lation during the act of extrusion, although the occurrence of prevalent and characteristic iddingsite rims around olivine phenocrysts in otherwise very fresh and young alkali basalts is one exception to this generalization. Mafic plutonic rocks which are inferred to have crystallized from generally water-poor magmas, generally show only slight effects of deuteric alteration. In mesozone granites, on the other hand, deuteric alteration is ubiquitous and may be marked, resulting in the recrystallization of albite-muscovite granites, for example, and grading to autometasomatism where migration and channeling of deuteric solutions has occurred in a regime of steadily declining temperature (see Chapter 7). An early "pneumatolytic" substage of deuteric alteration is sometimes distinguished, typically as a relatively high-temperature tourmalinization by inferred gaseous diffusion (a gaseous medium is connoted by the prefix pneuma-, but conditions were probably supracritical at this substage).

Secondary minerals, by contrast again, originate where the mineral parageneses resulting from pyrogenetic processes, with or without any deuteric alteration, may be affected by still later processes of hydrothermal alteration, metamorphism, or weathering, all of which are of course unconnected with the cooling history of the rock. As pyrogenetic mineralogies were formed at high temperatures, any such later secondary alteration is retrograde, giving rise to the now well-appreciated phenomenon of **degradation** of igneous rocks (see Chapter 14), characteristically resulting, for example, in the widespread development of spilites and keratophyres from older volcanic rocks.

Clear as the distinction between deuteric and secondary alteration may appear in theory, attribution on the basis of mineralogy may be difficult in practice. In the case of the basaltic and gabbroic members of an ophiolite suite, formed at an extensional ocean-ridge environment under very high temperature gradients coupled with circulation of hydrothermal solutions, much alteration of a secondary appearance can be shown to have occurred essentially during the period of cooling of the complex as a whole and hence, strictly speaking, is deuteric.

5.3.2. Terms used to describe structural features of igneous rocks

In looking at igneous rocks in the field, as in many other subjects, it is useful to know what characteristic features to expect in context, all the more to keep an eye open for the unusual. Some relatively common structural terms are noted below and related to the type of rocks in which they commonly, though not necessarily exclusively, occur.

5.3.2.1. Mafic flows. Maffic flows, whether subaerial or shallow subaqueous, are commonly **vesicular**; in older flows the **vesicles** are commonly filled with secondary minerals such as zeolites, chlorite, calcite, quartz, epidote, albite, etc., in which case the rock is said to be **amygdaloidal.** Interestingly, individual **amygdales** (for practical purposes synonymous with amygdules, although some petrologists reserve the term amygdules for very small amygdales) may be concentrically zoned with different secondary minerals, or separate amygdales may be entirely filled with differing minerals (i.e. one may be filled by calcite, one next to it by chlorite, and so on), or the amygdales may be filled by secondary minerals in a more random manner.

Subaerial mafic flows (see Chapter 2, Section 2.2.1.1, for discussion of pahoehoe and aa structures, etc.) may exhibit varying proportions of massive, sometimes polygonally jointed, bottoms to blocky and scoriaceous tops, often weathered red. Subaqueous mafic flows are commonly in the form of pillow lavas of various shapes (see, e.g., Dimroth et al., 1978). Where extruded at relatively shallow depths, pillow lava units may be interbedded with **pillow breccias** and finer-grained **hyalotuff** material (see Carlisle, 1963; Honnorez and Kirst, 1975; and discussion of this phenomenon in Chapter 2, Section 2.2.1.3).

Some mafic lavas, generally those with highly alkaline affinities, may carry **ultramafic nodules** of presumed mantle provenance but of uncertain history (see Chapter 9). The nodules, which may range greatly in abundance up to as much as 10—20% in some highly explosive spatter cone products, are typically up to 10 cm in diameter.

5.3.2.2. Intermediate and acid flows. Intermediate and acid flows are commonly **autobrecciated** resulting in a monolithologic fragmental rock with a matrix of the same material, the fragments generally being smaller than in blocky basalts. Acid flows, in particular, commonly show **flow-banding** (syn. fluxion banding) and indeed only rarely are autobrecciation and flow-banding not evident in rhyolite flows. Some flows, especially mugearites and trachytes, may be **fissile** in outcrop due to a strong tendency towards a parallel arrangement of feldspar microlites in their matrices.

Ignimbrites (syn. ash-flow deposits) commonly reveal an **eutaxitic structure** in outcrop due to the flattening of discrete pumice fragments at the time of accumulation, resulting in a discontinuous layering effect quite unlike the fine-scale continuous lines of flow-banding in a rhyolite lava flow. Contraction during cooling often results in the formation of massive vertical **polygonal jointing** in ignimbrites, with individual columns ranging up to a metre or so in diameter, contrasting with the generally finer-scale (15—50 cm) polygonal joints that may be developed in some ponded mafic flows.

Also as a result of cooling, conchoidal **perlitic fractures** may be developed, generally on a microscopic scale in acid glasses; preferential devitrification along these curved cracks may result in a play of colours resembling that of pearl — hence the name.

The structures of fragmental volcanic rocks that are abundantly associated with intermediate and acid extrusive rocks are of very great interest and importance in interpretation as any field geologist will discover (see Chapter 2, Section 2.3).

5.3.2.3. Layered intrusions. **Igneous layering** in plutonic rocks (generally of basic magma parentage) refers to parallel layers of different mineral proportions such as are the hallmark of cumulates (see Wager and G.M. Brown, 1968). Individual layers that were horizontal, or near horizontal, at the time of formation can often be traced for long distances; sometimes, however, the layering may be complicated by sedimentary structures such as slumping and cross-bedding [see, e.g., the spectacular photographs in Irvine (1974)]. Associated with igneous layering, there may be an **igneous lamination** due presumably to the flat settling of tabular cumulus plagioclase crystals, or even an **igneous lineation**, picked out, for example, by prismatic cumulus pyroxene crystals, presumably due to magmatic currents at the time of deposition of the cumulus crystals.

5.3.2.4. Katazonal granitic rocks. Migmatites show distinctive structures, better appreciated in outcrop than in hand specimen, that are well illustrated and summarized in the definitive book by Mehnert (1968, ch. 2). Most of the ensuing structural terms are based on the relative configuration of a more or less granitic **neosome** (= leucosome) component and a darker **palaeosome** (= melanosome) component. Among the more widely used terms are **veined gneiss** (syn. lit-par-lit gneiss, injection gneiss) for parallel veins of neosome following a foliation in the palaeosome, **net-veined gneiss** where neosome veins anastomose, and **agmatite** where disorientated chunks of palaeosome material appear to be caught up in a "sea" of neosome. **Nebulite**, on the other hand, connotes a more intimate development of neosome material chiefly in the form of feldspar through the palaeosome. See Chapter 3, Section 3.5.6, for a discussion of the conditions of formation of these highly perplexing rocks.

5.3.2.5. Mesozonal granitic rocks. An **igneous foliation** (as opposed to any later penetrative metamorphic fabric) is rarely lacking from mesozone granitic bodies. The foliation may result from an obvious tendency to parallelism of platy or tabular minerals such as biotite or feldspar; sometimes, however, it may be very subtle and a geologist may have to give best to an experienced quarryman in picking it out by sight or touch. Igneous foliation is generally found to be parallel to the wall of intrusions and is commonly steep. A common problem, with an obvious bearing on the relative age of emplacement of a granite to that of tectonism (perhaps broadly synchronous), is the effective disentangling of any internal granite fabric from the effects of any penetrative post-emplacement deformation. This problem is frequently compounded by the variable development within a pluton of igneous foliation which may often be more pronounced near contacts. Aspects of this problem recur several times, for example, in the discussion of the Donegal granites by Pitcher and Berger (1972). **Igneous banding** is of a more general application (Wager and G.M. Brown, 1968, p.5):

"for any streaky or roughly planar heterogeneity in igneous rocks, whatever its origin"

and is commonly evident, for example, near some diorite and granodiorite contacts.

Some inclusions in intrusive rocks are clearly **accidental inclusions** of country rock, usually angular fragments of recognizable wall-rock lithologies. Another class of inclusions, found in many mesozone granites, comprises varieties of **cognate inclusions** which are usually rounded, up to 10–50 cm in diameter, and more melanocratic than the host granite. A long residence time in, and a mineralogical equilibration with the host granite has obscured the origins of these cognate xenoliths. Some may be in fact accidental inclusions that have been relatively thoroughly "digested"; others may have had an origin further back in the history of the evolution of the granite — hence the term "cognate" (see also Chapter 7, p.215).

Orbicular structures are rarely found in some granitoid rocks, typically those of a dioritic–granodioritic range in composition. The structures consist of concentric shells of relatively leucocratic and relatively melanocratic mineralogies. The orbicles have dimensions in the range of centimetres to decimetres in diameter. The origin of these structures is problematical (see, e.g., Leveson, 1973; J.G. Moore and Lockwood, 1973; T.B. Thompson and Giles, 1974).

Jointing is common in intrusive rocks, particularly in granitic plutons, where it may be associated in a regular fashion with igneous foliation, aplite and other veins, etc., in mesozone bodies. In complex epizone granite terrain, different intrusive units can sometimes be distinguished on the basis of their differing joint patterns.

5.3.2.6. Epizonal granitic rocks. In contrast to mesozonal granites, many epizone granite outcrops are devoid of foliation or any other directional mineralogical structure; a useful term to describe this negative character is **homophanous**.

Some epizone granitic intrusions may have vesiculated, resulting in the formation of **drusy cavities** (syn. **druses; miarolitic cavities**); these are not spherical as their perimeter is generally much modified by the presence of small ingrowing crystals of feldspar and quartz, more rarely of topaz and other late-stage minerals.

Some intrusive rocks, particularly shallow fine-grained dacite and rhyolite porphyries formed from degassed viscous magmas, may undergo **autobrecciation** (just as corresponding flows) resulting in the formation of **protoclastic breccia**, particularly near their margins, where they may also, in addition, be flow-banded.

5.3.2.7. Lamprophyres. **Ocellar structures** or **ocelli** (singular, ocellus) are pale, subspherical to ellipsoidal bodies the size of birds' eggs, characteristically found in lamprophyres and related rock types. Mineralogically, they are much more felsic than the host lamprophyre, often containing notable amounts of analcite or calcite, and are thought to have been formed by the splitting of the magma into two immiscible fractions (see also discussion of lamprophyres in Chapter 4, Section 4.4.8.; and immiscibility in Chapter 7, Section 7.2.2.).

5.3.3. Terms used to describe mineral shapes

In holocrystalline rocks, the juxtaposition of crystals, obviously geometrically demands that not all can have boundaries conforming to the outlines of their own crystal faces. Some mineral grains, notably phenocrysts in fine-grained rocks and some accessory minerals, do tend to be bounded by their own crystal faces; this condition is best called **euhedral**, literally "good faces", (syn. automorphic, idiomorphic). On the other hand, specific crystal faces may be entirely lacking and a mineral grain in an igneous rock may have boundaries that are the crystal faces of adjacent minerals or that are irregular; this condition is best called **anhedral**, literally "without faces", (syn. xenomorphic, allotriomorphic). Another condition is where a mineral species in a particular rock may show a combination of some well-developed crystal faces and some boundaries which are not crystal faces. Biotite in many granitic rocks, for example, has straight crystal faces parallel to (001) and ragged ends, and hornblende similarly may have straight boundaries parallel to its own prism faces but uneven or irregular boundaries where dome or pyramid faces would be expected. Again, a crystal may have a shape closely approaching its own crystal form but irregular in minor boundary detail. These possibilities are encompassed by the term **subhedral** (syn. hypautomorphic, hypidiomorphic).

Whatever the degree of perfection of crystal face may be, as expressible by the above terms, a specific mineral grain or alternatively the grains of a particular mineral species in a rock may exhibit a certain characteristic overall shape. Isometric minerals and (paramorphed) high-temperature quartz are common examples of **equant** (syn. equidimensional) minerals with no tendency towards elongation in any direction. **Prismatic** crystals are elongated in one direction, e.g., hornblende and augite, the latter characteristically in rather stubby prismatic crystals. Where an elongation in one direction is very pronounced, let us say with a length/breadth ratio of 5 or greater, crystals are described as "elongate prismatic" ranging to **acicular** or "needle-like", e.g., apatite microlites, and tourmaline microlites in "sunburst" tourmaline. **Tabular** crystals are flattened in appearance, for e.g., the flat pseudo-hexagonal tablets of biotite, some larger feldspar crystals, and feldspar microlites generally.

Skeletal crystals, generally found in situations where an origin by quench crystallization seems probable, e.g., olivine crystals in high-magnesium basalts, plagioclase microlites in pillow basalts (see Section 6.2.1.4 in Chapter 6), are of varied morphology but, as their name implies, are not "solid". They may have a central cavity or a more complex skeletal morphology.

In employing the term **corroded** crystal we are implying an origin, i.e. this is an example of a descriptive term carrying a genetic connotation, an implication that is often better avoided in petrographical description (after all we could have called skeletal crystals quench crystals). The term corroded is applied to phenocrysts in a fine-grained rock that show lobate inward extensions of groundmass in a manner suggesting partial resorption by the liquid magma of a former more euhedral phenocryst grain. Quartz phenocrysts in rhyolitic and dacitic rocks show this feature so commonly that one suspects, in fact, that it may not be a corrosion feature at all but rather a reflection of an innate tendency of growing quartz crystals to fail to develop euhedral faces during growth, particularly in the region where the two sets of pyramid faces join. Some plagioclase phenocrysts in andesitic and related rock types may show one or more concentric sets of small blebby groundmass inclusions within their boundaries and are also commonly referred to as corroded crystals. In some subalkaline olivine-bearing cumulates with intercumulus orthopyroxene, a reaction between olivine crystals and liquid during intercumulus crystallization has clearly resulted in corrosion of olivine cumulus crystals as shown by carious embayments at their contacts with intercumulus orthopyroxene.

The term **microlite** refers to a small crystal visible only microscopically. It may be a small crystal in the holocrystalline matrix of a basalt, for example, or a small crystal embedded in a glassy mesostasis as is the habit of small oligoclase crystals in the groundmass of many andesites and related rocks. Although microscopic in dimensions a microlite is large enough to show birefringence. **Crystallites**, by contrast, are yet smaller crystals too small to show birefringence. They are discernible under the microscope as tiny black objects embedded in a glassy matrix, as in most obsidians for example. Crystallites are minutely dendritic and occur in curious hair-like forms, or curved, spiral, branched (trichites), or feathery (scopulites) shapes.

5.3.4. Terms used for overall mineralogical features

The **mode** of an igneous rock refers to the proportions, usually expressed in percentages by volume, less commonly by weight, of the different pyrogenetic minerals present in the rock, either estimated by visual inspection of a rock or a thin section, or more precisely measured by using a point counter. These minerals include: (1) **essential minerals**, generally present in large or at least significant amount, the presence, absence and abundance of which determine the position of the igneous rock in a scheme of classification; and

(2) **accessory minerals** such as apatite, zircon, opaque minerals (accessory in most igneous rocks, but obviously essential in a magnetite cumulate, for instance), etc. Accessory minerals are generally found in small amounts of a few per cent or less and their presence, absence, or abundance does not affect the classification of the rock. The terms essential and accessory refer to the pyrogenetic mineralogy, and are therefore ideally exclusive of any deuteric or secondary alteration products; exceptions have to be made however, e.g., much muscovite, an essential mineral of two-mica granites and muscovite granites, originated under subsolidus conditions.

Note that the terms **mafic** and **felsic** strictly refer to dark and light minerals, respectively, i.e. mafic minerals include olivine, pyroxenes, amphiboles, biotite and opaque minerals, and felsic minerals include quartz, feldspars, feldspathoids and muscovite. Usage, however, has extended the term mafic to refer to basic rocks generally, and the term felsic to refer to acid rocks and indeed intermediate ones.

The **norm** of an igneous rock is derived from a chemical analysis and is explained and compared with the mode in Chapter 4. The terms **femic** and **salic** refer specifically to the dark- and light-mineral components, respectively, of the CIPW norm.

Most relevant grain-size terms have already been discussed in Chapter 4. Note that **aphanitic** is a useful term to refer to matrix grain size too fine to be discernible by eye even if aided by a hand lens. If, on the other hand, the crystalline grains in an igneous rock can be discerned the rock is said to be **phaneritic** — this term, however, is not widely used, as the more specific terms fine-, medium-, and coarse-grained can commonly be used in its place. **Hyaline** is a synonym of glassy — many acid flow rocks, for example, were partly or wholly hyaline when formed. **Holocrystalline** by contrast, as the name suggests, refers to an igneous rock that is wholly crystalline and thus devoid of any glassy mesostasis.

5.3.5. Terms of general application used to describe textures

The terms allotriomorphic, hypidiomorphic and panidiomorphic have already been given as (undesirable) synonyms for anhedral, subhedral and euhedral, respectively. The latter three terms refer only to the shapes of *minerals*, and it seems preferable to reserve the former three for the description of *textures* as follows.

Allotriomorphic refers to a texture in which all the component mineral grains are anhedral. **Hypidiomorphic** refers to a texture in which the grains of some mineral species are anhedral, those of others subhedral, and those of some may even be euhedral. The texture is typical of granitic rocks in many of which quartz and orthoclase tend to be anhedral, and plagioclase and biotite are subhedral to euhedral. **Panidiomorphic** refers to a texture in which, theoretically, all the component mineral grains are euhedral (a geometrical impossibility in a holocrystalline rock). The term is commonly applied to the texture of lamprophyres, which are characteristically densely and conspicuously crowded with euhedral mafic phenocrysts. However, one has to overlook the perforce anhedral nature of the groundmass mineral or minerals, although again the groundmass may be glassy, in which case all the crystalline components can in fact be euhedral.

Equigranular refers to an igneous rock texture in which the diameters of component minerals are comparable, allowing of course for the inherently tabular and prismatic, rather than equant, habits of some minerals. Porphyritic texture, by contrast, connotes one or more mineral species, or a generation of one or more mineral species, that are conspicuously greater in size than those minerals constituting the rest of the rock; these relatively larger crystals are termed phenocrysts (syn. "insets"). An aphyric rock is one devoid of phenocrysts. The term aphyric is generally applied only to aphanitic or fine-grained rocks; thus an equigranular granite, for example, devoid of phenocrysts would not be termed aphyric*.

Phenocrysts are generally conceived of as relatively large crystals which were crystallized slowly from magma under intratelluric conditions (the term primocrysts is applied to them at this early stage of crystallization) occurring in a finer-grained groundmass (syn. matrix) that reflects a subsequent more rapid crystallization of the remaining melt consequent, for example, on extrusion of a lava or emplacement of a hypabyssal rock. However, an amusing example where phenocrysts of early crystallization are actually eclipsed in size by minerals formed during later sequential crystallization processes within the groundmass of granophyres is described by C.J. Hughes (1972a).

Some crystals that may be of comparable size to phenocrysts may actually be xenocrysts derived from the incorporation and fragmentation of a relatively coarse-grained country rock within a magma that later crystallized with a fine-grained groundmass. Xenocryst species will not, in general, be those appropriate to an early stage of crystallization and may reveal signs of reaction with the enclosing magma.

Some relatively large crystals, for example, some alkali feldspars in certain mesozone granites, may in fact have had an origin by growth essentially in the solid state with, or without, accompanying metasomatism; they may be in fact porphyroblasts rather than phenocrysts. However, where the origin may be debatable, and in view of the specific genetic connotation of the term porphyroblast, relatively large crystals of whatever origin in a rock conventionally regarded as igneous may be termed phenocrysts for descriptive purposes.

Exceptionally large porphyritic crystals are termed megacrysts. Some

---

*N.B.: the terms aphyric and aphanitic are not, however, synonyms; many, indeed most, aphanitic rocks are porphyritic, not aphyric.

notable examples are plagioclase megacrysts in some diabase dykes (presumably reflecting the long continued suspension of growing plagioclase primocrysts within a mafic magma of density nearly equal to the plagioclase), and
large alkali feldspar crystals (possibly in part of subsolidus metasomatic
origin as mentioned on p. 137) in some deep-seated granites.

Microphenocrysts, as the name implies, are small phenocrysts in rocks
with an aphanitic groundmass. Microphenocrysts may occur to the exclusion
of larger phenocrysts, or they may occur in addition to an earlier generation
of larger phenocrysts. An upper limit in grain size of 1 mm is sometimes
arbitrarily applied for the term microphenocryst, but it is the texture of the
rock in question rather than arbitrary limits that governs the choice of the
term. Whereas a complicated history of previous fractionation and/or accumulation [see Cox and Bell (1972) and Krishnamurthy and Cox (1977) for
discussion of possible "compensated crystal settling"] may well be suspected
in porphyritic rocks generally, especially in the relatively less viscous mafic
compositions, the small size of microphenocrysts may well have precluded
the operation of these effects in their case (see also discussion of the significance of fractionation in Chapters 6 and 7).

Glomeroporphyritic texture results when phenocrysts aggregate in groups;
sometimes only some of several phenocryst species may be so aggregated,
suggesting that those species not involved belong to a later period of intratelluric crystallization [quartz, for example in the porphyritic felsites of Rum
described by C.J. Hughes (1960)]. Some glomeroporphyritic (?) clusters
represent the incorporation of texturally more complex material, perhaps
consolidated early fractionated material. Some included clots of crystalline
material have a yet more problematic origin in that they may be mineralogically as well as texturally distinct from the phenocryst assemblage in the
rock (see discussion of ultramafic nodules, pp. 309 and 310).

Seriate porphyritic texture describes a situation where there is a continuous range in grain size of one or more mineral species from that of
phenocryst to groundmass size, and in which crystals of progressively smaller
sizes are increasingly numerous. This texture is commonly shown by plagioclase in some andesite porphyries, for example. An explanation of this texture
in terms of cooling rate and kinetics of crystallization is given on p.162.

Vitrophyric texture refers to a rock ("vitrophyre") which is composed of
phenocrysts of intratelluric crystallization in a glassy matrix. The above is
the one occasion where the terms groundmass (syn. matrix) and mesostasis
could perhaps be used interchangeably, but it is better to reserve the term
mesostasis to refer to the glassy component of an otherwise microcrystalline
matrix. This mesostasis component, commonly devitrified in older rocks,
often proves to have a composition close to that of ideal granite whatever
the bulk composition of the rock. Many tholeiites, for example, commonly
have a small proportion of glassy mesostasis of thoroughly acid composition
(although complicated in some by the presence of small amounts of a second,

iron-rich glassy phase) giving rise to intersertal texture (see p.142). Most andesites typically contain a prominent early set of calcic plagioclase and mafic phenocrysts (sometimes seriate) in a groundmass composed of oligoclase and mafic microlites set in an abundant glassy mesostasis. Many latites are comparable to andesites in these respects, but typically contain a much higher proportion of mesostasis to microlites, as of course befits their inherently more alkali-feldspar rich range of compositions. In a continuation of this trend, rhyolites typically have glassy or very fine-grained groundmasses throughout.

Minerals of igneous rocks can commonly be intergrown on various scales. Specific textures of a type, generally very fine-grained, that results from exsolution are discussed separately below. **Poikilitic** texture, however, is a general term describing a texture in which one or more mineral species may be partly or wholly enclosed by another mineral species. Although conventionally and generally correctly thought of as indicating a crystallization sequence of minerals in the igneous rock in question, kinetic factors may come into play in the formation of some poikilitic textures [see, e.g., the dicussion of ophitic texture in Bowen (1928, pp. 68—70)].

Sieve texture, a term sometimes reserved for metamorphic rock textures, refers in an igneous context to numerous small inclusions of one component within a crystal, a common example being the numerous inclusions of glassy groundmass material within "corroded" plagioclase phenocrysts of some andesites and related rock types.

Symplectic intergrowth connotes an intimate fine-grained intergrowth of two minerals; although some symplectic intergrowths may well be due to exsolution and others may reflect post-crystallization reaction or metasomatism, some are of perplexing origin and the non-genetic connotation of the term symplectic is thus useful for descriptive purposes.

A tendency towards a *concentric* relationship of certain minerals in igneous rocks has given rise to a plethora of terms, not always consistently defined or employed. A **reaction rim** strictly results where an early formed mineral later reacts with the still crystallizing magma. A conspicuous example is the occurrence of prominent "black rims", composed of small granules of pyroxene and opaque oxide mineral, around hornblende and biotite phenocrysts in andesite and latite flows. Presumably these hydroxyl-containing phenocrysts grew intratellurically in equilibrium with a water-rich melt and consequent on extrusion, accompanied by a loss of volatiles from the melt, they became no longer in equilibrium with a relatively water-poor lava melt (these rims are not seen in *intrusive* andesite porphyries, etc.). Another example of reaction is that of early-formed forsteritic olivine with magma to form orthopyroxene, typical of the crystallization trend of tholeiitic magmas. Direct textural evidence of this reaction is rarely visible in volcanic rocks; in some plutonic rocks, however, such as appropriate levels of the cumulate sequence of the Bushveld and Stillwater intrusions, reaction has clearly occurred between cumulus olivine crystals and intercumulus liquid, resulting in large oikocrysts of orthopyroxene surrounding embayed carious remnants of olivine crystals.

The term **overgrowth** describes the partial mantling of one mineral by another, presumably conforming to a sequence of crystallization and without conspicuous reaction. For example, alkali feldspar may nucleate and grow around parts of plagioclase phenocrysts in some latites and trachytes, and hornblende may similarly nucleate and grow around augite crystals in some diorites (reflecting in the latter instance the progressive buildup of H_2O concentration in the crystallizing magma, and thus contrasting with the sudden loss of H_2O accompanying extrusion of andesites, etc., referred to on p.319). The term overgrowth carries the additional connotation of crystallographic continuity between the two participating minerals insofar as their differing crystal structures permit this. Where the overgrowth forms a more or less continuous rim around the enclosed mineral, such as, for example, the rim of plagioclase around a proportion of orthoclase phenocrysts in some textural varieties of rapakivi granite, it may be termed a **mantle**.

In describing a **zoned crystal**, commonly plagioclase, olivine, augite, or hornblende, one may also speak of *core* and *rim* in a relative geometrical sense although neither conspicuous reaction nor a different mineral species is involved. **Normal zoning** connotes the gradual transition during the growth of a crystal (from core to rim) to a relatively low-temperature composition in a crystalline solution series. It is the anticipated result of fractional crystallization where equilibration has failed to keep up with falling liquidus temperature and concomitant changing liquidus composition. **Reverse zoning**, in contrast, connotes the transition, generally abrupt, to a higher-temperature outer zone in a crystal. Some hiatal event such as an accession of fresh magma to a magma chamber undergoing fractional crystallization or sudden loss of volatiles from a sub-volcanic magma chamber is responsible for reverse zoning. **Oscillatory zoning** generally accompanies reverse zoning and refers to a succession of normally zoned shells in a crystal each separated by a sharp reverse zone. Oscillatory zoning is often evident in the plagioclase phenocrysts of andesite, originating intratellurically in high-level magma chambers where a complex crystallization history is inferred. Note that in conceptualizing zoning, it is important to realize that random orientations in thin section, in general cutting through only an outer portion of a zoned crystal and not necessarily through its centre, can result in a very different appearance of the width, number, and compositional range of zones from crystal to crystal, i.e. in this respect, as in many others, it is essential to maintain a three-dimensional mentality while looking at the very two-dimensional object of a thin section.

The description of alteration processes affecting many igneous rocks demands another set of terms. A **pseudomorph** is a secondary mineral (or group of minerals) that partly or completely replaces a pyrogenetic mineral while preserving the characteristic crystal outline of the mineral being replaced. Students will soon appreciate the characteristic mesh textures of serpentine minerals replacing olivine, the pseudomorphing of orthopyroxene by bastite (a relatively highly birefringent chlorite in optical continuity), the saussuritization of plagioclase, the replacement of biotite by lamellae of chlorite, etc. **Devitrification** refers in general to the transformation with time of originally glassy groundmass or mesostasis material to a fine-grained cryptocrystalline or microcrystalline product. The patterns of devitrification textures can be quite varied and superimposed, for example, on primary features such as perlitic fractures or the outlines of shards in an ignimbrite, where varying degrees of "primary devitrification" may have occurred varying with height in the cooling unit (see p.143 and other references to ignimbrites in Chapter 2, Section 2.2.3.3; and references to crystallization in Chapter 6, Section 6.2.2.2).

Relict textures refer to such features as preservation of shard outlines and perlitic cracks in some devitrified ignimbrites and glassy rhyolite flows respectively, or the presence of small kernels of fresh olivine within a crystal that may be nearly completely pseudomorphed by a mesh of serpentine.

Kelyphitic rims composed of secondary hydrous minerals such as hornblende may develop around the boundaries of crystals in altered coarse-grained igneous rocks reflecting the partial degradation of the pyrogenetic mineralogy under low-grade metamorphic conditions. Metamorphism at higher temperatures may result in spectacular **corona textures** comprising concentric rims of pyroxene, spinel, amphibole, etc., produced by reaction between original olivine and plagioclase during the metamorphism, for example in gabbroic rocks (see Griffin and Heier, 1973). However, in the cases of both kelyphitic rims and corona structures, we are well away from the igneous history of the rock and into the domain of subsequent metamorphism.

5.3.6. Terms used to describe specific intergrain textures

A reasonable selection of useful, commonly-used terms is given below, in the context of the rock types to which they most commonly refer.

Basaltic flows generally have a microcrystalline matrix consisting of an unoriented (syn. **felty**) array of tabular plagioclase microlites and grains of augite. Where the grains of augite and accessory opaque mineral are small enough to fit between the plagioclase microlites the texture is said to be **intergranular**. Kinetic factors affecting crystallization often result in the augite nucleating less readily than the plagioclase but forming relatively larger, anhedral grains which thus tend to partly or wholly enclose the plagioclase tablets. Where the dimensions of the augite crystals are substantially larger than those of the plagioclase, several or even numerous plagioclase tablets may come to be included within each grain of augite; this is known as **ophitic texture.** Where, however, the augite grains are somewhat smaller, and not much larger than the plagioclase, they will thus only partly enclose individual plagioclase tablets, resulting in **sub-ophitic texture.** Whereas intergranular and sub-ophitic textures are common in basaltic flow rocks, ophitic texture is relatively rare in them, and where it is found the rock is generally an alkali basalt rather than the commoner tholeiite. Ophitic texture, however, is very common in the more slowly cooled hypabyssal rocks of all basaltic compositions such as sills and some thick dykes. For this reason, ophitic texture is sometimes referred to as **diabasic texture;** the two terms are not quite synonymous, however, because in diabasic textures both orthopyroxene and opaque mineral as well as augite may develop a comparable poikilitic habit with respect to the plagioclase, whereas ophitic texture specifically connotes the poikilitic enclosure of plagioclase by augite alone. **Intersertal texture** is characteristic of tholeiite flows where it may be present in addition to intergranular or sub-ophitic textures; intersertal texture denotes the presence of small, disconnected, patches of a glassy mesostasis of acid composition, often containing numerous tiny inclusions of opaque mineral, and universally altered in older flows generally to varieties of chlorite such as highly coloured chlorophaeite. Intersertal texture is not present in all tholeiite flows but it is diagnostic of tholeiite where it is seen. The amount of glassy mesostasis is small, generally no more than a few per cent by volume. In some low-magnesium flood basalts and in basaltic andesites generally (both also of subalkaline affinity) the amount of glassy mesostasis is greatly increased attaining modal amounts of 5—10 vol.%.

In the groundmass of andesites (and in most dacites and latites) there is generally a continuum of a pale-brown glassy or cryptocrystalline mesostasis, in which embedded microlites of plagioclase predominate over dark-mineral microlites producing characteristic hyalopilitic and pilotaxitic textures. The strict definitions of these two textures unfortunately involve overlapping criteria: glassiness of the mesostasis, and the amount and degree of fluxioning of plagioclase microlites. **Hyalopilitic texture** refers to a texture in which plagioclase microlites are set in an abundant glassy mesostasis (but the plagioclase microlites may and generally do show a degree of fluxioning); **pilotaxitic texture** connotes abundant plagioclase microlites prominently fluxioned in an overall sub-parallel manner and locally around phenocrysts (but strictly in a holocrystalline non-glassy matrix). **Orthophyric texture,** also quite common in andesites and related rocks, results where the plagioclase microlites have the form of stubby rather than flat tablets so that fluxioning of these relatively more equidimensional tablets does not occur and they exhibit squat unorientated rectangular sections in thin section; there is no requirement for the mesostasis to be glassy or not in the definition of orthophyric texture. Indeed, in all these andesitic textures the mesostasis component is likely to be cryptocrystalline or extremely fine-grained microcrystalline rather than glassy, due to devitrification with age.

Spherulitic and variolitic textures result from the crystallization of super-cooled eutectoid melts of rhyolitic and basaltic composition respectively; these textures are found in some glassy rhyolite flows and shallow intrusions, and basaltic pillow lavas respectively, and represent the product of extreme supercooling (see pp. 163 and 164). Felsitic texture, on the other hand, results from slow devitrification over geologically long periods of time of rhyolitic material originally cooled to a glass; in felsitic texture aggregates of cryptocrystalline or very fine-grained microcrystalline material extinguish together in small patches throughout the rock.

Ignimbrites, the more characteristic volcanic product of eruptions of rhyolite (see pp. 39—42) contain a groundmass of glassy shards that show varying degrees of flattening and welding resulting in eutaxitic texture (paralleling the eutaxitic structure displayed in outcrop by larger flattened pumice fragments if present). Devitrification processes in ignimbrites occur in two stages: (1) a "primary devitrification" while the ignimbritic "cooling unit" is cooling, commonly results in a complete fine-grained devitrification of the groundmass of the entire upper portion of the ignimbrite cooling unit, often associated with vague spherulitic devitrification textures within pumice fragments; and (2) a "secondary devitrification" with the prolonged passage of time will come to affect all glassy material not devitrified by the primary devitrification. The pattern of devitrification textures in ignimbrites is highly variable. Sometimes it partly or totally obscures original shard outlines; more rarely it may accentuate shard outlines by forming two rows of centrally-directed tiny fibrous crystals, arranged with their long axes perpendicular to the axis of the shard, axiolitic texture.

Many trachytes and some phonolites are characterized by a high proportion of tabular sanidine microlites in the groundmass that are very flat in shape and thus prone to strong fluxioning; the resultant texture is known as trachytic texture.

Considering now intrusive rocks, reference is sometimes made to "cumulus texture". However, consideration of the various textures shown by cumulate rocks shows that there is no one textural feature in common; indeed cumulate rocks are best recognized by their distinctive layering in outcrop. Where there is only one cumulus phase, the cumulate nature of the rocks is generally evident from the (varied) nature of intercumulus growth such as oikocrysts and adcumulus growth (see discussion of this on pp. 235—237). Where the cumulate contains several species of cumulus minerals, a typically gabbroic texture results where all the grains tend to be anhedral with intergrown irregular boundaries but may retain an overall subhedral outline (i.e. each cumulus grain has acted as a convenient nucleation centre for intercumulus growth).

Granitic texture connotes a granitic rock which is both equigranular and hypidiomorphic — the common condition. Monzonitic texture is a specific type of poikilitic texture in which orthoclase (often perthitic) poikilitically encloses crystals of plagioclase (that often display limpid albite rims in this situation) and mafic minerals. **Aplitic texture** refers to a fine-grained rock, often occurring as narrow aplite veins, generally of ideal granite composition, in which all the felsic minerals are equigranular and anhedral. Much has been written about **granophyric textures**. In general, these are delicate intergrowths, often with an overall radiating habit, of quartz and alkali feldspar in ternary minimum proportions. The texture often coarsens radially outward and passes into a microgranitic texture. This transition may be apparent on a small scale within the area of one thin section, and may be found also on a larger scale within intrusive bodies (typically small epizone intrusions emplaced below critical vesiculation depth) which may have granophyric margins and microgranitic interiors. **Micropegmatite** (syn. micrographic intergrowth) is compositionally akin to granophyric intergrowth, but the term is generally used to refer to the small interstitial patches of quartz and alkali feldspar that characterize quartz-diabases of tholeiitic affinity.

Myrmekite is a symplectic intergrowth of quartz and oligoclase occurring in small cauliflower-shaped embayments into microcline at oligoclase—microcline boundaries in katazonal granitic rocks and migmatites, or more rarely at oligoclase—orthoclase boundaries in some mesozonal granitic rocks. Unlike granophyric and micropegmatitic textures which are produced by crystallization from a silicate melt, myrmekitic textures reflect subsolidus replacement effects.

Pleochroic haloes are very obvious in many biotites of granitic rocks, much less commonly seen in hornblende. Where the plane of the thin section transects the centre of the sphere (halo is an inappropriate two-dimensional term) a tiny zircon may be seen. Radiation from the uranium and thorium contained in trace amounts within the zircon crystal has resulted in the prominent dark sphere of metamict material in the host mineral (incidentally, this metamict material is generally *not* pleochroic).

Lamprophyric texture refers to the voluminously porphyritic, apparently panidiomorphic texture that is common in lamprophyric rocks.

5.3.7. Terms used to describe intragrain textures produced by exsolution

These textures are highly specific and distinctive and merit separate consideration, particularly as they are common in the two most abundant igneous rock mineral genera, namely feldspars and pyroxenes. The principles of *crystalline solution* (syn. *"solid solution"*), whereby increased amounts of other component elements may enter into a given crystal lattice at elevated temperatures, have been discussed in Chapter 1 (see also discussion of solvus in Chapter 6). What interests us as igneous petrographers are the textures produced by **exsolution** (syn. "unmixing") which is the partial or complete response to equilibrium at a later set of lower-temperature conditions than the temperature of formation of the mineral crystal in question.

Where exsolution occurs the element present in minor amount typically enters a phase that forms small **exsolution lamellae** or **exsolution blebs**, within the host mineral. Due to the observed extreme sluggishness of exsolution in silicate minerals (reflecting extremely slow diffusion rates within the lattices of silicate minerals), the relatively short cooling history of lava flows and small minor intrusions inhibits exsolution. In large plutonic bodies, however, the effects of exsolution are commonly conspicuous.

In orthopyroxenes "of Bushveld type", for example, slow cooling has resulted in the exsolution of lamellae of diopsidic augite, "Henry lamellae", parallel to (100) of the host orthopyroxene. These lamellae are so thin as to be optically irresolvable, but give rise to an unmistakable extinction effect in thin section (and the typical "bronzite" appearance in hand specimen), and can be shown by X-ray techniques in fact to be augite. A complementary exsolution phenomenon is the exsolution of thin lamellae of orthopyroxene, often altered, along regularly spaced (001) planes in host augite, giving rise to the well-known salite parting.

Plagioclase only rarely shows exsolution visible on a microscopic scale, although plagioclases in the compositional range $An_3 - An_{22}$ are generally sub-microscopic intergrowths of relatively albite-rich and relatively anorthite-rich phases with slightly differing structures; such intergrowths are known as peristerite, and may result in some plagioclase crystals of this compositional range showing a characteristic schiller effect. Plagioclase of anorthosites within the Grenville province that have been subjected to presumably prolonged high temperatures during regional metamorphism subsequent to emplacement of the intrusions commonly reveal **antiperthite** which consists of numerous tiny blebs and rods of orthoclase within the host plagioclase. Antiperthite is not commonly seen in igneous rocks; although included here for the sake of completeness it would appear to be strictly a texture produced during metamorphism; it is also common in granulites.

Perthite, however, is very common in igneous rocks and consists of quantitatively minor lamellae, shreds, patches and rims of an albite component within and around host orthoclase or microcline. Whatever the orientation in this section, the albite component always has the higher birefringence and appears brighter under crossed nicols — a useful feature in identification, as the exsolution lamellae are generally far too small to show any diagnostic multiple twinning. **Microperthite** refers specifically to exsolution textures that are visible only on a microscopic scale (some perthitic intergrowths are sufficiently coarse to be conspicuous in hand specimen), and **cryptoperthite** refers to even finer-grained intergrowths that are only resolvable by X-ray methods. The apparently optically homogeneous sanidines of older volcanic rocks are in fact cryptoperthites (as quickly revealed by their distinctive values of 2V), their submicroscopic pattern of exsolution having been effected during the passage of time rather than during the cooling of the igneous rock. **Mesoperthite** refers to the typical perthite of epizone granitic bodies wherein the albite component is comparable in abundance to the orthoclase component (see explanation in Chapter 6, Section 6.3.9.3); in mesoperthites the albite component may come to form an anastomosing host around remnants of a (generally still relatively more abundant) orthoclase component. Even coarser-grained granular exsolution textures formed during the slow cooling of large epizone plutons are discussed and figured by Tuttle and Bowen (1958).

PHYSICAL PROPERTIES AND PHYSICAL CHEMISTRY OF MAGMAS

What we know about the physical properties and physical chemistry of magmas, direct observation being impossible, is derived from legitimate inferences based on observations of volcanic activity, the physical properties of minerals, rocks and their melts, and experimental work on synthetic and natural silicate systems.

6.1. PHYSICAL PROPERTIES OF MAGMA

6.1.1. Temperature

Estimation of lava temperatures can be based upon measurements of the incandescent colour value with an instrument, utilized in steelworks, coking plants and the like, known as an optical pyrometer. Qualitatively speaking, following Bullard (1962, p. 53) categories for estimation of lava temperatures are:

		Temperature (°C)
Incipient red heat	corresponds to	540
Dark-red heat	corresponds to	650
Bright-red heat	corresponds to	870
Yellowish-red heat	corresponds to	1,100
Incipient white heat	corresponds to	1,260
White heat	is attained by	1,480

Basalt is found to erupt at temperatures generally in the neighbourhood of 1100°C although values as much as 100°C higher or lower than this have been reported. Aerial colour photographs of Surtsey, Iceland, for example, when actively erupting lava in 1964, reveal a yellowish-red vent area of fountaining lava and flows showing deepening reddish colours and increasing proportions of dark congealed crust away from the vent. In Hawaii, it has been possible to approach the barely moving, steeply fronted, distal, aa-type toe of a basalt flow, and similarly estimate a temperature of ~900°C on freshly exposed, red-glowing lava surfaces. Clearly, solidification of basalt takes place over an appreciable temperature range represented, as we shall see, by the liquidus—solidus interval. The pyroxene andesites of the famous eruption in 1943 of Parícutin, Mexico, with silica contents in the range of 55—60 wt.%, were similarly estimated to have been erupted at temperatures in the range of 1020°—1110°C (Zies, 1946). Rhyolite temperatures are virtually impossible to measure in this way, as extrusive rhyolite is too viscous to fountain or flow readily and thus expose measurable surfaces. Nuées ardentes, however, have been recorded as having a dull-red colour, correlatable with temperatures of ~600°—700°C.

It is possible to venture out onto the congealed surface of basaltic lava lakes in Hawaii and insert a thermocouple into the still partly molten lava below. Temperature values so obtained are consistently a little higher than those obtained by pyrometer and are in the range of 1160°—980°C, the lower temperature corresponding with nearly completely crystallized material. Still higher temperatures, up to 1200°C, have been recorded directly above molten lava surfaces, where temperatures may likely have been raised by combustion of gases. The nature of andesitic and rhyolitic volcanism, of course, precludes this type of measurement.

Liquidus temperatures of examples of common lava types, determined in an open crucible at atmospheric pressures, have been found to be as follows:

	Liquidus temperature (°C)
Basalt	1,060—1,090
Andesite	1,200—1,250
Rhyolite	~950 (difficult to measure)

One notes (a) the rather good correspondence of these values with those determined by other methods for basalt; (b) the surprisingly high liquidus temperature of the measured andesite where a more intermediate value in line with the intermediate composition of andesites might have been anticipated. Andesites, of course, not only embrace a spectrum of compositions but also commonly contain some 20 vol.% or more phenocrysts of intratelluric crystallization and relatively refractory compositions so that the *extrusion* temperature of an andesite would not in fact be expected to equal its *liquidus* temperature (other considerations, such as the liquidus-depressing effect of dissolved water under pressure, apart); (c) the rather high liquidus temperature of rhyolite at atmospheric pressure as compared with nuée ardente eruptions. This, as we shall see, is a real difference and not solely attributable to cooling within the nuée ardente during eruption.

Phenocryst compositions are a guide to temperature of formation, but their magmatic crystallization temperatures compared to those determined in pure simple systems may be considerably reduced by the interplay of other constituents in natural magmas. The compositions of plagioclases from different igneous rocks, for example, are a guide to *relative* magma temperatures rather than their *absolute* temperatures. However, the compositions of magnetite—ilmenite solid-solution pairs (Buddington and Lindsley, 1964) when coexisting as phenocrysts have indicated fairly consistent temperatures (and incidentally oxygen fugacity values) in the range of 1000°—750°C for the compositional range andesite to rhyolite.

From the above data, a reasonable consensus clearly emerges on a temperature spectrum paralleling the compositional spectrum of common lava types. In view of the somewhat imprecise nature of the data, it would be unwise to put undue reliance on any particular numerical value. In passing, one may remark that a degree of imprecision often characterizes **quantitative** work in geology*.

6.1.2. Gas content of magmas

Juvenile volcanic gas is that originally contained in magma, of which it is often a significant component, almost entirely lost during the eruption and crystallization of lavas. The estimation of the amounts and compositions of juvenile gases is a very difficult and hazardous procedure. The great majority of reliable published results refer to juvenile gases derived from basaltic lavas, owing to the very great practical difficulties of sampling from other lava types. Sampling of escaping gas has been achieved, for example, at vesiculating basaltic lava pools from the insecure stance of the partly congealed surface of a basalt lava lake. Obvious possible contaminants are the atmosphere and meteoric water that may have become incorporated at some stage within the magma—lava system. High and varying amounts of nitrogen would indicate the former and $^{18}O/^{16}O$ ratios may give an indication of the latter. On these bases and also on consideration of the isotopic ratios of contained Ar and C, corrections can be applied to the compositions of the sampled and analysed gases. A further difficulty is the frequently observed rapid combustion of escaping volcanic gas on contact with the atmosphere, resulting in the collection of a more oxidized assemblage of gases than that originally contained in the magma.

Although the collection and adjustment procedures are thus delicate, a clear consensus emerges that water is the predominant constituent of juvenile volcanic gas and hence of magma (excluding kimberlitic magmas that may contain relatively more CO_2). On the average, water accounts for some 90 vol.% of total gas content, incidentally justifying the simplifying assumption made in direct comparison with much experimental work involving dissolved water in silicate melts. Other gases present, in approximate order of abundance, include the carbon-containing gases CO and CO_2, sulphur-containing gases H_2S and SO_2, hydrogen, and halogen gases of which HCl is predominant. An overall reducing primary mixture of juvenile gases is evident, and vigorous burning would indeed be anticipated on contact with atmospheric oxygen. This factor may account for wide variations in reported CO/CO_2 ratios, hydrogen content, etc.

---

*Students are occasionally led to remark that geology is not "scientific" on this account. This, however, is not true: the proper handling of available data, and the formulation and testing of hypotheses in geology exemplifies scientific method not only as well as in other disciplines but also frequently demands, for this same reason, a high quality of scientific ability and impartiality.

In fumarole areas associated with andesitic and rhyolitic volcanism contamination with meteoric water usually proves to be gross. The HF/HCl ratio increases with lower temperatures and more acid compositions, and the H_2S/SO_2 ratio also increases with lower temperature of emission, the H_2S being responsible for the familiar rotten egg smell of fumarolic areas.

The *theoretical* maximum amount of water that can be dissolved in silicate melts at different pressures has been investigated by several workers including Goranson (1936) for albite, Tuttle and Bowen (1958) for granite ternary minimum compositions, D.L. Hamilton et al. (1964) for basaltic and andesitic melts. The amount proves to increase greatly with pressure (Fig. 6.1), and a similar pattern of decreasing rate of increase of dissolved water content with increase in pressure is common to all. Ascending magma containing water in solution, not necessarily saturated at depth, will eventually rise to crustal levels where diminishing confining pressure will at first equal and then be exceeded by the vapour pressure of the contained water in the rising magma. The confining pressure could be the lithostatic pressure of overlying country rock, or the hydrostatic pressure of overlying magma if it is in a vent in continuity with the surface, or the hydrostatic pressure of water in the case of a submarine eruption. At levels shallower than this critical **vesiculation depth**

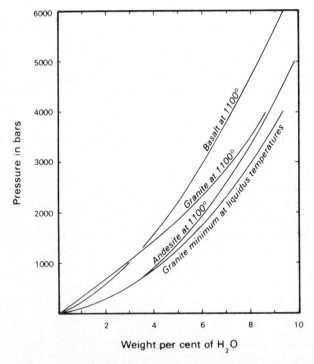

Fig. 6.1. Solubility of water in a typical basalt and andesite (after D.L. Hamilton et al., 1964, p. 29, fig. 1) and in granitic melt (after Tuttle and Bowen, 1958, p. 38, fig. 28).

[the **saturation level** of McBirney (1974)] gas will tend to vesiculate and escape from the melt. Retention of gas and consequent development of **magmatic overpressure** (C.J. Hughes, 1971) may, however, develop transiently within high-level magma chambers bounded by country rocks of high strength and low diffusivity.

The *actual* content of water and other volatile constituents in magmas is an interesting question leading to several ingenious and necessarily oblique lines of attack.

By correlating the size and amount of vesicles present in recent submarine pillow lavas off the southeast coast of Hawaii with ocean depth and therefore confining pressure, J.G. Moore (1965) showed that this oceanic-island tholeiitic basalt magma likely contained 0.45 ± 0.15 wt.% H_2O, plus much smaller amounts of another, less soluble gas, CO_2 (?), to account for the persistent pressure of some very small vesicles even in the deepest lavas sampled. Subsequent work of a similar nature, (J.G. Moore, 1970) on samples of oceanic basalts from other areas and of different compositions has resulted in the following estimates of initial magmatic water content:

	Initial magmatic water content (wt.%)
Low-potassium oceanic tholeiite	0.25
Oceanic-island tholeiite	0.5
Oceanic alkali basalt	0.9

In line with this apparent progressive increase of H_2O content with alkalinity, the commonly diatremic habit of eruption of nephelinites and other more alkalic mafic magmas suggests a yet higher content of volatile constituents in them.

By correlating experimental investigations on an actual andesite from Parícutin, Mexico, at various pressures and H_2O contents with the observed phenocryst assemblage, Eggler (1972) showed that the phenocrysts likely crystallized from a magma with a water content of 2.2 ± 0.5 wt.%, and incidentally at a temperature of 1110° ± 40°C.

An ingenious evaluation taking into account welding temperatures of shards in ignimbrite, magma temperature, and magma cooling due to loss of water (Boyd, 1961) suggests that the water content of the magma that erupted to form the Yellowstone Tuff ignimbrite was not greater than an inferred maximum figure of ~4 wt.%, although it might well have approached this figure.

It is possible, indeed probable, that H_2O contents of magmas of intermediate to acid compositions in high-level magma chambers may have been subject to an increase on parental magma compositions due to crystallization of anhydrous phases and to fluctuations consequent on upward volatile diffusion processes and intermittent escape of gas.

A consensus emerges, however, from the above lines of evidence that water content of magmas, although modest, is significant and is broadly related to composition.

6.1.3. Viscosity

The viscosities of magmas will clearly affect their mode of intrusion, crystallization, and possible crystal fractionation processes (see Chapter 7). The unit of measurement of viscosity is a **poise** (P)*. Water, for example, has a viscosity of $\sim 10^{-2}$ P at room temperature, glycerine 10 P, and pitch $\sim 10^8$ P. Some idea of the viscosity of magmas can be inferred from experimental work on silicate melts, from observation of the flow behaviour of lavas, and from theoretical considerations (see Bottinga and Weill, 1972). In general, it can be said that magma viscosity will depend on composition, temperature and dissolved gas content.

In the range of magma compositions, acid melts are several orders more viscous than basic melts. For example at 1200°C, a typical basalt has a viscosity of ~ 500 P, a typical andesite $3 \cdot 10^4$ P, and a typical rhyolite 10^7 P, all measured on "dry" melts, i.e. open to the atmosphere and lacking dissolved water.

As might be expected, viscosity also varies significantly with temperature. Fig. 6.2 shows viscosity plotted against temperature for three basalts determined by: (1) measurements of an observed flow in Japan (a pahoehoe type changing to aa distally); (2) measurement by rotation viscometry on a tholeiitic lava lake in Hawaii; and (3) a laboratory determination. Note the steepening of slope below liquidus temperature, and the lower viscosities of the flows in the natural temperature range, probably reflecting some retained volatile content. All three curves exhibit a comparable trend of increasing viscosity at lower temperature. A dry melt of rhyolite at its liquidus temperature of ~ 950°C would thus be expected to possess an enormously high viscosity, much higher than that quoted above at 1200°C.

It is believed that just as the molecular structure of silicate minerals is dominated by ordered SiO_4 tetrahedra, so magmatic silicate melts consist largely of SiO_4 tetrahedra that are polymerized to some extent. Polymerization and consequent viscosity should consequently be: (1) *increased* by the melt having a composition close to SiO_2, and the acidic magmas are indeed very viscous; and (2) correspondingly *decreased* by the melt having a chemical composition relatively rich in basic cations that would insert themselves between SiO_4 tetrahedra and thus prevent polymerization. Ugandite, a

---

*"According to Newton's law of fluid friction, viscosity is defined as the ratio of the shearing stress to the rate of shear. If the stress is measured in dyn cm^{-2} and the rate of shear in s^{-1}, the viscosity is given in dyn s cm^{-2}, or equivalently in g cm^{-1} s^{-1}. This unit is called the poise" (following Clark, 1966, p. 292).

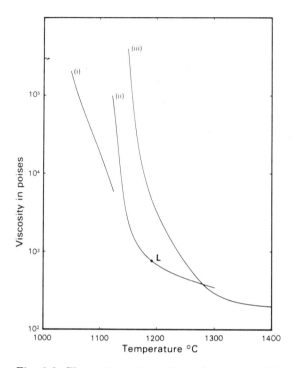

Fig. 6.2. Illustration of variation of viscosity of three natural basalts with temperature: (*i*) computed from field measurements of flowing basalt lava. Minakami, 1951, *Bulletin of the Earthquake Research Institute, Tokyo University*, 29, p. 491 (quoted by Clark, 1966, p. 293); (*ii*) measured by rotation viscometry on Hawaiian tholeiite by H.R. Shaw et al., 1968, p. 257, fig. 15, showing liquidus temperature (*L*) inserted; and (*iii*) measured in the laboratory by K. Kani, 1934, *Proceedings of the Imperial Academy, Tokyo*, 10, pp. 27 and 79 (quoted by Clark, 1966, p. 293).

basic rock rich also in alkalis and with a very low silica activity, has been calculated to have the lowest measured viscosity of any igneous rock, ~400 P at a liquidus temperature of just under 1100°C.

A pertinent discussion of Bingham and Newtonian models of liquid behaviour is incorporated in H.R. Shaw et al. (1968), prefacing data of field measurements of viscosity of tholeiitic basalt in Hawaii. A concise account of a direct measurement of viscosity is given by Scarfe (1977) who reports a viscosity of ~10^4 P for a pantellerite in the temperature range of 1300°– 1450°C. This value of viscosity is significantly lower than obsidian but higher than basalt values, and in line with predicted values following Bottinga and Weill (1972). For recent details on field measurements of the rheology of lava, see Pinkerton and Sparks (1978).

Over and above these considerations based upon temperature and major-element chemistry, dissolved water content has a very great effect in lowering the viscosity of silicate melts. This is partly because of the low molecular weight of water which means that for a given weight per cent there are correspondingly a larger percentage of water molecules present. Because of their strong polarity, dispersed water molecules are very effective at preventing polymerization of SiO_4 tetrahedra. Furthermore, water in significant amounts behaves as an extra component of very low freezing point temperature compared to those of the silicate components in silicate systems. Fig. 6.3, reproduced from H.R. Shaw (1965), indicates that, given reasonable estimates of water contents in the range say of 1.5—4 wt.%, granitic magmas near the ternary minimum composition in the system Or—Ab—SiO_2—H_2O with liquidus temperatures in the range of 880°—760°C would have viscosities in the order of 10^7—10^8 P. It should be noted that although this estimate represents a considerable reduction on anticipated "dry" melt viscosities, it is also considerably higher than inferred values for basic magmas.

Experimental work reported by Kushiro et al. (1976) shows that the

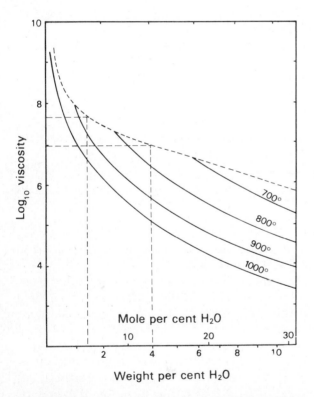

Fig. 6.3. Variation of viscosity with H_2O content for melts having bulk compositions approximating those in the low melting regions of the system Ab—Or—SiO_2—H_2O. The *dashed curve* represents viscosities for temperatures and H_2O contents of liquids along the minimum melting curve of the system Ab—Or—SiO_2—H_2O saturated with H_2O (after H.R. Shaw, 1965, p. 121, fig. 1).

viscosity of a water-free olivine tholeiite from Kilauea, Hawaii, decreases markedly with increasing pressure (and concomitant increase in temperature) along its liquidus to the very low value indeed of ~10 P at 30 kbar. (Low viscosities of this order would permit ready fractionation by crystal settling — see Chapter 7.) The same workers report viscosities of ~1100 P for a Crater Lake Oregon, andesite at 1350°C at pressures of 15 and 20 kbar (values straddling the liquidus). The same andesite, but with 4 wt.% H_2O added, was found to have the much lower viscosity of 45 ± 5 P at 15 kbar and 1350°C.

6.1.4. Density

Density becomes an important parameter when considering such phenomena as the behaviour of a body of magma with respect to country rocks and the possible movement of crystals within a magma chamber. Common fresh igneous rocks have densities in the approximate range 2.65 (granite, s.s.) to 2.97 g cm^{-2} (diabase). Some average densities for Coast Range rocks, for example, are as follows: biotite granite, 2.63; biotite granodiorite, 2.67; hornblende granodiorite, 2.71; hornblende quartz-diorite, 2.77; and hornblende diorite, 2.84 g cm^{-3}. An idea of the densities of more unusual rocks, ultramafics for example, can be obtained from the densities of their component minerals (Table 6.1), some of which, it will be noted, are presented as theoretical end-members of solid-solution series. For comparative purposes, some other approximate average densities in g cm^{-3} are given in Table 6.2.

The coefficient of thermal expansion is such that for a temperature increase from 20° to 1000°C there is a volumetric expansion of between 1.45

TABLE 6.1

Densities (in g cm^{-3}) of some common rock-forming minerals and of end-member components of some solid-solution series

Light (coloured) minerals:		Diopside	3.28
Leucite	2.48	Enstatite	3.20
Nepheline	2.62	Hedenbergite	3.55
β-Quartz (high-temperature quartz)	2.53	Common pyroxenes	3.3—3.4
α-Quartz (low-temperature quartz)	2.65	Common hornblende	~3.2
Orthoclase	2.55	Phlogopite	2.79
Albite	2.62	Annite	3.32
Anorthite	2.76	Common biotite	3.1—3.2
Muscovite	2.83	Pyrope	3.56
Jadeite	3.31	Almandine	4.32
		Spessartite	4.19
Dark minerals:		Common chromite	~4.7
Forsterite	3.21	Common magnetite	~5.2
Fayalite	4.39	Common ilmenite	~4.8
Common forsteritic olivine	~3.33		

TABLE 6.2

Average densities (g cm^{-3}) of miscellaneous rocks

Shale	2.1, rising to ~2.65 with decreasing porosity
Sandstone	2.2, rising to ~2.6 with decreasing porosity
Limestone	~2.75
Gneiss	equivalent to igneous rock of comparable composition, many acid gneisses ~2.7
Mica schist	~2.8
Orthopyroxene bearing granulite	~2.9
Garnet lherzolite	~3.35
Eclogite	3.4—3.5

and 3.3 vol.% for the common rock-forming silicate minerals, with most values lying around 2 vol.%. In the case of igneous rocks, of course, a significant consideration is the equivalent contraction during cooling through a comparable temperature interval.

Volumetric expansion on melting is somewhat greater; artificially prepared glasses are ~4—10 vol.% less dense at room temperature than the corresponding igneous rock of the same composition. Natural basaltic glass at room temperature has a density of ~2.77 and rhyolite glass of ~2.37 g cm^{-3}. Basalt glasses at between 1200° and 1250°C have yielded densities in the range of 2.60—2.79, average 2.65, all values except one high value close to 2.62 g cm^{-3}; for basalt at liquidus temperature a density of 2.65 g cm^{-3} would be a fair approximation. From less abundant data, the density of a rhyolite glass also at 1100°C would seem to be ~2.3 g cm^{-3} (allowing for a small but significant quantity of dissolved gases), and therefore fractionally higher at its lower magmatic liquidus temperatures.

6.2. KINETICS OF CRYSTALLIZATION OF MAGMA

6.2.1. Introduction

It is evident that the *pyrogenetic* texture and mineralogy of most igneous rocks reflect one physical phenomenon — namely, the solidification of a melt usually resulting in the formation of an aggregate crystalline product, more rarely a partly or wholly glassy rock. Of course, the rock so produced may be considerably modified by subsolidus recrystallization, deuteric alteration, later hydrothermal, metamorphic and metasomatic processes, and certainly most katazonal and mesozonal granites never originated from a high proportion of melt in the first place. Nevertheless, the crystallization of a silicate melt is the point at which the petrographer must attempt to begin an understanding of the textures of many igneous rocks.

An enormous range of detailed experimental studies on liquid-crystal *equilibria*, summarized in Section 6.3 of this chapter, both on natural rocks and appropriate synthetic compositions has yielded among other things fairly precise prediction of what mineral phases and their solid-solution compositions should crystallize at what temperatures from various magma compositions. One notable field of success, as we shall see, has been the correlation of experimental work with the observed succession and composition of "cumulus" minerals crystallizing from large chambers of basic magma under conditions closely approaching equilibrium.

However, superimposed on these considerations, and indeed causing departures from predictions based on considerations of equilibria alone, there is also the question of *kinetics*, i.e. the rate at which the crystallization process occurs. This, in fact, becomes a prime consideration in the interpretation of textures of volcanic and hypabyssal rocks, and the concepts involved also find application even in some of the more slowly cooled intrusive rocks where equilibrium conditions might have been anticipated. The writer finds it surprising that this question of kinetics, so fruitfully applied, for example, to metamorphic rocks where crystallization in the solid state is concerned, is so commonly neglected, even ignored, in texts on igneous petrology. One notable exception is the very thorough quantitative treatment of this subject in Carmichael et al. (1974, pp. 147—168). What follows here is a more qualitative approach with relatively greater emphasis on observed rock textures.

6.2.1.1. Phenomenon of supercooling. The formation of a crystal from its melt should, under equilibrium conditions, take place at the melting point (= freezing point), which is the temperature below which, at any given pressure, the solid is thermodynamically more stable than the liquid. In the case of a pure substance, and ignoring those which undergo incongruent melting, this fixed temperature at which the solid changes entirely into liquid (provided only that sufficient heat is provided for the latent heat of fusion) can be accurately ascertained. However, on cooling it is frequently found that the same substance may remain in a liquid condition as a **supercooled liquid** at temperatures below the anticipated freezing point. Silicate melts, with their high viscosity, are very prone to exhibit this phenomenon. Crystallization of a supercooled melt can be thought of as involving two steps, first **nucleation** and then **growth**; these two steps operate at varying rates at temperatures significantly below the freezing points of pure silicates and the liquidus temperatures of more complex silicate melts.

6.2.1.2. Nucleation of crystals. The commencement of crystallization depends on the formation of a nucleus of the crystal, a random event resulting from thermal motions within the liquid. Surface energies cause small nucleii to be resorbed, so that it is also a matter of a nucleus attaining a certain critical size before it becomes viable and subsequent growth can occur. At small degrees of undercooling the necessary critical size of the nucleus is large and thus unlikely to form by random groupings. At higher degrees of undercooling the critical size is much less, so that a **nucleation rate,** expressed for example as the number of viable nucleii forming per unit volume of melt per unit of time, is very much increased under these conditions of greater supercooling. At some still higher degree of supercooling however, greatly reduced thermal motions at lower temperatures result in a reduced nucleation rate despite the smaller viable size necessary. These considerations are best shown in the form of a graph (Fig. 6.4), from which it will be seen that the "kinetic" nucleation rate for nepheline reaches a broad maximum at 45°—80°C below the "thermodynamic" freezing point (H.R. Shaw, 1965, p.134).

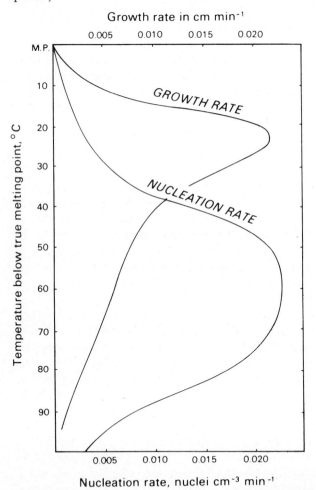

Fig. 6.4. Curves showing rates of nucleation and of growth of nepheline from its melt at degrees of supercooling up to 100°C. After H.R. Shaw, 1965, p. 154, fig. 5 (after Winkler, 1947).

6.2.1.3. Growth of crystals. A second consideration is the growth rate of viable nucleii. This also is found to vary with temperature. In a melt just below the freezing point the energy difference between the liquid and crystalline state is small, so there is little energy potential to drive crystallization and hence the growth rate is small. This potential increases rapidly with falling temperature, so that the growth rate increases. However, with still further fall in temperature, the crystallization process slows, and the growth rate falls again. Fig. 6.4 shows a fairly sharp peak in the growth rate of nepheline ~ 20—$25°C$ below freezing point. This consideration is enhanced in a natural rock melt, in which at near-liquidus temperatures the crystallizing phase or phases do not generally have the same composition as the melt, necessitating for their growth also therefore a degree of diffusion of material through the melt, a process that also slows with falling temperature.

Experimental work on melts of other compositions has yielded comparable results. It should be noted, however, that a low latent heat of crystallization, other things being equal, results in a somewhat faster nucleation rate and slower growth rate. Silica and the feldspars have lower latent heats of crystallization than the dark minerals.

6.2.1.4. Quench crystals. The term "quench crystal" originated with experimental work on crystal—liquid equilibria where quenching from the experimental temperature should result in the preservation of material that was liquid at equilibrium temperatures as glass in the quenched product. Unfortunately, this glass is commonly found to contain not only (the desired) crystals grown at equilibrium but also some additional quench crystals (extraneous to the equilibrium conditions) that grew even during rapid quenching. It is obviously important for the correct interpretation of the equilibrium study carried out at high temperature to be able to differentiate between these two possible sets of crystals. It is generally found that the equilibrium crystals are euhedral whereas the quench crystals exhibit distinctive morphologies (see, e.g., the illustrations in D.H. Green and Ringwood, 1967a, p. 127). These morphologies include: (1) skeletal forms and (2) unusually acicular or bladelike forms, sometimes with spiky protuberances, grading to (3) various dendritic morphologies. They can be envisaged as being the product of conditions of a rapid potential growth rate exceeding the rate at which diffusion processes within the liquid can bring the appropriate constituents to the growing crystal faces, so that the growing crystal, instead of acquiring a euhedral layeritic habit, extends as far as possible into the surrounding liquid during growth and hence adopts these morphologies.

An interesting experimental study on plagioclase crystals grown from heated plagioclase gels (Lofgren, 1974) showed that their morphology

". . . is markedly dependent on the degree of supercooling, ΔT, changing from tabular crystals at small ΔT to skeletal crystals, dendrites, and spherulites with progressively increasing ΔT."

The results of this experimental work indeed parallels the observed textures in igneous rocks deduced from their field occurrences to have crystallized at different degrees of quenching. Kirkpatrick (1975) similarly reviewed aspects of crystal growth theory and figures fine examples of experimentally-produced plagioclase crystals of tabular, skeletal, dendritic, and spherulitic habits grown respectively at 50°, 100°, 200°, and 430°C below liquidus temperatures (Kirkpatrick, 1975, p. 811, plate 1).

In the case of naturally occurring examples, the term **quench crystal** is an example of the habit petrologists have of applying a term with a genetic connotation to textures, etc., the description of which might often be better served by a more descriptive term without genetic implications. However, the application of the term does in fact seem reasonable, given the locale of rocks displaying such textures.

It is important however to appreciate that quenching in igneous rocks can be effected in different ways:

(1) *Supercooling* leads to scattered nucleation but rapid growth of one, some, or all of the appropriate phases to be expected on crystallization. The occurrence of skeletal and/or spiky plagioclase microlites in pillowed basalts, often in a zone just within the glassy rim, is a familiar example; the frequent absence of microlites of other anticipated phases from pillow lavas with such a texture is in line with the relatively greater facility with which plagioclase nucleates as mentioned on p. 159. Morphologies of quench crystals of olivine and augite as well as plagioclase from MORB pillow basalts are, however, described, figured and discussed by Bryan (1972).

(2) A more subtle possibility is where, for example, ascending magma, without necessarily undergoing any violent temperature quenching, under-goes *changes in its pressure regime* such that it rapidly attains below-liquidus temperatures for the new conditions. A fairly common example is the occurrence of quench crystals of skeletal olivine in some basalts of relatively high magnesia content (Clarke, 1970). In contrast to strictures expressed below on the validity of analyses of porphyritic volcanic rocks, analysis of a rock containing skeletal crystals may well represent a magma composition, as the crystals have grown very quickly and may not therefore have undergone significant relative movement with respect to the enclosing magma. Sinking of quench crystals in some lunar basalts and komatiites has, however, demonstrably occurred (Pyke et al., 1973).

(3) A *sudden loss of water* will raise liquidus temperature so that a magma melt may undergo sudden quenching on this account. This kind of situation occurs most frequently in very shallow chambers of acid magma that, as we have seen, tend to be richer in contained H_2O than other magmas in the first place. This kind of quenching does not usually result in the formation of skeletal crystals but more commonly in the formation of a glass or a crypto-crystalline or very fine-grained microcrystalline product from viscous, degassed acid magma. See, for example, the inferred complex sequence of events related to textures of minor acid intrusions at Yosemite, California (Fournier, 1968).

6.2.2. Effect of kinetic factors on igneous rock textures

6.2.2.1. Conditions of slow cooling (mainly plutonic rocks).

Very slow cooling should result in the formation of a few nucleii followed by slow layeritic growth to produce euhedral crystals. This condition is realized in slow *intratelluric crystallization* in sizable magma chambers, giving rise to *primocrysts* in equilibrium with the melt. Theoretically, once formed they should go on growing. However, in practice this consideration will be balanced against the formation of a number of competing nucleii per unit volume and a limiting value of diffusion rate within the magma of material to and from the site of the growing crystal faces, particularly in the relatively more viscous acid magmas. Furthermore, in the less viscous basic magmas, these growing primocrysts may move in relation to enclosing magma of differing density and may eventually settle to the floor of the magma chamber to form cumulus crystals and the observed types of *cumulate rocks**.

Again, primocryst growth may be halted by the magma with some included growing primocrysts being injected to form a minor hypabyssal intrusion or extruded as lava. In these cases, the liquid matrix will undergo rapid crystallization with the production of a finer-grained product surrounding the primocrysts, which would then be termed *phenocrysts* in a *porphyritic rock*. In passing, one may remark the inherent danger of analysing such a porphyritic rock and assuming the analysis to represent a former magma composition. This is because of possible relative movement of primocrysts within a melt of differing density during the intratelluric stage of crystallization and the consequent physical enrichment or impoverishment in any particular part of a magma chamber of crystal phases of which the composition is generally not equal to that of the original melt. In porphyritic acid rocks, where the equivalent plutonic rocks of the same composition rarely show evidence of cumulus textures, grain size of phenocrysts is variable, although quartz crystals seldom exceed 5 mm and feldspars 10 mm in diameter, and commonly both have about half these linear dimensions.

With an appreciable intratelluric cooling rate resulting from heat loss from the magma chamber exceeding the latent heat of crystallization of growing

---

*Cumulus crystals commonly range in size up to a few millimetres across, but tend to be much smaller in cumulates from komatiite magmas which were very hot and fluid. In some anorthosites, on the other hand, the grain size is characteristically variable and very coarse with some crystals up to 50 cm across. This variation reflects the length of time a growing primocryst is in contact with magma before accumulation in a crystal pile. Cumulus grain size thus ranges from a minimum in some komatiite sill rocks, where high temperatures and low viscosities would permit quick settling of crystals, to a maximum for the plagioclase of anorthosites, where a very small density difference between plagioclase crystals and parent magma melt could lead to long periods of possible suspension of plagioclase crystals in the magma before the eventual apparent upward accumulation of crystals to form a crystalline aggregate.

crystals, nucleation should occur at a progressively increasing rate. This condition is realized by *seriate porphyritic texture* shown by some porphyritic rocks, where phenocrysts show a range in size from the largest downwards towards matrix size, with phenocrysts of smaller size being progressively numerically more abundant. This situation is more likely to occur in small high-level magma chambers.

The interpretation of textures of many granitic rocks in pyrogenetic terms is difficult. One obvious reason is that many granitic bodies were likely never more than partially molten during their entire period of formation and emplacement. A second reason is considerable subsolidus recrystallization, particularly involving recrystallization of feldspars in response to unmixing and metasomatism during long cooling periods. Some rocks of dioritic compositions preserve features indicative of crystallization from a melt such as complexly zoned plagioclases, and others of granodioritic and monzonitic compositions reveal a sequential pyrogenetic crystallization of poikilitic alkali feldspar after plagioclase, preserved as monzonitic texture. The shapes of grains of quartz and feldspar in some hypersolvus epizone granites closely resemble those in compositionally-equivalent hypabyssal porphyry intrusions modified only by later overgrowths of quartz and alkali feldspar equivalent to the porphyry matrix. It comes as a shock therefore to realize that many alkali feldspar "phenocrysts" of mesozonal granitic rocks, sometimes anhedral, sometimes remarkably euhedral, are probably nevertheless in fact porphyroblasts rather than phenocrysts (the term phenocryst has an implied pyrogenetic connotation although it can also be used purely descriptively in a size connotation). The whole question of granitic petrography cannot be divorced from petrogenetic considerations related mainly to depth of emplacement. Aspects of this problem are reviewed in Chapters 3, 5 and 7.

6.2.2.2. Conditions of supercooling (mainly extrusive rocks). Turning now from considerations of intratelluric crystallization to those of matrix textures in extrusive (and some hypabyssal) rocks where cooling is relatively rapid, there is a considerable variety and a range of grain size downwards to rocks that are wholly or partly glassy, i.e. in the latter case those which failed to nucleate at all, owing to rapid supercooling to temperatures below those of measurable nucleation rates.

The textures of fine-grained and aphanitic basic rocks illustrate this variability well. At significant degrees of supercooling, the relatively higher nucleation rate but slower growth rate of plagioclase as opposed to augite results in the familiar range of *sub-ophitic* to *ophitic* textures of plagioclase microlites partially or wholly enclosed within augite, typically developed in diabases. Hypersthene, where present, and opaque minerals often have a similar textural relationship towards plagioclase microlites in diabase. It can easily be demonstrated that most diabases are close to cotectic compositions from which plagioclase and pyroxene would be expected to crystallize together under equilibrium conditions. An appreciation that ophitic and diabasic textures do not mark sequential equilibrium crystallization of the plagioclase followed by the dark minerals, but are merely the result of kinetic factors, was well summarized by Bowen (1928, pp. 68 and 69). Under apparently higher degrees of supercooling, e.g., in many basaltic flows, the lava seems to have readily nucleated pyroxene in addition to plagioclase, resulting in characteristic *intergranular* textures with squat prisms of pyroxene and grains of opaque minerals between tabular microlites of plagioclase*. Quench crystals, common in pillow lavas, have been discussed on p. 160. Very pronounced supercooling of mafic melts in contact with air (e.g., "Pele's hair") or water (e.g., hyaloclastites and the outer rim of pillow lavas) or more rarely, country rock (e.g., the chilled contacts of some dykes) results in the formation of a basaltic glass, *"tachylite"*. This incidentally is commonly altered to and pseudomorphed by a hydrated weakly-birefringent material, "palagonite", in the geological record, even in young Pleistocene volcanics as, for example, in Iceland (G.P.L. Walker, 1965, p. 35).

Granophyric texture deserves mention here, a good pyrogenetic quench texture that often suffers the indignity of being lumped for descriptive purposes in many texts with myrmekitic texture of a quite dissimilar metamorphic/metasomatic origin. Granophyric intergrowth is a mixture of quartz and alkali-feldspar in ternary minimum proportions, and of characteristically variable grain size tending to increase radially outwards to microgranitic varieties from centres of nucleation which are often phenocrysts if present. It represents quenching under shallow epizonal conditions of a magma that retained water in solution (see C.J. Hughes, 1960, 1971). The comparable "micropegmatitic" texture refers to the few per cent of intergrowth of quartz and alkali feldspar, also in ternary minimum proportions, found interstitially between plagioclase and pyroxenes in some tholeiitic diabases. It is the more slowly cooled compositional equivalent of the acid mesostasis of tholeiites with intersertal texture.

The *spinifex* textures of many komatiite flow rocks are characterized by the presence of remarkably coarse, bladed and branching crystals of olivines, formed under conditions of quench crystallization of unusually high-temperature, high-magnesium lavas.

At some appreciable temperature below the freezing point, some natural glasses may crystallize radiating aggregates of fine-grained material of the same bulk eutectoid composition as the glass. The commonest examples are *spherulites* composed of silica and alkali-feldspar from acid glass, and *varioles*

---

*These are frequently mistermed "laths" rather than tablets, indicative of a practically irremediable habit of petrographers of applying two-dimensional terms — and worse, a two-dimensional mentality — based on features seen in the area of a thin section to what, of course, is a three-dimensional fabric. What is sometimes interpreted as a square or rectangular microphenocryst of plagioclase in such a rock proves to be merely a matrix crystal of average size that happens to lie in the plane of the thin section.

composed of plagioclase and pyroxene from basic glass. Spherulites, as the name implies, are commonly spherical in shape; varioles also tend to be spherical (weathering as whitish bumps on exposed surfaces of some pillow lavas, with a fanciful resemblance to the scars left by smallpox or variola), but variolitic texture also occurs as sheaf-like aggregates between and around plagioclase microlites that are often skeletal. Spherulites in rhyolitic obsidian flows generally overgrow an earlier flow-banded fabric of microlites and crystallites, the latter commonly of dendritic morphology. Sometimes, however, flow-banding is flattened around spherulites. Spherulites may also develop in some ignimbrites in a layer in the upper parts of a welded glassy base and below a lithoidal devitrified upper part. The inferred temperature of accumulation of ignimbrites is, of course, well below liquidus temperatures, at atmospheric pressures. Experimental work has indicated a temperature of formation of spherulites below crystallization temperatures but at sufficiently elevated temperatures that they obviously originated during the cooling process (and not during any later secondary devitrification), and therefore should be mentioned here.

The somewhat spherulite-like but distinctive textures of *chondrules* have been reproduced by drastic supercooling of droplets of high-temperature magnesium silicate melts to 400°—750°C below their equilibrium liquidus temperatures (Blander et al., 1976).

6.3. STUDIES IN THE EQUILIBRIUM CRYSTALLIZATION OF SYNTHETIC MELTS AND THEIR BEARING ON MAGMATIC COOLING HISTORY, IGNEOUS ROCK COMPOSITIONS AND TEXTURES

6.3.1. Historical review

Early work took a little of the mystery out of igneous rocks. J. Hall, for example, in 1798 melted a diabase and noted that on cooling quickly it produced a glass, but on being allowed to cool more slowly a crystalline product of basaltic appearance resulted. In 1878, F. Fouquet and A. Michel-Lévy synthesized a basaltic product from an appropriate synthetic mixture of major oxides by fusing and cooling. Most significant experimental work has however been confined to the twentieth century.

One pioneer was J.H.L. Vogt who, around the turn of the century, fused simple mixtures of oxides in known proportions and investigated their freezing points and first crystalline products, a technique which has been followed ever since. For example, Vogt identified eutectic mixtures and related them to some common igneous textures, wherein certain major mineral phases commonly occur in the same proportions, e.g., "graphic granite" and the matrices of most basaltic rocks. Although much of the later sequence of experimental studies seems fairly obvious to us now in retrospect,

Vogt's work was timely and highly significant in that he demonstrated that crystallization equilibria could be attained in highly viscous silicate melts and that correlation between simple systems and observed igneous rock compositions and textures was indeed possible.

The most important single contribution to these studies began with the setting up in 1907 of the Geophysical Laboratory in Washington, D.C., a branch of the Carnegie Institute. Geophysics has, of course, nowadays acquired a more specialized meaning, but the continuing work of this laboratory has largely been in the field of equilibrium studies of melts and their crystallization. Early work by N.L. Bowen, J.F. Schairer, and others was carried out at atmospheric pressure in open crucibles, that is to say essentially on water-free melts. This work was extremely productive and successful in relating experimental studies to the crystallization behaviour of basic magmas that are in nature relatively hot and water-poor. An important milestone relating to this period was the publication by Bowen (1928) of the *Evolution of the Igneous Rocks*, a masterpiece, the contents and philosophy of which have greatly influenced igneous petrologists. It is only fair to add, however, that the temptation to facilely explain the genesis of many deep-seated intermediate and acid plutonic rocks solely in terms of magmatic crystallization and fractionation pressures should be succumbed to only if other working hypotheses have been tested.

Later work after the Second World War developed techniques for studying the crystallization of melts under pressure with water as a possible additional component. The somewhat paradoxical term *dry melt* refers to one in which water is absent, and the term *wet melt* refers to one where excess water is present, ensuring that the melt is saturated with water at the pressure and temperature conditions prevailing during the experiment. This latter condition perhaps relates more closely to granitic magma, and certainly experimentally speaking more easily to melts of acidic compositions. These are extremely viscous and notoriously difficult to crystallize under open-crucible conditions, necessitating impracticably long runs and with the often uncertain prospect of melt—liquid equilibrium being obtained. Another important milestone relating to this period was the publication by Tuttle and Bowen (1958) of the *Origin of granite in the light of experimental studies in the system: $NaAlSi_3O_8$—$KAlSi_3O_8$—SiO_2—H_2O.*

Work is proceeding at many centres at the present day, often involving advanced experimental techniques. One refinement involves the difficult task of working at high pressures with a determinable partial pressure of H_2O, a condition approximating probable magma conditions still more closely. The development of very high-pressure equipment now enables experimental work to simulate the probable conditions of *generation* of basic magmas, thus adding another dimension to the contribution of experimental work to petrology, which classically was more concerned with the *crystallization* of magma. An important tool now available is the electron probe which allows determination of the exact composition of the often small crystals produced in experimental work, for example, of such an important parameter as the Mg/Fe ratio of mafic crystals.

6.3.2. Equilibrium and the phase rule

The number of different pyrogenetic crystalline phases in any igneous rock is small, commonly no more than six, seven, or eight, even including those accessory minerals which may be present in small amounts such as apatite and zircon. Given the relatively small number of major oxides and their natural groupings allowing for diadochy (see Chapters 1 and 4) and the dispersed nature of nearly all trace elements, this phenomenon is predictable from Gibbs' phase rule, whereby for a system in equilibrium:

$$P = C + 2 - F^{(*)} \tag{6.1}$$

For many groups of igneous rocks with comparable compositions and mineralogies that clearly may have crystallized over an appreciable pressure and temperature field implying two degrees of freedom, $P = C$, i.e. the number of phases is equal to the number of components, the so-called **mineralogical phase rule**, as first indicated by Goldschmidt (1954). This phenomenon, together with the observed comparable mineralogy of igneous rocks of comparable compositions, indicates that equilibrium was approached during crystallization and that conclusions based on equilibrium studies of crystallizing silicate melts may indeed be valid.

By contrast, as a result of deuteric and later alteration processes a considerable number of extra phases such as sericite, epidote, chlorite, etc., may be present in an igneous rock in which they may only partially replace the pyrogenetic mineralogy. The existence of this increased number of phases implies that a new equilibrium assemblage appropriate to lower-temperature conditions has not been fully attained in such a rock.

A system at equilibrium is in its lowest energy state given the prevailing conditions of pressure and temperature and has no tendency to change spontaneously. In a qualitative way, **Le Châtelier's principle** indicates how a system at equilibrium will react to a change in these conditions:

> "if a system is in equilibrium a change in any of the factors determining the conditions of equilibrium will cause the equilibrium to shift in such a way as to tend to nullify the effect of the change."

For example, raising the temperature will favour an endothermic reaction such as melting (if the system previously were wholly or partly solid). Raising the pressure, however, will favour the formation of denser phases such as the production of solid phases by freezing (if the system were wholly or partly

---

*See Frye (1974, pp. 215—218) for a lucid account of the phase rule and its importance.

liquid). The interplay of the effect of variations in these two determining conditions is given by the **Clausius—Clapeyron equation**:

$$\frac{dP}{dT} = \frac{\Delta H}{T\Delta V} \tag{6.2}$$

where ΔH represents the change in heat content of the system; T is the absolute temperature; and ΔV represents the volume change in the system, all expressed in the appropriate units.

The phases present at equilibrium in the unary system, SiO_2, for example, occupy various fields in a pressure—temperature diagram with their boundaries defined by equations of the above type (Fig. 6.5), wherein thermodynamic calculations can be verified by experimental observation. Note how (1) any of the phases can exist over an appreciable **field** of different temperatures and pressures [2 d.f. (degrees of freedom)] ; (2) certain pairs of phases can coexist but only under certain boundary conditions shown by **univariant lines** (1 d.f., i.e. the choice of a certain value of a possible range of pressures predetermines the temperature and vice versa); (3) certain sets of three phases can coexist but only at an **invariant point** with fixed coordinates of pressure and temperature (no d.f.).

This whole topic of equilibrium is treated at a more fundamental level in such texts as Ehlers (1972) and Ernst (1976).

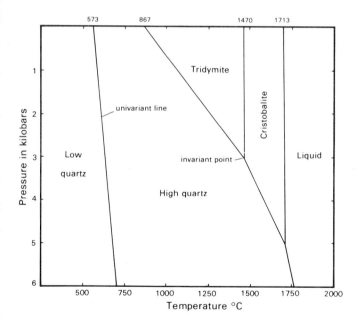

Fig. 6.5. Stability relationships of polymorphs of SiO_2 (after Tuttle and Bowen, 1958, p. 29, fig. 12).

6.3.3. Melting points of the pure end-members of some common rock-forming mineral species

This might seem a logical place to begin in an attempt to comprehend the crystallization of natural silicate melts; some values are given in Table 6.3. Note (1) the high melting points compared to observed lava temperatures; (2) some curious relative positions, forsteritic olivine is a common phenocryst phase in many basic lavas, and therefore of a relatively early crystallization that is qualitatively consistent with its high melting point; the same, however, cannot be said of quartz; (3) the great temperature range between melting points of end-members of the plagioclase series and the even greater range within the olivine series; and (4) the curious "incongruent melting" of enstatite and orthoclase, and the apparent anomaly of the coexistence in equilibrium of the undersaturated minerals forsterite and leucite with liquids that are oversaturated with respect to silica.

Clearly, the crystallization temperatures of these pure mineral species can be related only in a complex way to magma temperatures. Presumably, the interplay of the several major and other constituents within a natural magma lowers melting points drastically; this is apparent from consideration of the following simple systems.

TABLE 6.3

Melting points (°C) of the pure end-members of some natural silicate melts

Forsterite	1,880	
Quartz	1,713	
Enstatite	1,557	(producing a slightly oversaturated liquid + a few per cent of forsterite crystals)
Anorthite	1,553	
Diopside	1,391	
Fayalite	1,200	
Orthoclase	1,150	(producing an oversaturated liquid + leucite crystals)
Albite	1,090	

6.3.4. The system diopside—anorthite, a binary system with a eutectic

This is a convenient starting point as a common augite not far removed from diopside in composition together with a calcic plagioclase, usually of labradorite composition, are the essential minerals of basalt, the commonest extrusive igneous rock. Furthermore, the system illustrates a eutectic, a very common phenomenon. The discussion of this and the following systems will also serve to define the terms commonly used in phase diagrams of this type.

Pure diopside crystallizes under equilibrium conditions at 1391°C. A melt composed of a mixture of 90 wt.% diopside (Di) and 10 wt.% anorthite (An) begins to crystallize at 1370°C, the crystals formed being pure diopside; the amount of diopside crystals formed at this temperature is very small but more diopside crystallizes with falling temperature. Similarly, melts composed of mixtures of diopside and anorthite by weight in the ratios 80:20, 70:30 and 60:40 begin to crystallize pure diopside in small amounts at temperatures of 1345°, 1315° and 1275°C, respectively.

Considering now the more anorthite-rich compositions, pure anorthite crystallizes at 1553°C. A 90 wt.% An—10 wt.% Di mixture begins to crystallize small amounts of pure anorthite at 1520°C. Similarly, melts composed of mixtures of anorthite and diopside in the ratios 80:20, 70:30, 60:40 and 50:50 begin to crystallize, again only anorthite in small amounts, at temperatures of 1490°, 1450°, 1400°, and 1330°C, respectively.

It seems, therefore, that the presence in the melt of increasing amounts of the one mineral component progressively lowers the melting point of the other mineral component; each, it will be noted, has a dissimilar crystalline structure. Plotting of these values and values determined from other intermediate compositions (Fig. 6.6) shows a smooth curve known as the **liquidus**. A liquidus temperature is defined as the temperature at which under

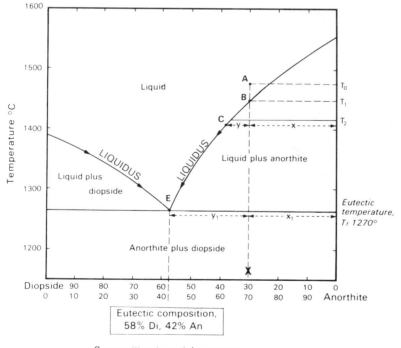

Fig. 6.6. The system diopside—anorthite at 1 bar (after Bowen, 1915, p.164, fig. 2).

equilibrium conditions crystallization begins from a melt of specified composition*. Where there is, as here, a spectrum of compositions to be considered, the liquidus becomes an envelope of varying crystallization temperatures with composition, separating a liquid field from lower-temperature fields of liquid + crystals in the appropriate phase diagram.

The two converging parts of the liquidus in the system diopside—anorthite intersect at a point corresponding to a composition of 58 wt.% diopside and 42 wt.% anorthite and a temperature of 1270°C. A melt of this composition will crystallize diopside and anorthite together in the same proportions as itself, namely 58 wt.% Di and 42 wt.% An, and furthermore, unlike other mixed compositions, will complete crystallization at the one fixed temperature of 1270°C. This special composition is called a **eutectic composition** (eutectic means literally "well-built") and the temperature is called the **eutectic temperature.** No liquid can exist in equilibrium below this eutectic temperature for any composition in the system. For any given pressure, the phase rule shows that the coordinates of the eutectic are fixed, as it is an invariant point:

$P + F = C + 2$ (phase rule)   (6.1a)
$P + F = C + 1$ (where pressure already specified, here atmospheric)
$P = 3$ (liquid, diopside crystals, anorthite crystals coexisting)
$C = 2$ (anorthite, diopside)

Therefore:

$F = 0$

It is worthwhile considering the sequential behaviour of a progressively cooling wholly liquid melt in the general case where it does not have the composition of a pure compound or that of the eutectic, but a composition, say, of X and at an initial temperature T_0. The melt, originally at point A, will start to crystallize anorthite at the appropriate liquidus temperature T_1, point B in Fig. 6.6. Further cooling to a temperature of say T_2 will result in further crystallization of anorthite and a consequent shift in the remaining melt composition towards a more diopside-rich composition, C. A **tie-line** links the composition of the liquid (here C) with the composition of the solid phase (here anorthite) in equilibrium at a specified temperature (here T_2). The relative proportion of anorthite crystals and melt at temperature T_2 is necessarily related to the original bulk composition of the closed system, X; it is a simple arithmetical procedure to show that the proportions are in fact related by the useful **lever principle.** In the specific case under discussion, the proportion of liquid of composition C to anorthite at T_2 is given by the ratio of the length x to length y measured along the tie-line.

---

*The liquidus temperature can also be defined as the temperature above which no solid phase can exist in a system of specified composition.

With falling temperature, it is obvious that the liquid composition must follow the liquidus curve in the direction of the arrow. Crystallization at C, for example, results in crystallization of anorthite — this causes a resultant shift in liquid composition to the left, where the liquid would be above its liquidus — further fall in temperature causes it to intersect the liquidus, resulting in further crystallization of anorthite — and so on in a circular infinitesimal calculus like reasoning. The liquidus in fact is a univariant line; application of the phase rule as above but to a system with two phases (anorthite crystals + liquid) shows one degree of freedom, i.e. for any given temperature which may vary between the melting points of the pure compounds and the eutectic, the composition must be fixed.

Given any liquid starting composition (except that of either pure diopside of anorthite), loss of heat from the system will cause crystallization and eventually result in a residual liquid of eutectic composition, E, plus crystals of either diopside or anorthite (for our starting composition X, crystallization will lead to a liquid of eutectic composition plus anorthite crystals in the proportion $x_1:y_1$). With further loss of heat, the liquid of eutectic composition itself crystallizes in the eutectic proportions at the constant eutectic temperature and is all used up.

$A-B-C-E$ is the **crystallization path** of the liquid. A horizontal line through E in this phase diagram is the **solidus** defined as the locus of temperatures below which the system is entirely solid for given compositions under equilibrium conditions. T_1-T_E is the **liquidus—solidus temperature interval** (for composition X) over which crystallization occurs; all intervening temperatures result, as we have seen, in varying proportions of liquid and crystals.

If the liquid, again of composition X, were left to cool by conduction through its container walls and a thermometer were inserted, the graph of temperature against time would look something like the solid line $A-B-C-E-F-G$ in Fig. 6.7 in which the points A, B, C and E correlate respectively with those in Fig. 6.6. The cooling of liquid in the temperature interval $A-B$ would be continued by the slightly curved dashed line, reflecting a gradual smooth lowering in the rate of conduction of heat from the system with falling temperature if crystallization did not intervene. However, crystallization begins at B and the rate of loss of heat is partially counterbalanced by the latent heat of crystallization of the crystallizing anorthite in the temperature interval represented by the $B-E$ portion of the liquidus in Fig. 6.6, resulting in a reduced rate of fall of temperature with time. At E, eutectic crystallization begins and is completed at the constant eutectic temperature over a time interval $E-F$, the loss of heat from the system by conduction being balanced by the latent heat of crystallization so that the temperature remains constant during this time interval. At F, the system is entirely solid and begins to cool by conduction in a regular manner ($F-G$ and on down to ambient room temperature) again uncomplicated by further crystallization.

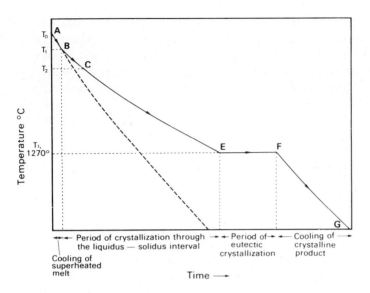

Fig. 6.7. Diagrammatic graph of temperature against time for a cooling melt of random composition in a binary system with a eutectic. See text for explanation; the lettering and specific temperatures that have been inserted refer to the same composition, X, considered in Fig. 6.6.

This procedure affords the experimenter with a quick way of determining the eutectic temperature, T_E, using any random intermediate composition.

Comparing this rather simple system with actual basaltic rocks, it will be seen that:

(a) Although still on the high side, the eutectic temperature of $1270°C$ offers a much closer approach to inferred magma temperatures than the melting points of either diopside or anorthite.

(b) The eutectic composition, 42 wt.% An and 58 wt.% Di, although a little high in colour index, is not far removed from basalt matrix mineralogy.

(c) Unless the melt happened fortuitously to be of eutectic composition, there would be a significant temperature interval over which the **fractional crystallization** of one phase would occur before eutectic composition was attained; similarly, in the more complex case of a magma, one phase could crystallize until a cotectic (see p. 179) was attained; then two phases could crystallize together cotectically until a ternary eutectic or cotectic was attained, and so on. In principle, therefore, for basaltic rocks there may be time available for the slow crystallization of one or more phenocryst phases before the more rapid crystallization of a matrix of eutectic composition. In nature the timing of eruption with respect to the liquidus—solidus interval would be a determining factor in phenocryst/matrix proportions and number of phenocryst phases present, but in general, different basalts might be expected to have differing phenocrysts but a more convergent matrix mineralogy.

6.3.4.1. Fractionation. If early forming crystals, intratelluric primocrysts for example, *were somehow separated from the magma,* the resultant melt composition would change to that of a lower-temperature liquidus composition than that of the original magma that crystallized the primocrysts, and would eventually crystallize to give a product with a composition different to that of the original magma. This process, known as **fractionation** (syn. **crystal fractionation**), could be, and indeed is believed to be, a most effective mechanism for producing variation in resultant igneous rock compositions. Note that the process of *fractionation* depends on the prior *fractional crystallization* of phases which do not have the same composition as the crystallizing liquid, but the two terms are *not* synonymous. The compositions of aphyric lavas or hypabyssal rocks or the groundmass compositions of porphyritic fine-grained rocks that are closely associated in space and time may thus indicate the **crystallization path** (syn. **liquid line of descent**) of a parental magma, initially usually mafic, undergoing fractionation.

The opposing concept of **equilibrium crystallization** supposes that the system is closed, that crystals and liquid remain together throughout the crystallization process, and that reaction may take place between crystals and the cooling liquid to produce equilibrium at lower temperatures. Having regard to specific gravity differences between crystals and liquid, viscosity, and time available for crystallization within large magma chambers, it would seem that equilibrium crystallization, particularly in the less viscous basic magmas, becomes, in fact, an improbable situation. Field evidence, the final arbiter of all geological debate, indeed supports the widespread operation of fractionation processes.

6.3.4.2. Melting processes. What would happen on heating up a solid mixture, again in the general case of a composition X in our diopside—anorthite system? Melting would begin at temperature T_E and continue at this temperature while heat sufficient for the latent heat of fusion was added to the system, producing a melt of eutectic composition plus residual anorthite in the proportion x_1 of liquid to y_1 of anorthite (see Fig. 6.6). If this eutectic melt were removed in its entirety, further melting would not take place until the melting point of the residual anorthite was reached, when a second melt of pure anorthite composition would be produced. This is known as **fractional melting.** If, on the other hand, the eutectic melt remained in contact with the residual anorthite, further heating would result in a gradually more anorthite-rich melt on the liquidus curve in equilibrium with a steadily decreasing proportion of anorthite (in the proportion $x:y$ at T_2, for example), until a 100% melt of the original composition X resulted at a temperature T_1, point B. This sequence of possible events, in part or in whole, is known as **equilibrium melting.** The distinction between equilibrium melting and fractional melting in more complicated systems is particularly clearly made by Presnall (1969), who includes reference to natural basalts. For modest proportions of melt, until one of the component solid phases is exhausted, the composition of first melt will have the same eutectic composition whether produced by fractional melting or equilibrium melting.

It might be supposed that, as the mantle is essentially solid, melting has to occur there to produce primordial basaltic magma. If mantle were to a first approximation homogeneous, basalts might therefore be expected to lie at or close to some eutectic composition with compositional variations depending on the prevalence of either fractional or equilibrium melting processes and the proportion of melt produced. The general truth of this supposition is borne out by the overall similarity in composition of many basalts and their approach to a eutectic composition. However, there are significant differences between basalts, both distinct and subtle, which as we shall see would seem to be related not only to the degree of partial melting, but also the varying pressures at which this may occur. A variation in pressure, being another degree of freedom, may result in a change in eutectic composition. Superimposed on these general considerations, there are several additional factors that may affect basalt composition — a compositionally and thermally inhomogeneous mantle, the likely effect of fractionation processes operating on ascending magma at various depths prior to eruption at crustal levels, and complications due to "zone refining" and "wall-rock reaction" processes (see Chapter 13).

6.3.5. The plagioclase solid-solution series

In attempting to match basalt with the synthetic system Di—An, one obvious difference is the relatively more sodic composition of the plagioclase in naturally-occurring basic rocks. For example, bytownite occurs as an early cumulus mineral in layered basic intrusions and as phenocrysts in some basalts and andesites; labradorite is a common matrix composition in many basalts and diabases; andesine occurs in some classes of basaltic rocks such as hawaiite, even oligoclase in mugearite. The plagioclase system is therefore of considerable interest to petrologists. To a first approximation, it can be said that natural plagioclases in igneous rocks exhibit a continuous range in optical properties, cell size, and recorded compositions, i.e., a **solid-solution series** (syn. **crystalline-solution series**). This is not unexpected as albite and anorthite are not structurally dissimilar as are diopside and anorthite. What then is the pattern of crystal—melt equilibria in the plagioclase system?

It has already been noted that pure anorthite melts (and also crystallizes under equilibrium conditions) at 1553°C, and pure albite at 1090°C. A liquid melt of 50:50 composition by weight begins to crystallize at 1450°C, producing a small amount of crystals of composition An_{81}*; a solid crystalline

---

*The useful notation An_{81} denotes a composition of 81 wt.% anorthite, and correspondingly 19 wt.% albite in the pure An—Ab system. In naturally occurring plagioclases, however, there are also significant amounts of K_2O and Fe_2O_3.

material, again of a 50:50 composition, begins to melt at 1285°C, producing a small quantity of a liquid of composition An_{14}! The appropriate tie-lines linking these and other respective liquidus and solidus compositions are incorporated in Fig. 6.8, which shows the complete experimentally determined liquidus and solidus for the plagioclase solid-solution system. Two salient contrasts to the Di—An system are: (1) the absence of a eutectic, rather a continuous fall in liquidus temperature from anorthite to albite, the two end-members; and (2) a continuum in solidus compositions, rather than the crystallization of pure end-members or of eutectic mixtures. What are the implications of this diagram in interpreting igneous rocks?

If *equilibrium crystallization* obtained, a cooling liquid of An_{50} composition would begin to crystallize at 1450°C, producing crystals of An_{81} composition; then, with falling temperature, the liquid composition would move down temperature along the liquidus in the direction of the arrow, *and* a similarly progressively changing solidus composition would necessitate continuous reaction of the first formed crystals with the liquid in order that equilibrium be maintained. For example, beginning again with an An_{50}—Ab_{50} composition, and applying the lever principle to tie-line compositions, at 1400°C there should result x wt.% of crystals of composition An_{72} and y wt.% of liquid of composition An_{35}. Similarly at 1350°C, the same 50:50

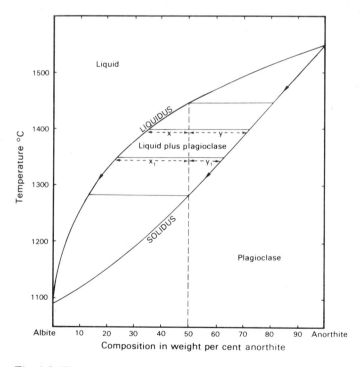

Fig. 6.8. The system anorthite—albite at 1 bar (after Bowen, 1913, p. 583, fig. 1).

compositional mixture would consist of x_1 wt.% of crystals of composition An_{63} and y_1 wt.% of liquid of composition An_{23}. Finally, at $1285°C$, all the liquid would be used up, leaving a crystalline product of composition An_{50}. The continuously changing composition of the solidus crystals necessitated by this model would have to be effected by diffusion processes within the crystals, processes that are slow in the crystalline state.

Suppose this does not happen, and that the existing crystals in the crystallizing liquid offer a ready growing site for further crystal growth at a rate faster than diffusion processes can re-equilibrate their composition. Then the crystals will be compositionally zoned from a relatively anorthite-rich, more calcic, core to a relatively albite-rich, more sodic, rim — the well-known *normal zoning* commonly displayed by igneous plagioclase. **Normal zoning** is defined in the general sense as compositional variation in a solid-solution series from a relatively high-temperature crystallization product to a relatively low one as crystallization proceeds. The effective selective removal from the system of relatively anorthite-rich crystals protected from reaction by an armour of more sodic outer zones again results in visible evidence of *fractional crystallization* of the liquid and the last crystalline product will correspondingly be more albite-rich than An_{50}. Fractionation can thus be effected by the physical removal of early formed crystals in the plagioclase solid-solution system as in the case of the Di—An eutectic system*.

Phase relations in the olivine solid-solution series are very similar to those in the plagioclase series, resulting in a comparable normal zoning of natural olivines in basic rocks from relatively Mg-rich cores to Fe-rich rims, more generally noticeable in alkali basalts for reasons that are explained in Chapter 7.

Students sometimes wonder why normal zoning is easily seen in plagioclase but not so in olivine. The reason has to do with the lower triclinic crystal symmetry of plagioclase, wherein the optical indicatrix can and does lie at varying angles to the crystallographic axes for different plagioclase compositions. Hence, such an obvious parameter as extinction measured to the fixed composition plane of ubiquitous albite twinning varies with composition and is the basis of the familiar Michel-Lévy and Carlsbad-albite twin methods of optically determining plagioclase compositions. In orthorhombic olivine, on the other hand, the optical indicatrix is necessarily fixed with respect to the

---

*Notice again the difference in connotation between *fractional crystallization* and *fractionation*. Normally zoned plagioclase crystals within a rock that had formed from a magma that had crystallized as a compositionally closed system would constitute an obvious example of the *chemical process* of fractional crystallization. If, on the other hand, early formed plagioclase crystals of a relatively anorthite-rich composition had been *physically removed* from a crystallizing magma, thus effecting a change in its bulk composition and that of its eventual crystallized rock product, this would constitute fractionation. Again, if a porphyritic basalt had a large proportion of *unzoned* plagioclase phenocrysts, this would strongly suggest that their presence is attributable to accumulative fractionation, *not* to fractional crystallization.

crystallographic axes because of the orthorhombic symmetry, so that any difference in extinction position of zoned segments of an olivine crystal is precluded. What does vary significantly in olivine, however, is birefringence, so that the relatively fayalite-rich rims of normally-zoned olivine phenocrysts found in alkali basalts characteristically show a higher birefringence than a more forsterite-rich core.

6.3.6. The ternary system diopside—anorthite—albite

This system which combines the first two studied should result in a yet closer approximation of a simple synthetic system to a natural basalt.

How can compositions and liquidus temperatures conveniently be shown on a planar phase diagram for a three-component system? Compositions can be represented as points within an equilateral triangle; it can be demonstrated very simply geometrically that for any point within an equilateral triangle (Fig. 6.9), $a + b + c$ = constant h, the distance from any apex to the opposite side. If we scale h to be 100, then a, b and c can represent the percentage amounts by weight of components A, B and C, respectively, and the point thus derived uniquely represents this composition. Note that two values only

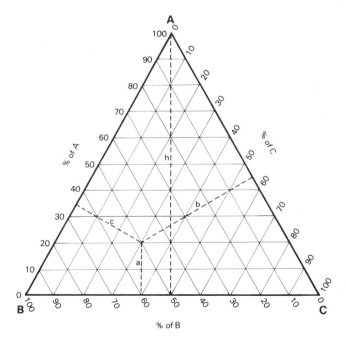

Fig. 6.9. Diagram to illustrate the plotting of the relative amounts of three components within an equilateral triangle. The point X uniquely defines a composition of 20 wt.% of component A, 50 wt.% of component B and 30 wt.% of component C.

serve to define the point; the third serves as a useful check to help eliminate plotting errors. If temperature were considered as a perpendicular axis arising vertically out of the plane of the triangle, liquidus temperatures for various compositions would form a surface above the triangle. Now, if this liquidus surface were contoured in appropriate temperature intervals, it could easily be shown in the plane of the triangle just as surface relief is shown by contours in a topographical map. Of course, the smooth contours are constructed, to pursue the topographical analogy, from selected "spot heights", each in our phase diagram resulting from an individual observation of equilibrium liquidus temperature in a series of experiments on melts of differing compositions.

Projected in this way, Fig. 6.10 portrays the system Di—An—Ab. It incorporates the eutectic E_1 between diopside and pure anorthite that we have already considered, and another eutectic E_2 between diopside and pure albite, that lies very close to albite in composition. Just as in the plagioclase

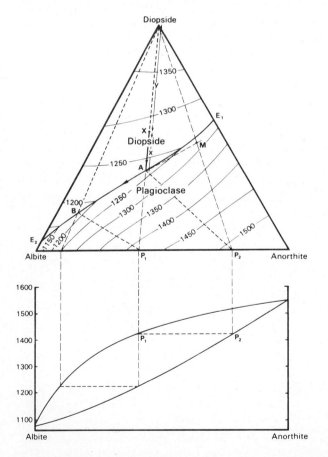

Fig. 6.10. The system diopside—anorthite—albite at 1 bar (after Bowen, 1915, p.167, fig. 5).

solid-solution series, where the liquidus temperature varies continuously with composition between anorthite and albite, so E_1 and E_2 are linked by a continuous line, representing a steady fall in liquidus temperature. This line separates two fields on the diagram, an upper diopside field where diopside is the first phase to crystallize from a cooling melt at liquidus temperatures, and a lower plagioclase field where a plagioclase solid solution is the first to crystallize.

Consider the crystallization path of a melt initially above liquidus temperature and of a composition represented by the point X. On cooling, the liquidus is intersected in the diopside field at a temperature of 1280°C. Applying the lever principle, crystallization of diopside will cause the remaining liquid to change in composition in a straight line away from the diopside apex until it intersects the line E_1—E_2 at A, at which point the system will consist of x wt.% diopside crystals and y wt.% liquid at a temperature of 1230°C. Any further crystallization of diopside would put the liquid in the plagioclase field, whereupon crystallization of plagioclase would bring the composition back to the line E_1—E_2, along which it is thus apparent, again in an infinitesimal calculus like manner of reasoning, that both diopside and plagioclase must in fact crystallize together. The line E_1—E_2 is a **cotectic line**; **cotectic crystallization** therefore connotes the simultaneous crystallization of two (or more) phases in a system in which the crystallizing liquid is undergoing continuous change in composition and temperature. A solid-solution series, as here, does not necessarily have to be involved in cotectic crystallization. The composition of the plagioclase crystallizing at A can be crudely estimated as follows (assuming that the presence of a component of diopside has no effect on the An—Ab system, an assumption which is not strictly tenable as diopside contains CaO, itself a component of anorthite but not of albite): the melt at A can be considered as a mixture of diopside and plagioclase of composition P_1; in the pure An—Ab system the latter would crystallize plagioclase of composition P_2, as indicated by the appropriate tie-line. The crystallizing melt at A, therefore of a cotectic composition, is linked by two tie-lines to the two crystallizing phases, diopside + plagioclase of composition P_2. Furthermore, these two phases must crystallize in the proportion represented by a bulk composition M, found by the intersection of the line Di—P_2 with the tangent to the cotectic at A. This must be so as M represents the only proportion of diopside + plagioclase of composition P_2 that can crystallize from melt A and send the liquid, for reasons given above, along the cotectic. The removal of crystals of bulk composition M drives the crystallizing liquid along the cotectic line in the direction of the arrow, i.e. down the "thermal valley" in which the cotectic line lies. Under equilibrium conditions, crystallization of diopside and plagioclase (accompanied by continuous re-equilibration of plagioclase compositions) proceeds until the system becomes a totally crystalline mixture of diopside + plagioclase of composition P_1, by which time the liquid, again following our simplified illustrative reasoning with reference to the simple An—Ab system, will have reached composition B and have a temperature of 1180°C. Removal of the first forming plagioclase crystals from the system or their failure to equilibrate during further crystallization will of course result in the liquid composition moving further along the cotectic towards E_2 than B. The actual tie-lines linking compositions of crystallizing plagioclase with melt compositions in the plagioclase field or on the cotectic can, of course, be determined by experiment and analysis; they do, in fact, lie pretty close to the ones predicted by the simplified treatment above.

The position of the diopside—plagioclase cotectic relates quite well to rocks of the basalt family. Lunar basalts, for example, in which the plagioclase is of high anorthite content have high colour indices. Many terrestrial basalts have a colour index somewhat over 40, corresponding to a similar value obtained from the cotectic curve for plagioclase of around An_{65}, which is close to the observed labradorite composition of many basaltic groundmass plagioclases. Within the alkali basalt series, rocks such as hawaiite and mugearite have colour indices ranging down to 25, again corresponding with the more albitic plagioclase that they contain and in line with an origin by fractionation following a cotectic comparable to that of our simple ternary system.

Note also how liquidus temperatures are considerably depressed and condensed when compared with the pure plagioclase system: a liquidus temperature range of 1553°—1090°C in the plagioclase system contrasts with the range of 1270°—1085°C of the diopside—plagioclase cotectic. This is a yet closer approach to observed basalt eruption temperatures, and incidentally also allows considerable compositional variation by fractionation over a relatively small temperature interval.

6.3.7. Other simple systems relevant to an understanding of the crystallization of mafic magmas

Considering first binary systems with intermediate compounds, two common situations are illustrated by the nepheline—silica and forsterite—silica systems, respectively. Each system contains a saturated silicate mineral, albite or enstatite, and an undersaturated silicate mineral, nepheline or forsterite.

6.3.7.1. The system nepheline—silica.
Silica, nepheline, and the compositionally intermediate compound albite melt congruently at 1713°, 1524°[*] and 1090°C, respectively. Simple eutectics divide nepheline from albite and albite from silica as shown in Fig. 6.11.

---

*This temperature is actually the melting point of a polymorph of nepheline, carnegieite. To avoid using unfamiliar high-temperature polymorph names, familiar mineral names are often retained in phase diagrams with an implied compositional notation, although use of a chemical name or formula would obviate any ambiguity. Silica, for example, is a handy chemical name and is thus preferable to quartz in this context.

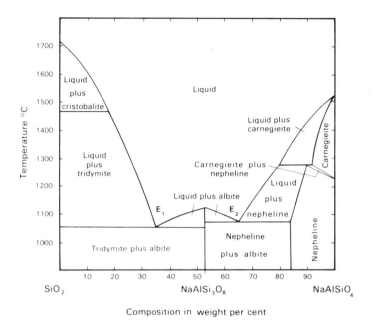

Composition in weight per cent

Fig. 6.11. The system nepheline—silica at 1 bar with the intermediate compound, albite (slightly simplified, in part after Grieg and Barth, 1938, p.94, fig. 1).

In geological terms it would appear therefore that fractionated melts poor in anorthite component might trend towards either a decisively oversaturated end-member or an undersaturated one, corresponding to these two eutectics, respectively. Rhyolites, peralkaline rhyolites, and phonolites do indeed characterize fractionated end-members of the tholeiitic, transalkaline, and basanitic series, respectively. Many alkali basalts, however, are associated with trachytes, generally presumed to be fractionation products, that are not far from being just saturated with silica (see also Section 6.3.9 of this chapter).

6.3.7.2. The system forsterite—silica. This system is of considerable interest to igneous petrologists as it affords an insight into the crystallization behaviour of tholeiitic magmas at high crustal levels. Whereas silica and forsterite melt congruently at 1713° and 1880°C, respectively, enstatite (or more strictly speaking the monoclinic polymorph clinoenstatite at the temperatures in question) melts **incongruently** at a temperature of 1557°C to give a large proportion of a slightly oversaturated liquid of composition R (i.e. not of the same composition as the enstatite) plus a much smaller proportion of residual solid forsterite crystals, as shown by the solid tie-line at this temperature in Fig. 6.12. There is a eutectic E between a melt of this composition and silica. Melts of more magnesian compositions have progressively higher liquidus temperatures up to the crystallization temperature of pure forsterite

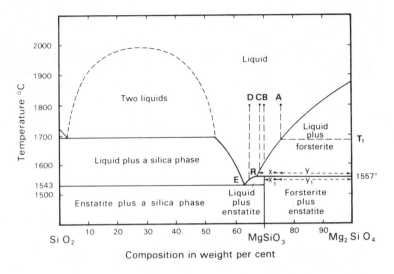

Fig. 6.12. The system forsterite—silica at 1 bar with the intermediate compound, enstatite, showing the incongruent melting of enstatite at 1557°C (after Bowen and Andersen, 1914, pp. 487—500; and Grieg, 1927, p.15, fig. 3).

as shown. Melts of highly siliceous compositions in this system display a field of immiscibility between two liquids that is imprecisely known in detail and is therefore indicated by a broken line in Fig. 6.12.

Let us consider briefly the crystallization path and crystalline products of different melts in this system, broadly comparable to mafic igneous rock compositions, beginning with an undersaturated melt of composition A. This will begin to crystallize forsterite at T_1, and continue to crystallize forsterite until the liquid reaches R at a temperature of 1557°C, at which point, by application of the lever principle, x wt.% of forsterite crystals will coexist in equilibrium with y wt.% of liquid of composition R. With further loss of heat from this system and if equilibrium is maintained, the reverse reaction of the incongruent melting of enstatite occurs: liquid of composition R reacts with the forsterite crystals to form enstatite, resulting in complete crystallization at this temperature to produce a mixture of x_1 wt.% forsterite crystals and y_1 wt.% enstatite crystals, again necessarily related by the lever principle to the initial composition A.

If, on the other hand, forsterite had been fractionated, that is to say removed from the system, for example by sinking, the liquid at R would have no forsterite to react with, and would commence to crystallize enstatite in place of forsterite, eventually reaching a eutectic composition E at a temperature of 1543°C, where the system would finally solidify as a mixture of enstatite and quartz. Under conditions of fractionation, an undersaturated melt A, a just saturated melt B, a slightly oversaturated melt C (but not a more oversaturated melt D), could each in this way first crystallize the undersaturated mineral forsterite, followed by the eventual crystallization of an oversaturated product containing quartz.

This reaction of early formed olivine to orthopyroxene is the mineralogical hallmark of the large group of basaltic rocks known as tholeiites which in nature indeed comprise both undersaturated and oversaturated parental magma varieties. The cumulus mineral sequences of such classic intrusions as the Bushveld and Stillwater complexes among layered basic intrusions, the mineralogies of Palisades Sill rocks among differentiated diabase sills, and the presence of olivine phenocrysts in tholeiitic basalts that may contain an intersertal oversaturated glass, all illustrate this. In the latter case, equilibrium was not attained for kinetic reasons — the tholeiitic basalt on extrusion cooled too quickly for olivine to react with the enclosing liquid; in the intrusive rocks olivine was in fact fractionated by gravity settling. As an illustration of this distinctive trait of tholeiites, analysis 4 of Table 9.2 shows ~2 wt.% of normative quartz in a submarine Hawaiian tholeiite which modally consists of 2 wt.% forsteritic olivine phenocrysts (plus a few weight per cent of plagioclase and augite phenocrysts) in a glassy matrix. Furthermore, analyses 5—9 of the same table show the progression in a Kilauean tholeiite series from olivine-normative tholeiite to more evolved tholeiites which contain increasing amounts of normative quartz.

Although a discussion of this phase diagram began with mention of the incongruent melting of enstatite, igneous petrologists have classically been more concerned with crystallization processes rather than with melting. The point R therefore where the crystallizing liquid reacts with first formed forsterite is generally known as a **reaction point**. Like a eutectic, it is an invariant point where three phases may coexist in a two-component system at a predetermined pressure (here atmospheric). The reaction of liquid plus forsterite to produce enstatite is exothermic, so that if the temperature of a forsterite—silica system that was steadily losing heat by conduction were monitored in the same manner as that described above for a cooling mixture of diopside and anorthite, a flattening of the temperature—time graph would be observed at the reaction point, again paralleling the behaviour of a eutectic. However, given an appropriate starting composition, e.g., that of C in Fig. 6.11, crystallization would not end at the temperature of R but proceed towards E, so in this observational feature and, of course, from the evidence of igneous rocks and experiment, the reaction point obviously differs from a eutectic. It should be noted, however, that the descriptive term reaction point generally employed by petrologists is also known to physical chemists as a **peritectic**.

6.3.7.3. The ternary system forsterite—diopside—silica. This portrays the crystallization from appropriate melts of olivine and two pyroxenes, a calcium-rich one and a calcium-poor one, the compositions of which are complicated by a degree of possible solid solution, indicated by the continuous lines along the dashed line joining (clino)enstatite and diopside (Fig. 6.13). The reaction point R_1 and the eutectic E_1 in the forsterite—diopside—silica system correspond with R and E in Fig. 6.12, respectively. A reaction relationship holds along the line R_1—R_2 where olivine is resorbed and clinoenstatite precipitated, E_1 is joined by a cotectic line to a second eutectic E_2 between diopside and silica. A third eutectic E_3 occurs between forsterite and diopside. Paths of cotectic crystallization are indicated by arrows. Note the thermal high at A. The field of olivine crystallization at liquidus temperatures occupies the area Fo—R_1—R_2—E_3; the (clino)enstatite (solid solution) field, the area R_1—E_1—B—R_2; the silica phase field, the area E_1—SiO_2—E_2—B; and the diopside (solid solution) field, the area Di—E_3—A—R_2—B—E_2. Given this information, crystallization paths for melts of compositions contained in the diagram can readily be traced. The compositions of rocks of tholeiitic affinity which will first crystallize olivine at liquidus temperatures (and that may be, as we have seen above, either oversaturated, just saturated, or undersaturated in composition with respect to silica) can be envisaged as lying in the area Fo—R_1—R_2—A. Note the existence of a more undersaturated area Fo—A—E_3 where early crystallization of olivine leads the liquid to the line A—E_3 where olivine and diopside crystallize together. This offers, in contrast

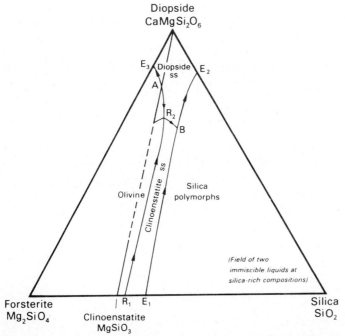

Fig. 6.13. The system forsterite—diopside—silica at 1 bar (after Bowen, 1914, p.217, fig. 6; Kushiro, 1972b, p.1263, fig. 1).

to the tholeiites, an analogy to the group of alkali basalts in which there is no reaction of an olivine to an orthopyroxene and in which no calcium-poor pyroxene is present.

6.3.7.4. The basalt tetrahedron. Following Yoder and Tilley (1962, see pp. 349—352 for rationale), this system of four components, forsterite, diopside, nepheline, and quartz, represented by the four corners of a tetrahedron respectively, introduces nepheline (and albite) to the components (and compounds) considered in Fig. 6.13, and offers an easy conceptual analogy to natural basalt groups (Fig. 6.14). Basalt compositions can be represented within the tetrahedron by plotting relative amounts of the necessarily restricted combinations of normative Q, en, fo, di, ab, and ne. No attempt is made in this diagram to show the curved three-dimensional bounding surfaces between the fields of initial crystallization of different phases. Two-dimensional representations can be made by projecting points onto a plane within the diagram, e.g., projecting from the diopside apex onto the triangle ab′—fo′—en′, for example. Within the tetrahedron, the important plane Di—Ab—Fo is a critical *thermal divide* that separates alkali basalts and more alkaline mafic rocks from tholeiites in much the same way as does the line Fo—A in Fig. 6.13. The plane Di—Ab—En is the *plane of silica saturation* that divides

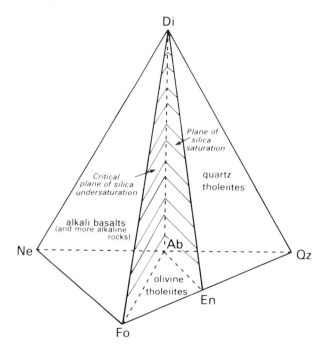

Fig. 6.14. The "basalt tetrahedron" (after Yoder and Tilley, 1962, p.350, fig. 1). See text for explanation.

compositions that are technically oversaturated and undersaturated with respect to silica, and merely divides olivine tholeiites from quartz tholeiites. It can be seen that cutting of the large tetrahedron along these two planes defines three tetrahedral wedges that are correlatable with the compositions of: (1) *alkali basalt* that would remain undersaturated whatever proportions of forsterite, diopside, albite, or nepheline were to crystallize from it, (2) *olivine tholeiite*, and (3) *quartz tholeiite*; the two latter are both capable of fractionation to give an oversaturated product within the quartz tholeiite field by early fractionation of a proportion of olivine. The diagram shows how alkali basalt is a good descriptive term as alkali basalt (abbreviated from the former term "olivine alkali basalt") is, in terms of this diagram, relatively richer in albite and nepheline components than is tholeiite. To a first approximation also occurring in nature, tholeiites are hypersthene-normative whereas alkali basalts are nepheline-normative. An old somewhat vague term "olivine basalt" is no longer used in modern igneous rock series nomenclature as it could be taken to refer equally to alkali basalt or olivine tholeiite, both of which contain olivine, but which are the progenitors of two distinct fractionation sequences.

6.3.7.5. Fractionation trends of olivine and pyroxene. With iron-rich compositions there is no field of pyroxene crystallization separating the fields of olivine and a silica mineral; iron-rich olivines thus crystallize cotectically with a silica mineral eventually leading to a fayalite—silica eutectic (Fig. 6.15).

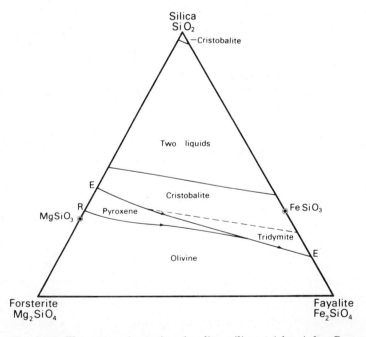

Fig. 6.15. The system forsterite—fayalite—silica at 1 bar (after Bowen and Schairer, 1935, p.159, fig. 5).

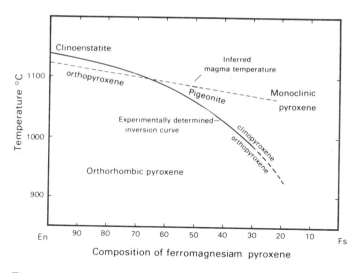

Fig. 6.16. Diagram showing the fields of stability of monoclinic and orthorhombic poly-morphs of ferromagnesian pyroxene (after Bowen and Schairer, 1935, p.164, fig. 8). Inferred temperature of crystallization of fractionating subalkaline mafic magma inserted to show the changeover from crystallization of orthopyroxene to pigeonite at about En_{70}.

Fayalitic olivine and quartz (or tridymite) are, of course, commonly found together in the highly-fractionated late crystallizing residua of tholeiitic diabase sills and layered basic intrusions. Fayalite and quartz are also found in some epizone bodies of granitic composition and occur as phenocrysts in some rhyolites that also quite possibly may have had a history of fractional crystallization.

Mention should also be made of the (experimentally determined) inversion temperatures related to (inferred) magmatic crystallization temperatures in the ferromagnesian pyroxene series (Fig. 6.16). From field evidence provided by fractionated tholeiitic intrusions, it is apparent that early crystallization of primocrysts of a magnesian orthopyroxene (succeeding the still earlier crystallization of magnesian olivine) is consistently followed by the later crystallization of a more iron-rich monoclinic pigeonite. The point where the two curves intersect in Fig. 6.16 provides a theoretical indication of magmatic temperature at this stage of fractionation*. In contrast to the low CaO con-tent of orthopyroxene formed under these conditions (~3 wt.% Wo compo-nent), pigeonite contains appreciable CaO (~8—10 wt.% Wo component), as the larger-sized calcium atom can be more easily accommodated in the monoclinic lattice than in the orthorhombic one. Pigeonite may persist metastably in lava flows and in fairly quickly cooled intrusions, such as, for

---

*For a more elegant treatment of this taking account of liquidus—solidus relationships in naturally occurring magmas, see G.M. Brown et al. (1957, pp. 530 and 541).

example, the Palisades Sill, but in larger intrusions such as the Bushveld layered intrusion there is time during cooling for equilibrium to be attained and the pigeonite to invert to the stable orthorhombic form; this inversion is accompanied by the copious exsolution of augite which, of course, is relatively richer in the Wo component that was originally present in crystalline solution in the pigeonite and that cannot all be incorporated in the orthopyroxene formed by inversion.

For practical purposes common igneous pyroxenes of non-peralkaline rocks can be thought of as falling in the lower trapezium-shaped area of Fig. 6.17 (igneous wollastonite is known but is extremely rare). Average compositional trends of pyroxenes produced under equilibrium conditions during fractional crystallization of tholeiitic magmas are indicated by arrows in Fig. 6.17. This is not a phase diagram, but the result of plotting analyses of actual cumulus crystals from large layered intrusions. There are two well-defined series of pyroxenes, a calcium-rich one and a calcium-poor one linked by tie-lines, indicating a limited degree of solid solution between a pure ferromagnesian pyroxene and a pure diopside—hedenbergite range in composition. Groundmass pyroxenes of volcanic rocks, however, commonly have intermediate, presumably metastable, compositions reflecting crystallization under quench conditions. This widely used diagram does not allow for Na, Ti, Al and Fe^{3+}, often present in significant amounts in pyroxenes.

The two ends of the pigeonite trend in the calcium-poor series reflect, as we have seen above, the change from the stability field of ortho- to clino-pyroxene during crystallization, often around En_{70}, and the later cessation of crystallization of ferromagnesian pyroxene in favour of fayalitic olivine, at around En_{30}.

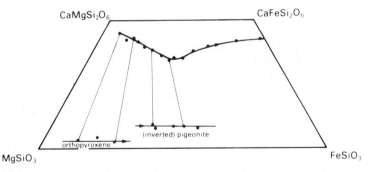

Fig. 6.17. Compositions of cumulus pyroxenes from the Bushveld and Skaergaard intrusions. *Tie-lines* link coexisting pairs of calcium-rich augites and ferroaugites above to calcium-poor orthopyroxenes and pigeonites below. *Arrows* show trend of compositional variation with fractionation and increasing height in the intrusion (after G.M. Brown et al., 1957, p.519, fig. 3; G.M. Brown and Vincent, 1963, p.181, fig. 3; and Atkins, 1969, p.239, fig. 3).

In the far less common alkali basalts, fractionation leads, as we have seen indicated by the experimental work, to the cotectic crystallization of olivine and clinopyroxene. With falling temperature, both phases become more iron-rich and there are no complications introduced by ferromagnesian pyroxene species which are conspicuously lacking from alkali basalt series.

6.3.7.6. Overall fractionation trends of mafic magmas. The last system we shall consider in connection with basaltic rocks is the ternary system for-sterite—diopside—anorthite (Fig. 6.18). This contains a *ternary eutectic, E*, and a significant field of primary crystallization of spinel, not a true inter-mediate compound between forsterite and anorthite as it contains no lime or silica. Consider the crystallization of a melt initially above liquidus tempera-ture and of composition X. The sequence of fractional crystallization would be forsterite (during liquid path X—B), followed by forsterite + spinel (along cotectic B—C), followed by forsterite + anorthite (along the cotectic C—E) and, finally, eutectic crystallization of forsterite + anorthite + diopside at the eutectic E.

Now if we take into account that in naturally-occurring basic magmas Na will be present in addition to Ca, and Fe in addition to Mg, and if in addition we assume that the magma is of a tholeiitic affinity and will thus reach at

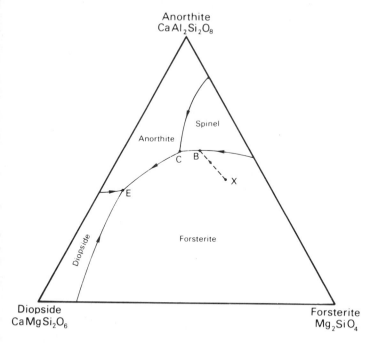

Fig. 6.18. The system forsterite—diopside—anorthite at 1 bar (after Osborn and Tait, 1952, p.419, fig. 5).

some stage the olivine to ferromagnesian pyroxene reaction boundary, we arrive at the following possible fractional crystallization sequence:

(1) Forsteritic olivine
(2) Forsteritic olivine + spinel (chromium-rich in naturally occurring rocks)
(3) Orthopyroxene + spinel
(4) Orthopyroxene + calcic plagioclase
(5) Orthopyroxene + plagioclase + augite
(6) Cotectic crystallization of increasingly iron-rich augite + increasingly soda-rich plagioclase + increasingly iron-rich pyroxene in the lime-poor series including pigeonite and the eventual replacement of the latter phase by fayalite plus silica

As will be seen from the descriptive data in Chapter 7, we thus obtain a remarkable approximation to the observed fractionation sequence of tholeiitic layered basic intrusions. An initial magma composition that, although still tholeiitic, had a lower silica activity could affect the above sequence slightly: depending on the relative timing of the cessation of spinel crystallization and the olivine > pyroxene reaction, the combination of olivine + plagioclase could replace (3) above. With parental magma of still lower silica activity, no ferro-magnesian pyroxene would ever appear during fractional crystallization, and an increasingly iron-rich olivine would crystallize cotectically with augite + plagioclase — paralleling the observed fractionation sequence of alkali basalt. With more undersaturated, highly alkaline parental magmas of very low silica activity, feldspathoids accompanying or even totally taking the place of plagioclase would occur. Of course, other important variables like pressure might be expected to affect equilibrium both in synthetic melts and in nature. A high oxygen fugacity, f_{O_2}, might be expected to raise the Fe^{3+}/Fe^{2+} ratio and hence favour early crystallization of magnetite; a high partial water pressure, P_{H_2O}, could increase the potential field of amphibole crystallization, and so on.

6.3.8. Effect of pressure on crystallization temperatures

For the great majority of substances and including all rock-forming silicates the solid is denser than the liquid at melting point temperatures*. The effect of increased pressure can therefore be predicted to raise melting-point temperatures, the actual slope of the graph of melting-point temperature against pressure being given by the Clausius—Clapeyron equation:

$$\frac{dP}{dT} = \frac{\Delta H}{T\Delta V} \tag{6.2}$$

where ΔH = latent heat. In practice, values of melting-point gradients for

---

*The familiar substance water constitutes a rare but notable exception in that the solid is *less* dense than the liquid. Skating on ponds depends therefore not only on the fact that, in the first place, ice floats on the surface of denser water, but also on the considerable pressure exerted on the ice by the skater's weight via the narrow skate blade, which *lowers* the melting point of the ice, causing it to melt transiently under the blade, i.e. one skates therefore on a thin film of water. When the skater has passed and the pressure is removed this water immediately refreezes.

silicate minerals are found to be not far from 10°C/kbar. It would seem probable that for multicomponent silicate melts liquidus temperatures would behave similarly, and at increasing depths in the lithosphere therefore one might expect liquidus temperatures to increase at a rate of ~10°C/kbar, equivalent to ~3°C/km, other factors such as eutectic compositions remaining more or less the same.

However, these considerations do not take account of the role of water in magmas and silicate melts. Water as we have seen is capable of being dissolved in increasing amounts in silicate melts as pressure increases. Although the great majority of natural magmas are probably not saturated with respect to water, their water contents are nonetheless significant. Not only is water an extra component of relatively very low freezing point but its presence also reduces polymerization and lowers viscosity markedly, and because of its low molecular weight a modest weight per cent content of dissolved water implies a much higher molecular per cent content. It follows that the effect of dissolved water under pressure on liquidus temperatures of water-saturated silicate melts is to lower them drastically; indeed this lowering more than counteracts the temperature increase that might have been expected to result from the effect of increased pressure alone. In the diopside—anorthite system, for example, already reviewed at some length (Fig. 6.6), the nett effects of increasing amounts of dissolved water under high pressures on liquidus temperatures are very marked (Fig. 6.19). Note also that not only eutectic temperatures but also eutectic compositions are greatly dependent on P_{H_2O} in this system.

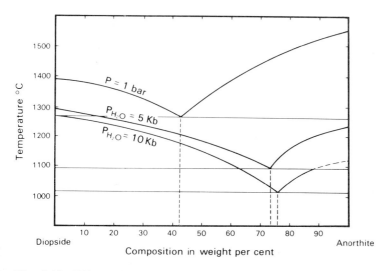

Fig. 6.19. Effect of P_{H_2O} on liquidus temperatures in the system diopside—anorthite, showing the marked shift in both eutectic temperature and eutectic composition with increasing water pressure (after Yoder, 1965).

6.3.9. Experimental systems relating to felsic rock melts

So far our illustrations of experimental work have been taken in the main from systems relating to basic rock compositions on the grounds that these are not only the commonest extrusive rocks and thus undoubtedly represent a common liquid magmatic composition, but are also the most likely "parental magmas" of fractionation sequences, to a first approximation synonymous with igneous rock series (see Chapter 8). It is appropriate now to turn to systems relating to the granite family, the most common intrusive rocks in continental crust, and including compositions that together with nepheline syenites can be regarded as "residual liquids", capable of being produced by fractional crystallization processes and thus complementary to basic rocks. However, it must be emphasized that field relations unequivocally show that by no means all granites can be interpreted to have been formed in this way. In the simplest terms, what might be a final eutectic reached after prolonged fractionation might equally well be close to a first product produced by anatexis. Within the broad spectrum of granitic rocks, the nearer the composition to granite (sensu stricto), the greater the content of quartz and alkali feldspars and the lower the colour index, so that the simple system $Or-Ab-SiO_2$ very closely approaches the composition of at least some granitic rocks.

6.3.9.1. The binary systems $Ab-SiO_2$, $Or-SiO_2$, and $Ab-Or$.

Considering first "dry" melts at 1 bar, there is a eutectic between albite and silica at 1065°C with a composition of 32 wt.% silica and 68 wt.% albite. Similarly, there is a eutectic between orthoclase and silica at 990° ± 20°C (the wide error limit here reflects the extreme difficulty of determining equilibrium crystallization temperatures in these very viscous melts) with a composition of 42 wt.% silica and 58 wt.% orthoclase. This system is complicated by the incongruent melting of orthoclase at 1150°C to produce an oversaturated liquid + leucite crystals, in an analogous manner to the forsterite—silica system. Liquidus—solidus relationships in the third binary system, orthoclase—albite reveal a **minimum**, M (Fig. 6.20), that has some of the features of a eutectic and solid solution combined and is thus a little different to anything we have so far seen. This reflects the fact that orthoclase and albite have somewhat similar crystalline structures that, however, are not as close as those of albite and anorthite (orthoclase is monoclinic, albite triclinic). The leucite field, present because of the incongruent melting of orthoclase, is drastically reduced under natural "wet-melt" conditions of P_{H_2O} (see Tuttle and Bowen, 1958, pp. 41 and 42).

Let us, therefore, consider the crystallization of a cooling melt of composition X under a modest P_{H_2O} of 1 kbar and initially above liquidus temperature. This begins to form crystals of composition S_1 at A. Under equilibrium conditions, with further cooling the proportion of crystalline material in the

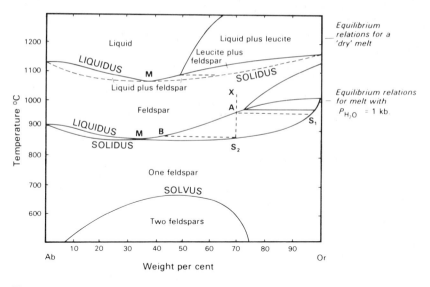

Fig. 6.20. The system albite—orthoclase showing solvus and liquidus—solidus relationships at 1 bar and at $P_{H_2O} = 1$ kbar (after Tuttle and Bowen, 1958, p.40, fig. 17).

system increases, and the compositions of the liquid and crystals change progressively to B and S_2, respectively, at which temperature the system becomes entirely crystalline with composition S_2 necessarily equaling initial melt composition X. Given fractionating conditions by the separation of crystals from the melt or the failure of early formed crystals to equilibrate with changing liquidus compositions, the liquid composition may reach M where crystals of the same composition as the liquid will crystallize. The minimum M is thus akin to a eutectic, but the system differs from a eutectic in that the early formed crystals do not have the composition of either of the pure end-member compositions. This implies, inter alia, that fractionation is not as pronounced as in a eutectic system where early fractionating phases of a more contrasted end-member composition are removed from the liquid.

6.3.9.2. The ternary system orthoclase—albite—silica. This includes the two eutectics and the minimum referred to above. The two eutectics are joined by a cotectic line dividing the fields of primary crystallization of a silica mineral and feldspar (Fig. 6.21). Note the arrows showing crystallization paths along this cotectic towards the point T. Another trend line can be drawn from M to T, representing the crystallization path with falling temperature of liquids of minimum composition in the feldspar field; this line is not strictly speaking a cotectic as only one phase is crystallizing, namely an alkali feldspar solid solution, but for purposes of tracing the crystallization path it fulfils a similar role, and divides a field of initial crystallization of a relatively albite-rich alkali feldspar from one of a more orthoclase-rich composition.

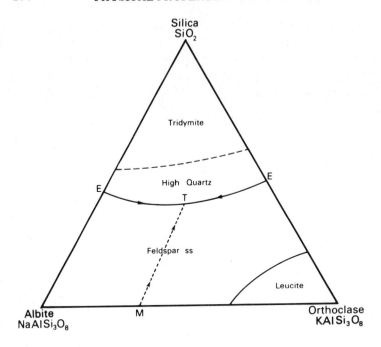

Fig. 6.21. The system orthoclase—albite—silica at 0.5 kbar water pressure (after Tuttle and Bowen, 1958, p.54, fig. 23).

The point T, where the liquidus temperature is 950°C, the lowest in the system, is called the **ternary minimum**, where alkali feldspar and a silica mineral crystallize together in the same proportions as the melt composition.

It is of some interest that this composition has the lowest liquidus temperature of any we have so far considered. This being so, prolonged fractional crystallization in natural magmas containing the components of free silica, albite and orthoclase might therefore be expected to yield finally some small proportion of liquid of this composition. This prediction is indeed realized by the intersertal glass in tholeiitic lavas, the interstitial micropegmatite found in the upper parts of some fractionated tholeiitic sills, and the acid rocks occurring in the upper parts of some fractionated layered basic intrusions of tholeiitic affinity; the prediction is realized more problematically by discrete intrusions of epigranite not physically connected with basic rock but showing geochemical indications (e.g., high concentrations of residual trace elements) of being the product of strong fractionation. Rock compositions closely approaching the ternary minimum composition are alternatively known as **ideal granite** compositions. This composition is also referred to as **petrology's residual system**, although it is but one of two such minima (see p.198).

6.3.9.3. The solvus between plagioclase and K-feldspar. A further complication within the alkali feldspars is limited solid solution with falling temperature. Under equilibrium conditions at surface temperatures, for example, albite can only contain a fraction of 1 wt.% of orthoclase in solid solution, and orthoclase only slightly more of albite in solid solution. Alkali feldspar of all intermediate compositions is thus not in equilibrium and, given time, should unmix to nearly pure albite and orthoclase. With rising temperature the field of possible solid solution increases, and above ~680°C, at atmospheric pressure, there is complete solid solution for all compositions between albite and orthoclase. The bounding curve limiting the field of solid solution and thus dividing (under equilibrium conditions) the field of one feldspar solid solution from a field of two feldspars is known as the **solvus** and is shown in Fig 6.20.

Homogeneous alkali feldspars with compositions falling more or less midway between albite and orthoclase (sanidines) do occur in nature, only because equilibrium has fought a losing battle with kinetics. That is to say lavas containing sanidine phenocrysts or microlites cooled too quickly for exsolution processes necessitating intra-lattice diffusion to have occurred. The rate of exsolution, at best very sluggish in most silicates, becomes like all chemical reactions very much slower with falling temperature so that the high-temperature solid-solution equilibrium, although technically unstable, may become frozen in consequent on quick cooling.

In older volcanic rocks, however, alkali feldspar, originally sanidine, is typically in the form of **cryptoperthite,** where exsolution on a sub-microscopic scale (but clearly shown by X-ray diffraction) has occurred with the passage of time.

Epigranites typically contain **mesoperthite,** generally a fine-grained intergrowth of albite and orthoclase in comparable proportions, formed by exsolution during cooling from a homogeneous alkali feldspar that had a composition like sanidine, typically intermediate between albite and orthoclase. In this kind of rock, exsolution textures can become quite coarse and approach an irregular mosaic of discrete grains of orthoclase and albite.

All of the above examples reflect a **hypersolvus** condition, wherein crystallization occurred at liquidus temperatures above the solvus. Rocks of this category that approach an ideal granite composition near the ternary minimum will thus contain one (original) feldspar only (±exsolution effects as noted above).

With plutonic granitic rocks of mesozonal and katazonal affinities a complication arises in that increasing pressure actually raises the solvus appreciably (to 730°C at 5 kbar; Morse, 1970) and P_{H_2O}, which may come to equal P_{total} during final emplacement and consolidation of the granite, decreases the liquidus drastically. At $P_{H_2O} = P_{total} = 5$ kbar, for example, the liquidus— solvus relationship is as shown in Fig. 6.22. Note the disappearance of the leucite field and, most significantly for feldspar textures, the intersection of

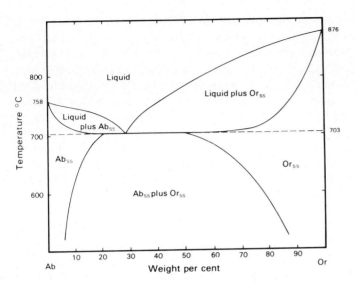

Fig. 6.22. The system albite—orthoclase—water at 5 kbar projected onto the anhydrous join (after Morse, 1970, p.234, fig. 2).

the solidus with the solvus, so that at lowest liquidus temperature two feld-spars, not one, are in equilibrium with melt. Leaving aside the question as to whether these deeper-seated granitic rocks were ever wholly liquid (and thus whether we are dealing with (1) crystallization from a liquid or (2) equilibra-tion in the solid state ± silicate pore fluid), it is clear that equilibration at these lowest liquidus temperatures, however achieved, will result in the formation of two feldspars. This is known as a **subsolvus** condition. Further subsolidus exsolution, generally solely within the orthoclase-rich feldspar which shows a greater compositional gradient against temperature than does the albite-rich feldspar, typically results in a mesozonal granitic rock contain-ing a sodic oligoclase and a slightly perthitic orthoclase (as opposed to the mesoperthitic intergrowths in hypersolvus epizone granitic rocks). At con-tacts of plagioclase with orthoclase, exsolved albite from the orthoclase preferentially forms around the convenient plagioclase nucleii thus forming the typical clear **albite rims** very noticeable where plagioclase is included in poikilitic orthoclase.

   Under katazone conditions, equilibration at still higher pressures and high P_{H_2O} (be it in an incipient silicate melt phase only) typically results in a sodic plagioclase and a non-perthitic microcline equilibrating at lower temperatures and thus further down the solvus. As well as considerations of pressure and water content, mesozonal and katazonal granites, of course, have progressively greater periods of time to equilibrate and re-equilibrate than do the relatively quickly and continuously chilled epizonal granite intrusions.

   At first sight paradoxically, but on consideration quite logically therefore, the highest-temperature feldspars have crystallized from intrusions emplaced at highest crustal levels whereas the lowest-temperature feldspars have equilibrated at lower crustal levels.

Volcanic rocks and high-level intrusive rocks of compositions significantly more intermediate than ideal granite may also contain two feldspars, a sanidine (±exsolution products) and a plagioclase in the oligoclase to labradorite compositional range. These rocks too are technically subsolvus owing to the fact that the solvus between orthoclase and plagioclase compositions occurs at significantly higher temperatures as the plagioclase component becomes increasingly more calcic than albite. The spectrum of feldspar compositions in Fig. 6.23 signals in a general way the intersection of liquidus with solvus surfaces in the feldspar system under differing conditions of P_{H_2O} at high crustal levels. Higher P_{H_2O} will shift this compositional spectrum further towards the An—Ab and Ab—Or lines, eventually as we have seen intersecting the Ab—Or line in two places in the case of the subsolvus granites of near ideal compositions.

It is apparent that for an understanding of granitic rocks the historically later experimental work at elevated P_{H_2O} is essential, thus contrasting with the effective approach to the crystallization behaviour of basic, generally water-poor, magmas afforded by the classical work on water-free systems at 1-atm. pressure.

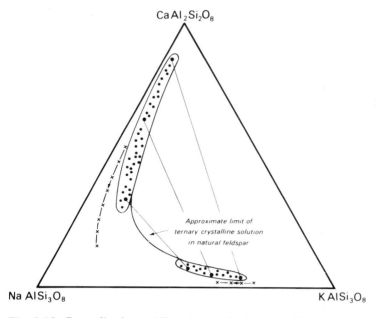

$CaAl_2Si_2O_8$

Approximate limit of
ternary crystalline solution
in natural feldspar

$NaAlSi_3O_8$ $KAlSi_3O_8$

Fig. 6.23. Generalized trend lines (*crosses*) of compositions of feldspars in rhyolites, and generalized field of compositions of feldspars in trachytes and phonolites, showing also tie-lines between coexisting sanidine—plagioclase pairs (simplified after data in Carmichael, 1962, 1963; and Carmichael et al., 1974, pp. 218—250).

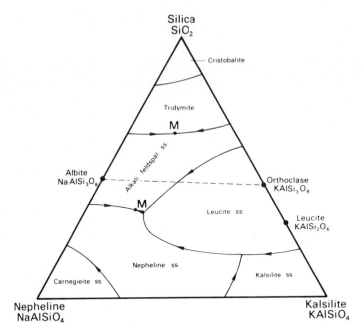

Fig. 6.24. The system silica—nepheline—kalsilite at 1 bar (after Schairer, 1950, p.514, fig. 1).

6.3.9.4. The system $NaAlSiO_4$—$KAlSiO_4$—SiO_2. The legs of the SiO_2—Ab— Or compositional triangle can be extended to the compositions of nepheline and kalsilite and thus encompass the compositional range of undersaturated felsic rocks such as nepheline syenite. Fields of crystallization in this extended compositional range are shown in Fig. 6.24. Significant water pressure would, of course, lower liquidus temperatures and decrease the size of the leucite field. Note the second minimum in the lower undersaturated trapezium, coincidentally comparable in temperature with the ideal granite ternary minimum. Between these two minima the liquidus is rather flat. Strong fractionation of parental basic magma could therefore be expected to produce end-products either of granitic (rhyolitic) or nepheline syenite (phonolitic) compositions, or conceivably in the absence of a well-defined oversaturated or undersaturated condition fractionation could produce syenite (trachyte). The observed rock associations seen in inferred genetically related igneous rock series (see Chapter 8) do fit these predictions remarkably well. Tholeiitic magmas are associated with rhyolite as are rocks of the calc-alkaline association. Alkali basalts, however, appear to yield trachyte often associated with very small amounts of even more strongly fractionated phonolite or (peralkaline) rhyolite, the latter more characteristic of transalkaline parental basic magma between tholeiite and alkali basalt in silica activity. More decisively undersaturated alkaline parental basic magmas such as basanites occur, where fractionated, with phonolite.

Chapter 7

DIFFERENTIATION OF IGNEOUS ROCKS

7.1. INTRODUCTION

Having surveyed the mineralogical and textural variety of igneous rocks and having attempted to classify them, the scientist in us wants to know if and how they may be interrelated. Questions to ask in this regard would include the following. Are there any consistent associations of varieties of igneous rocks from place to place? Can evidence of the production of different rock types from one parent body be found in the field? How might this have occurred? Do some or all varieties of igneous rock spring from one or more parental magma types? Can any working hypotheses in these respects be extracted from the overlapping complexities in igneous terrain? Can these hypotheses be tested by experiment or by prediction and comparison? This is the stuff of scientific enquiry which constantly strives to contradict Henry Adams' tongue-in-cheek assertion,

"Anarchy is the law of nature, and order is the dream of man."

and make that dream an acceptable reality.

The observed differentiation in igneous rocks is in fact the keystone of petrology. It is important to the field petrologist as a study of it leads on as to what to map and collect in the field. In general, lava flows have solidified too quickly to show any internal differentiation (and thus may have preserved magma compositions less only volatile constituents), but a study of plutonic rocks, however, reveals that few in fact are free from the effects of differentiation.

Of great significance in petrogenesis is the fact that a study of differentiation leads back to the characterization of parental magma types. To anticipate what is to come, we may say that, deep-seated granitic rocks apart, there are indeed recognizable *parental magmas* of basic compositions, which have differentiated, predominantly by crystal fractionation processes, to yield distinct *igneous rock series*, commonly characteristic of distinct plate-tectonic environments. Ideal materials to work with in this regard are fresh recent volcanic rocks, as analysis of aphyric varieties or of the groundmass of porphyritic varieties reveals the *liquid line of descent*. For the present, however, we must follow the historical path of discovery and examine first the field evidence of differentiation as seen in intrusive bodies, as interpretations based on analyses of discrete lava flows involves assumptions themselves based on this incontrovertible evidence of differentiation.

Historically, in considering the study of igneous rocks, we note the successive "heroic ages" of petrography in the 19th century, followed by that of petrology in the second quarter of the 20th, which was largely concerned with the elucidation of differentiation processes and tentative recognition of provinces and associations, now in turn largely superseded by questions of the petrogenesis of recognizable igneous rock series. This is not to say that former workers may not have observed differentiation processes and speculated on petrogenesis at much earlier dates than those indicated by the above. Classic examples of early perceptive deduction well worth the reading include reflections on differentiation by Pirsson (1905, pp. 183—186) based on field observations of the differentiated Shonkin Sag laccolith and other minor intrusions in the Highwood Mountains area, Montana, and the description of gravitational settling and floating of primocrysts in the Ilimaussuaq Intrusion, southern Greenland, by Ussing (1912). However, for the full flowering of any scientific subdiscipline the time must be ripe in that parallel subdisciplines must be advanced and man's minds prepared and attuned to countenance new hypotheses.

Magmatic differentiation is defined as the processes of developing more than one rock type in situ from a common magmatic source during the time of emplacement and cooling*. It refers to chemically distinguishable products and not mere textural variants, so that, for example, a glassy tachylitic margin of a basalt dyke is not necessarily differentiated with respect to a crystalline core. Note that we are *not* here strictly concerned with the hypothetical processes that lead to the production of magmas of different compositions prior to emplacement at accessible crustal levels, although such processes need to be included in the case of the deep-seated granites within the framework of differentiation processes.

Are magma chambers ever homogeneous to begin with? Even in rapidly emplaced basic dyke systems and in fissure eruptions there could be some temporal variation in the composition of magma supply, and there is strong evidence that many large basic plutonic bodies were supplied with magma in several pulsatory events. The relative constancy of composition of discrete magma batches as preserved in historic and recent lava flows, at Kilauea, Hawaii (T.L. Wright, 1971), for example, does however seem to suggest that broad assumptions of initial homogeneity in the supply of magma may not be unreasonable, at least in the environment of eruption of voluminous basic rock. It should be noted also that there is a fair constancy in composition

---

*In Cox et al. (1979, p.2), the term "fractionation" is used in the sense that "differentiation" is used here. Historically, however, the term differentiation has always had the connotation for igneous petrologists that is assigned to it here, and furthermore the term fractionation has, rightly or wrongly, come to be used synonymously by many petrologists with crystal fractionation, itself but one of several kinds of differentiation processes. Hence, the term differentiation is retained here. See, however, the pertinent discussion on these matters in Cox et al. (1979, ch. 1).

between discrete eruptions from "coherent" andesitic volcanoes such as Merapi, Java, again suggesting the possibility of homogeneity between different magma batches. However, magma at any one time and place is likely either crystallizing or resorbing included crystals. Thus, in the history of evolution of porphyritic magma there is the possibility (probability in the case of the less viscous basic magmas) of relative movement between the liquid and crystals, which in the general case are of a different density and composition to that of the liquid, thus resulting in differentiation. The convenient assumption that the contents of any given magma chamber were ever homogeneous is therefore probably only tenable to a first degree of approximation. Another rider is that katazone and mesozone granites were likely never homogeneous or more than slightly liquid. Emplacement processes may mix them up more thoroughly and result in the production of volumes of relatively homogeneous material — a process of "undifferentiation"! Petrology has often suffered from attempts to apply a unifying philosophy to embrace both (essentially liquid) mafic magmas and (essentially solid) granitic "migmas".

However, given a reasonably homogeneous body of magma as a starting point, whether the magma be liquid or partly, or even mainly, composed of crystalline material, what processes could affect it to produce differentiated solid end-products, and what evidence is there for the operation of these processes in nature?

Just as a modern study of igneous petrogenesis encourages us to take a global look at igneous activity with respect to new global tectonics, so too have many distant igneous provinces and intrusions become widely known as classic examples of differentiation. Selection is difficult and in part reflects the desirability of alluding here to accessible references in English, with a lesser emphasis on localities known personally to the writer.

7.2. DIFFERENTIATION PROCESSES WITHIN LIQUID MAGMA

7.2.1. Diffusion and gaseous transfer processes within liquid magma

We are here concerned not with the small-scale diffusion around growing crystals inevitably associated with the crystallization process, but rather with the possibility of large-scale diffusion within a magma chamber as a possible cause of differentiation. One theoretical possibility is that diffusion may occur in response to a temperature gradient thus tending to establish an equilibrium compositional gradient, the *Soret effect*; another is diffusion in response to vertical increase in pressure, setting up a density and compositional gradient in the liquid magma. Authorities appear to agree, however, that the order of magnitude of these effects and the rate at which the necessary diffusion processes could operate in "dry" silicate melts are so small that they can be neglected, and indeed there is no compelling field evidence for them.

G.C. Kennedy (1955), however, has drawn attention to the anticipated behaviour of dissolved water in a silicate melt, and has suggested that a sufficiently large potential would exist within magma chambers with a large vertical extent to result in a concentration of H_2O in those parts of the magma chamber at relatively lower temperatures and pressures (i.e. upwards) within geologically reasonable periods of time. Higher concentrations of dissolved water produced in this manner would reduce magmatic crystallization temperatures thus accentuating the temperature gradient before or during crystallization, and adding further potential to the migration of dissolved water. Kennedy included a graph (Fig. 7.1) showing equipotential concentrations of water against depth, based on the classic work of Goranson

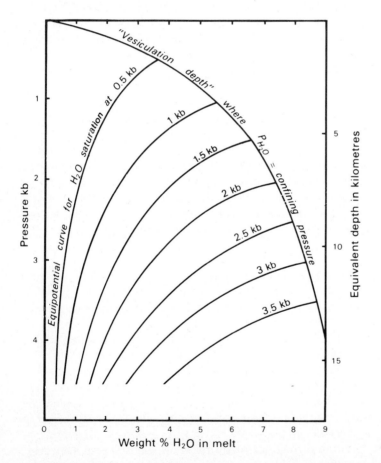

Fig. 7.1. Equilibrium distribution of water in a column of melt of albite composition at 1000°C (after G.C. Kennedy, 1955, p.493, fig. 1).

(1936); the assumptions involved in this in turn have been fully discussed by Burnham (1967) and several modifications suggested. Kennedy further suggested that alkalis and certain metals will coordinate with the water and, similarly, be concentrated in the regions of lowest pressure and temperature. If this were true then indeed significant major-element differentation could occur within the magma.

What could be interpreted as geological evidence for the efficacy of this process is provided by ash-flow deposits, many of which are zoned and within one cooling unit show a continuous subtle compositional variation upwards from relatively acid to relatively basic. The inference is that the source material, catastrophically erupted within a very short period of time from one interconnected magma chamber, was compositionally differentiated from relatively more acid and highly volatile-charged material at the top, whence the eruption was initiated, to relatively hotter, more basic liquid below. Wherever reported, zoning in ash-flow deposits is always in the same sense, thus agreeing with the hypothesis of a zoned magma chamber in the sense predicted by Kennedy. Very interesting work by Allègre and Condomines (1976) on the *Fine chronology of volcanic processes using $^{238}U-^{230}Th$ systematics* lends some support to this view in that their work indicates the existence of a differentiating magma chamber below the andesitic Irazú stratovolcano for a period of no less than some 0.14 Ma, a surprisingly long time-span and one long enough one would think for differentiation processes to operate.

On a smaller scale and shorter time interval, the well-known recurrent volcanism of Hekla, Iceland, affords a parallel. In some twenty recorded eruptions within the last millennium, the first products are gassy and relatively high in silica, changing rapidly to a uniform basaltic composition. The longer the time interval between successive eruptions, the higher the silica content of the initially erupted material (Thorarinsson, 1968, pp. 5—9).

It is the combination of inferred high water content with felsic differentiation products in these instances which lends support to Kennedy's proposed model of differentiation within a magma chamber. Nevertheless, it should be noted that fractionation processes (see pp. 225—249), acting concurrently with or independently of any processes producing differences in volatile content in the magma, could result in the production of bodies of more felsic liquids in the upper parts of a column of fractionating magma.

Appeal has been made to the possible role of gaseous transfer of alkalis within high-level parental tholeiitic magma chambers to account for the eruption of alkali basalts late in the volcanic cycle of Hawaiian volcanoes (G.A. Macdonald, 1949, pp. 88—92). However, other compositional parameters of the alkali basalts, distinctive trace-element fingerprints, and a worldwide similarity of alkali basalts in many terrains (Schwarzer and J.J.W. Rogers, 1974) point inexorably to a separate distinct origin for this parental magma type at depth in the mantle, and the observed distinctive association of differentiated rock types associated with many occurrences of alkali basalt can reasonably be accounted for by processes of fractionation of a parental alkali basalt magma.

It is in connection with the small but highly distinctive group of peralkaline rocks that alkali transfer seems to have occurred with the strong possibility of transfer of material both into and out of partially open magma systems. See, for example, a review by D.K. Bailey (1975) who concludes that:

> "the mobility of alkalis (and iron) must figure in any realistic scheme of peralkaline petrogenesis."

Whether this phenomenon may be the cause or the effect of a peralkaline condition is of course another question. It is conceivable, for example, that continued strong fractionation of plagioclase (always more aluminous as well as more calcic than the crystallizing magma) could result in fractionated peralkaline liquids [the *plagioclase effect* of Bowen (1945)], and that subsequent migration of alkalis could occur as a result of the solubility of an alkali metasilicate component in water.

So, although migration of dissolved water within magma chambers seems to have occurred, the case for diffusion processes vs. fractionation processes, as a probable mechanism of significant magmatic differentiation has not been established except for peralkaline rocks where it seems to be demanded. Perhaps, however, this latter instance should alert us to the possibility of some degree of diffusive transfer of water, alkalis, and possibly complex ions within magma chambers even in situations where other mechanisms of differentiation appear to predominate. Thus an element of uncertainty is added to rigid closed-system models, envisaging only crystal fractionation, for example. See below also for the role of pore-fluid diffusion in auto-metasomatism.

7.2.2. Liquid immiscibility

If magma were to split into immiscible liquids of differing composition the aggregate crystalline product of each would be different and hence liquid immiscibility could be a potent cause of differentiation of igneous rocks. Faced with the diversity of igneous rocks sometimes occurring together in sharp contrast (e.g., many acid—basic associations) postulates invoking liquid immiscibility, not an a priori impossible or even unlikely happening, enjoyed a vogue in the early years of this century.

Early experimental work on silicate melts related to common magma compositions of the type reviewed in Chapter 6, however, revealed no examples of immiscibility. Some melts very rich indeed in silica, and thus well outside the range of any igneous rocks, were known to show a limited field of immiscibility between a nearly pure, very viscous silica-rich liquid and one containing silica plus FeO, MnO, MgO, etc., at elevated temperatures

of the order of 1700°C. Furthermore, the addition of even small amounts
of alkalis to any of these systems caused the immiscibility field to disappear.
Armed with this kind of experimental data petrologists could therefore well
afford to ignore the possibility of liquid immiscibility in natural silicate
melts (see Bowen, 1928, pp. 7—19) — not the first time they have been
beguiled by the lucid prose and authoritarian arguments of the master!

Cumulative geological evidence, however, from such diverse instances
as the occurrence of sulphides in basic rocks, from magnetite—apatite
bodies, from rocks of the broad kimberlite—lamprophyre—carbonatite
grouping, and more recently from lunar basalts and fractionated terrestrial tholeiites has demonstrated liquid immiscibility in magmas beyond
doubt. This in turn has led to further experimental work paying particular
attention to these kinds of melts. An important discovery was the demonstration of the existence of an unsuspected additional immiscibility field in
the leucite (orthoclase)—fayalite—silica system (Fig. 7.2) at temperatures
as low as 1100°C completely surrounded by fields of crystallization of
fayalite and silica (Roedder, 1951). Liquid immiscibility thus became an
important reality for the igneous petrologist when dealing with rocks of
comparable compositions, and at the time of writing it is receiving considerable reinvestigation and re-evaluation. See, for example, the reviews by
Philpotts (1976) and Roedder (1978) and a timely warning against the

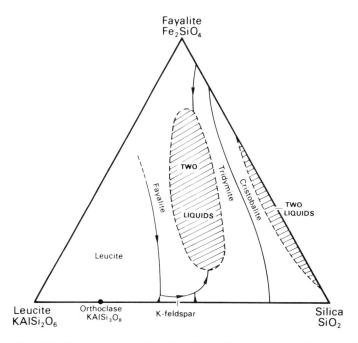

Fig. 7.2. The system leucite—fayalite—silica at 1 bar (after Roedder, 1951, p.283, fig.1;
1978, p. 1601, fig. 2).

possible overenthusiastic embracing of immiscibility models (Philpotts, 1977).

Criteria for immiscibility include both textural and mineralogical considerations: texturally, immiscibility will result in rounded, sometimes coalescing globules of one liquid in another; mineralogically, an important consideration is that crystallizing phases in equilibrium with one liquid phase must also be in equilibrium with the other liquid phase.

Considerable impetus to the revival of interest in immiscibility in igneous rocks came from the discovery of textures attributable to immiscibility in lunar basalts and in some terrestrial basaltic rocks (Roedder and Weiblen, 1970). For example, in some specimens of tholeiitic flood basalts from the Deccan Traps, India, the intersertal glassy mesostasis comprises minute globules of a dark-brown glass in a matrix of pale glass with a much lower refractive index (De, 1974). This evidence of glassy globules in glass is the best kind for proving immiscibility, and fulfills in this regard one of the criteria set out by Bowen half a century earlier. This demonstration of immiscibility on a microscale is perhaps paralleled on a much larger scale but in chemically analogous rocks from the Skaergaard Intrusion, Greenland, where an immiscible relationship has now been postulated to account for the contrasted granophyres and extremely iron-rich rocks represented by the last crystallizing rocks of the intrusion (McBirney, 1975). A similar phenomenon on a small scale is indicated by the (rare) occurrence of globules of plagiogranite in ferrogabbro belonging to the gabbroic unit of some ophiolites.

The case for low solubilities of sulphides in naturally-occurring silicate melts (much lower than that indicated by experimental work on some simple systems), is convincingly made by the study of textures in the Skaergaard Intrusion (see Wager and G.M. Brown, 1968, pp. 55—57; 188—190). The basic rocks contain 0.05 vol. % overall of ubiquitous microscopic globules mainly composed of copper sulphides, and interpreted on the basis of their shape and distribution as having formed from an immiscible sulphide liquid. At a very late fractionation stage ($\sim$98% crystallized), the immiscible sulphide phase is mainly represented by iron sulphides and is relatively more abundant. On a much larger scale, the economically very important sulphide ores of the Sudbury Complex, Ontario, with a mineralogical assemblage consisting mainly of chalcopyrite—pentlandite—pyrrhotite—magnetite with accessory cubanite and pyrite, are interpreted as having formed from segregated bodies of immiscible sulphide liquid generated at the same time as the basic rocks. In their final emplaced structural position, the sulphide bodies appear to be associated with relatively young marginal norite intrusions and present complicated field relationships with them, country rocks, and other igneous units (Naldrett and Kullerud, 1967).

Philpotts (1967) showed that a magnetite—apatite mixture roughly in a 2:1 proportion constitutes a eutectic mix and that this liquid composition is immiscible with certain silicate magma compositions at reasonable magmatic temperatures. Among several field areas where immiscibility has been claimed for magnetite—apatite deposits is that of the Camsell River area, Northwest Territories (Badham and Morton, 1976). Spectacular evidence of the existence of melt of somewhat comparable composition comes from the discovery of flows of almost pure magnetite—hematite with minor amounts of apatite (Park, 1961).

Several examples have been recorded of the occurence of carbonate-rich ocelli in kimberlite, and of the association with some kimberlites of discrete veins, etc., of carbonate formerly attributed to alteration processes but now believed to be genetically associated in that carbonatite liquids may have formed by immiscible segregation from kimberlite magmas (Janse, 1969, 1971). Subsequent experimental work has demonstrated the capacity of undersaturated melts coexisting at upper-mantle pressures with olivine and two pyroxenes to dissolve up to 40 wt.% of CO_2 (Wyllie and Huang, 1976). The fractional crystallization of such a liquid with a high amount of dissolved CO_2 yields a residual liquid that eventually splits to form two immiscible liquids corresponding roughly with kimberlite and carbonatite. This work thus provides an experimental demonstration of the hypothesis that carbonatites have been formed by the immiscible splitting of a carbonatite liquid fraction from parental undersaturated alkalic mafic melts (not necessarily all kimberlitic) during their crystallization and ascent to crustal levels.

Ocelli* attributed to liquid immiscibility are common in many lamprophyres. An excellent definitive account of ocelli in lamprophyres with many references to similar occurrences is .that by J. Ferguson and Currie (1971), who show that:

"(1) droplets obeying Bowen's (1928) criteria for liquid immiscibility are present in lamprophyre dikes around the Callander Bay alkaline carbonatite intrusion, Ontario; (2) these rocks melt into two or three immiscible liquids under geologically plausible conditions; (3) the compositions of the ocelli and matrix suggest that the ocelli-forming process was a major factor in the genesis of the alkaline igneous rocks; (4) the Callander Bay ocelli are identical to features observed in lamprophyric rocks of restricted composition the world over."

Another interesting and well-illustrated account of lamprophyres with ocelli attributed to liquid immiscibility is that by Strong and A. Harris (1974). What emerges from this geological deductive work is an apparent range in composition of the two complementary immiscible liquids (if the aggregate crystalline products in each apparent phase can be taken as the equivalent

---

*Ocellar structure has previously been described in a different fashion in the A.G.I. *Glossary of Geology* (Gary et al., 1972) as the tangential arrangement of smaller crystals around large eye-like phenocrysts, and ocellus would then refer to a phenocryst in such a rock. The term ocelli, as most commonly used at the present day, however, usually in this plural form (but note singular, "ocellus"), refers not to phenocrysts but to rounded globules of leucocratic material enclosed in a more mafic host rock that is often kimberlitic or lamprophyric. Crystals in the host rock can sometimes be found arranged tangentially around the ocelli (as in ocellar structure as defined above), but sometimes project into them.

of liquid compositions, an assumption that may be complicated by the existence and possible loss of a separate gas phase within ocelli). J. Ferguson and
Currie (1971, p. 565), for example, record carbonate ocelli in some olivine-
rich dykes, and ocelli composed mainly of K-feldspar and zeolites in kaer-
sutite-bearing lamprophyres. Favorable conditions for the development of
immiscibility in lamprophyres include the presence of CO_2 (but not necessarily so) and a high potassium content.

Certain structures in granite pegmatite (Jahns and Burnham, 1969) and
in a granophyre body (C.J. Hughes, 1971) have been attributed to immiscibility late in the crystallization history of water-rich granitic melts near
the ternary minimum composition. What seems to have happened in these
instances is the local development of pockets containing a water-rich liquid
coexisting with water-saturated silicate melt. Partitioning between the two
liquids has resulted in giant quartz crystals crystallizing from the hydrothermal liquid at the same time as granitic material with feldspar-rich bands
has crystallized from the silicate melt. The structures produced are striking
but the scale of the differentiation is not great.

In sum therefore, the possibility of significant immiscibility cannot be
ignored in: (1) crystallizing volatile-rich magmas, particularly alkalic ones
(here questions arise too of the rate of diffusive escape of H_2O and CO_2
vs. crystallization rate, or alternatively of the explosive escape of these
volatile constituents); (2) the occurrence of sulphides in basic rocks; and
(3) fractionated iron-rich liquids of tholeiitic affinity.

7.3. HYBRIDIZATION

Hybridization is defined as the mixing of two magmas of differing composition, thus resulting in a product which is compositionally distinct from
either and hence in differentiation. A priori, it might seem to be an unlikely
process as it depends on the intimate juxtaposition of two magmas of
different compositions in a critically short time interval before either had
completely crystallized. Furthermore, given the considerable and differing
viscosities of silicate melts (see Chapter 6) the appropriate physical conditions for mixing are hard to visualize. Indeed composite dykes and associated flows from eastern Iceland containing basalt and rhyolite show no
evidence of hybridization — even small blobs of basalt show sharp glassy
selvedges against inferred coexisting liquid rhyolite (Gibson and G.P.L.
Walker, 1963).

Field evidence in support of limited hybridization has, however, come
from the early Tertiary Hebridean province of western Scotland where
composite sill and cone-sheet intrusions are common. Wherever these are
seen, an intrusion of basic composition has preceded one of more acidic

composition*. This typically results in a sill or sheet with basic selvedges of fine-grained diabase chilled against country rock surrounding a central layer of a relatively much more acid composition such as rhyolite porphyry, often called felsite in Hebridean accounts. Usually, the basic selvedges are thin and the central layer is relatively thick. The interior contact relationships between the acid rock and the basic rock are highly variable. In some of the composite intrusions, the acid rock is chilled in a normal manner against the basic. In others, although the interior contacts are sharp, chilling is absent and it may be presumed therefore that the time interval between the successive emplacement of magmas of basic and acid compositions was so short that the acid magma was emplaced while the basic rock although crystalline was still hot, thus accounting for the absence of chilling. Yet, again, and these are the instances which concern us here, in some of the composite intrusions in addition to an absence of chilling a compositional gradation takes the place of a sharp contact between the acid and basic members. Here, one argues, the time interval between the emplacement of the two intrusive members was still shorter, so that the acid magma was intruded before its basic predecessor had completely crystallized and there was consequently a degree of intermingling or hybridization of the two magmas either in situ or during emplacement resulting in the observed compositional gradation of the crystallized product. These phenomena are shown, for example, by composite sills in the island of Arran, and an example of hybridization within a cone sheet ~5 m in thickness in the Isle of Skye is figured and described in *The Tertiary Volcanic Districts Regional Geology Handbook* (Richey et al., 1961, p.93), following an account by Buist (1959).

Also in the Isle of Skye on the mountain of Marsco, hybridization on a somewhat larger scale has been claimed between magmas of ferrodiorite and granite compositions (Wager et al., 1965). The petrographic interpretation there depends critically on both the assumed parent magmas having been porphyritic, and observation of the presence in the resultant hybrid rock, "marscoite" (subsequently then itself intruded to its present position), of two sets of phenocrysts: one set of plagioclase, An_{50}, derived from the relatively basic magma, and the other set of quartz and alkali feldspar derived from the acid magma. The marscoite approximates in major-element composition to some weighted arithmetical average of the two presumed end-members, a relationship that would be expected to ensue from hybridization.

From time to time, petrologists have recorded an apparent small-scale "streakiness" related to compositional variation of some lavas of dacitic to rhyolitic compositions, and/or the coexistence in some of them (or in associated minor intrusions) of apparently distinct sets of plagioclase pheno-

---

*An apparently unique exception is the Creag Dubh composite sill, Whiting Bay, Arran (N.W. Rogers and Gibson, 1977).

crysts of differing compositions (C.J. Hughes, 1960, pp. 117 and 118; Curtis, 1968; Eichelberger, 1975; Carey and Sigurdsson, 1978). These observations tend to suggest intimate mingling (although not necessarily complete hybridization) of magmas of contrasted compositions. However, what may well be happening in these instances (as is indeed also hinted at in the case of Marsco) is the physical incorporation during eruption of material from one compositionally zoned magma chamber rather than necessarily postulating two originally distinct magmas. That is to repeat that complete homogeneity of a crystallizing magma chamber, although a convenient assumption, may be unlikely due to the operation of processes leading to zoning described above and the processes of fractionation to be described below. See also in this respect an interesting discussion by Simons (1963) of a composite andesite—rhyolite dyke; hybridization is not claimed here, rather the successive intrusion of andesitic and rhyolitic material, arguably not from the same magma chamber. A.T. Anderson (1976) records numerous examples of apparent magma mixing primarily from evidence of the compositions of glass, phenocrysts, and glass inclusions within phenocrysts, within glassy tephra. The implication is that this process could go further under intratelluric conditions and produce a homogeneous magma, the final crystallization of which would obscure evidence of its mixed origin by this process of differentiation. Anderson concludes that evidence of magma mixing is particularly prevalent where the parental magmas were saturated in water vapour and had compositions that place them in the orogenic andesite suite. It is possible, of course, that some of the quoted examples may represent merely the local mechanical juxtaposition of the erupted products of a magma chamber that was *already differentiated* under these conditions by other processes. In any event, it is important to note that the evidence of compositions of glass inclusions in phenocrysts must be handled very carefully to allow for a component of nucleation and crystallization of the trapped liquid on the enclosing host crystal (see Watson, 1976).

In sum therefore it seems that hybridization between basic and acidic magmas is negligible, probably because of significantly differing viscosities and temperatures, although mixing, as opposed to hybridization, may be a mechanism for triggering acid explosive eruptions (Sparks et al., 1977b). Apparent widespread mixing and hybridization among melts of intermediate and acid compositions, typically of calc-alkaline affinities, may reflect inhomogeneity in high-level magma chambers (in which case the apparent mixing is itself a result of earlier differentiation rather than a significant cause of it). In any event it would be fair to say that the great wealth of field observations of igneous rocks has not produced compelling evidence for hybridization, as strictly defined, having played a quantitatively significant role in processes of magmatic differentiation.

7.4. ASSIMILATION

The term **assimilation** refers to the incorporation of country rock in magma thus resulting, in the general case where the country rock is of a different composition, in a change in composition of the magma and hence differentiation. The term **contamination** is also used in describing the same geologic phenomenon but in a different voice, e.g., one may say that a certain granite has assimilated pelitic sediments, or alternatively that it has been contaminated by them.

Intuitively it would not seem to be an unlikely process, particularly in connection with the relatively hotter magma compositions. For example, basic magma (specific heat of ~ 0.25 cal. g^{-1} $^{\circ}C^{-1}$) at near-liquidus temperatures in the vicinity of $1100^{\circ}C$ might be expected to melt some amount of granitic gneiss (latent heat of fusion ~ 75 cal. g^{-1}) which fuses at around $950^{\circ}C$ (or less under P_{H_2O}) and perhaps incorporate the anatectic melt so produced. A granitic magma, on the other hand, emplaced at significantly lower temperatures would not be expected to cause melting of country rock and thus there would be no resultant assimilation. This approach, based solely on considerations of anatexis and temperature, is in fact mistaken, and needs considerable modification in the following respects: (1) a more thorough appreciation of the principles involved in fusion and reaction phenomena between silicate melt and solid rock; and (2) an outline of what is actually observed.

7.4.1. Principles

Following the very clear treatment of the subject by Bowen (1928, pp. 175—223) the following principles emerge:

(1) A magma may partially or completely melt a rock that has a temperature range of melting partly or wholly below that of the temperature range of crystallization of the melt. The total amount of melt in such a system cannot be increased because, lacking superheat (a majority of magmas are porphyritic and thus below their liquidus temperatures), the heat necessary both to raise the temperature of the country rock from ambient temperature to melting point and then to cause melting (the latent heat of fusion) can only be supplied by the comparable latent heat of crystallization of the magma and the cooling of the magma through a limited temperature range above the temperature of melting of the country rock.

(2) Relatively refractory minerals in the country rock, however, cannot be melted but may be converted by chemical reaction to minerals in equilibrium with the magma. This can be illustrated by reference to the continuous and discontinuous chains of Bowen's reaction series: for example, the relatively calcic plagioclase of a basic rock immersed in a granitic magma crystallizing oligoclase, quartz, alkali feldspar and biotite will become con-

verted by chemical reaction in the solid state to oligoclase; and, in similar circumstances, forsteritic olivine will become converted stepwise via pyroxene and hornblende to biotite.

7.4.2. Limited extent of assimilation by mafic magmas at upper crustal levels

Melting is in fact very seldom observed at the exterior contacts of mafic igneous rocks with country rock, even, for example, at the contact of a large basic intrusion with granite gneiss. The reason for this is that as heat is conducted away into the country rock there is a concomitant crystallization and cooling of the marginal part of the magma chamber. After a short interval of time after emplacement one could imagine a thin layer of crystallized igneous rock — a typical chilled margin in fact — and an adjacent thin layer of the country rock both at some temperature necessarily intermediate between the original temperatures of the magma and country rock. Further continued conduction of heat has to take place across an increasing thickness of congealed, cooled igneous rock and heated-up country rock and a temperature gradient is established. It can be shown from calculations that for a planar contact and utilizing reasonable values of specific heat, latent heat of crystallization, and conductivity, the country rock, even that right at the contact, never attains a temperature of more than ~60% of the combined initial temperatures of the magma and country rock [see F.G. Smith (1963, pp. 477—482), and Jaeger (1968, with references to his previous classic work) for mathematical treatment of this topic]. This is not a high enough temperature to cause fusion at upper crustal levels, and indeed in line with these theoretical calculations the typical observed high-grade hornblende hornfels to pyroxene hornfels facies rocks were recrystallized in the approximate temperature range of 500°—700°C only.

Exceptional instances where some anatexis has occurred include the following. In the Isle of Rum, Scotland, arkose in contact with a large steep gabbroic ring-dyke intrusion has been partially melted and contains paramorphed tridymite which is stable only at temperatures above an inversion temperature of 870°C, itself a minimum figure applicable at 1 atm. and rapidly increasing with pressure at a rate of ~190°C/kbar. The inference here is that the basic magma in the ring-dyke intrusion was not static but moving vigorously and could therefore replenish the marginal part of the magma chamber with fresh hot magma and thus vitiate the theoretical calculation assuming a static crystallizing magma. Nonetheless, the arkose retains a sharp contact with the gabbro and there is no suggestion of assimilation of arkose by the gabbro. Also in the volcanic hearth of Rum, at some contacts of a similar gabbro with a porphyritic felsite, the latter appears to send a reticulate network of veins up to a few centimetres wide into the gabbro. Closer examination of the vicinity of such outcrops shows, however, that the gabbro is unequivocally the later intrusion as it truncates cone sheets that in turn had intruded the felsite. This is an example of anatectic *back-veining* produced by the localized remelting of the acid rock; the phenomenon incidentally has the paradoxical effect in outcrop appearance of apparently reversing the true age relationship of the two igneous rocks (C.J. Hughes, 1960, p.114). Here, perhaps, the acidic country rocks, although crystalline, may still have been hot at the time of emplacement of the gabbro thus resulting in above normal temperatures in the contact rocks. It is worth noting that these back-veins have sharp contacts against the gabbro which is of a chilled marginal facies at the contact, and again there is no suggestion of assimilation.

The Nodre Bumandsfjord peridotite pluton of northern Norway has converted a wide aureole of gabbroic country rock to granulite facies. Within a zone of intimate intrusive contact produced by hydraulic fracturing, partial melting of the gabbro along grain boundaries and physical inclusion of fragmented gabbro within the peridotite have occurred (B.A. Sturt, pers. commun., 1978). Peridotite melts, with necessarily extremely high liquidus temperatures, are of course rare beasts at crustal levels.

The thermal history of inclusions within mafic magma, however, can be quite different as it is possible for them to attain considerably higher temperatures. For example, gneiss inclusions up to 1 or 2 m in diameter are found in the lower parts of the layered series of the Skaergaard Intrusion; they show clear evidence of intense thermal metamorphism leading to partial melting and the local development of vesicular structures. Above a certain level in the layered series none are seen, suggesting that after a time represented by this level they may have become completely melted and incorporated in the magma. The quantitative effect of assimilation in this way can only have been small as acid rocks account for only a small fraction ($\sim$1—2%) of the rocks of the complex; furthermore, the acid rocks of the complex have chemical features which suggest that they owe their origin to processes of strong fractionation, possibly augmented by this possible small amount of assimilation of anatectic material. Diabase sills intruded into argillaceous sediments in Mull, western Scotland, have included xenoliths of the country rock that have undergone melting; crystallization from the anatectic melt has produced such phases as sanidine, tridymite, corundum and mullite (Richey et al., 1961, p.75). Some of the xenoliths are surrounded by a reaction zone of spinel, anorthite and cordierite but there is no evidence of physical incorporation of the anatectic melt within the basic magma.

In these instances involving the anatexis of country rock in contact with relatively hot basic and ultrabasic magmas at accessible crustal levels, we seem to be witnessing petrological curiosities rather than the operation on any significant scale of assimilation differentiation by way of anatexis.

Although the evidence at exposed crustal levels is thus meagre, there is still the possibility of assimilation by processes of anatexis at deeper levels in the crust where country rock temperatures are higher. Flood basalts in continental areas, the so-called continental tholeiites, are characterized by high

K and high initial $^{87}Sr/^{86}Sr$ ratios. It has been suggested that both these features indicate assimilation of crustal material into basic magmas. It is, however, difficult to envisage crustal assimilation uniformly affecting large amounts of mafic magma that from the field relationships apparently reached the surface quickly and in large batches. Furthermore, along with high K there are high contents of a suite of other typically incompatible elements, not all of which can readily be attributed to contamination by crustal material, and the overall composition may alternatively be explicable in terms of partial-melting processes, etc., within the mantle. The high initial $^{87}Sr/^{86}Sr$ ratios, although unusual, are not unique to this class of mafic volcanic rocks but are found in addition in some alkaline mafic rocks for which no crustal contamination is invoked, and may be related to the residence time of underlying lithospheric mantle beneath stabilized continental crust.

Geophysical work has indicated pronounced sharp gravity highs below several of the Scottish Tertiary volcanic complexes, and it seems reasonable to correlate these highs with the presence of large bodies of basic rocks emplaced within the underlying crust. Dunham (1970) has shown that the heat produced by the crystallization and cooling of bodies of basic magmas of the size indicated by the gravity data would be sufficient to cause crustal melting in significant amounts; this is in line with earlier conclusions by G.M. Brown (1963), and Moorbath and J.D. Bell (1965), based respectively on compositional and Sr-isotope considerations, that some at least of the acid rocks of the Scottish Tertiary province were produced by crustal remelting. Anatexis of this type, of course, is a possible source of magma *generation*, but there is little evidence in the province of rocks resulting from *assimilation*.

All in all, one is led by the available field evidence to agree with Bowen (1928) that assimilation at upper crustal levels does not play a significant role in the differentiation of mafic magmas, although these hot magmas may cause local anatexis of crustal material. There remains, however, the probability of greater anatexis and possible assimilation at lower crustal levels (see pp. 219 and 458).

7.4.3. Assimilation phenomena involving granitic rocks

In the case of deep-seated granitic rocks the field evidence is far less strongly weighted against assimilation. This is perhaps paradoxical in view of their lower temperatures of emplacement. However, a factor favoring assimilation by reaction is the long time-span afforded by the evolutionary development of deep-seated granitic rocks during orogenesis. One must therefore re-examine the old controversy, of which Bowen (1928) and Daly (1933) were proponents half a century ago, as to whether assimilation or fractionation may have played a major part in their development. This is perhaps best done by including direct references to some well-documented granitic provinces.

7.4.3.1. Skarns. In passing, one notes the marked reaction that occurs between granitic rocks and country rock carbonates, a special case consequent on the escape of CO_2 from the heated-up carbonates and the formation of highly reactive bases, e.g., periclase, MgO, and brucite, $Mg(OH)_2$. These rocks of extremely low silica activity thus come to exist next to silica-saturated granitic rocks with the result that silicon and other elements diffuse into the contact rocks. This has the effect of typically producing: (1) contact skarns in which carbonate rock is replaced by Ca-, Mg- and Fe-silicate minerals such as garnet, pyroxene and hornblende, often carrying payable values of scheelite, base metal sulphide, and precious metals and (2) adjacent desilicated varieties of granitic rock such as quartz syenite, syenite, and even nepheline syenite. These obviously differentiated igneous rocks are, to be precise, not the result of assimilation but the reverse, i.e. the selective *escape* of silicon and other elements from the granite. However, they may profitably be included here, lest they be overlooked, within the broad rubric of phenomena involving reaction. Some score examples of varieties of desilicated granite intruding dedolomitized limestones occur to the north of Ottawa in the Grenville province, Ontario.

7.4.3.2. Cognate inclusions. Leaving aside this rather special case generally found in upper crustal rocks where carbonates have not previously been metamorphosed to amphibolite, let us turn to the consideration of deeper-seated granitic rocks and particularly to the **cognate inclusions** (syn. *autoliths*) so commonly found in them. These puzzling inclusions, which are distributed in variable proportions throughout many mesozonal granitic plutons, are rounded in outline, subelliptical in shape, variable in colour index, and generally distinct from sharp angular *accidental inclusions* of country rock that by contrast occur near contacts, particularly upper stoped contacts, and whose presence is readily understood. Generally, mineralogical equilibration is complete, and the cognate inclusions are found to be composed of the same mineral phases as the enclosing granitic rock, thus fulfilling the prediction of Bowen (1928) regarding the end-product of reaction. The higher proportions of mafic minerals distinguish them from the host granite. Some have a relatively high colour index and are consequently dark in appearance, others are paler and include varieties that in texture and composition closely resemble the host granite. In some instances, porphyroblast alkali feldspars or porphyroblast biotites are developed both in the granitic host rock and in the cognate inclusions alike, thus proving that these crystals cannot have been phenocrysts of early crystallization from a melt, and alerting us to the far-reaching possibilities of subsolidus recrystallization and possible metasomatism in granitic rock bodies. In addition to this mineralogical equilibration is the demonstration by Farrand (1960) of a remarkable similarity even in trace-element content between the same mineral phases present in cognate inclusions and their host granite. This is a striking demonstration of the efficacy of reaction processes in attaining equilibrium again in the manner predicted by Bowen's principles.

If we are to regard the variations often seen within one exposure between cognate inclusions, from dark and relatively mafic to pale and granitic in appearance, as an evolutionary sequence, we could further postulate that the observed surprisingly complete mineralogical equilibration is a step towards a more complete compositional equilibration involving necessarily reciprocal reaction and hence assimilation of material that was originally either accidental or conceivably part of a parental migmatite complex from which the granite was derived. Assimilation could thus be inferred to have been more important as a factor in the production of certain granites than would be apparent from the evidence actually available at any one outcrop.

7.4.3.3. The Coast Range batholith. Turning now to specific examples, within the Coast Range batholith, field evidence strongly indicates the formation of granitic bodies by the granitization processes described by Hutchison (1970). As outlined in Chapter 4, a range in structural styles encompasses allochthonous and parautochthonous diapiric bodies ranging downwards to autochthonous bodies that apparently root in an older migmatite complex. A marked reduction in xenolith content in the more evolved, structurally higher bodies has apparently been achieved by assimilation processes in the manner described above for cognate xenoliths resulting in homogenization. This, of course, does not necessarily imply differentiation unless an overall compositional change occurred. The most evolved allochthonous bodies, however, as well as being more homogeneous, are relatively more acidic and are characterized by higher silica contents, greater proportion of alkali feldspar to plagioclase, lower An content of plagioclase, and higher biotite/hornblende ratios. Their formation has apparently involved not merely homogenization processes but also the concentration of some least refractory material of ternary minimum composition; the proportion of liquid existing at any stage of this evolution is unknown; however, even if it were small, at least incipient anatexis probably occurred to facilitate the diffusion of alkalis and silica under chemical potential gradients towards the structurally higher parts of the intrusive complexes (cf. Marmo, 1967). The range in composition — diorite to granite, quartz diorite and granodiorites greatly predominating, gabbro and granite notably rare — favours the operation of these processes of assimilation, homogenization, and diffusion acting concomitantly on starting material of migmatitic nature and overall dioritic composition.

7.4.3.4. Birrimian granites of Ghana. The Birrimian granites of Ghana, formed during the early Proterozoic Birrimian orogeny, show suggestive indications of the operation of assimilation. On a broad scale, the "G_2-type" granites are hornblende-bearing and relatively rich in lime, and are invariably spatially associated with the predominantly basic metavolcanic rocks of the

Upper Birrimian. By contrast, the "G_1-type" granites are biotite-bearing, often peraluminous with muscovite- and/or almandine-bearing varieties, poor in lime, and are invariably associated, often intimately in intermixed migmatite zones, with the predominantly pelitic Lower Birrimian. The field evidence from the plutonic rocks is compatible with a progressive physical incorporation of xenolithic material derived from migmatite. On a smaller scale, it is not uncommon to find, on outcrop scale, apparently homogeneous, hornblende-rich varieties of the otherwise predominantly biotite-bearing G_1 granite where there are amphibolitized inclusions of calcareous greywacke present (C.J. Hughes and Farrant, 1963). Field and petrographic evidence for the granitization of metabasites in Birrimian G_2 granitic terrain is lucidly presented by Lobjoit (1969). The following rock types were apparently produced from metabasites during the progressive evolution of granitization processes, involving, like the Coast Range rocks, eventual assimilation, homogenization and metasomatism: trondheimite-veined metabasites, agmatites, patchy hornblende quartz-diorite, biotite-hornblende quartz-diorite and adamellite.

7.4.3.5. Nova Scotia batholith. The granitic rocks of the southern part of the Nova Scotia batholith, outcropping over a large contiguous area $\sim 10^4$ km^2, are predominantly granodiorite and adamellite; their genesis has been interpreted in terms of the fractionation of plagioclase and biotite from a parental magma of granodiorite composition (McKenzie and Clarke, 1975) or by varying degrees of partial melting of metasediments (de Albuquerque, 1977). The rocks are highly silicic and are peraluminous throughout, carrying the relatively uncommon igneous mineral andalusite in some adamellites (Clark et al., 1976). In the field, however, it is impossible not to be impressed with the abundant evidence of the physical incorporation of xenoliths apparently largely derived from the intruded, very thick, Meguma Group; this, suggestively in view of the compositions of the batholithic rocks, comprises varying proportions of pure orthoquartzites and pelites. Within a few metres of the upper contact, accidental xenoliths of Meguma rocks lose their sharp outline, and what appear to be comparable recrystallized inclusions are common in many of the igneous rocks of the complex. The quartzites sometimes contain small calcareous nodules; where included in xenoliths within the granite these nodules sometimes project from the host xenolith in a manner suggesting significant preferential incorporation of the quartzite into the granite. It is, of course, entirely possible that differentiation by a degree of assimilation could be superimposed on the one produced by fractionation. Certainly it is not clear how fractionation of any reasonable proportion of plagioclase and biotite (with an overall highly aluminous composition) from granodiorite magma could alone account for the production of large volumes of peraluminous adamellite.

Although hard chemical evidence of assimilation is not always to hand, one cannot help being impressed by the suggestive nature of field observations of granitic rocks such as the above. One wonders if the volume of studies as detailed as some of those in support of a fractionation model (a viable, indeed probable, assumption for behaviour of mafic magma) might not, if differently inspired, have yielded quantitative data compatible instead with assimilation for some granitic complexes.

7.4.3.6. Sierra Nevada batholith. It is interesting in this context to trace an evolution over the last few years in some ideas on the petrogenesis and mechanisms of differentiation of the rocks of the composite and polycyclic Sierra Nevada batholith. The arguments adduced are richly illustrative of the complexities presented by mesozone granites, and although they range well outside assimilative phenomena, it is not inappropriate to allude to them briefly here. Very broadly stated, are the igneous rocks of the Sierra Nevada merely the plutonic equivalents of effusive andesites and their associated rock types, or are they remobilized lower crustal rocks, a by-product of the heat resulting from magmatic intrusion and crustal thickening? One possibility that seems to be excluded on account of their enormous volume is that they could simply be the products of fractionation of basic magma. Field evidence (Bateman, 1961) reveals sequences of intrusions ranging from granodiorite to alaskite in composition, emplaced in order of increasing acidity; granodiorite and adamellite* greatly predominate and are about equally abundant. There is evidence of differentiation within some plutons, always in the sense of from relatively basic marginal facies to more acidic centrally, presumably by fractionation due to "congelation crystallization" (see p.225). However, at least local evidence of assimilation has been recorded (Crowder et al., 1973). The possibility of crustal anatexis producing parental magma has been explored (Presnall and Bateman, 1973), and indeed given certain assumptions on crustal thickness, temperature gradient, and heat added by access of andesitic magma, refusion can be made to appear inevitable. Equilibrium melting (if the lower crust in question was obliging enough to stay put for the necessary time required) could result in a magma of granodiorite composition which could then fractionally crystallize to produce adamellite and granite. However, granodiorites, although abundant, are not the basic high-temperature end of the observed intrusive sequences: quartz-diorite and more basic rocks are. Furthermore, the comparable Southern California batholith (E.S. Larsen, 1948), possibly representing merely a lower level of dissection than that of the Sierra Nevada, contains ~14% by area of horn-blende gabbro. An overall enrichment in K_2O towards the east within the granitic rocks of the Sierra Nevada has been demonstrated despite a non-

---

*Many published descriptions of Sierra Nevada rocks use quartz-monzonite in the sense of adamellite in the proposed I.U.G.S. classification (see Chapter 4); adamellite, which, of course, is synonymous with the former usage of quartz-monzonite, is used here to avoid any possible confusion.

regular variation of longitude of intrusion with respect to time along any particular latitude. This could be compatible with a deeper Benioff zone to the east on the andesite model, or with anatexis of different material on the crustal remelting model, but there are difficulties with both of these simple models (Bateman and Dodge, 1970). Early strontium isotope data showed most initial $^{87}Sr/^{86}Sr$ ratios to lie within 0.7073 ± 0.0010. These values were considered too high to be equatable with magma of mantle origin and too low to be equatable with remelted average crustal material; rather, they favoured anatexis of young greywackes with comparable ratios, or some degree of contamination of mantle-derived melts (Hurley et al., 1965). Subsequent work, however, revealed a range of initial ratios from 0.7030 to 0.7075, the lower end of the range apparently too low for an anatexis model to be universally applicable (Kistler et al., 1971). Later work (Kistler and Peterman, 1973) showed a distinct regional pattern in initial ratios with a three-fold grouping as follows: (1) values greater than 0.7060 in the east overlying Precambrian basement (2) values between 0.7040 and 0.7060 in a central zone overlying Palaeozoic geosynclinal rocks; and (3) values less than 0.7040 in the west, further west than the eastern limit of ophiolites, overlying presumed oceanic crust. These zones are transected by geosynclinal trend-lines and the belts of granite intrusions themselves, and thus appear to be a fundamental parameter. The data indeed appear to confirm the role of incorporation of lower crustal rocks, and it would seem therefore that a zone of melting and/or incorporation intersected both upper mantle and lower crust to produce the parental magmas of the majority of the granitic rocks of the Sierra Nevada.

Thus this oblique isotopic evidence favours an origin involving the assimilation of some crustal material at depth by mantle magmas, a compromise between the two contrasted possibilities mentioned at the beginning of the preceding paragraph. Superimposed on this inferred process to produce the parental magmas are the further processes of fractionation and assimilation for which there is more direct field evidence at the higher crustal levels at which emplacement occurred.

7.4.3.7. Granites of southwest England. In this connection also, a salutary review, including much interesting detail, of the classic Hercynian granitic rocks of Cornwall and Devon (Exley and Stone, 1964) sounds a note of caution in that a complex spectrum of mechanisms of differentiation is proposed, within which not one simplistic hypothesis of differentiation can satisfactorily be maintained to the exclusion of others:

"We envisage a palingenetic origin of the granites which is consistent with their regional setting and with the 'granite series' of Read (1957). Selective fusion would result in a liquid having, initially, a composition close to the 'natural' ternary minimum, 'contaminated' by solid material not taken into solution, as well as xenolithic fragments. The density difference between the magma and its mantle would aid the upward rise of the former, partly by plastic deformation of the country rocks, partly by assimilative granitization, and partly by mechanical stoping. . . . At higher levels, crystallization of the bulk of the magma would have occurred; magma would be subordinate and interstitial and would pass continuously into interstitial aqueo-silicate fluid containing alkalis. Differentiation could have occurred in two ways:

(a) by crystal fractionation together with the gravitational removal of 'contaminants', resulting in a liquid having a composition near the natural ternary minimum, and

(b) by differentiation of the interstitial aqueo-silicate fluid.

The latter could result in the marked lithium and fluoride enrichment found in some of the granites, together with the removal of potassium and enrichment in sodium.

Whilst megacrysts of potash feldspar were growing, largely as a result of internal metasomatism as the temperature was falling, it is considered that the granites were emplaced into their present positions as sensibly solid bodies containing interstitial lubricating fluids.

Thus, it is reasonable to suppose that the granite variants have arisen in part from the magmatic differentiation of a contaminated liquid—crystal mush, and in part from ion-exchange reactions between the rock and late-stage fluids derived from magma. The occurrence of some features suggesting replacement and others pointing to a magmatic origin are thus resolved as resulting from dependent processes: replacement follows magmatism in the 'granitization' of granites."

A fitting epilogue to this attempt to survey the phenomena involved in granite genesis is the essay by Krauskopf (1968) entitled *A Tale of Ten Plutons*. Krauskopf raises the disturbing question how much we can ever know with certainty about the evolution of a granite body, even allowing for the possible results of the 'future work' to which many writers ingenuously appeal at the end of their own contribution. This most readable essay provoked a riposte (Shea, 1968). The reader is heartily recommended to read the exchange (discussion and reply) and to consider his own philosophy.

7.5. AUTOMETASOMATISM

The extract on the granites of southwest England quoted above leads naturally to a consideration of autometasomatism as a mechanism of differentiation. This kind of differentiation is produced during the cooling of an intrusion, essentially in the solid state, by the migration of pore-fluid material. The evidence for the operation of autometasomatism at the lower temperature end of the temperature spectrum, where the pore fluid was hydrothermal solutions, is undeniable. The possible operation of autometasomatism in certain katazone and mesozone granitic rocks at higher temperatures where the diffusive medium could be a water-rich or water-saturated silicate pore-fluid [the possibility alluded to by Exley and Stone (1964)] is a more controversial subject (see Marmo, 1967); certainly it is a possible, indeed probable, mechanism for some of the phenomena accompanying the formation of late K-feldspar crystals and megacrysts. However, as with all interpretations appealing to metasomatism, it is legitimate to raise the query how does one know what the original composition was, and at times this question cannot be unambiguously answered, and at best is generally capable of only circumstantial proof.

The entrapment of late-stage magmatic water in intrusive igneous rock bodies is shown by the formation of several characteristic hydrous minerals that clearly post-date the development of the pyrogenetic mineral fabric. This phenomenon is known as **deuteric alteration,** and is generally found in the more slowly cooled of the larger granitic bodies, in which independent evidence indicates a significant content of volatile constituents (see Chapter 6). Minerals formed by deuteric alteration, for example, include (1) the partial or complete replacement of biotite (H_2O content ~4 wt.%) by chlorite (H_2O content ~12 wt.%) containing included blebs of sphene (reflecting the original presence of significant amounts of Ti in biotite, but not in the later chlorite); and (2) the alteration of plagioclase to yield plagioclase of a more albitic composition plus included blebs of zoisite—epidote and tiny flakes of sericite; in this process of **saussuritization** the zoisite—epidote and sericite contain much of the Ca and K, respectively, that were originally present in crystalline solution in the igneous plagioclase. Some secondary deuteric minerals such as muscovite, epidote, and sphene* commonly nucleate apart from the minerals being altered and form relatively large anhedral grains in some deuterically altered granitic rocks. There is nothing in this process that demands metasomatism; on the contrary, it would seem to reflect merely the *retention* of original magmatic water and hence *no* differentiation due to the movement of any original magmatic constituents.

However, in some granitic rocks in particular, processes of deuteric altera- tion are so pronounced that there *has* been a resulting change in the compo- sition of the igneous rock and hence a differentiation by autometasomatism, caused presumably by the *migration* of fluids at the deuteric stage.

Hydrogen-ion metasomatism operating at low temperatures to produce **kaolinite** from feldspar has resulted in large bodies of **china clay** (essentially kaolinite + quartz) in cupolas of the St. Austell granite in Cornwall, England, by reactions such as the following:

$$4H_2O + 4KAlSi_3O_8 \rightleftharpoons Al_4Si_4O_{10}(OH)_8 + 8SiO_2 + 2K_2O \tag{7.1}$$
$$\ 72 \qquad 1184 \qquad\quad 516 \qquad\qquad 480 \qquad 188 \quad \text{(molecular weights)}$$

Note that only ~6 wt.% of water is necessary for this reaction to proceed to completion, although more than this may be needed to carry away all the released K_2O in solution. An old controversy whether this kaolinization was due to near-surface weathering or to autometasomatism during cooling of the intrusion has been decisively settled in favour of the latter explanation (see Edmonds et al., 1975).

---

*Sphene, of course, can also be a mineral of early pyrogenetic crystallization; for example, it forms neat euhedral microphenocrysts in some andesites.

Another higher-temperature kind of alteration involves the formation of mica rather than kaolin:

$$H_2O + 3KAlSi_3O_8 \rightleftharpoons KAl_2AlSi_3O_{10}(OH)_2 + 6SiO_2 + K_2O \qquad (7.2)$$

18   852   398   360   94   (molecular weights)

Local veins, alteration haloes around mineralized quartz veins, and more generalized areas of **greisen** (essentially mica + quartz) reflect the operation of this process of **greisening** for which still less water is required. The mica in greisen is commonly yellowish in colour, and on analysis is generally found to contain significant amounts of Li and F (e.g., zinnwaldite); other fluorine-bearing minerals commonly found in greisen include topaz, $Al_2SiO_4(OH, F)_2$, and fluorite.

Large bodies of **china stone,** also exploited, for example, in the St. Austell granite, are composed essentially of albite, muscovite and quartz, plus varying proportions of kaolinite and fluorite; the relative amounts of the latter two minerals give rise to such commercial terms as hard white, soft white, hard purple and soft purple. These varieties reflect the operation of varying proportions of the processes of greisening and kaolinization.

Tourmalinization affecting both country rock and granite is another clearly metasomatic process closely related to the late-stage cooling history of some granites. Field evidence indicates that tourmalinization predates kaolinization, and both predates and overlaps with greisening.

In **porphyry coppers** it would appear that anomalously large amounts of Cu and Mo, and other metals of presumed igneous provenance have been concentrated by autometasomatic processes within some epizonal plutons and deposited in a telescoped geothermal system in and around the margins of the pluton (see Whitney, 1975), probably involving also an element of groundwater circulation. A continuum from magmatic crystallization through deuteric alteration to hydrothermal alteration has been demonstrated by Robertson (1962) in the northwestern part of the Boulder batholith. See also a discussion of models of fluid circulation and their magnitude resulting from emplacement of magmas in the crust (Norton and Knight, 1977).

The formation of *simple pegmatites*, typically occurring in lensoid and irregular bodies around the margins of some deep mesozonal granites, can be interpreted as the crystallization of a more or less closed system of ternary minimum composition melt from which, under conditions of very high P_{H_2O}, muscovite has joined microcline, albite, and quartz as primary crystallizing phases. The occurrence, generally in small amounts, of such relatively rare minerals as beryl, apatite, columbo-tantalite, spodumene, etc., reflect the concentration of elements such as Be, P, F, Nb, Ta, Li, etc., that are incompatible with granite mineralogy into a residual pegmatite-forming melt (concentrated in part by filter differentiation, see p.248). However, it is abundantly clear that the interconnected "plumbing systems" of originally simple pegmatite have often acted as convenient channelways for metasomatizing fluids on a gross scale to form complex pegmatites formed during protracted cooling commonly in an environment of regional metamorphism. In these **complex pegmatites** a former simple pegmatite mineralogy and texture has been replaced on greatly varying scales and in a highly variable manner (ranging from mimetic replacement to obliteration of original texture) by mineral assemblages characterized by albite, muscovite, quartz, lepidolite, spodumene, etc., in the formation of which metasomatism has clearly been involved (see Jahns, 1955, C.J. Hughes and Farrant, 1963). Sugary or platy albite (var. cleavelandite) is a common mineralogical indicator of this kind of autometasomatism. It is found not only in complex pegmatites but also within certain granites, for example, the albite-bearing granites of northern Nigeria (Jacobson et al., 1958).

In a completely contrasted igneous setting to the granitic rocks hosting the above instances of autometasomatism, the well-known **platiniferous pipes** of the Bushveld Intrusion, similarly owe their origin to an autometasomatic replacement during the late cooling history of the intrusion. The Onverwacht pipe in the eastern Bushveld, for example, only ~15 m across at surface, has been mined out down to 330 m, and had the form of an elongate parsnip-like body tapering irregularly downwards. It was composed mainly of horto-nolite olivine with high platinum values that had clearly replaced more or less horizontally layered ultramafic cumulate rocks containing olivine of a more magnesian composition (Cameron and Desborough, 1964, pp. 215—222). Some of the pipes in the western Bushveld are localized at the inter-section of major vertical joint systems apparently developed during cooling and are thus obvious channelways for the upward migration of late-stage liquids within the intrusion.

In yet another geologic setting some autometasomatism may accompany the primary devitrification that occurs during the cooling of ignimbrite, although a more pronounced metasomatism is commonly associated with later secondary hydration, devitrification and leaching processes (Lipman, 1965).

Many texts dealing with igneous rocks are reticent about the operation of these "messy" autometasomatic processes of differentiation which inci-dentally commonly have considerable economic significance.

7.6. CRYSTAL FRACTIONATION

Crystal fractionation has long been claimed by petrologists to be by far the most important mechanism in producing differentiation in igneous rocks, certainly in those that have passed through a wholly or largely liquid mag-matic stage in their development such as the mafic eruptive rocks and their associated igneous rock series. Nearly a century ago, W.C. Brögger, for example, pointed out that much magmatic differentiation is governed by the laws of crystallization. As we have seen in Chapter 6 fractional crystallization is a necessary consequence of the slow cooling of most naturally-occurring silicate melts. The effect is evident, for example, in the presence of zoned plagioclase crystals. However, differentiation will not ensue from fractional crystallization until some mechanism operates that tends to separate primocrysts from the crystallizing magma. If and when this happens differentiation is inevitable in all melts except those of eutectic or minimum compositions because the resultant liquid magma is depleted in primocrysts of which the bulk composition is not the same as the original magma. This is what is termed **crystal fractionation**, or generally just **fractionation**.

The various ways in which fractionation might occur include *flow differentiation, congelation crystallization, gravity settling* (evidence of the latter from large layered basic intrusions indeed provides the most spectacular and compelling demonstration of the efficacy of fractionation in producing differentiation in igneous rocks), *filter press action, autointrusion,* and as *a consequence of nuée ardente eruption.* An element of inference is added by ascribing the compositions of many lava series to fractionation processes, and an important foundation for the concept of an igneous rock series (see Chapter 8) is laid.

7.6.1. Flow differentiation

It has been established that a liquid containing a suspension of solid particles moving between walls will tend to concentrate particles in the more central parts of the flow away from the walls; in this way differentiation of a porphyritic magma could theoretically occur.

The effect depends in a complex manner on such variables as velocity, viscosity and relative density of the liquid, size and concentration of the particles, and channel width. No less than three separate forces have been identified: the poorly understood **"wall effect"** tending to repel grains from the wall, and vanishing a few grain diameters from the wall, the **"Magnus effect"**, a transverse force due to particle rotation which gives rise to an inward translation of suspended particles; and the **"Bagnold effect"**, likely two orders more important than the Magnus effect, a grain dispersive pressure due to the shear of the suspension. The Bagnold effect would be expected to be strongly operative in relatively narrow dykes where the phenocryst concentration exceeds ~8 vol.%, but the importance of even this, the strongest effect, should decline to zero in the case of larger dykes or pipes, even if appropriately porphyritic, if their width or diameter exceeds 100 m (Barrière, 1976).

Flow differentiation is often discernible in porphyritic basaltic dykes producing noticeable increases in the concentration of plagioclase phenocrysts towards the centre of the dyke, for example. The classic account by Bowen (1928, pp. 148—159) of the famous peridotite dykes of Skye suggests that it was operative there too, resulting in places in a non-porphyritic selvedge to a dyke containing numerous olivine crystals in its central portions [see Gibb (1968) for a more detailed account of these same dykes].

Flow differentiation has also been invoked to account for a high concentration of olivine primocrysts in central portions of some Archean high-magnesium flows and sills. Here, however, the problem of interpretation is compounded by the great abundance of apparently rapidly crystallizing olivine showing skeletal and dendritic growth and overgrowth from super-cooled melts and the relatively rapid sinking of primocrysts and skeletal crystals under the influence of gravity in hot melts of inferred low viscosity.

Perhaps a less happy invoking of flow differentiation is to attempt to explain transverse variations in composition of relatively large bodies such as, for example, the Alaskan type pipes, as Barrière demonstrates, in connection with these relatively large masses, it seems that:

"geologists must return to more classical explanations."

In sum, therefore, it seems that flow differentiation is a relatively unimportant mechanism of effecting the fractional crystallization of a crystallizing magma, although its possible operation during the emplacement of relatively small and quantitatively minor hypabyssal intrusions should not be dismissed.

7.6.2. Congelation crystallization

This could be anticipated in large plutons of liquid magma undergoing relatively slow crystallization. Considering the progress of crystallization with time, first one can imagine the solidification of a thin marginal envelope of supercooled liquid — the chilled margin — commonly presumed by petrologists to be indicative of the composition of the parental magma of the intrusion. Within this congealed envelope the rate of loss of heat and the rate of crystallization, buffered by the rate of conduction of heat into the country rock and the latent heat of crystallization, will gradually become less rapid. It will involve the congealing of crystalline mineral phases in equilibrium with the liquid magma at, or near, the walls of the magma chamber. In general, these first crystallizing minerals will not be of the same bulk composition as the liquid magma but would in the case of a basic magma, for example, essentially comprise relatively calcic plagioclase and relatively magnesian augite for reasons discussed in Chapter 6. The crystallizing aggregate will not be 100% solid but will include a proportion of interstitial liquid. If this liquid were trapped and subsequently completed crystallization as a closed system the resultant rock would have a chemical composition similar to that of the parental magma and thus not be differentiated. However, there is another possibility. The early crystals offer favourable growth sites for further crystallization from the interstitial liquid magma initially at, or near, the same magma temperature at which the crystals formed. The crystals could thus begin to enlarge with little or no compositional zoning. As a result of this further additive crystallization of the first products of fractional crystallization, a compositional gradient would be established between the liquid network within the crystal mush and the large volume of the liquid magma remaining within the magma chamber. In response to this gradient there could possibly be time for diffusion processes throughout the liquid continuum so as to replenish the interstitial liquid from the magma chamber with elements appropriate to the relatively refractory mineral phases continuing to crystallize and, in a complementary manner, to remove concentrations of those elements not required for their crystallization. Diffusion processes in this manner need not go to completion, for even if they are only partially operative, their continued aggregate contribution will result in a perceptible degree of fractionation inevitably resulting in a compositional gradation towards lower-temperature liquid magma within the interior of the intrusion and hence in an igneous differentiation visible in the finally crystalline rock product.

So much for theory, what about the evidence?

This seems pretty compelling for the pattern of differentiation observed within thick diabase sills. Within homogeneous upper and lower chilled margins such classic examples as the tholeiitic Palisades sill reveal a pattern of gradual differentiation, devoid for the most part of layered cumulate structures (the famous olivine layer being an obvious exception to this), with the most refractory mineralogies at the top and bottom and the lowest temperature mineralogy not exactly in the middle but about three quarters of the way up. As cooling and amount of crystallization would be approximately equal across both the upper and lower contacts, this geometry suggests that some of the crystallizing material forming at the upper magma—solid interface may have quietly sunk, perhaps partly in polycrystalline aggregates, perhaps slightly concentrating relatively heavier mafic minerals, to the lower interface without necessarily forming distinctively layered rocks such as those which distinguish the larger basic intrusions (see p.232). As an index of fractionation in the sill rocks one could use bulk chemistry, or perhaps better (as, apart from the chilled margins, no one rock in the sill need ever have corresponded to a liquid of the same composition), one of a number of mineralogical parameters such as: (1) An content of plagioclases throughout the sill; (2) Mg/Fe ratio of augites; (3) composition and nature of the ferromagnesian minerals; as well as a continuous range in composition these display interesting phase changes, so that one passes inwards from an early forsteritic olivine to magnesian orthopyroxenes to less magnesian pigeonites to relatively iron-rich fayalitic olivines in a manner completely equatable with the experimental work reviewed in Chapter 6; (4) presence and amount of micropegmatite, occurring as an interstitial intergrowth giving rise to a distinctive white-speckled appearance in rocks from the more central parts of the sill sandwich. (Apart from a gradual coarsening in grain size centrally, this is the only easily visible evidence in hand specimen of differentiation in the sill rocks, which only in thin section are seen to belie their rather monotonous field appearance.) In the Palisades sill all these parameters run parallel in conclusively demonstrating an overall gradual compositional variation towards lower-temperature varieties centrally, as shown by the classic work of F. Walker (1940) and the later account by K.R. Walker (1969).

A comparable pattern of compositional variation is displayed by the rocks of the marginal and upper "border groups" of the Skaergaard Intrusion, east Greenland, although the intrusion is, of course, better known for its remarkable cumulate sequence of layered rocks. The border group rocks are inhomogeneous in appearance, variable in grain size, sometimes crudely banded parallel to the walls of the intrusion, and in places contain large dendritic, inwardly directed, crystals of plagioclase, the "perpendicular feldspar rock". In terms of mineral compositions these border group rocks include, among a zoned sequence to lower-temperature varieties inwards, relatively refractory material of early crystallization in their outermost part composed of even higher-temperature members of solid-solution series than those exposed in the cumulate sequence of which the lower part is hidden below sea level. All the features that the border group rocks display are readily explicable in terms of congelation crystallization.

The Binneringie Dyke, Western Australia (McCall and Peers, 1970) up to 3 km in width, is of overall basic composition and characterized by consistently vertical layering accompanied by normal* compositional mineral variation inwards, particularly in a marginal border zone extending some hundreds of metres or so in from the chilled margin. As there is no evidence of later tilting, the vertical disposition and composition of these layered rocks suggests an origin by congelation rather than by gravity settling. The rocks have been termed **congelation cumulates** which the authors state, in a very full account,

"may be a far more significant feature of layered intrusions than has hitherto been realised."

It will be apparent that fractionation by simple congelation crystallization is much more likely to develop in an intrusion emplaced as liquid magma rather than a crystal mush, and, other things being equal, in a large intrusion where the rate of fall in temperature is relatively slow. Many large epizonal granitic plutons of inferred liquid magmatic origin indeed show some effects, and the literature abounds with descriptions of streaky dioritic contacts to granitic bodies of an overall more acidic composition. For purposes of illustration we shall consider two **zoned plutons**: (1) the Shaler pluton, which

---

*The term "normal" usefully refers to a continuous compositional variation in a mineral belonging to a solid-solution series, *either* to zoning within one crystal *or* overall variation in composition of the mineral in a cumulate sequence, in the sense of variation from a higher-temperature to a lower-temperature composition with time of formation.

shows a very pronounced compositional zoning; and (2) the Tunk Lake pluton, which shows a relatively subtle variation in composition, necessarily so in fact because the parental magma appears to have been much nearer the residual ternary minimum composition.

The Shaler pluton of Miocene age is emplaced into Lower Miocene andesites and derived sediments that overlie older Tertiary pillow lavas on the island of Unalaska, Aleutian Islands (Drewes et al., 1961, pp. 609—634). The pluton is sub-elliptical in plan with a maximum diameter of ~25 km. With an outcrop area of over 250 km², it is technically a batholith according to definitions based on area, although this term continues to conjure up the connotation of a subjacent mesozonal body for many petrologists, whereas the affinities of the Shaler pluton are unmistakably epizone. A flattish roof to the intrusion is visible in parts of the mountainous interior of the island where dissection ranges up to ~1 km. The overall average composition of the exposed rocks of the pluton is granodiorite and ~70% of the outcrop area is underlain by granodiorite. This average, however, comprises a remarkable compositional variation (with no visible internal contacts despite good exposure) from marginal pyroxene diorite to biotite adamellite centrally. One of the components to have concentrated centrally was water, as crystallization of anhydrous primary pyroxene gives way centrally to hydrous dark minerals. Isopleths of such mineralogical parameters as the outgoing of pyroxene, presence of more than 20 wt.% quartz, presence of more than 20 wt.% K-feldspar, and a curious central concentration of allanite (Fig. 7.3.) are sensibly concentric and reveal the pattern of differentiation within this pluton, for which congelation crystallization seems to provide an explanation.

The Tunk Lake pluton, southeastern Maine, is also quite large with an outcrop area of 175 km² (Karner, 1968). A marginal chilled zone and a calculated average composition based on outcrop area are similar and of a granitic (sensu stricto) composition. This inferred bulk composition of the pluton is in fact not far removed from that of the ternary minimum, although SiO_2 (~74—75 wt.%) is a little lower, and K_2O (~5 wt.%) exceeds Na_2O (3.8 wt.%) and is a little higher than ternary minimum composition. What is observed are zones (again lacking any sharp internal contacts) comprising, from the outside inwards, hornblende granite, hornblende-biotite granite, biotite granite and biotite adamellite. The latter most central variety of rock is not relatively calcic as the name alone might imply but the reverse, possessing the lowest colour index and highest silica content of the complex and also the least perthite and markedly less albite component in the perthite, that is to say, a lower-temperature subsolvus mineralogy. The effect of increasing amounts of trapped water within the crystallizing magma within a seal of congealed rocks (attested by the common occurrence of miarolitic cavities in the central part) has been to reduce crystallization temperatures. Dark minerals include magnetite and aegirine-augite of early crystallization in the chilled margin; hornblende is seen to replace the pyroxene and then become

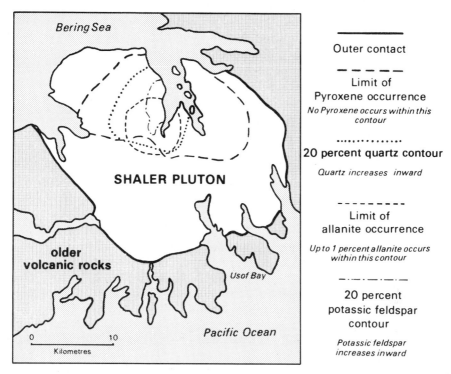

Fig. 7.3. Simplified sketch map of the Shaler pluton, Unalaska Island, Aleutian Islands, showing compositional zoning (after Drewes et al., 1961, p.632, fig. 98).

predominant in the outer zone, and in turn is replaced by biotite centrally, an example of Bowen's (1928) discontinuous reaction series. Accessory minerals include apatite zircon, sphene, allanite-epidote, and fluorite. Although the compositional range displayed by the rocks of the Tunk Lake pluton is small, Karner infers a nett inward migration of aqueous, silica-rich fluids during crystallization. The intrusion thus apparently affords another example of differentiation by congelation crystallization, in which an increasing concentration of water also played a significant role in reducing magma temperatures and affecting the dark-mineral phases and the feldspar mineralogy.

7.6.3. Gravitational crystal fractionation

7.6.3.1. Introduction. Gravitational crystal fractionation describes the sinking or rising of primocrysts within magma as a result of their being in the general case either more or less dense than the enclosing silicate melt. The terminal velocity of particles (assumed to be spherical for ease in calculation) within a liquid, if the spheres are not above a certain size, is given by Stokes' law, expressed in the equation:

$$v = Kgr^2(d_1 - d_2)/n \tag{7.3}$$

where v = terminal velocity; g = acceleration of gravity; r = radius of particle; d_1 and d_2 = densities of particle and liquid, respectively; n = viscosity of liquid; and K = a constant depending on units employed.

It is apparent that, other things being equal, larger particles would move faster with a velocity proportional to the square of their radius, and that a small difference in density between particle and liquid or a high liquid viscosity will curtail relative movement. Given observed or deduced values of mineral and magma densities (see Chapter 6), primocryst size, and magma viscosity, it appears inevitable that all dark-mineral primocrysts will readily sink in mafic magma, that calcic plagioclase will barely sink in mafic magma, that the behaviour of plagioclase of intermediate compositions will depend critically on the composition and hence the density of the fractionating magma, that leucite would rise in the relatively very rare potassic mafic magmas from which it crystallizes, but that relative movement of any primocrysts in acid magmas may be very slow or negligible on account of their generally greatly higher viscosities, viscosities that are, however, known to be significantly reduced by the presence in some granites of large amounts of dissolved H_2O, HF, etc. (see Chapter 6).

Early forming primocrysts from basic magmas, as already discussed in connection with congelation crystallization, are relatively high-temperature members of solid-solution series, and their aggregate composition in any proportion is not the same as that of the magma crystallizing them. Thus the physical removal of primocrysts by sinking or rising results in profound changes in composition of the remaining liquid fraction, and hence in significant differentiation.

Because of the *field evidence* from large layered basic intrusions (where a *sequential solidus series* of primocrysts is preserved as *cumulates*), and of the *chemical evidence* from volcanic rocks of various igneous rock series of basaltic parentage (where *sequential liquidus compositions* are presumably preserved by the compositions of *aphyric lavas* or alternatively of the groundmasses of porphyritic ones), crystal fractionation is assumed by petrologists to play a major role in the patent differentiation of these rocks. The major-element chemistry of volcanic rock series in particular lends itself to exact models of crystal fractionation by the successive subtraction from magmatic liquids (beginning with a parental magma) of specified proportions of primocrysts of a composition equivalent to observed (or inferred) phenocryst phases. Ideally, these models are testable, as although it is possible to ring the changes between phenocryst phases and the relatively few major oxides to produce more or less whatever variation one wants, any specified crystal fractionation should also have a predictable effect on the concentrations of numerous trace elements in successive liquids resulting from the fractionation. Without wishing to sound unduly cynical, it is worth noting that although fractionation is often claimed, indeed assumed, the instances in which a reasonably complete balance sheet of major and trace elements in successive liquid fractions has been worked out in this way are still relatively rare. Note also that although the physical removal of primocrysts by sinking is a very effective method of fractionation, the nett effect is comparable to that of congelation crystallization; the effectiveness of each is enhanced by a component of what has now come to be termed *adcumulus* growth (see p.235).

The notion of gravitational crystal fractionation is not new; it occurred to Charles Darwin during the voyage of H.M.S. Beagle when he observed porphyritic andesites and associated lavas of a range of compositions in the Chilean Andes. The relative rising in magma of leucite crystals, commonly concentrated in the first products of the cyclic eruptions of Vesuvius, had long ago been deduced. Ussing (1912), in a beautifully written description of the highly alkalic Ilimaussuaq Intrusion, southern Greenland, reported the rising of sodalite crystals to form naujaite, and the sinking of arfedsonite and eudialyte crystals, etc., to form layered kakortokites and stated that:

"the differentiation may partly be explained on the old [sic] principle of mechanical separation by gravity of crystals from their magma."

Daly (1933, pp. 334–337) listed no less than 54 concordant layered intrusions in which gravitational fractionation had been proved, the great majority formed from mafic parental magmas.

7.6.3.2. The Skaergaard Intrusion. It was, however, the somewhat later account of the **Skaergaard Intrusion**, east Greenland, by Wager and Deer (1939) — the most cited reference in igneous petrology — that marked an important landmark, parallel to the theoretical work of Bowen (1928), from which time on it could be truly said that both the efficacy and the validity of crystal fractionation models was enthusiastically accepted by petrologists. It is still worth the time of every student of igneous petrology to read this classic account, incorporating intrepid fieldwork, many illustrations, and accurate analytical work. The later more precise and widely adopted cumulate nomenclature (Wager et al., 1960) is employed in a relatively condensed account of the Skaergaard and other layered intrusions by Wager and G.M. Brown (1968).

What factors resulted in the Skaergaard, allowing for the hiatus caused by the Second World War, becoming an instant and continuing classic, given the intrinsic excellence of the work of Wager and Deer? First, the tie-in of the sequence of observed crystallized mineral phases and their compositions with the classic experimental work at 1 atm. of Bowen and his associates (reviewed in Chapter 6) is uncannily precise. Then the rocks of the intrusion served as a testing ground for the principles of V.M. Goldschmidt governing diadochy and the ensuing partitioning of trace elements during fractional crystallization (Wager and Mitchell, 1951). This avenue of work has been followed up by a host of papers, using advanced analytical techniques dealing with specific elements and relating their abundance to later concepts governing diadochy. The intrusion served as an early testing ground for the behaviour of strontium isotopes during fractionation (E.I. Hamilton, 1963). Latterly, the development of two immiscible liquids at a very late stage in the fractionation of the Skaergaard magma has been proposed (McBirney, 1975). The Skaergaard magma is thus an outstanding example of the tendency for an intrusion or province, once classic, to remain so and become a battle ground for subsequent concepts and fashions in petrology. A striking example of this is the detailed reinterpretation by McBirney and Noyes (1979) of the mechanism of crystallization of the Skaergaard Intrusion. Based on field and textural evidence and on theoretical considerations of viscosity, heat transfer, crystallization rate and densities (particularly the embarassingly low density of plagioclase that appears to be a cumulus mineral), a relatively greater role is advocated for congelation crystallization in the layered series rocks themselves, in addition to the dominant role congelation crystallization played in the formation of the border-group rocks.

The intrusion is superbly exposed in recently glaciated coastal mountains rising to 2000 m. Structurally, the intrusion reveals, within a tholeiitic chilled margin and an envelope of border-group rocks composed of congelation cumulates (see p.227), a series of layered rocks originally accumulated on flattish saucer-shaped surfaces (Fig. 7.4) representing the progressively rising lower magma—solid interface of a crystallizing magma chamber. There is suggestive evidence from amazing sedimentary-like structures such as gravity bedding, cross-bedding, locally developed inwardly directed trough-like structures, relative concentration of dark minerals within layers traceable across the intrusion, that powerful convection currents were operative within the magma chamber; it is these, in addition to a postulated simple gravitational

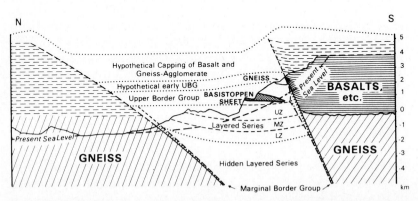

Fig. 7.4. Cross-section of original Skaergaard Intrusion restored to attitude at time of crystallization after allowing for effects of later tilting (after Wager and G.M. Brown, 1968, p.20, fig. 8).

settling of primocrysts, that seem to have conspired to produce the observed layering in the complex. In addition to the considerable relief of the area underlain by the layered series, a post-intrusion tilting of ~20° associated with the East Greenland Crustal Flexure has had the happy result of increasing the exposed thickness of the layered series to ~2600 m.

The layered series thus comprises a series of solidus crystals that show a progressive and systematic change in composition upwards whatever their exact mode of formation may have been. The crystals represent primocrysts, generally belonging to solid-solution series, that were crystallizing in equilibrium with a magma that was itself undergoing strong fractionation due to the effective removal of these primocrysts from the liquid magma as a result of their accumulation on the rising floor of the magma chamber. This compositional variation of the crystals upwards in the layered series was first termed *cryptic variation*, a term now superseded by the term **cryptic layering**. The term cryptic, of course, refers to the fact that this subtle variation in mineral composition is not apparent in hand specimen, but can only be appreciated by variation in optical parameters in thin section, or more precisely by careful analysis of the individual crystals.

This cryptic layering is the key to understanding the progress of fractionation with falling temperature in the crystallization history of the intrusion. The cryptic layering is quite independent of such accidental features as the actual proportions of minerals present in the various layers, it remains steadily and relentlessly *normal* upwards, i.e. what is observed is the presence of lower-temperature varieties of solid-solution series upwards in the cumulate sequence (just as they occur inwards in the border groups). This phenomenon is particularly well shown by the continuous pattern of cryptic layering against structural height in the layered series displayed by the plagioclase and calcium-rich pyroxenes of the intrusion (Fig. 7.5).

However, there are some apparent discontinuities in the upward sequence of primocrysts: early magnesian olivine, for example, gives place to a ferromagnesian pyroxene which in turn gives place to fayalitic olivine upwards in the layered series. This apparent discontinuity is, however, in fact the predicted result of continued fractional crystallization. We have seen in Chapter 6, for example, how the early crystallization of forsteritic olivine from tholeiitic melts is superseded via a reaction point by crystallization of ferromagnesian pyroxene, and that with more iron-rich compositions at falling temperature the pyroxene field is eliminated in favour of the coprecipitation of fayalite and quartz. Opaque minerals and apatite also become cotectically crystallizing phases in upper levels of the layered series rocks when their respective solubility products are attained in the crystallizing magma. The mineralogy of the opaque minerals in the intrusion is elegantly treated by Vincent and Phillips (1954), but in the simplest terms, the crystallization of magnetite, $FeO \cdot Fe_2O_3$, depends on a sufficiently high concentration of ferric iron being reached in the crystallizing magma; this is presumably attained by

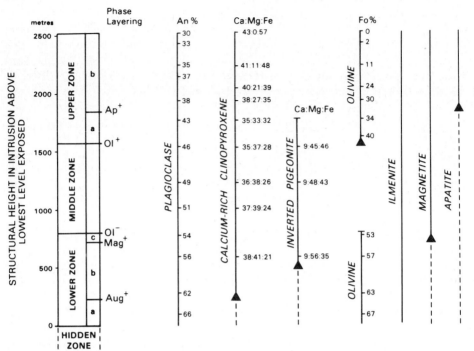

Fig. 7.5. Minerals present in the layered cumulate rocks of the Skaergaard Intrusion, east Greenland, and their compositions. *Continuous vertical lines* indicate cumulus minerals; *broken vertical lines* indicate that a mineral is present as an intercumulus phase only (slightly simplified after Wager and G.M. Brown, 1968, pp. 26—28, figs. 14 and 15).

the continued fractionation of mafic silicates containing no essential ferric iron, possibly aided by a buildup of f_{O_2} by the selective diffusive escape of H_2 from the magma chamber; the hydrogen would originate from the dissociation of water, $2H_2O \rightleftharpoons 2H_2 + O_2$, itself originating as juvenile water or absorbed meteoric water. The commencement of crystallization of apatite presumably reflects the progressive buildup during strong fractionation in the crystallizing magma of P_2O_5 which is not an essential component of any other primocryst phase.

These changes in the primocryst assemblage in what is essentially a continuum of crystallization are obvious mappable features of the accumulated rocks of the layered series and are examples of what has been termed **phase layering**.

7.6.3.3. Cumulates. The remarks in the above four paragraphs refer to the accumulated primocrysts of the rocks of the layered series, known as **cumulus crystals**, and **cumulate rocks** or **cumulates**, respectively (Wager et al., 1960), a nomenclature that has been widely adopted in dealing with layered igneous rocks of this type.

Cumulate rocks exhibit a distinctive range of textures because they were apparently formed in two stages: first, the cumulus crystals settled to form a rather loosely-packed crystal mush of roughly 60% crystals and 40% liquid as indicated by their textures, by theoretical considerations on packing in a viscous liquid of low density contrast, and by experimental observations. The trapped **intercumulus liquid,** with a composition of the magma at a liquidus temperature corresponding to the fractionation stage reached, must crystallize later. Considerable interest and significance attaches to this crystallization; two theoretical and contrasted possibilities are as follows.

The principle of the first, so-called **orthocumulate** crystallization, intuitively perhaps the most likely, is that the products of crystallization of the intercumulus liquid sum to its composition that it had when it became entrapped as a liquid network within a cumulus crystal mush. Intercumulus crystallization would be expected to begin by additive growth around cumulus crystals and progress to form compositionally normally zoned rims of solid-solution series around cumulus grains, in contrast to the virtually unzoned character of cumulus crystals themselves (see Carr, 1954). Additional phases may result from the crystallization of the intercumulus liquid: for example, given a parental tholeiitic magma, intercumulus plagioclase, augite and opaque mineral would crystallize between first-formed olivine cumulus crystals; at a later stage of crystallization, represented by a higher level in the cumulus pile, small amounts of quartz or micropegmatite might be anticipated in the interstices of a plagioclase-augite cumulate, and so on. A distinctive texture results where an additional intercumulus phase nucleates only slowly and at scattered centres during the slow cooling of the intercumulus liquid. These nuclei then serve as growing centres, sometimes spaced as much as 10 cm apart for poikilitic intercumulus crystals, **oikocrysts,** each of which may include up to a very large number of cumulus crystals of other mineral species. This ability of material to diffuse through the intercumulus liquid to form such oikocrysts should alert us to another possibility, already referred to in the case of congelation cumulates, and contrasting sharply with orthocumulate crystallization.

The formation of **adcumulates,** the contrasted possibility, happens because the cumulus crystals in the crystal mush act as ready nuclei for the commencement of crystallization of the intercumulus liquid just as in the first possibility above. Initially, this crystallization will result in additive growth of the same composition as the cumulus crystals. Concomitantly, however, diffusion between the crystallizing intercumulus liquid and the overlying magma may occur so as to equilibrate their compositions. Because of this a process of additive growth of cumulus crystals may ensue and complete solidification may result at temperatures at, or only very slightly below, the ambient magmatic temperature of cumulus crystal formation. Adcumulate growth would be indicated by the presence of unzoned minerals in the final rock, and also bands of virtually monomineralic rocks such as dunite, anorthosite, etc., developing from layers of monomineralic cumulus grains, phenomena which would not occur if the intercumulus liquid has in fact crystallized as a closed system as in the theoretical case of orthocumulates.

The further possibility of **heteradcumulus** growth, a textural variety arising under adcumulus growth temperature conditions, occurs where a phase (or phases) different to any of the cumulus phases in a particular layer may nucleate from the intercumulus liquid, but the phase(s) so nucleating have a composition compatible with crystallization at the same temperature of formation as the cumulus minerals. An example would be a sequence of plagioclase-olivine-augite cumulates containing a layer of plagioclase-olivine cumulate rock itself containing intercumulus augite of the same composition as the cumulus augite in adjacent rock layers. It is this compositional feature that distinguishes heteradcumulate growth from orthocumulate textures that may superficially resemble it. Oikocrysts, as described above, are in fact relatively more common in heteradcumulates than in orthocumulates, presumably because a slow rate of cooling favours both adcumulus growth and slow nucleation rate, both a reflection of a very low degree of undercooling in the intercumulus liquid (see discussion on kinetics of crystallization, Section 6.2, in Chapter 6).

The mineralogy, textures and composition resulting from intercumulus crystallization of most cumulate rocks reveal in fact a compromise between these two extreme possibilities of ortho- and adcumulate crystallization, and can conveniently be termed for this reason **mesocumulates**. A consideration of cooling rates suggests that the cumulate layers in a small, layered intrusion accumulated relatively rapidly and thus favoured a highly orthocumulate component in intercumulus material (as shown by many Skaergaard rocks); on the other hand, in a large intrusion a slower accumulation rate would favour adcumulus growth (as shown by many Bushveld rocks). However, generalizations and preconceived hypotheses tend to be dangerous in geology (where the rocks themselves are the final arbiters of scientific truth), and in fact different proportions of ortho- and adcumulate growth are commonly seen in layered rocks of the same intrusion. In the Skaergaard Intrusion, for example, there seems to be a higher component of adcumulus growth at higher levels in the exposed layered series.

The explanation above follows the simple "crystal settling" approach of L.R. Wager and coworkers. McBirney and Noyes (1979) in their re-interpretation object to the cumulus nomenclature and regard it as an unfortunate imposition of a "genetic" assumption, a tendency all too common in igneous petrology. Nevertheless the nett chemical effects on the remaining magma (our present concern) of contrasted "adcumulus" and "orthocumulus" crystallization is the same, whether the rocks originated in the main by either crystal settling or by congelation crystallization, and the terms are thus retained here.

It will be appreciated that adcumulus growth implies a perfect separation of solidus material from crystallizing magma at its liquidus temperature and thus represents the *most efficient possible fractionation*. This is not to deny, however, that mesocumulus growth or even orthocumulus growth, in which a complete separation of solidus material from the liquid is not achieved, is also a potent ongoing mechanism of fractionation over the period of crystallization of an intrusion.

It is worth noting that the most distinctive cumulate textures of all in thin section are found where there happens to be only one cumulus phase in the rock: in this event both the contrasted ortho- and adcumulus mechanisms produce distinctive and readily identifiable products such as obvious intercumulus textures, often with oikocrysts on the one hand or monomineralic rocks on the other hand. By contrast, where there are several cumulus phases, for example plagioclase and two pyroxenes, intercumulus crystallization tends to add on to and modify the shape of all the primocryst cumulus crystals, resulting in a typically gabbroic texture and masking the cumulate nature of the rock in hand specimen or even thin section (although a cumulate origin, of course, may very well have been surmised from layering and other structures visible in the field). It should also be noted that in some of the very large layered basic intrusions such as the Bushveld Complex and the Stillwater Intrusion which have cooled very slowly, a subsolidus *recrystallization* has affected parts of the ultramafic cumulate rocks sequences. This produces an equigranular texture with straight-line boundaries and triple-point junctions, thus masking the original cumulate texture. This well-documented phenomenon (see, e.g., Jackson, 1961) becomes an important consideration when petrologists attempt to apply facile textural criteria to the interpretation of the textures of ultramafic nodules in alkaline mafic lavas.

The analysis of cumulate rocks, although still indulged in, is futile; belonging as they do to generally conspicuously layered sequences, individual hand specimens collected for analysis, albeit from the same structural level in the intrusion, may vary widely or subtly in the proportions of different cumulus species that they contain, a purely accidental attribute, but one that would be reflected in fortuitous compositional differences. Furthermore, cumulate rocks comprise a high proportion of intercumulus material itself made up of an imprecisely known ratio of orthocumulus to adcumulus crystallization products that differ in composition. The composition of cumulus rocks is thus never equatable with the composition of a magma and is only a crude parameter of fractionation. The important parameter to determine is cumulus mineral compositions by optical or by analytical methods. It is these compositions which reflect the fractionation of the magma.

The compositions of cumulus plagioclase and olivine from the lowest level exposed of the Skaergaard layered series are An_{66} and Fo_{67}. These are considerably more soda-rich and iron-rich, respectively, than phenocrysts in basalt or earliest crystallized congelation cumulates in the Skaergaard marginal border group. Assuming a basaltic parental magma, as indicated by the mineralogy and composition of chilled marginal rocks of the intrusion, these compositions imply that a considerable amount of fractionation had already occurred before their formation. Lower levels of the Skaergaard layered series are regrettably hidden below sea level, but presumably they are composed of cumulates of earlier crystallization and more refractory mineral compositions. The amount of the "hidden layered series" has been ingeniously calculated based on chemical parameters as 70% (Wager, 1960) or perhaps much less (McBirney, 1975).

7.6.3.4. Other tholeiitic layered basic intrusions. Other layered basic intrusions of similar tholeiitic affinities fill this gap in the Skaergaard cumulate sequence. The large **Stillwater Intrusion,** Montana, fully described by Hess (1960), has an exposed base and a continuous layered sequence upwards until it is cut off by overlying nonconformable sedimentary rocks. Coincidentally, the cumulus mineral compositions at this uppermost level correspond rather closely with those of the lowest exposed Skaergaard cumulates, so that the layered sequences of the two intrusions, although the one is two orders larger in size than the other, can be pieced together to reveal a continuous whole in the pattern of cryptic layering.

A whole continuum is provided in one intrusion by the **Bushveld Intrusion,** South Africa, the largest basic intrusion in the world, ~400 km across its maximum diameter, and attaining a staggering maximum thickness of 9 km. Fig. 7.6 shows the cumulus minerals and their compositions of this remarkably complete fractionated sequence, the present-day accepted interpretation of which had to await the classic work on the Skaergaard, Stillwater, and other intrusions (see Wager and G.M. Brown, 1968).

The Skaergaard Intrusion has a visible roof of border-group congelation cumulates, and a pattern of fractionation that suggests that, apart from a small and questionable degree of assimilation of gneiss inclusions, it crystallized as a *closed system* of fractionating basic magma in situ. Fairly efficient mesocumulate fractionation gave rise to no more than some 1—2% (depending on assumptions made in calculating the bulk of the hidden layered series) of acid granophyric rocks. In the very latest stages of fractionation it is possible that an acid magma of this composition may have coexisted in an immiscible relation with an iron-rich liquid that crystallized the uppermost ferrogabbros. However this may be, it is clear that the field evidence places constraints on the amount of residual acid rock of granitic composition that can be generated from basic magma of appropriate tholeiitic composition by fractionation processes. Field relationships of acid rocks in the larger Bushveld Intrusion are confusing because of the apparent generation of some acid rocks by remelting of the roof of the intrusion and the presence of unrelated later acid intrusions (von Gruenewaldt, 1968). Nevertheless, the total volume of acid rocks attributable to the period of emplacement of the Bushveld layered intrusion whether produced by fractionation or by anatexis, is also relatively small.

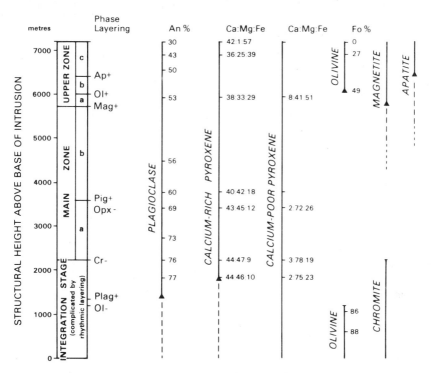

Fig. 7.6. Minerals present in the layered cumulate rocks of the Bushveld Intrusion, South Africa, and their compositions. *Continuous vertical lines* indicate cumulus minerals, *broken vertical lines* indicate that a mineral is present as an intercumulus phase only (after Wager and G.M. Brown, 1968, p.351, fig. 192).

Some layered basic intrusions appear to have been *open systems* in that their pattern of cryptic layering suggests that magma chambers underwent periodic partial evacuation of partly fractionated magma and replenishment by fresh magma, i.e. they were subterranean staging points during an interrupted ascent of basic magma to the surface. In the **Muskox** Intrusion (Irvine and C.H. Smith, 1967), although cryptic layering in cumulus olivine is not pronounced, there are subtle reversals in Fo content coupled with pronounced increases in Ni content of olivine. As nickel is strongly partitioned into early crystallizing olivine, these data strongly indicate replenishment of the crystallizing magma chamber with fresh additions of magma. In the Isle of **Rum**, Hebrides, western Scotland, a small high-level layered intrusion was formed from a magma chamber within a volcano. Major rhythmic units, defined by conspicuous changes in the relative proportions of cumulus olivine, augite, and plagioclase, typically some tens of metres thick, olivine-rich at the base and plagioclase-rich towards the top, have been held to reflect intermittent accessions of fresh magma (G.M. Brown, 1956), a view subsequently corroborated by work on the nickel content of the cumulus olivine crystals (Henderson and Gijbels, 1976).

Also in Rum, a curious texture has been attributed to the temporary cessation of the fall of cumulus crystals, resulting in the upward growth of coralline crystals of olivine and plagioclase up to 20 cm in height from the lower magma—crystalline mush interface. This was originally termed *harrisitic structure* after Glen Harris in the centre of the isle; now the more general term **crescumulate** is applied to this type of crystal formation. It has been interpreted as a type of congelation cumulate appearing intermittently within a layered cumulate sequence that accumulated in the main by gravity settling (Wadsworth, 1961).

7.6.3.5. Layered intrusions of compositions other than tholeiite. The above examples of observed sequences of solidus minerals have for the greater part apparently been derived from parental tholeiitic magmas. Most of the world's layered intrusions and all of the larger ones are in fact tholeiitic. Interesting examples of other sizeable intrusions of differing magma compositions that have apparently undergone gravitational crystal fractionation include the following intrusive rock bodies, with their probable parental type in square brackets:

(1) Blue Mountain Complex, New Zealand [alkali basalt] (Grapes, 1975).

(2) Great Dyke of Rhodesia [high-magnesium basalt] (Worst, 1958; C.J. Hughes 1977).

(3) Michikamau anorthosite, Labrador [?high-alumina basalt] (Emslie, 19   upward accumulation of plagioclase crystals on massive scale).

(4   ssaluk granite, Ivigtut region, west Greenland [fluorine-rich epi-granite*] (Harry and Emeleus, 1960).

(5) Kungnat Intrusion [fayalite quartz-syenite*] (Upton, 1960).

(6) Ilimaussuaq Intrusion, south Greenland [peralkaline nepheline sye-nite*] (Ussing, 1912; Sørenson, 1958; J. Ferguson, 1964).

See Wager and G.M. Brown (1968, ch. 15—17) for accounts of these and other layered intrusions.

7.6.3.6. Fractionation in sill rocks. So far discussion of gravitational crystal fractionation has been presented in the light of the compelling field evidence from large layered intrusions. Some thick sills also show evidence of differentiation by crystal settling.

Considering first sills of overall tholeiitic composition, the Whin sill of northern England, thicknesses generally of 25—60 m, does not show crystal settling; presumably, congealing in this relatively thin body proceeded too fast for any settling of discrete primocrysts to occur. Crystallization of the considerably thicker Palisades sill (thickness up to 300 m) has been described above in terms of congelation crystallization. Some 15 m above the base, how-

---

*Major rock type present and (immediate) parental magma, almost certainly derived by strong fractionation of unexposed, relatively more basic (ultimate) parental magma.

ever, in southern exposures of the sill, the well-known olivine layer shows in thin section clear evidence of the concentration of cumulus crystals of olivine. This has recently been re-interpreted not in terms of the crystallization of a static body of magma, but in terms of crystallization ensuing from the accession of a new magma pulse intermingling with the still crystallizing old magma, and resulting for a short period of time in the formation within the body of the intrusion of olivine primocrysts that had time to accumulate towards the base of the intrusion (K.R. Walker, 1969). Where the olivine layer is developed there is a slight reversal within the layer in the upward cryptic layering of plagioclase which in habit also changes upwards from tabular to oikocrystic, suggesting that a partial hiatus in the crystallization of the sill rocks accompanied the second pulse of magma and provided time for some olivine primocrysts to accumulate. Even in some very thick tholeiitic sills belonging to the Karroo, there is rather little evidence of crystal settling (F. Walker and Poldervaart, 1949, pp. 651 and 652) and the prevalent method of fractionation in them appears to be by means of congelation crystallization. Further interesting examples of fractionated tholeiite sills include the Red Hills diabase, Tasmania, (McDougall, 1962), Antarctic sills (Gunn, 1966), and the Tumatumari—Kopinang diabase sill, Guyana, (Hawkes, 1966).

Sills crystallized from more alkaline mafic magmas, however, tend to display gravitational crystal fractionation on a more appreciable scale. Pronounced gravity settling within the Shiant Isles sill, Hebrides, ~1 0 r in thickness, for example has led to the formation of a lower layer at le   m thick of accumulative picritic rocks containing over 60 vol.% o ivine crystals (Drever, 1953). This has a sharp upper contact with overlying olivine diabase containing titaniferous augite passing upwards into crinanite (olivine—analcite diabase). The olivine of these diabases and crinanites shows a remarkable degree of normal zoning from cores not much more iron-rich than the accumulative olivines in lowest rocks (Fo_{81}) to rims of fayalite-rich compositions in uppermost rocks (Johnston, 1953). It seems clear therefore that after early dumping of olivine crystals the sill crystallized in congelation cumulate manner in a strongly orthocumulate fashion. Very likely the accumulative olivines were already present as phenocrysts when the sill magma was emplaced.

Remarkably complete exposures of the Shonkin Sag sill (incorrectly termed a laccolith in older literature) show it to have a flat lens-shaped form, ~3 km in diameter and up to 80 m maximum thickness in the middle. Within a chilled margin of leucitite (not "mafic phonolite" as in some accounts) containing phenocrysts of augite, leucite (generally replaced by pseudoleucite) and rare olivine, crystal settling of augite has produced the accumulative rock type known as shonkinite. The shonkinite occurs in two flat layers near the lower and upper contacts, each shows increasing amounts of euhedral accumulative crystals of augite upwards, a curious feature attributed to congealing outpacing a more complete fallout of augite primocrysts probably present as phenocrysts on intrusion (Hurlbut and Griggs, 1939). The two shonkinite layers sandwich between them a more fractionated non-accumulative nepheline-augite syenite itself auto-intruded (see p.248) by veins of a more iron-rich aegirine syenite.

Accumulative rocks within sills are thus on the whole relatively rare; where they occur they reflect one or more of the following: (1) unusually thick sills, (2) presence of phenocrysts at the actual time of intrusion; (3) more alkaline parental magmas with relatively low viscosities because of their composition (see Chapter 6); and (4) komatiite sills with relatively low viscosities because of their high temperatures.

7.6.3.7. Fractionation in flows. Examples of gravity settling in lava flows are rare. One amusing example is the concentration of olivine phenocrysts towards the base of individual pillows of basaltic flows of transalkaline parentage in southern Iceland (Mathews et al., 1964). Considering the rapid crystallization of a pillow lava environment, the sinking of olivine was here achieved remarkably rapidly. Gravity settling is more prevalent in peridotitic komatiite flows. In these high-temperature, ultrabasic, high-magnesium melts congealing has been rapid as shown by the remarkable examples that they contain of skeletal and dendritic crystallization habits (see Chapter 6); spectacular examples of the latter are the well-known spinifex textures of olivine, augite and opaque mineral. Despite this rapid crystallization, some larger flows contain olivine cumulates near the base and augite cumulates near the middle, a differentiation produced by gravity settling. The numerous small cumulus crystals in these rocks suggest rapid nucleation and easy settling, features to be anticipated in a hot and relatively fluid magma. There is evidence too in some flows of the accumulation of much larger bladed olivine crystals of dendritic crystallization habit.

Olivine settling within the prehistoric Makaopuhi lava lake of Kilauea, Hawaii, has been documented in detail by J.G. Moore and Evans (1967); the scale, however, a vertical section close to 70 m in thickness, is more analogous to that of sills than of lava flows*.

---

*Note in passing the demonstration by Moore and Evans of considerable re-equilibration of the compositions of olivine phenocrysts of the Makaopuhi lava lake during crystallization of the groundmass and later subsolidus cooling down to temperatures at least as low as 800°C by the interdiffusion of Fe, Mg, Ni and Mn. They conclude that:

"the compositions of olivines in basaltic rocks only reflect those of the intratelluric stage of crystallization when cooling is extremely rapid."

This is an important and disturbing consideration to be borne in mind in the analysis of igneous rock minerals. Somewhat paradoxically the other situation where olivine and other primocrysts may be expected to retain their original composition is under completely contrasted conditions to rapid cooling — namely the extremely *slow* cooling associated with adcumulus growth where solidification and equilibrium crystallization does in fact occur at ambient magma temperature.

7.6.3.8. The behaviour of trace elements during crystal fractionation. As we have seen, the pattern of evolution of major-element contents in a fractionating magma is governed by the sequence, composition and amounts of fractionating crystal phases, including the removal of one trace element, chromium, by the early fractionation of a separate phase, chromite. The great majority of trace elements, however, are dispersed in habit (see Chapter 1) and occur in generally small but highly variable concentrations within the crystal lattices of the fractionating mineral species. The concentrations of trace elements in fractionating magma will thus come to depend critically on their partitioning between the evolving sequence of sets of crystallizing mineral phases and continuously changing magma compositions.

An early study of this very significant phenomenon was made by Wager and Mitchell (1951) by comparing the concentrations of trace elements in minerals of the Skaergaard cumulate sequence with the inferred compositions of successively derived liquids. The compositions of these liquids were calculated from a study of the geometry of the intrusion, bulk cumulate compositions and parental magma composition; this calculation was subsequently refined by an ingenious and detailed consideration of Ti and P contents, and other parameters (Wager, 1960). A great deal of their overview and discussion, in which trace-element behaviour was related primarily to the classic principles of V.M. Goldschmidt, is valid today although principles governing diadochy have been refined by considerations of electronegativity and crystal field theory (see Chapter 1). It is instructive to plot and account for the trace-element compositions of successive Skaergaard liquids expressed as percentage of magma crystallized, using the data of Wager and Mitchell. Multiple band log paper should be used for this exercise, as this clearly reveals the *relative* patterns of behaviour of elements present in different orders of concentration.

Residual trace elements such as Rb, Y, and Zr, are those which are not readily incorporated into the lattice structures of crystallizing minerals and thus steadily build up in concentration in the melt as crystallization proceeds. The maximum extent to which this is possible is a two-fold increase in concentration after half the melt has crystallized, a four-fold increase after three-quarters has crystallized, and so on. This theoretical maximum gradient can be well shown as a straight bounding line, using log graph paper with concentration plotted against an appropriate abscissa representing proportion of the melt crystallized.

Ni, on the other hand, is strongly partitioned into crystallizing olivine and is therefore very strongly depleted in concentration in the melt during early fractionation stages. Cr shows a comparable depletion trend.

Ti, ferric iron, V, and P behave as residual elements and increase in concentration in the melt until ilmenite, magnetite and apatite become crystallizing phases after which the concentration of these elements shows a decrease in concentration in the melt.

Sr behaves as a residual element until plagioclase becomes a crystallizing phase after which the concentration of Sr decreases only slowly in the melt because Sr is not strongly fractionated by plagioclase.

The study of trace-element concentrations in igneous rocks provides much information in the delineation, distinction, differentiation, and petrogenesis of igneous rock series and further references to this important topic are made in Chapters 8—13.

7.6.3.9. The inferred operation of fractionation in the production of volcanic rock series, using Thingmuli as an example. With this strong field evidence of gravity settling as an agent of fractionation, plus cogent evidence of congelation crystallization which would have a similar effect and which might be expected to be of greater efficacy at greater depths where wall-rock temperatures are not far removed from magma temperatures, it is not surprising that liquid/crystal fractionation is considered to be an overriding factor in producing the observed differentiation in volcanic members of igneous rock series derived from mafic parental magmas. In the case of the discrete associated members of a volcanic rock series as opposed to that of a differentiated intrusive body, however, the demonstration of fractionation depends on inference and does not amount to outright proof however likely it might seem.

Layered intrusive rocks preserve, with some relatively minor and readily appreciated complications, a sequential series of observed solidus crystal products — the cumulus crystals. By contrast, a series of lavas erupted from a magma chamber crystallizing and fractionating in this way would presumably preserve an inferred *liquid line of descent.*

An example of this is provided by the dissected late Tertiary volcanic edifice of **Thingmuli,** eastern Iceland, (Carmichael, 1964; 1967a), situated amongst mainly tholeiitic basalt flows that are the products of fissure eruption. Mainly aphyric or sparsely porphyritic lavas and hypabyssal rocks belonging to the volcanic complex, however, range from olivine tholeiite through tholeiite and intermediate rock types to rhyolite in composition (see Table 9.3). Carmichael shows in a convincing manner how it would be theoretically possible to derive successive members of this series by the fractionation of primocryst phases observed to be actual phenocrysts in the less evolved members of the series.

Table 7.1. shows: (1) a petrographic grouping of what is actually a continuum of Thingmuli lava compositions; (2) the approximate abundance of these in that part of the volcanic pile attributable to the central volcano of Thingmuli; (3) their approximate silica content, (4) their differentiation indices; (5) their *M*-values, and (6) phenocryst phases. It should be noted that phenocrysts are generally sparse throughout the series, and that many lavas are aphyric, a useful feature because the bulk analysis of an aphyric lava presumably equates with that of a magmatic liquid less only volatile constituents.

TABLE 7.1

Petrological data on Thingmuli lavas, generalized from Carmichael (1964)

Rock type	Approximate relative abundance (vol.%)	Typical silica content (wt.%)	Differentiation index (range)	M-value (average)	Phenocryst phases
Picritic tholeiite (accumulative)	<1	~44		—	chromite, olivine (forsteritic)
Olivine tholeiite (parental)	7	46—48	19—27	52	olivine (forsteritic), plagioclase (bytownite)
Tholeiite	50	49—51	29—37	37	olivine (forsteritic), plagioclase (bytownite—labradorite), augite
Basaltic andesite ⎫	18	52—56	45—51	32.5	plagioclase (labradorite), augite, magnetite
Icelandite ⎭		59—65	61—72	19	plagioclase (An_{45-40}), ferroaugite, magnetite, hypersthene or pigeonite or olivine (fayalitic)
Rhyolite	21	69—75	74—94	13 (low silica) 4 (high silica)	plagioclase (An_{35-30}), ferroaugite, olivine (fayalitic), magnetite
Pyroclastic rocks (mainly rhyolitic)	3	—	—	—	

Phenocryst phases in boldface are almost always present in porphyritic samples of the designated rock type. Phenocryst phases in ordinary type may or may not be present in individual samples of the designated rock type, even when it is porphyritic.

In the pattern of compositions of observed crystallizing phenocryst phases, plagioclase shows a continuum, as would be expected of a solid-solution series, as does the calcium-rich pyroxene. Crystallization of oxides shows a pronounced gap between small amounts of early chromite and later opaque mineral. The ferromagnesian phenocrysts change from early forsteritic olivine through hypersthene and pigeonite (these two species are admittedly rare but they are present, not necessarily in the same specimen, in icelandites) to a fayalitic olivine. The pattern is typical of a tholeiitic fractionation sequence and can be closely matched in the cumulate sequences of tholeiitic layered basic intrusions.

The probable parental magma of the Thingmuli series needs some consideration. The most basic rock present, the "picritic basalt", with lowest SiO_2 and highest MgO, does not qualify, as it is an *accumulative* rock merely mechanically enriched in the primocryst phases, forsteritic olivine and chromite, presumably by gravitational differentiation in a magma chamber prior to extrusion. There was never necessarily a liquid equivalent in composition to the bulk composition of this rock. The *groundmass*, by contrast, of such a rock would not be far from the composition of parental magma to the series, as it presumably represents the liquid in equilibrium with these earliest crystallizing phases. Alternatively, the most magnesium-rich aphyric lava is a likely candidate for parental magma (as the predominant early crystallizing phase is forsteritic olivine leading to initial depletion in MgO of the liquid). These considerations point to the composition of olivine tholeiite as parental magma for the Thingmuli series. Olivine tholeiite is not, as some traditional petrologists would like to see of a parental magma, present in relatively large amounts in the lava pile — tholeiite is eight times more abundant; however, given the physics of a fractionation process, there is no reason why it should necessarily be the most abundant extruded rock type*.

The nomenclature employed in the description of the Thingmuli rocks is worth a note, as it typifies some of the difficulties involved in applying our inherited and changing web of petrographic nomenclature to members of the several major igneous rock series that are recognized today. Although here used adjectivally, the term picrite precisely defined has nothing to do with volcanic rocks but is an olivine-bearing plutonic rock, ultrabasic (SiO_2 content less than 45 wt.%), with more than 10 vol.% plagioclase; furthermore, picrite carries the connotation for many petrologists of a titanaugite-bearing alkalic rock. It is here used to connote a tholeiite rich in accumulative olivine primocrysts — an "oceanite" in fact. The two groups of basic rocks belonging as they do to a tholeiite magma series are thus appropriately named; one small possible confusion is that for many petrologists the term olivine tholeiite would imply the presence of olivine phenocrysts whereas it can also apply as in the case of some of the Thingmuli rocks to relatively magnesium-rich aphyric tholeiites with normative olivine as opposed to those with normative quartz. The next two groups are intermediate in composition (SiO_2 content between 52 and 66 wt.%); the only term for these rocks in an old-fashioned

---

*Inferred mafic parental magma, furthermore, may not always have been extruded at surface at all. An example of this is the pantellerite province within the rift valleys of Ethiopia and Kenya, the greatest grouping of extrusive peralkaline rhyolite on Earth. From their pattern of trace-element contents, Weaver et al. (1972) convincingly show that the rocks could have been produced by crystal fractionation from a basaltic parental magma, probably, by analogy with other pantelleritic rocks, one of transalkaline type, i.e. between tholeiite and alkali basalt in composition. It is indeed only to be expected that during a long ascent through relatively thick lithosphere a very great deal of fractionation could have occurred, and thus little or none of the mafic parental magma may have reached the surface.

petrography is andesite. However, as Carmichael (1964, 1967a) points out, they are quite different in their chemistry, phenocryst content and tectonic setting from the relatively abundant and well-known "orogenic" andesites. Carmichael therefore quite properly creates a new name, "icelandite", to denote an intermediate volcanic rock of a tholeiite fractionation series, but somewhat illogically retains the old term andesite in alluding to basaltic andesites (why not basaltic icelandite?) of the basic end of the range of icelandite. The proper nomenclature of intermediate members of different igneous rock series is a subject of current difficulty as there are not enough adequately defined traditional names to go round. See, for example, the discussion on this point in Cox et al. (1979, pp. 14 and 15).

The high relative abundance of acid rocks (some 20%) in the Thingmuli area is unexpected despite their chemical fit to a fractionation model, and Carmichael therefore left open the possibility of refusion of lower crustal material for them. In this connection, Sigurdsson (1977) has proposed that remelting of plagiogranites (themselves fractionation products of low-potassium tholeiitic magmas) due to "rift jumping" (eastern Iceland is apparently being pushed westwards over a mantle plume by the continued accretion of the African and Asian plates: see Chapter 9) has resulted in the generation of rhyolitic magmas in Iceland. Fractionation, therefore, frequently an attractive hypothesis, is not necessarily the sole candidate for processes producing the observed differentiation in lava series.

7.6.4. Filter differentiation

The idea of **filter differentiation** is that given a crystal mush, i.e. a body of crystals plus interstitial liquid, significant fractionation could be effected by (a) squeezing the system so that the crystals become more tightly packed and some liquid residue thereby concentrated ("filter press action"), or (b) the weight alone of the relatively heavier crystals leading to closer packing within the mush and upward concentration of liquid. The notion has had a certain vogue (see Emmons, 1940), and was invoked as a possible explanation of vertical mineral variation within the Glen More ring dyke of Mull (E.B. Bailey et al., 1924). Further convincing examples of differentiated mafic igneous rocks attributed to this mechanism of fractionation are, however, lacking. To be sure, during the actual accumulation of cumulus crystals accompanied perhaps by some small progressive degree of closer packing there should be a relative upward displacement of liquid. However, this much is part and parcel of a gravity settling process, and given such factors as the relatively small difference in density between crystals and magma, relatively high magma viscosities, and the role of a component of adcumulus growth in quickly reducing the amount of liquid in a crystal mush, there does not seem to be much call for this mechanism in producing fractionation in mafic rocks, which typically occur in anorogenic environments where they are unlikely to be "squeezed" during crystallization. The situation could be radically different, however, in the case of granite "migmas" emplaced under katazonal and mesozonal conditions under stress as shown by an igneous foliation that most of them possess. Under these conditions water-rich or water-saturated silicate pore fluids of composition approaching the ternary minimum could become squeezed off upward into relatively more felsic bodies or even separated from the parent body and concentrated in apical regions or to form marginal pegmatites, of which the mineralogy does reflect this overall composition, and hence a differentiation from the parent plutonic body.

A broad historical review of the gamut of possible processes leading to filter differentiation is given by Propach (1976).

7.6.5. Autointrusion

Autointrusion results from the brittle disruption of the solidified part of a crystallizing magma chamber in response to cooling or stress forces, whereupon interstitial liquid would tend to migrate to fill tensional fractures or other actual or potential cavities, forming while crystallization was still taking place. The writer has an amusing example of a vesicular tholeiite from Iceland in which some residual liquid appears to have oozed into small spherical vesicles and crystallized there to give material of a much more felsic nature than the average rock, thereby effecting differentiation, admittedly on a very small scale. On a somewhat larger scale, decimetre wide granophyre veins found cutting the marginal zones and wall rock of some tholeiitic sills such as the Palisades sill may well have originated by this process. Syenitic veins and patches in alkali basalt sills and in the Shonkin Sag sill (see p.241) which have a composition compatible with their being fractionated liquid products may also have formed in this way, and provide useful sampling material for illustrating the path of late-stage fractionation. On a larger scale again some of the relatively undifferentiated ferrodiorite and ferroadamellite bodies found within the upper parts of the large Michikamau anorthosite intrusion (Emslie, 1970) may represent the autointrusion of residual liquids into collapsed parts of the congealed roof of the complex; their compositions may similarly preserve the liquid line of descent of fractionating magma.

7.6.6. Fractionation in nuée ardente eruptions

One highly specialized and spectacular method of fractionation affects the products of large nuée ardente eruptions, in which there is apparently a tendency for the more finely particulate shard components to be concentrated in the large volumes of wind-blown ash accompanying the eruption (R.A. Bailey et al., 1976; Roobol, 1976; Sparks and G.P.L. Walker, 1977). This results in a dispersed thin ash layer relatively enriched in glass (representing the liquid fraction of the magma chamber), and an ignimbrite sheet relatively enriched in crystals (the crystallized fraction of the magma) and in lithic fragments. On this fundamental fractionation plus the frequent complication of zoned cooling units (see Chapter 2), factors of lateral and vertical compositional variation in ignimbrite sheets are superimposed, reflecting a tendency for relatively light pumice fragments to both rise up and travel further than lithic fragments during the ash-flow eruption. All in all, samples of ignimbrites (and some can be very rich indeed in crystal components) may be more highly fractionated than one may be led to believe on a simplistic view of the apparent overall homogeneity within an individual exposure of ignimbrite.

Chapter 8

IGNEOUS ROCK SERIES

8.1. HISTORICAL REVIEW

A preoccupation with the composition and distribution of recurring associations of igneous rocks has been a leitmotiv of igneous petrology, well summarized by Daly's (1933) observation that there *is* a degree of sameness in the midst of variety. The usage over the years of terms such as province, branch, suite, kindred, association, family, series, group, stem, clan, lineage and consanguinity reflects the groping by petrologists after an intuitively perceived order in these respects. Rephrased in contemporary terms this concern would appear as the recognition and definition of igneous rock series and the relationship of each series to global tectonics, a relationship that is in fact perceived in fairly sharp focus at the present time.

Illustrative of early concepts in distinguishing magmatic affinity are: (1) the recognition by J.P. Iddings in 1894 of "alkali" and "subalkali" groups, a concept not thought sufficiently definite to form a basis of classification at the time; (2) the distinction by H. Rosenbusch in 1907 of an "Alkaligesteine" (alkali series) and a "Kalk-Alkaligesteine" (calc-alkali series); and (3) the designation by Harker (1909, p.90) of "Atlantic" and "Pacific" branches to refer to rocks of alkaline and subalkaline affinity. These latter names related to the several series of alkalic volcanic rocks found on various of the smaller Atlantic islands and to the circum-Pacific orogenic andesite associations, respectively*.

To these were soon added an "Arctic" suite of continental flood basalts by F. von Wolff in 1914, and a "Mediterranean" suite of distinctly potassic flows by P. Niggli in 1921. The "spilitic" suite of H. Dewey and Flett (1911) was always regarded as of dubious validity by some petrologists as it specifically comprised only rocks of degraded mineralogy. Returning purely to petrographic criteria, Holmes (1920) distinguished "alkali rocks", in which a relative abundance of alkalis is reflected by the presence of essential soda-bearing inosilicates and/or feldspathoids, from the more abundant "calc-alkali" rocks, which specifically lack representatives of either of the above-mentioned alkali-rich mineral groups and which contain instead

---

*Note, however, that Iceland, the largest island in the Atlantic, is built mainly of tholeiitic basalt and related rock types that are *not* alkalic in nature, and that the Pacific contains several examples of distinctly alkaline suites, notably Samoa and Tahiti, as well as the massive Hawaiian archipelago which like Iceland is constructed largely of tholeiitic rocks.

essentially feldspars and common augite or hornblende.

The above concepts hovered uneasily between an overall chemical and/or mineralogical distinction between the several variously named groups, series, branches, or suites on the one hand, and an implication of some kind of geographic/tectonic distinction implied by the later names on the other (see the discussion of this topic in Loewinson-Lessing, 1954, p.23).

Tyrrell (1926) developed the idea of different "kindreds" comprising spilitic, Arctic, quartz dolerite and quartz gabbro, granodiorite—andesite, anorthosite—charnockite, trachybasaltic, ijolite—melteigite + nepheline syenite, and Mediterranean, all within a broad two-fold division into alkalic and calcic. Tyrrell considered that rocks within a kindred might be capable of further subdivision into recognizably distinct "tribes" or even further into "clans". Within each subdivision so delineated individual rocks would be shown to be genetically interrelated by acceptable processes of differentiation.

By this time the important experimental work of N.L. Bowen and other workers was widely disseminated and it became widely assumed that crystal fractionation was a dominant factor in producing differentiation. Curiously this realization of the efficacy of fractionation, which has been amply justified and which contains the key to linking the rocks in igneous rock series, impeded the progress of adequately delineating the series themselves. Bowen rightly recognized mafic rock as the parental magma to others in a fractionation series (see p.253 for definitions), but apparently missed the extra vital key of recognition of *different* parental mafic magmas. Time and time again Bowen (1928, pp. 78, 221, 235, 239 and 240) ignores the possibility of parental magmas other than tholeiite, granted that tholeiite is unquestionably the most abundant mafic magma type. In turn, the over-towering relevance of Bowen's synthesis itself acted as an impedence to further work on the actual distinction of different igneous rock series. Following the precepts of fractional crystallization and ensuing processes of fractionation and other concepts of differentiation processes so lucidly set out by Bowen, petrologists the world over busied themselves with the field evidence and elucidation of differentiation processes instead. Furthermore, the so-called "granite controversy" (see Read, 1957) was raging and absorbed petrologists' energies*. Faced with the variety and interest of volcanic phenomena and igneous rocks in the field, it is not surprising that petrologists concentrated first on comprehending the emplacement and crystallization of lavas and magma, while tending to ignore the global significance of igneous activity. It was the scientific revolution of the new global tectonics in the late 1960's that created for the first time both a need and an adequate basis for the distinction of igneous rock series. The nomen-

---

*Many granitic rocks of the "plutonic association" have never passed through a largely liquid stage and are not relatable to any parental mafic magma and are thus distinct from the igneous rock series considered here. Their study constitutes a distinct major branch of igneous petrology.

clature of these series as we shall see cuts across a traditional classification such as that developed in Chapter 4 (but does not replace it for much field and other use).

8.2. DEFINITION OF TERMS

A **province** (alternatively an *igneous province* or a *petrographic province*) is a recognizably distinct area where igneous activity has occurred over a recognizably coherent period of time. The term carries the strong implication, but does not necessarily demand, that the igneous rocks contained in any one province are somehow genetically related; certainly petrologists would look for some genetic relationship. To give some idea of space and time scales involved, examples of provinces include such classic areas as: (1) the central part of the North Island of New Zealand from late Pliocene time to the present; (2) the Hawaiian archipelago with a history of active constructive volcanism over the last 10 Ma; (3) Iceland with a history extending over the last 16 Ma (note that present-day volcanism in Iceland can be further subdivided into "subprovinces" based on systematic and coherent differences in composition of parental magma type in different areas); (4) the Tertiary volcanic districts of the Hebrides and Northern Ireland; and (5) the Oslo region of Norway in Permian time.

A **suite** is merely a collection of igneous rocks from a particular province with no more and no less of a genetic connotation than that applicable to the term province.

An **igneous rock series** is an association of igneous rocks that can reasonably be inferred to be genetically interrelated by known processes of differentiation. If the same association recurs in several or more provinces the assumption that it constitutes a genetically interrelated igneous rock series and not merely a fortuitous juxtaposition is, of course, considerably strengthened. Whereas a province is factually definable (subject only perhaps to subdivision or arguable limits in space and time), the concept of an igneous rock series represents an hypothesis which must run the customary gauntlet, familiar to geologists, of analogy, testing, time, and adoption. More than one igneous rock series may occur within a single province: for example, the Hawaiian and the British Tertiary provinces each include two major series, a tholeiite series and an alkali basalt series.

Parental magma is defined as that which is capable of producing by sequential processes of magmatic differentiation all the other rock types in any specified igneous rock series. If these processes of differentiation are primarily fractionation processes, as seems most probable for all except the plutonic granite association, then the parental magma must satisfy certain conditions. Accumulative rocks apart, it should possess the highest liquidus temperature of any in the series. It must also have a major- and trace-element composition such that the progressive subtraction from it, of the compositions of visible or inferred fractionating crystal phases, may yield the compositions of the remaining rocks in the series. Mafic magma types do indeed fulfill these conditions, and most igneous rock series bear the name of their mafic progenitor for this reason. Another attribute of a parental magma considered desirable by some petrologists, that it be seen to be present in quantity in any particular province, is not essential, and indeed may be demanding of nature what just does not happen in all cases. For example, strong fractionation may have occurred within the crust resulting in the eruption of relatively greater amounts of evolved members of a series, conceivably even to the exclusion of representatives of crystallization products of accepted parental magma at or near the surface. Alternatively, denudation in older provinces may have partly or completely removed volcanic rocks with the inevitable result of drastically increasing the relative abundance of more evolved felsic members of a series in any accompanying high-level intrusions (see pp. 69 and 70 for the explanation of this phenomenon).

Usage of the term parental magma does not imply that it be equivalent to a "first melt" produced by processes of partial melting within the mantle, nor to a "primary magma" formed by ensuing processes of segregation into discrete liquid bodies. On the contrary, parental magma is defined in more pragmatic terms as that which arrives at crustal levels wherein documentable differentiation at relatively shallow levels occurs to produce an igneous rock series. Between generation and segregation to produce primary magma in the mantle and ensuing arrival in the upper crust as parental magma it is apparent that primary magma could itself undergo fractionation, possible wall-rock reaction, etc., at elevated but varying pressures and temperatures resulting in differentiation (see, e.g., Jamieson, 1970). Assuming in the simplest case progressive cooling and continuous fractionation during ascent, the exact definition of when fractionation of primary magma becomes fractionation of parental magma may become semantic. Indeed, the above-mentioned demand to see abundant representatives of the proposed parental magma among exposed rocks of a series is understandable, for such material at least provides a tangible starting point for the series*.

8.3. THOLEIITE AND ALKALI BASALT

As already mentioned, tholeiites are by far the commonest mafic extrusives in crustal rocks, and probably exceed all other mafic rocks in abundance by

---

*The term parental magma is also commonly applied in a slightly different context by igneous petrologists, to refer to the original magma of a particular differentiated intrusion often inferred from the composition of chilled margins. In this sense, the term parental magma refers to a magma belonging to some igneous rock series but not necessarily to the parental magma of that series.

two orders. Of the remainder, alkali basalts greatly exceed in abundance yet more alkalic mafic rocks. In view, therefore, of the absolute abundance of tholeiite and alkali basalt, and their resultant predominance in the historical evolution of ideas on igneous rock series, it is appropriate to consider these two groups here in a historical context.

One of the first differentiated magma series to be recognized in volcanic and hypabyssal rock compositions was in Mull, Scotland, (E.B. Bailey et al., 1924), where a "plateau magma type" represented by plateau-forming lava flows was distinguished from a "non-porphyritic central type" represented by flows within a caldera. A third proposed basaltic magma type, the "porphyritic central type" was less happily conceived and illustrates the difficulty posed by the possible accumulation of phenocrysts in porphyritic fine-grained rocks, the bulk composition of which may not, therefore, be equivalent to any liquid line of descent. The "non-porphyritic central type" was shown by Harker variation diagrams to be plausibly linked by inferred fractional crystallization processes to oversaturated intermediate and acid rocks. This series of differentiated rocks, which was termed the "Mull normal magma series", would now be termed a tholeiite magma series. One would nowadays also recognize the "plateau magma type" to be similarly linked with mugearites and trachytes in a distinct alkali basalt magma series [perhaps even in two distinct subseries (R.N. Thompson et al., 1972)], though this was not explicitly stated by the authors of the Mull memoir, who preferred to envisage the derivation of the one major basalt type from the other.

A formal chemical and mineralogical distinction of these two major parental basaltic magma types and their associated magma series coupled with recognition of their occurrence on a global scale was made by W.Q. Kennedy (1933), who referred to them as tholeiitic and olivine-basalt magma types.

Further refinements on the same theme were made by Tilley (1950), in an historically interesting state-of-the-art address, in which he (1) employed chemical discriminants in addition to mineralogical ones; (2) recognized the critically higher alkali content of Kennedy's olivine basalt magma type and renamed it and its associated series "alkali olivine basalt"; and (3) recognized that tholeiites occur in abundance in an oceanic environment and are not merely restricted to areas of continental crust.

Experimental petrologists subsequently demonstrated (Yoder and Tilley, 1962) that basaltic compositions of these two types could be conceived of as lying within the now well-known "basalt tetrahedron", based on the relative CIPW-normative proportions of di, ab, Q, en, fo and ne, and separated compositionally by a critical thermal divide at low pressures (see Fig. 6.14 and accompanying discussion).

Continuing work on the voluminous fresh volcanic rocks of Hawaii clearly delineated the two contrasted basalt types petrographically, along with their associated fractionation series: tholeiite—andesitic rocks—rhyolite, and alkali basalt—hawaiite—mugearite—benmoreite—trachyte. A simple Harker variation diagram plotting total alkali content [weight per cent $(Na_2O + K_2O)$] against silica content (G.A. Macdonald and Katsura, 1964, following Tilley, 1960) showed that a straight line on the diagram served effectively to separate analyses of basic members of the two series (Fig. 8.1). Thus a very widely used chemical discriminant was introduced to divide tholeiite from alkali basalt (the latter term is now preferred to the longer "olivine alkali basalt", itself as we have seen, derived from the "olivine basalt" of Kennedy). In a more general context, the same dividing line can be used to separate subalkaline rocks from alkaline, each group being capable of subdivisions of which tholeiite and alkali basalt series are examples. This simple diagram was a forerunner of an increasing trend towards employing chemical discriminants, conveniently expressed graphically, to separate igneous rock series one from another. Note that in the diagram of Macdonald and Katsura the boundary line extends only up to 52.4wt.% silica content. Some later workers have arbitrarily extended the boundary line in a straight line to higher silica contents.

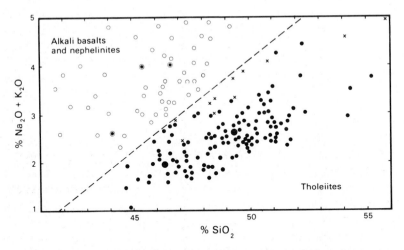

Fig. 8.1. Alkalis vs. silica diagram of analyses of Hawaiian basaltic rocks showing an empirical boundary (*diagonal dashed line*) between rocks of tholeiitic mineralogy (*dots*) and rocks of alkali basalt and nephelinite mineralogies (*circles*) (after G.A. Macdonald and Katsura, 1964, p.87, fig. 1). Average Hawaiian tholeiites (*large dots*) and alkali basalt series rocks (*large circles with dot*), after data in G.A. Macdonald (1968). Plots of analyses of Thingmuli series tholeiites from Iceland (*crosses*) inserted for purposes of comparison (after Carmichael, 1964, p.448, fig. 2).

8.4. THERMODYNAMIC BASIS OF CLASSIFICATION BASED ON SILICA ACTIVITY

As we have seen in Chapter 4 a relatively higher alkali content within the otherwise compositionally roughly comparable class of fine-grained mafic rocks has a drastic effect on silica saturation, reflected in a greater abundance of olivine in alkali basalt than in tholeiite and the occurrence of olivine plus feldspathoid in yet more alkaline types. It is the silica content taken in conjunction with the abundance and composition of other major-oxide components of a silicate magma that determines how "undersaturated" it is. Quantitative expression of this concept of the effective concentration of silica in the liquid can be represented as the silica activity (Carmichael et al., 1970). **Silica activity** is defined as the ratio (at constant temperature) of the fugacity of SiO_2 in the liquid phase of the magma to a standard state which can be stipulated, for example that of a silica glass (Carmichael et al., 1974, p51):

> "In other words, the activity of silica measures the difference between the chemical potential of silica in the liquid lava phase and that in the standard state, silica glass"

The silica activity may be derived by thermodynamic calculation, though for natural magmas this calculation would be enormously complex. Silica activities can, however, be more readily calculated for simple reactions that bound the mineral assemblages observed in distinct igneous rock series. The results of these calculations, based on Carmichael et al. (1970, 1974), are shown in simplified from in Fig. 8.2. Reactions (*i*) to (*iva*) or (*ivb*) thus serve as bounding reactions and form a thermodynamic basis for the delineation of the five fields of the following igneous rock series, designated by parental mafic magma type as presented in Table 8.1. In this table, in which the order of increasing alkalinity is similar to the outline petrographic classification of mafic rocks adopted in Chapter 4 (Table 4.2), olivine denotes a forsteritic olivine, and ferromagnesian pyroxene denotes either a separate ferromagnesian pyroxene phase or the ferromagnesian component of a subcalcic augite. The use of the perovskite $\rightleftharpoons$ sphene reaction as the lower limit of basanite is based on the observation that perovskite and feldspar have never been observed together in volcanic rocks. Two possible reactions involving end-members of the melilite crystalline solution series are shown, (*iva*) and (*ivb*).

Of course, this can serve only as an approximate guide to the silica activities of natural magmas and lavas as the quoted reactions are based on end-members of crystalline solution series. In practice, as well as the obvious diadochy in olivine, pyroxene, plagioclase, and melilite, there are also more subtle possibilities such as the presence of the undersaturated jadeite ($NaAlSi_2O_6$) and Tschermak's molecule ($CaAl_2SiO_6$) components in the augites of alkali basalts and more alkaline mafic rocks. A further complication is that potassium, in addition to occurring diadochically in plagioclase

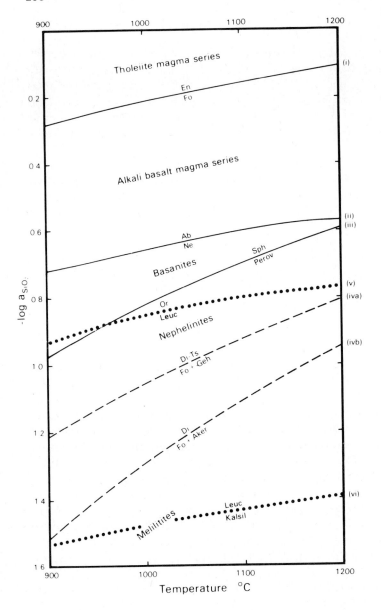

Fig. 8.2. —log a_{SiO_2} plotted against temperature for the following reactions which define fields of designated magmas and their series (see also text for explanation):

(*i*) $Mg_2SiO_4 + SiO_2 \rightleftharpoons 2MgSiO_3$
  forsterite   enstatite

(*ii*) $\frac{1}{2}NaAlSiO_4 + SiO_2 \rightleftharpoons \frac{1}{2}NaAlSi_3O_8$
  nepheline   albite

(*iii*) $CaTiO_3 + SiO_2 \rightleftharpoons CaTiSiO_5$
  perovskite sphene

TABLE 8.1

Name and modal mineralogy of mafic member of igneous rock series

Tholeiite	olivine (in reaction relationship to ferromagnesian pyroxene), plagioclase, augite
Alkali basalt	olivine, plagioclase, augite
Basanite	olivine, nepheline, plagioclase, augite
Olivine nephelinite	olivine, nepheline, augite
Olivine melilitite	olivine, nepheline, augite, melilite

and nepheline, may assume major quantitative importance in some alkaline mafic rocks and separate potassium mineral phases may be present. The silica activities of reactions (v) and (vi) divide the fields of the appropriate potassium-bearing phases, orthoclase, leucite and kalsilite.

It is re-assuring to discover that the sequential steps in dealing with under-saturated rocks in the calculation of the CIPW norm (viz. the successive conversion of hypersthene to olivine, albite to nepheline, orthoclase to leucite, diopside to calcium orthosilicate, and leucite to kalsilite) thus apparently have a sound thermodynamic basis; these steps were, of course, based on observation of common mineral parageneses in the first place.

8.5. BASIS OF CLASSIFICATION OF IGNEOUS ROCK SERIES FOLLOWED IN THIS BOOK

The problem is that we are not dealing solely with a single continuous range of igneous rock series as might be construed from a superficial appreciation of the thermodynamic continuum of silica activities so elegantly demonstrated by I.S.E. Carmichael and coworkers. For example, three obvious examples of distinct groups of series of overlapping silica activities would be (1) tholeiites of ocean ridges as contrasted with those of oceanic islands; (2) a spectrum of SBZ volcanic rock series ranging from tholeiitic to alkali basalt differing from a parallel range of tholeiites and alkali basalts

---

Fig. 8.2. (*continued*)

(iva) $\frac{2}{3}Ca_2Al_2SiO_7 + \frac{1}{3}Mg_2SiO_4 + SiO_2 \rightleftharpoons \frac{2}{3}CaMgSi_2O_6 + \frac{2}{3}CaAl_2SiO_6$

   gehlenite   forsterite   diopside   Ca-Tschermak's molecule

(ivb) $\frac{2}{3}Ca_2MgSi_2O_7 + \frac{1}{3}Mg_2SiO_4 + SiO_2 \rightleftharpoons \frac{4}{3}CaMgSi_2O_6$

   åkermanite   forsterite   diopside

(v) $KAlSi_2O_6 + SiO_2 \rightleftharpoons KAlSi_3O_8$ (vi) $KAlSiO_4 + SiO_2 \rightleftharpoons KAlSiO_6$

   leucite   orthoclase   kalsilite   leucite

(After Carmichael et al., 1970, p.250, fig. 1; 1974, p.52, fig. 2-3.)

in oceanic crustal areas; and (3) relatively potassium-rich series amongst alkaline rocks of both cratonic and SBZ settings.

This same inherent defect affects three other widely quoted single "spectral" parameters of alkalinity in which alkalis are measured against CaO, SiO_2, and $(Al_2O_3 + CaO)$, respectively:

(1) The *suite index* of Peacock (1931) is based on a simple Harker variation diagram in which total alkalis and lime are plotted against silica. Since with increasing silica content the alkali content increases in any igneous rock series and lime decreases, there will be a silica content within any given analysed igneous rock series where a crossover point is attained, i.e. where,

$$\text{weight per cent of } (Na_2O + K_2O) = \text{weight per cent of } CaO \tag{8.1}$$

The silica content in weight per cent so defined is known as *alkali-lime index*, usually quoted to the nearest integer. Other things being equal, an igneous rock series that is intrinsically more alkaline (i.e. a parental magma richer in alkalis) will achieve this crossover point at a lower silica content and thus be represented by a lower alkali-lime index. Several series yielding different alkali-lime indices were recognized by Peacock and grouped into four as follows:

	Alkali-lime index
Alkali series	<51
Alkalic-calcic series	$51-56$
Calc-alkalic series	$56-61$
Calcic series	<61

(2) The *alkalinity index* of A. Rittmann, σ, relatively more popular on the European continent, is:

$$\sigma = (Na_2O + K_2O)^2 / SiO_2 - 43 \tag{8.2}$$

Series of higher alkalinity will have higher σ-values. The parameter is devised empirically so as to remain more or less constant when calculated for any of the rocks of one particular series for any silica content. However, J.B. Wright (1969a) has shown that resolution using this index is not good at higher silica contents.

(3) J.B. Wright (1969a) proposed a *"simple alkalinity ratio"*:

$$Al_2O_3 + CaO + Na_2O + K_2O/Al_2O_3 + CaO - (Na_2O + K_2O) \tag{8.3}$$

and graphically showed a fair resolution into fields designated as calc-alkaline (+ tholeiite), alkaline, and peralkaline. The expression is of the form $(1 + x)/(1 - x)$, where $x = (Na_2O + K_2O)/(Al_2O_3 + CaO)$, and thus has a multiplier effect on values of x which is less than unity.

Given then that we cannot place all our series in one simple sequential order, how do we go about the job of describing them? Do we rely on observed mineralogical differences between series in the manner of classical petrography, or should we rely on compositional distinctions, as, for example, in the review by Irvine and Baragar (1971)? Alternatively, should we follow apparent natural groupings that we now reasonably discern to be related to global tectonics, or is there a sufficient consensus to put forward a genetic taxonomy based on considerations of petrogenesis of magma in the mantle?

The solution adopted here is to present an apparent natural grouping of igneous rock series related to global tectonics, an approach presaged in many chapters of Holmes (1965), as a framework to describe the mineralogy, chemistry, and distinctive characters of each in Chapters 9—11, to defer a consideration of Precambrian igneous rocks, some of which do not have contemporary analogues, until Chapter 12, and to reserve petrogenetic discussion in the main until Chapter 13. This kind of grouping by natural association is not new as can readily be seen by a perusal of at least some of the chapter headings in F.J. Turner and Verhoogen (1960), Hyndman (1972) and Carmichael et al. (1974).

8.6. USEFUL PARAMETERS, INDICES AND VARIATION DIAGRAMS

A selection of numerous parameters, indices and variation diagrams based on chemical composition are employed in the following four chapters. Some of the more common ones that index variations *within* igneous rock series are noted for ease of reference below.

It is important to note that compositional criteria are used for two major purposes: (a) to indicate the stage of differentiation *within* a given series; and (b) to discriminate *between* different series. Some variation diagrams can usefully serve and portray both functions where the context is appropriate, and it is important to bear in mind what is being accomplished in a given context.

As we have seen in Chapter 7, the liquid line of descent, i.e. the changing liquidus composition of differentiating magma, is best indicated by analyses of aphyric lavas. Petrologists customarily attempt to portray this kind of systematic chemical variation within a series by variation diagrams. The closer the individual points lie to a smooth connecting curve and the more numerous they are, the greater the confidence one has that there is some systematic cause of the variation such as fractionation.

One such variation diagram is the traditional **Harker diagram** [with a history of employment going back to Harker's (1909) classic *Natural History of Igneous Rocks*] wherein the different major oxides are plotted against silica as the abscissa (Fig. 8.3). The dual rationale is that silica content generally shows the greatest arithmetical variation in amount in any igneous rock series and that silica apparently increases with fractionation. A shortcoming of this variation diagram, however, is that a basaltic rock can undergo

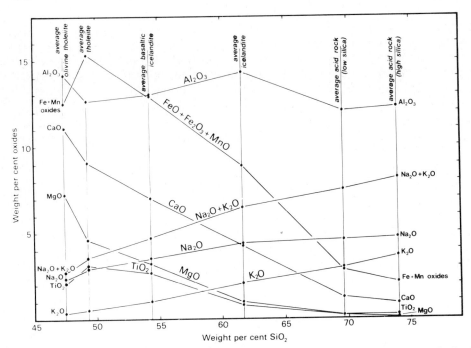

Fig. 8.3. Harker variation diagram for the tholeiitic lava series of Thingmuli, Iceland, constructed by grouping and averaging the analyses given in Carmichael (1964).

a great deal of early fractionation, resulting from say the cotectic crystallization of olivine + plagioclase + augite, without materially affecting the silica content of the liquid. The reason for this can be seen from their stoichiometric silica contents listed in Table 8.2. Early cotectic crystallization and fractionation of comparable proportions of olivine, say Fo_{90}, plagioclase, say An_{75}, and diopsidic augite with say 20wt.% hedenburgite component will result in the removal from the magma of crystals containing ~41.5, 50, and 54.5 wt.% SiO_2, respectively. Their joint fractionation in comparable proportions therefore will not have a significant effect on the silica composition of a magma originally containing say 48 wt.% silica. It is worth noting that in early fractionation stages only removal of olivine, an oxide mineral, nepheline, or amphibole can raise the silica content of basic magma.

An index which gives a more effective spread on the abscissa for the basic end of the spectrum of an igneous rock series is the **Larsen index** [Larsen (1938), with reference to earlier work on the San Juan volcanic series] equal to:

$$\tfrac{1}{3}SiO_2 + K_2O - (FeO + MgO + CaO) \tag{8.4}$$

A **modified Larsen index** (Nockolds and Allen, 1953) is just:

$\frac{1}{3}$Si + K − (Ca + Mg)   (8.5)

this recognizes the fact that whereas K increases and Mg and Ca decrease with early fractionation, Fe does not show a marked similar decrease (in fact, Fe content actually increases in basic to intermediate members of tholeiitic series).

Another popular variation diagram uses rectilinear coordinates to plot major oxides, not against silica content or a comparable index as in the Harker or Larsen variation diagrams, but against a **differentiation index, DI**:

DI = Q + or + ab + ne + lc   (8.6)

i.e. the sum of CIPW-normative Q + or + ab + ne + lc. Any one norm will, in fact, contain only three of these normative minerals as normative feldspathoid only appears when there is pronounced deficiency in free quartz, and the norm calculation procedure converts all albite to nepheline before orthoclase is converted to leucite in even more silica-undersaturated rocks (see Thornton and Tuttle, 1960; and pp. 97—99). The differentiation index used as an abscissa is an effective "spreader" in the representation of major-oxide variation within any one igneous rock series particularly at the basic end. It should be noted, however, that the parental mafic magmas of the more alkalic igneous rock series will have different differentiation indices than parental tholeiites owing to an intrinsically higher alkali content being

TABLE 8.2

Stoichiometric silica contents of minerals of mafic rocks

Mineral	Idealized end-member of solid-solution series	Formula	SiO_2 content (wt.%)
Olivine	forsterite	$2MgO \cdot SiO_2$	43
	fayalite	$2FeO \cdot SiO_2$	29
Augite	diopside	$CaO \cdot MgO \cdot 2SiO_2$	56
	hedenbergite	$CaO \cdot FeO \cdot 2SiO_2$	48
Ferromagnesian pyroxene	enstatite	$MgO \cdot SiO_2$	60
	ferrosilite	$FeO \cdot SiO_2$	45
Plagioclase	anorthite	$CaO \cdot Al_2O_3 \cdot 2SiO_2$	43
	albite	$Na_2O \cdot Al_2O_3 \cdot 6SiO_2$	69
Opaque mineral	chromite	$FeO \cdot Cr_2O_3$	nil
	magnetite	$FeO \cdot Fe_2O_3$	nil
	ilmenite	$FeO \cdot TiO_2$	nil
Nepheline		$Na_2O \cdot Al_2O_3 \cdot 2SiO_2$	42
Leucite		$K_2O \cdot Al_2O_3 \cdot 4SiO_2$	55
Amphibole	—	—	~38—42

offset by the occurrence of some soda in normative nepheline rather than albite (stoichiometrically relatively heavier than nepheline per unit soda content). Continued fractionation within a series leads inexorably to rocks with very high DI-values approaching 100.

A modification of the DI is a *fractionation index*, FI, designed specifically to cope with oversaturated peralkaline rocks (R. Macdonald, 1969):

$$FI = Q + or + ab + ac + ns \tag{8.7}$$

i.e. similarly the sum of CIPW-normative $Q + or + ab + ac + ns$. This reflects

"the observation of D.K. Bailey and Schairer (1964) that peralkaline felsic liquids fractionate not towards petrogeny's residua system but towards eutectics enriched in alkali silicate relative to that system."

With parallel rationales and application, but not as widely employed as the DI, are two indices with confusingly similar names: (1) the *crystallization index*, CI, (Poldervaart and Parker, 1964) equal to the sum of CIPW-normative anorthite, magnesian diopside calculated from total normative diopside, normative forsterite plus normative enstatite converted to forsterite, and magnesian spinel calculated from any normative corundum; and (2) the *solidification index*, SI, (Kuno, 1959):

$$SI = 100 \, MgO/(MgO + FeO + Fe_2O_3 + Na_2O + K_2O) \tag{8.8}$$

As fractionation of plagioclase leads to enrichment of potassium and sodium relatively to calcium in the liquid, and fractionation of dark-mineral silicates similarly to enrichment in iron relatively to magnesium, an often revealing variation diagram is of the type of Fig. 8.4, reproduced from Carmichael (1964, p.452), showing progressive relative enrichment in these two ratios. The distinct steepening of the slope equates with magnetite joining the cotectically crystallizing phases. The large number of points on, or close to, a continuous curve, each representing the analysed composition of a lava (ideally aphyric), lends strong support to the implicit assumption that they are indeed related in a coherent manner.

The *AFM* (syn. *FMA*) variation diagram uses a triangular representation to show the relative proportions of alkalis ($Na_2O + K_2O$), iron oxides ($FeO + Fe_2O_3$), and magnesia (MgO). Fig. 8.5 compares the Thingmuli trend with the (calculated) trend of the Skaergaard liquid. This *AFM* variation diagram illustrates well the marked *iron enrichment* which is found to characterize intermediate rocks belonging to tholeiitic fractionation series, presumably because strong fractionation of iron from the liquid has to await the crystallization of magnetite, a relatively late event in the crystallization history of tholeiitic melts which characteristically contain low initial Fe_2O_3/FeO ratios. Care is necessary in making comparisons based on this variation diagram, as some workers, in order to avoid fortuitous variation due to

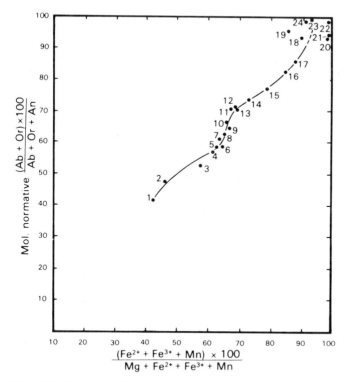

Fig. 8.4. The molecular normative feldspar ratio, $[(Ab + Or) \times 100]/(Ab + Or + An)$, plotted against the atomic ratio $[(Fe^{2+} + Fe^{3+} + Mn) \times 100]/(Mg + Fe^{2+} + Fe^{3+} + Mn)$ for analyses of the Thingmuli tholeiite series (see text for discussion) (after Carmichael, 1964, p.452, fig. 5).

weathering, record total iron as ferrous oxide [i.e. they would use $(FeO + 0.9Fe_2O_3)$ for the F component], and yet others use proportions of elements instead of weight per cent oxides (viz. $(Na + K):Fe:Mg$). Use of these varying components in different diagrams results, of course, in relative distortion of superimposed compositional variation trends.

Another parameter of differentiation within a series is based on Mg/Fe ratios alone, the main rationale being that fractionation of dark silicates, which are always relatively richer in their Mg/Fe ratio than the crystallizing silicate melt, results in a marked progressive lowering of this ratio with evolution in a series. Another rationale for using this ratio is that, as the work of Roeder and Emslie (1970) has indicated, there is a determinable partition coefficient between the Mg/Fe ratio of a crystallizing olivine (an early crystallizing phase in virtually all mafic magmas) and the Mg/Fe ratio of the crystallizing mafic liquid. Consequently, given an approach to homogeneity in the upper mantle there should be some *common* upper limiting value of Mg/Fe ratio in derived mafic melts that have been *in equilibrium with*

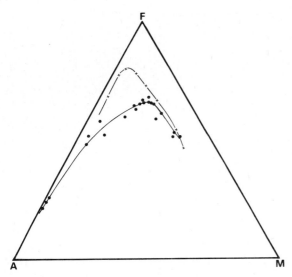

Fig. 8.5. *AFM* plot of Thingmuli lava compositions (*dots*) and calculated Skaergaard liquid compositions (*crosses*). $A = Na_2O + K_2O$; $F = FeO + Fe_2O_3$; $M = MgO$ (all in weight per cent) (after Carmichael, 1964, p.448, fig. 3).

average upper mantle (although a local high proportion of partial melting could result in a primary magma with significantly higher Mg/Fe content). Many variations on a Mg/Fe ratio have been proposed; some of these are reviewed by C.J. Hughes and Hussey (1976), who propose (1) that the Mg ratio should be defined as:

$$\text{Mg ratio} = 100\ Mg^{2+}/(Mg^{2+} + Fe^{2+} + Mn^{2+}) \tag{8.9}$$

after appropriate correction of Fe^{3+}/Fe^{2+} ratios in rocks that may have undergone post-crystallization oxidation, and (2) that the *M*-value should be:

$$M = 100\ Mg^{2+}/(Mg^{2+} + Fe^{2+} + Mn^{2+} + Fe^{3+}) \tag{8.10}$$

It is the Mg ratio so defined in using ferrous iron only that should be of greater relevance to arguments concerning compositions of olivine phenocrysts, magmas, and mantle related by partitioning of Mg^{2+}/Fe^{2+}. The *M*-value, of course, makes no assumptions involving correction of ferrous/ferric oxide contents of raw analytical data and thus has some pragmatic use for comparative purposes. Fig. 8.6 is a graphical representation of Mg ratios for analysed MgO and FeO contents. The *M*-value can be obtained from the same graph by adding $0.9Fe_2O_3$ content to the content of (FeO + MnO). The numbers at the upper and right margins give the Mg ratios of olivines coexisting in equilibrium with melts of varying Mg ratios according to the equation of Roeder and Emslie (1970).

Variation diagrams of different kinds can be used to show the **control** exerted by the fractionation of specific primocryst compositions upon the liquid line of descent. For example, Fig. 8.7, in which MgO content is used as the abscissa, indicates the effect of removal or accumulation of a determined proportion of olivine and augite crystals (of determined composition) for a suite of analysed lavas from Kartala volcano, Grande Comore, Comores archipelago (after Strong, 1972). The compositions of accumulative rocks (ankaramites) plot closer to a point on the olivine—augite join, and the compositions of fractionated magma compositions trend linearly away from this same point for as long as olivine and augite are the sole fractionating phases in the same more or less fixed proportion. Making reasonable allowance for the noise of the data such as analytical error and for some variation in the proportion of phenocrysts added or subtracted from rock to rock, the plots in Fig. 8.7 are consistent with a simple fractionation model. Note that models of this type must perforce fit every chemical element to be valid.

The most fundamental index of fractionation of all is quite simply the percentage of original magma that has crystallized, necessarily a matter of inference [but see treatment of Skaergaard liquid compositions by Wager (1960)]. The content of a highly residual element such as Zr, which systematically increases in concentration in the melt phase as crystallization proceeds, can also usefully be employed as an abscissa in variation diagrams to portray variation in other components (see, e.g., Pearce, 1975).

A combination of the above two approaches is employed by Barberi et al. (1975), where the fraction, f, of the parental magma represented by any residual melt composition is estimated by comparison of contents of Ce, selected from the residual elements Ce, La, Zr and Hf.

8.7. COMPLICATIONS AFFECTING SIMPLE FRACTIONATION MODELS

A complication that could affect the trace-element compositions of liquid lines of descent is the effect of replenishing a fractionating magma chamber with renewed accessions of parental magma. Within a layered cumulate sequence this could result in a horizon of sharply reversed cryptic layering (often in practice over a very small compositional range) and a concomitant duplication in the compositions of sections of the layered cumulate sequence, often accompanied by rhythmic layering as in the lower parts of Bushveld and Great Dyke sequences. The nickel content of fractionating olivine crystals would, however, reveal a much greater change (see discussion of the Muskox and Rum complexes on p.239). In a complementary manner to the selective removal of Ni from the liquid by the crystallization of olivine, those trace elements *not* preferentially entering the lattices of fractionating crystals could undergo an *increase* in concentration in the liquid magma. As a consequence of repeated accessions of fresh magma to a fractionating magma chamber, these trace elements could attain concentrations not only

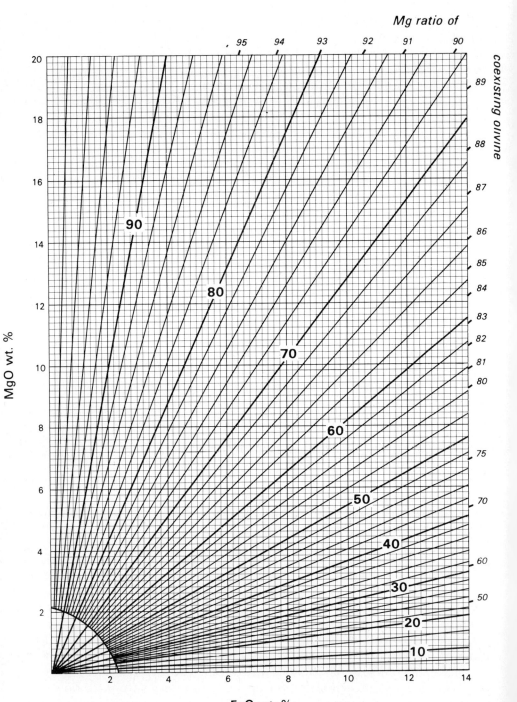

Mg ratio of

coexisting olivine

MgO wt. %

FeO wt. %

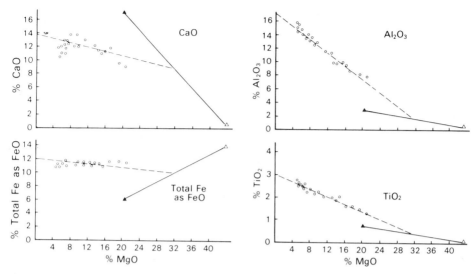

Fig. 8.7. Variation diagrams of selected major elements plotted against MgO content for analyses of rocks of the alkali basalt series of Kartala volcano, Grande Comore. *Solid lines* link compositions of olivine and augite phenocrysts. *Dashed lines* are apparent control lines drawn from average olivine/augite phenocryst ratio in accumulative ankaramite members of the series through the compositions of other rocks in the series (after Strong, 1972, p.194, fig. 6).

higher than that of the original magma (a normal result of fractionation) but also, for a given stage of fractionation as indicated by the major-element compositions of crystallizing minerals, higher than those they would have attained by continuous fractionation of an unreplenished magma chamber. A mathematical computation of this possibly significant effect is given by O'Hara (1977), and may have to be borne in mind when making use of trace-element "signatures" to help distinguish between different igneous rock series [see also discussion by Pankhurst (1977)].

Another complication is suggested by the different settling rates of primocryst phases of different densities, notably the anticipated slow settling of plagioclase crystals relative to the more rapid settling of mafic crystals. To some extent magma composition could be buffered by crystals settling

Fig. 8.6. Graph giving Mg ratio (*diagonal lines radiating from origin*) for given contents of MgO and FeO. The M-value can similarly be obtained by first adding $0.9Fe_2O_3$ to the FeO content. The numbers in the upper and right margins give the Mg ratios of olivines coexisting in equilibrium with melts of various Mg ratios according to the equation of Roedder and Emslie (1970). Note that the Mg ratio of an olivine differs from, and is always greater than, the commonly employed Fo content of an olivine expressed in weight per cent. For example, Fo_{90} (weight per cent) equates with an Mg ratio of 92.9, and Fo_{80} similarly equates with an Mg ratio of 85.4.

from it being compensated by crystals settling *into* it from higher levels of the magma chamber. The selective fractionation of primocryst settling at different rates could, however, lead to enrichment in plagioclase crystals in the upper parts of a fractionating magma chamber. These two processes seem to be demanded by the results of a perceptive study of porphyritic basaltic rocks from the Solomon Islands (Cox and J.D. Bell, 1972; see also Krishnamurthy and Cox, 1977), and indeed there are numerous instances of "feldsparphyric" mafic rocks that do not seem to fit simple fractionation models.

These possibilities involving some degree of "open-system" conditions are probably more realistic than simpler models of fractionation, although correspondingly more complex to model. It is likely that more attention will be paid to these considerations in future years.

The last few pages have concentrated in the main on means of portraying differentiation *within* igneous rock series. The distinction *between* different igneous rock series involves relatively more specific features, the discussion of which is deferred until appropriate positions in the following chapters.

Chapter 9

IGNEOUS ROCKS OF OCEANIC AREAS

9.1. OCEANIC CRUST

With a piece of string and National Geographic Society's excellent globe one can quickly arrive at a rough estimate of the world's ocean ridge system as 60,000 km in length. From a crude sector averaging of vectors of movement from data by Le Pichon (1968) coupled with crustal ages from later compilations one obtains an average spreading rate of 2.6 cm/a. Taking an average thickness of oceanic crustal material as 6.5 km, one obtains then a figure of $6 \cdot 10^4 \times 5.2 \cdot 10^{-5} \times 6.5 = 20.3$ km^3 of new basic oceanic crust being formed each year along the ocean ridge system. The spreading rate is conventionally taken as *half* the total extensional rate which in fact usually appears to be symmetrical. Note also the comparable estimate of the length of the world's active ridge system to be 53,700 km, with an average half spreading rate of 2.74 cm/a (D.L. Williams and von Herzen, 1974).

Back-arc spreading behind the Scotia, Lesser Antilles, Tonga, Izu-Bonin, Japan, and Aleutian arcs (Moberly, 1972) is less well defined and documented. Nevertheless, over a possible maximum strike length of ~7000 km, and assuming an average lower spreading rate of 1.5 cm/a, one similarly obtains an additional figure of 1.4 km^3/a of new oceanic crust being formed in this tectonic environment.

At this total rate of 21.7 km^3/a, say $21\frac{1}{2}$ km^3/a, it would take little over 100 Ma to create all of the present oceanic crust, some 65% of the surface of the Earth. Such a figure accords well with the oldest known oceanic crust being Triassic and allows for both the continuing assumption of some oceanic crust of different ages and the fact that most is much younger than Triassic in age.

This total average figure of yearly emplacement of basic magma to form ocean crust is substantial, more than an order greater than the comparable rate of visible subaerial volcanic activity. It necessitates the emplacement at crustal levels, principally along the ocean ridge system, of an average amount of 690 m^3 of basic magma per second! It is all the more enigmatic because it is unwitnessed by man and was unsuspected by all but a very few geologists until the 1960's. Compared with continental crustal areas the density of sampling of oceanic crustal material is ridiculously low. Even where samples have been obtained they literally represent merely a scratching of the surface at the crust–water interface.

This inadequate sampling of fragments of pillow lavas recovered either by dredging, shallow drilling, or, more recently, from submersibles gives one a considerable feeling of uneasiness in presenting data. Acceptance of the correlation of ophiolite suites with oceanic crust provides an invaluable extra dimension; however, it must be borne in mind that the mafic parts of ophiolite successions are commonly either tectonically incomplete or overlain by volcanics of possibly different affinity. Furthermore, the study of these mafic rocks of ophiolites is universally complicated by degradation consequent on burial metamorphism and prone to debate on magmatic affinity, as for example in the classic ophiolite sequence of the Troodos Massif, Cyprus, [see Gass et al. (1975), with references to five preceding papers and discussions].

9.2. MID-OCEAN RIDGE BASALTS (MORB)

Petrographically, recovered samples of mid-ocean ridge basalts (MORB) are generally either aphyric, or contain olivine or plagioclase phenocrysts, or both. The great majority are olivine tholeiite in composition. Early workers (e.g., Engel et al., 1965) tended to be impressed by a remarkable uniformity of MORB. While this impression has to a significant extent been borne out by additional sampling and analysis (e.g., Cann, 1969, 1971; Kay et al., 1970), it has also, as we shall see, been modified by additional data.

Three average chemical analyses are presented in Table 9.1. Standard deviations were found to be small in compiling these averages which are obviously very similar. The greatest variation between individual analyses was found to be in FeO and Fe_2O_3 figures. Despite efforts to collect and analyse only fresh material it is probable that some late-stage and post-crystallization oxidation had occurred in some samples resulting in increases of Fe_2O_3 contents that may originally have been as low as 1.5 wt.% (Kay et al., 1970, p.1590), and approximating an Fe_2O_3/FeO ratio of 0.15 (C.K. Brooks, 1976).

Assuming that these average compositions of lavas collected from the rock–water interface have some validity, what do they tell us in terms of major-element composition? Compared with basaltic rocks generally, most major-oxide contents are not unusual. MgO and CaO are generally high; TiO_2, however, is a little low, P_2O_5 is low, and K_2O is very low.

Trace-element contents have come to be an important discriminant between different igneous rock series, particularly in degraded rocks (see Chapter 14). They also place constraints on source rock composition and processes of partial melting (see Chapter 13). The low to very low contents of Ti, P and K signal the even more marked depletion in MORB relative to other mafic rocks of a group of trace elements that have come to be known

TABLE 9.1

(a) Average chemical compositions (in wt.%) and CIPW norms of some mid-ocean ridge basalts compared with average basalt, and (b) abundances (in ppm) of selected trace elements of mid-ocean ridge basalts compared with average basalt

	(a) Average chemical compositions (wt.%)					(b) Abundances (ppm)		
	1	2	3	4		5	6	7
SiO_2	49.34	49.13	49.11	49.5	Rb	<10	~2	39
Al_2O_3	17.04	16.31	15.85	15.9	Sr	130	169	544
Fe_2O_3	1.99	2.41		3.0	Ba	14	—	303
FeO	6.82	7.85	11.38[*1]	8.0	Y	43	50	31
MnO	0.17	0.20	0.18	0.17	Zr	95	131	116
MgO	7.19	7.82	7.76	6.6				
CaO	11.72	10.84	11.21	10.0	Ni	97	253	90
Na_2O	2.73	2.92	2.73	2.7	Co	32	80	40
K_2O	0.16	0.21	0.22	1.0	Cr	297	360	168
TiO_2	1.49	1.61	1.42	1.9				
P_2O_5	0.16	0.07	0.14	0.33				
H_2O^+	0.69	0.56		0.9				
H_2O^-	0.58	0.32						
Total	100.08	100.25	100.0	100.0				
or	1.0	1.3	1.3	6.0				
ab	23.4	24.9	23.1	23.1				
an	34.2	31.0	30.1	28.6				
di	19.2	18.5	19.9	15.8				
hy	11.9	8.8	9.5	15.0				
ol	4.6	9.4	9.6	3.9				
mt	2.6[*2]	3.0[*2]	3.4[*2]	3.2[*2]				
il	2.9	3.1	2.7	3.6				
ap	0.4	0.2	0.3	0.8				
M-value	59.3	57.7	54.5	52.0				

(a) *Major-element compositions:* 1 = average composition of oceanic tholeiitic basalt (Engel et al., 1965, p.721, table 2); 2 = average composition of five fresh basalts from 5°30′N on the Carlsberg Ridge (Cann, 1969, p.9, table 3, average of analyses 1 and 3—6); 3 = mean water-free composition of 94 selected ocean-floor basalt analyses [Cann, 1971, p.497, table 2 (recast to 100 wt.%)]; 4 = average basalt (Manson, 1967, p.226, table VI, analysis 1).
(b) *Trace-element contents:* 5 = average oceanic tholeiitic basalt (Engel et al., 1965, p.721, table 2); 6 = fresh basalt from the Carlsberg Ridge (Cann, 1969, p.4, table 2, average of analyses 3, 5 and 6); 7 = arithmetic mean of all basalts (Prinz, 1967, pp. 278—280, table II).

[*1] Total Fe reported as FeO.
[*2] $Fe_2O_3/(FeO + Fe_2O_3)$ recalculated to 0.2.

as the incompatible trace elements*[1] (see D.H. Green and Ringwood, 1967a).
The contents of these elements and ratios such as K/Na and Rb/K are on the
whole low in MORB and some are remarkably low; some comparative data
are included in Table 9.1.

Anticipating later petrogenetic discussion (p.285), it would appear that
the low incompatible trace-element content of MORB correlates with
their ancestral primary melts having been produced from significantly
"depleted" mantle sources. A depleted mantle refers to one which may
hypothetically have undergone one or more previous cycles of partial
melting, during which its tenor of incompatible elements would have been
significantly reduced by their preferential incorporation into earlier removed
liquids.

Another distinctive trait of MORB is held to be a low content of rare-
earth elements (REE) with a more or less "flat" distribution and a sporadic
slight "negative europium anomaly" when "normalized"*[2] to the abundance
of REE in chondritic meteorites. If this generalization be valid, the data are,

---

*[1] Incompatible trace elements are generally those of large ionic radius (the so-called
LIL trace elements, "large-ion lithospheric") but include others of high valency that are
not readily accommodated into the lattice sites of upper-mantle minerals. These minerals
comprise olivine, orthopyroxene, clinopyroxene, and spinel or garnet, and offer a preponderance of normally divalent sites of small ionic radius. For this reason, the incompatible
elements would tend to be more or less strongly partitioned into a melt phase resulting
from the partial melting of mantle. Incompatible elements as thus defined are found to
comprise the larger alkali elements K, Rb and Cs, the larger alkaline earth elements Sr
and Ba, the rare-earth elements (REE), a group of large heavy elements such as Pb, U and
Th, and a group of elements of high valency such as Ti, Zr, P, Ta and Nb.
Incompatible trace elements thus form a set that is nearly the same as that of the *residual*
trace elements discussed in Chapter 7 that tend to become concentrated in the melt phase
during the fractional crystallization of magma. The reason is the same — incompatibility
with the lattice sites of the solid phases in equilibrium with the melt phase. However, in
the case of the residual elements, the set of solid phases crystallizing cotectically from
basic magmas eventually comes to include plagioclase, ilmenite, magnetite and apatite in
addition to ferromanganese silicates, so that the behaviour of Sr, Ti, V and P comes to
depart significantly from that of a residual element, whereas all of these elements are
incompatible as defined above.
*[2] The lanthanides, atomic numbers 57—71, commonly equated by petrologists with the
REE (see Chapter 1), are closely similar in their chemical properties but vary greatly in
abundance. The reason for presenting data on the contents of REE in igneous rocks in
this way (see Fig. 9.6) is as follows. The abundances of REE in certain chondritic
meteorites that are believed to represent primitive solar system material (see Chapter 12),
presumably equate with REE abundances in the primordial Earth. These abundances
would all be equally enhanced in the first silicate mantle after initial mantle/core fractionation by a factor of ~ 2. Any subsequent partial melting of mantle would be expected to
result in a marked increase of the concentration of REE in the melt fraction (as with all
the group of incompatible elements), and indeed in higher overall concentrations the
smaller the proportion of melt. The *relative* abundances to each other of the REE (which
the chondrite-normalized presentation is designed to reveal) should, however, be
unchanged by partial melting or other processes unless other, possibly petrogenetically

to say the least, noisy; see Puchelt and Emmermann (1977), for a discussion of the "surprising variability" found in the chondrite-normalized REE patterns of some ocean-floor basalts. The flat distribution of REE in MORB suggests that they have not had a history of equilibration with garnet as a residual mantle phase, and have thus been derived by partial melting of mantle at shallow depths (see Chapter 13). Furthermore, some MORB reveal a slight relative depletion in La and immediately adjacent light REE (Kay et al., 1970); this, coupled with a low overall REE content, further supports derivation from a depleted mantle source, as no mantle mineral composition could consistently produce this pattern of LREE depletion by crystal/liquid fractionation alone.

Because the content of REE and other incompatible elements in MORB is low and more closely akin to chondritic meteorite abundance than in any other igneous rock series, MORB were once thought to be "primitive" in character, i.e. most closely akin to primordial mantle material. However, consideration shows that this rather loosely conceived speculation by no means implies a close genetic connection with primordial mantle. On the contrary, MORB may well owe their low content of incompatible elements to their having been derived from a depleted mantle at least one stage removed by partial-melting episodes from primordial mantle. This consideration assumes great importance when we consider the next series — the oceanic-island tholeiites.

MORB also have a distinctively low initial $^{87}Sr/^{86}Sr$ ratio ranging between 0.7023 and 0.7033. There seems to be in the main a systematic variation between the value of this ratio and proximity to oceanic islands on or near

---

significant, influences were at work, such as differential partitioning of REE according to their size between melt and mantle minerals. Garnet, for example, likely present at relatively high pressures equivalent to depths of over ~80 km as a residual mantle phase or as a component of eclogite fractionation, would strongly fractionate the heavier REE (HREE) (which have relatively smaller ionic radii due to the lanthanide contraction) with respect to a melt phase. The lighter REE (LREE) would thus be relatively enriched in a melt that had equilibrated with garnet, or to a lesser extent but for the same reason with clinopyroxene. Olivine and orthopyroxene have a negligible fractionating effect on REE.

Europium, atomic number 63, is the only REE that can be in part divalent under magmatic conditions, and in this oxidation state it tends to follow Sr in its geochemical behaviour. It is also just small enough to proxy for calcium and be accepted to a perceptible extent into the lattice of crystallizing plagioclase. Consequently, the fractionation of plagioclase from magma would result in the normalized abundance of europium in the resultant melt being slightly depleted with respect to its neighbouring REE, thus producing a *negative europium anomaly*. Note that the relative abundance of Eu^{2+} to Eu^{3+} in a magma, and hence the availability of Eu^{2+} during plagioclase crystallization, will depend on the oxidation state of the magma (in the same manner as does the proportion of Fe^{2+} to Fe^{3+}), so that this is a complicating factor in assessing the significance of europium anomalies.

See Hanson (1980) for a review of these and other aspects of REE geochemistry.

the mid-ocean ridge system such as Iceland and the Azores, sites of inferred mantle plumes (see p.280). MORB well away from these localities generally have low ratios in the range of 0.7023—0.7027, while MORB, for example, from the FAMOUS locality in the axial zone of the Mid-Atlantic Ridge between approximate latitudes $36°30'$ and $37°N$, ~225—275 km from the Azores, have initial ratios in the narrow range of 0.70288—0.70307 (W.M. White and Bryan, 1977), and the Azores lavas themselves have characteristically higher values clustering around 0.7034. Again these variations are consistent with MORB having been derived from a depleted mantle source with an anomalously low Rb/Sr ratio resulting from one or more episodes of partial melting in the past. See Faure and Powell (1972) and Faure (1977) for a definitive account of strontium-isotope geology and principles of isotope geology, respectively.

Chromium content in MORB is moderately high with an appreciable range in values: averages of 297 and 360 ppm, ranges of 160—460 and 220—460 ppm, in the ten and three analyses given by Engel et al. (1965) and Cann (1969), respectively. These data are correlatable with the early fractionation of a chrome spinel. The same general comment on abundances could be applied to nickel with averages of 97 and 226 ppm, in the ranges of 58—140 and 180—320 ppm from the same sources. Kay et al. (1970) found nickel contents in 33 MORB samples to lie generally between 70 and 170 ppm, higher nickel contents showing a high degree of correlation with increasing contents of MgO, and lowest nickel contents being found in some exceptionally iron-rich basalts from the Juan de Fuca Ridge. As nickel is strongly fractionated by forsteritic olivine, the earliest and generally the commonest of the observed silicate phenocryst phases in MORB, these data are compatible with some of the rocks representing fractionation products. Consequently, although mid-ocean ridge basalts apparently form a compact group in terms of silica content (commonly between 48.5 and 50 wt.% SiO_2), there are strong hints from this trace-element data that some have undergone significant fractionation.

In this respect, Hekinian et al. (1976), and also Bryan and J.G. Moore (1977), have described the petrography and chemistry of 80 samples of submarine basalt lavas collected from submersibles over a 4 km width of an 8 km long segment of the central rift valley portion of the Mid-Atlantic Ridge, near $36°49'N$. This represents the highest reported density of controlled sampling of the ridge system to date. Rocks collected from an axial zone 1 km in width are olivine basalts and picritic basalts containing phenocrysts of olivine and small phenocrysts of chrome spinel. These rocks have relatively high contents of MgO, Ni and Cr, as well as high M-values. Immediately flanking this axial region a greater variety of petrographical types are found, comprising a range from aphyric basalts to felsparphyric basalts (up to 35 vol.% of plagioclase phenocrysts). Some of the porphyritic basalts contain three phenocryst phases: olivine, plagioclase and augite

(augite phenocrysts are generally rare in MORB basalts). The evidence of fractional crystallization in these latter types, especially the presumptive evidence of appreciable cotectic crystallization in the rocks with three phenocryst phases, is reinforced by the relatively lower M-values of their glassy groundmass components. From thicknesses of palagonite and manganese oxide coating it appears that although relatively fractionated types are in general older than the relatively more parental types in the axial zone, yet they are significantly younger than the spreading age of the crust upon which they were erupted. Hékinian et al. (1976, p.84) conclude that:

> "The above relations indicate that the diverse lava types were erupted from a shallow, zoned magma chamber from fissures distributed over the width of the inner rift valley and elongate parallel to it. Differentiation was accomplished by cooling and crystallization of plagioclase, olivine, and clinopyroxene toward the margins of the chamber. The centrally located hills were built by the piling up of frequent eruption of mainly primitive lavas which also are the youngest flows. In contrast smaller and less frequent eruptions of more differentiated lavas were exposed on both sides of the rift valley axis."

Within the group of rocks sampled by Hekinian et al. the TiO_2 and K_2O contents of glassy groundmasses vary sympathetically with SiO_2 content and inversely with groundmass MgO content and M-values (Fig. 9.1) in a manner suggesting that fractionation of olivine, the most abundant overall phenocryst phase in all but most fractionated types, has played a major role in controlling successive liquid compositions. TiO_2 content, for example, doubles from ~0.7 to 1.4 wt.% in the group of MORB sampled from the one small area. Even if Ti behaved as a perfectly residual element during fractionation, these figures would imply 50% fractionation of phenocrysts from some of the liquids which still remain basic in their silica content and basaltic in their mineralogy when crystallized. Note especially how the MgO content, or better still the M-value, each serves as a more consistent index of differentiation than does SiO_2 content. The relatively greater scatter of K_2O contents in Fig. 9.1 might at least in part be due to a rela-

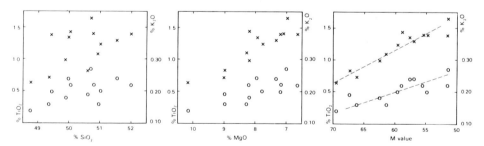

Fig. 9.1. Plot of TiO_2 contents (*crosses*) and K_2O contents (*circles*) against SiO_2, MgO, and M-value [$100Mg/(Mg + total Fe)$], respectively, for the glassy groundmass compositions of a closely spaced group of mid-ocean ridge basalts (constructed from data given by Hekinian et al., 1976, p.95, table 4).

tively greater analytical error in the small amounts reported to two signifi-
cant figures and probably also to a component of seawater metasomatism;
Ti is stable under low-grade degradation processes but K is highly mobile (see
Chapter 14).

It appears therefore that any generalizations about the chemistry of
MORB must take account (among other factors) of a significant degree of
fractionation that is clearly indicated by phenocryst assemblages, groundmass compositions, and aphyric rocks of differentiated compositions.
Philosophically, they must be approached as a *series*, with a restricted but
significant range in composition, *not* as some "average" rock type. As previously noted in Chapter 8, p.262, the silica content may not be greatly
affected during appreciable early fractionation and therefore it is not a
reliable guide as to whether fractionation may have occurred or not. It
is with this caution in mind that the interesting comparison of MORB with
oceanic-island tholeiite (see pp. 291 and 294) can better be appreciated.

Strong corroborative evidence that MORB are indeed generally not
unfractionated parental magma types comes from a study of the mafic rocks
in ophiolite complexes. As noted in Chapter 8, an *AFM* diagram is often
employed to portray the differentiation of tholeiites in which there is a
characteristic iron enrichment in the early to middle stages of fractionation.
A compilation (Fig. 9.2) has been made by Strong and Malpas (1975) who
were concerned with the possible shape of magma chambers under a spreading ridge system and the geometry of the spreading process, a timely dis-

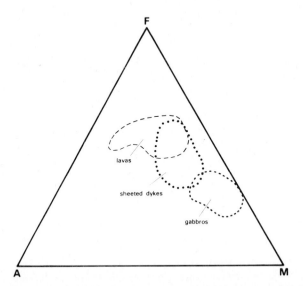

Fig. 9.2. *AFM* plot showing fields of composition of most analyses of gabbros, sheeted
dykes and lavas from several ophiolite associations. A = Na$_2$O + K$_2$O; F = total iron as
FeO; M = MgO (all in weight per cent) (based on Strong and Malpas, 1975, p.895, fig. 1).

cussion sparked by Brock (1974) and contributed to by Church and Riccio (1974). Based on the observation that the pillow lavas as a group appear to be more differentiated than the sheeted dykes which in turn are more differentiated than the underlying gabbros, Strong and Malpas propose that this apparent fractionation sequence developed as a result of numerous small and overlapping magma chambers rather than in a large continuous one (as proposed by Church and Riccio). Subsequently, a study of variation in major and trace elements in drilled basalts from the Mid-Atlantic Ridge has suggested that distinct cycles of low-pressure fractionation must have operated within a complex network of magma storage reservoirs beneath the crustal spreading axis (Flower et al., 1977). Flower et al. also state that a single extensive sub-axial magma chamber could not account for the observed chemical variation and moreover would probably be dynamically unstable. Thus the recognition and investigation of fractionation within MORB dovetails with an understanding of the ocean spreading process and generation of ophiolite.

One interesting observation concerns the very great rarity of intermediate and acid members among mid-ocean ridge extrusive rocks (see Pearce and Cann, 1973, p.291). G.P.L. Walker (1971) has pointed out that the elevated confining pressure due to depth of seawater would tend to inhibit the eruption of acid differentiates as ash flows — their typical mode of subaerial eruption. Small bodies of relatively evolved intrusive rocks such as diorite and trondheimite are found, however, within the gabbro layer of ophiolites. Coleman and Peterman (1975) have referred to these as the plagiogranite series, characterized by very low K_2O contents. To this extent the MORB volcanic series may be compositionally extended by this evidence presented by ophiolites.

Finally, not only are MORB a series rather than some average rock type as explained above, but also there is evidence (Erlank and Kable, 1976) of slightly but significantly *different* MORB *parental* magmas from different sections of the mid-ocean ridge system, and apparently *not* systematically related to proximity to inferred plumes (see pp. 290–291). Some MORB, for example, from 45°N in the North Atlantic, have a higher content of incompatible trace elements, a higher initial $^{87}Sr/^{86}Sr$ ratio, and a lower Zr/Nb ratio than "typical" MORB as described above. The Zr/Nb *ratio* in contrast to *concentrations* of residual elements, would scarcely be affected by early fractionation, and Erlank and Kable attach particular significance to this Zr/Nb ratio, not only as these two elements are considered to be "immobile" (see Chapter 14) and thus unaffected by possible processes of secondary alteration, but also because previous partial-melting episodes would be expected to raise the Zr/Nb ratio of residual mantle and hence that of any subsequently derived melt. Indeed all the trace-element and isotopic data cited by Erlank and Kable consistently indicate the derivation of some MORB from a mantle significantly less depleted than that producing typical

MORB. Thus some degree of continuum exists in MORB between the common extremely depleted varieties on the one hand and taken as a basis for description in this chapter, and apparently less common, less depleted varieties on the other hand, trending toward oceanic island tholeiite affinities.

9.3. OCEANIC ISLAND THOLEIITES (OIT)

It may seem curious to erect a series bearing a name connoting what might seem at first sight to be a purely accidental attribute, that of forming land above sea level as opposed, for example, to MORB occurring along the mid-ocean ridge system. However, a further consideration of the great length of the ocean ridge system and its nearly ubiquitous submarine character suggests that there might in fact be something genetically different about island tholeiites. The two prime examples of oceanic island tholeiite (OIT) are provided by the large islands of Iceland (straddling the mid-ocean ridge system) and the Hawaiian archipelago (with no links of any kind to a mid-ocean ridge). Notwithstanding this apparent difference in tectonic setting their tholeiites, the predominant rock type in both, are compositionally comparable to each other and distinct from MORB.

Both Iceland and Hawaii are held to be examples of hotspots, a term which is *factually* based, though perhaps with subjectively defined limiting parameters. A **hotspot** denotes a concise area of the Earth's crust at the present day which is distinguished by relative uplift or a topographic high over a diameter of up to some hundreds of kilometres, high heat flow, and generally by volcanism. The products of volcanism may vary from relatively sparse alkaline varieties to voluminous tholeiitic varieties (see also Chapter 10). A more detailed discussion of the characters, identification, and theory of hotspots is given in an invaluable paper by J. T. Wilson (1973, with references to earlier work), who identifies 32 examples of hotspots or, in some instances such as East Africa, groups of hotspots.

As there is no convincing explanation for the anomalous features of hotspots in terms of simple plate tectonics, Wilson postulates that hotspots are the surface expression of deep-seated **plumes** rising in the mantle. Here we are in the realm of *hypothesis* concerning the very existence of supposed plumes, their dimensions, their postulated rate of ascent, their possible differences in composition and temperature relative to other mantle material at the same depths, and possible processes of generation and segregation of magma within plumes. We can hardly therefore define that which nobody has seen and about which there is no general agreement at the present time. It can, however, be demonstrated that during a sufficiently great ascent under near-adiabatic conditions mantle will produce a partial melt (see Chapter 13), and a "plume", or for that matter any rising limb of a mantle convection system, therefore is a potential source of igneous activity which is what concerns us here. See also Gass et al. (1978) and Turcotte and Oxburgh (1978) for discussion of "intra-plate volcanism", syn. "mid-plate volcanism".

One interesting observation is that the great majority of hotspots (and their inferred underlying plumes) occur in areas of oceanic crust rather than continental crust, and no less than 16 of the grand total of 32 identified by Wilson occur on or close to mid-ocean ridges. This certainly suggests a causal relationship and leads naturally to the view that the positions of plumes may be at least one of the factors causing plate movement*. Certainly some mammoth event within the mantle seems to be required to initiate plate separation within old stabilized continental crust and thick underlying lithosphere, e.g., the breakup of Gondwanaland and the associated voluminous flood basalts. One might also envisage "weak" plumes with little associated volcanism, or considering an evolutionary timespan of the order of 100 Ma, one could also envisage "young" plumes (the presentation of cratonic igneous activity is presented below partly in terms of an evolutionary plume mechanism). However, putting aside controversy over possible relationships between hotspots and plumes and returning to our theme of oceanic island tholeiites, nobody can doubt the existence of anomalously voluminous tholeiitic volcanism in Iceland and Hawaii.

Accounts of the central Quaternary volcanic zone of Iceland (Fig. 9.3) include those of Thorarinsson et al. (1960), G.P.L. Walker (1965) and Thorarinsson (1968). Iceland has witnessed the most voluminous fissure eruption in history, that of Laki, 1783, producing an estimated volume of 12--15 km^3 of basalt, and the even greater post-glacial flow of Thjórsá. The continuing volcanic activity of Mývatn (1725--1729, etc.), Hekla (1947, etc.), Askya (1961, etc.), Surtsey (1963--1964), Heimay (1973), etc., is well documented. On a longer term view, within the last 16 Ma, which is the age of apparently the oldest volcanic rocks in Iceland situated on the extreme east and west coasts (Moorbath et al., 1968), volcanic activity has produced a subaerial mass of nearly 10^5 km^2 over a "strike" length of ~ 300 km of ridge, with an average height above sea level of ~ 500 m. This is accompanied by an even more massive submarine volcanic pedestal arising from shallower than average ocean depth itself reflecting a thickened basaltic crust up to a maximum of some 16—20 km in thickness, contrasting with the average thickness of oceanic crust of ~ 6.5 km (Fig. 9.3).

---

*Plate tectonics has been accepted by geologists before any adequate causal mechanism has been agreed upon. Wegener's philosophy was thus at last vindicated but not before his theories had been pilloried by many geophysicists and ignored by many geologists.

Fig. 9.3. Iceland in its oceanic setting showing place names mentioned in the text. Submarine bathymetry is indicated by 100-, 500- and 1000-fathom contours (1 fathom = 1.8288 m). Areas of ice caps are shown enclosed by *dashed lines*. The area of postglacial volcanic activity (currently subglacial in the case of Grímsvotn) is *shaded grey*.

The data from the Hawaiian archipelago, not complicated as in Iceland's case by occurring astride a mid-ocean ridge, are even more striking. The archipelago, over 2000 km in length, constitutes a remarkable submarine volcanic mountain chain, elongated in an approximately WNW—ESE direction of which only the peaks project above sea level. The only currently active volcanic island is Hawaii, area $\sim 10^4$ km^2, easily the largest in area of the archipelago and situated at its southeastern extremity. Hawaii has been built of five great coalescing shield volcanoes within a timespan of ~1 Ma (see G.A. Macdonald, 1968). Of these the only two active volcanoes are the enormous shield of Mauna Loa, 4170 m above sea level, which constitutes the south-western part of the island, and the volcanic area of Kilauea in southeastern Hawaii. To this bulk of volcanic material must be added a massive submerged pedestal of volcanic rocks, ~5 km in thickness, lying on older downwarped oceanic crust. Mauna Loa and the slightly higher, now extinct Mauna Kea thus have the highest constructional reliefs above base level of any other volcano or mountain on Earth, exceeding even the height of Everest above sea level, and they are, of course, merely part of the much greater elongate massif forming the archipelago.

Proceeding to the WNW (Fig. 9.4), one notes the contrast between (1) the constructional volcanic activity of Mauna Loa and Kilauea with their summit calderas and smooth slopes, with most extrusion taking place along the line of the East Rift of Kilauea and thus actively extending the land area of Hawaii to the southeast: (2) the recently extinct volcanoes of northern Hawaii; (3) the rocky nature of the cluster of extinct volcanic islands within ~150 km to the northwest of Hawaii including Maui, Kahoolawe, Lanai and Molokai; (4) the weathered rocks and erosional landscape of the Kauai group of islands and Oahu, on which Honolulu is situated; (5) the remnants of once much larger volcanic islands represented by Nihoa and Neckar islands; (6) the tiny pinnacle of volcanic rocks left projecting through coral reefs of the French Frigate Shoal; (7) the coral atolls of Midway and Ocean Islands, etc.; and finally, (8) still further to the NNW, the line of the Emperor seamounts which apparently form a yet older extension of the archipelago over a further distance of ~3000 km.

This geomorphological evidence strongly suggests that volcanism was sequential along the chain and has migrated progressively first to the SSE, then to the ESE. K—Ar ages of lavas from several of the younger islands indeed indicate that the apparent rate of migration of volcanic activity along the chain has been some 8 cm/a over the last 40 Ma (P.J. Smith, 1978, with references to earlier work on this important point). Recent work on the Emperor seamounts (Greene et al., 1978) indicates that they too are pro-gressively younger to the SSE.

The bulk, nature, age and distribution of volcanic rocks in the Hawaiian chain suggest that they reflect a moving hotspot formed by the overriding of a fixed mantle plume by the Pacific lithospheric plate (Morgan, 1972). This is not a forced postulate as, given the deduced pattern of plate move-ments relative to each other, most plates, if not all, must in fact also be moving relatively to mesospheric mantle beneath. Exceptionally, a plate could be stationary with respect to underlying mesosphere mantle as may possibly be the case with the African plate today. A plume could also remain fixed with respect to an overlying spreading ridge, a condition probably approximating to the Icelandic situation (although there too

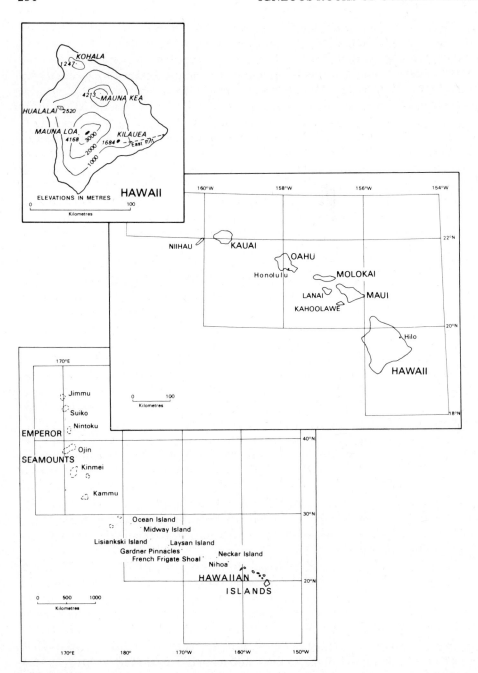

Fig. 9.4. Sketch maps of the island of Hawaii, the Hawaiian archipelago, and the Emperor seamount chain.

in common with other Atlantic volcanic islands, there is evidence that the steadily accreting African plate is slowly overriding Atlantic plumes westwards, resulting in an apparent eastward displacement of hotspots).

Petrographically, over 98% of Hawaiian volcanic rocks are tholeiites. Besides rather rare intermediate and acid rocks of tholeiitic affinity, the remainder comprise volcanic rocks of the alkali basalt series and still smaller amounts of relatively more alkaline mafic rocks. Iceland is also largely built of tholeiitic rocks although fractionated members of the tholeiitic series are relatively much more prominent than in Hawaii; rhyolitic flows, intrusions and pyroclastic rocks account for ~8% overall of surface exposures in Iceland. In Iceland too alkali basalts are found, but in an "off-axis" position such as the Snaefellsnes region in western Iceland where recent alkali basalts lie on an older basement of pre-Pleistocene tholeiites.

Phenocrysts in the Kilauea and Mauna Loa tholeiitic flows of Hawaii comprise olivine (the first to crystallize and often the sole phenocryst phase or most abundant), plagioclase, augite, and in some rocks orthopyroxene in place of olivine. In the less fractionated members of the Thingmuli lavas in Iceland the commonest phenocrysts are again olivine followed by plagioclase and augite. Carmichael (1967a, pp. 1829 and 1830) draws attention to an unexplained difference in groundmass mineralogy between the Thingmuli tholeiites and the Hawaiian tholeiites. The former contain two species of groundmass clinopyroxene, a pigeonite and an augite, whereas tholeiitic rocks of the latter commonly contain only one groundmass pyroxene, a sub-calcic augite.

Possible petrogenetic differences between MORB and OIT have been the subject of a classic debate between Schilling (1973b, c) and O'Hara (1973, 1975). Basically, Schilling attempted to demonstrate that Icelandic tholeiites not only have a significantly different chemistry to that of MORB, specifically MORB sampled at increasing distances to the south of Iceland along the Mid-Atlantic Ridge, but also that these differences could plausibly be related to partial melting within a mantle plume rising beneath Iceland. This "primordial hot mantle plume", PHMP, would be relatively undepleted by previous melting episodes. Consequently, it would be richer, or synonymously more "fertile" in contents of K_2O, TiO_2, P_2O_5 and incompatible elements, and would also have higher Na/Ca and Fe/Mg ratios than a relatively depleted low-velocity layer, DLVL, underlying the ridge system in general (Fig. 9.5). Thus the plume could generate a melt significantly enriched in these elements relative to MORB. O'Hara, on the other hand, pointed out that any direct comparison of the chemistry of the two groups is complicated by the operation of fractionation processes. Not only might fractionation be expected to be more pronounced in the Icelandic basalts which have to be emplaced through a relatively much greater thickness of crust but also the presence in many Icelandic basalts of olivine, plagioclase and augite as phenocryst phases suggests some appreciable degree of cotectic

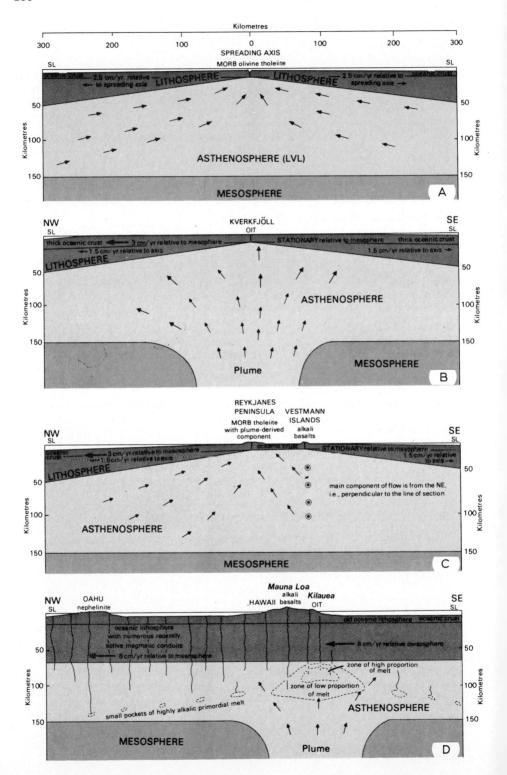

crystallization that itself would produce significant changes in Na/Ca and Fe/Mg ratios, and increased contents of minor and trace elements that are not incorporated into these phenocryst phases.

In the exchange of papers the arguments become complex, notably in the forced postulates of degrees of fractionation at different depths to account for the observed trace-element pattern in the two magma types. The point forthrightly made by O'Hara (1973, 1975) is valid, indeed a vital one*, but the balance of compositional evidence, taken in conjunction with that now available from other hotspots such as Afar (Schilling, 1973a), the Faeroe Islands (Schilling and Noe-Nygaard, 1974), and the Azores (W.M. White et al., 1976) gives strong support to accepting the idea of plume generation, at least as a credible working hypothesis, and consequently to the establishing of oceanic island tholeiite as a distinct series.

As well as basalt chemistry, the plume hypothesis also affords explanations of bathymetric, seismic, and gravity data, and is thus greatly strengthened thereby. These and other ramifications relating to the Iceland hotspot are discussed in Kristjansson (1974). A later review (Pearce and Flower, 1977) assesses the effects of possible mantle source heterogeneities and of

---

*The arguments adduced by O'Hara and Schilling are worth following in detail because they illustrate very clearly the vital need of igneous petrologists to distinguish variation *within* an igneous rock series from variation *between* different igneous rock series.

Fig. 9.5. Diagrammatic sketches to illustrate inferred differences between conditions of generation and eruption of MORB, OIT, and associated rocks. Cross-sections are natural scale with slight exaggeration of bathymetry and topography.
A. A typical mid-ocean ridge situation: "passive" flow of asthenosphere (LVL) to compensate for diverging movement of lithospheric plates that thicken away from the ridge; absolute motion of plates with respect to mesosphere immaterial; MORB erupted at spreading axis.
B. NW—SE section through Kverkfjöll hotspot, Iceland: "active" flow of plume material in LVL; note even *larger* components of flow (perpendicular to the line of section) towards Mývatn in the northeast and towards the Vestmann Islands in the southwest (see map, Fig. 9.3); OIT erupted at hotspot centre by virtue of high proportion of melt in the plume head.
C. NW—SE section through the Reykjanes peninsula and Vestmann Islands, Iceland; the line of section is parallel to, but 200 km to the southwest of the line of section B; note that the flow of plume-derived material in the LVL is mainly laterally from the northeast (perpendicular to the line of section) giving rise to the eruption of plume-derived alkali basalt in the Vestmann Islands (by virtue of a smaller proportion of melt generated away from the plume head centre), and to the admixture of a proportion of plume-derived melt at the spreading axis in the Reykjanes peninsula.
D. NW—SE section through Oahu and Hawaii (see map, Fig. 9.4): a plume underlies relatively rapidly moving, thick, oceanic lithosphere; voluminous eruption of OIT at Kilauea (and potentially much more to come to the southeast); alkali basalt reaching surface relatively easily at Mauna Loa (itself built mainly of OIT) through a young "plumbing system" of magmatic conduits; isolated eruptions of small quantities of nephelinites and related rocks at Oahu (see also Fig. 10.6).

TABLE 9.2

Analyses and CIPW norms of representative Hawaiian tholeiites

	1	2	3	4	5	6	7	8	9
SiO$_2$	49.36	49.40	49.44	50.32	49.20	50.10	50.45	50.69	51.24
Al$_2$O$_3$	13.94	11.69	12.98	13.61	12.77	13.78	14.01	13.51	13.60
Fe$_2$O$_3$	3.03	1.91	2.32	2.32	1.50	1.89	1.58	1.87	1.87
FeO	8.53	9.49	9.16	9.18	10.05	9.46	9.76	10.70	11.19
MnO	0.16	0.17	0.18	0.18	0.17	0.17	0.17	0.18	0.18
MgO	8.44	12.96	9.79	7.59	10.00	7.34	6.75	5.74	5.12
CaO	10.30	8.95	10.17	10.79	10.75	11.46	10.76	9.72	9.03
Na$_2$O	2.13	2.06	2.24	2.34	2.12	2.25	2.50	2.60	2.81
K$_2$O	0.38	0.40	0.47	0.49	0.51	0.57	0.60	0.72	0.83
TiO$_2$	2.50	2.16	2.47	2.65	2.57	2.71	2.97	3.56	3.75
P$_2$O$_5$	0.26	0.23	0.24	0.25	0.25	0.27	0.31	0.39	0.41
H$_2$O$^+$		0.45	0.45	0.32	0.09	0.02	0.08	0.08	0.03
H$_2$O$^-$		0.03	0.03	0.03	0.02	0.03	0.01	0.00	0.00
Total	99.03	99.90*[1]	99.94*[1]	100.07*[1]	100.00*[1]	100.05*[1]	99.95*[1]	99.76*[1]	100.14
Q	1.3	—	—	1.9	—	1.1	1.5	3.7	4.2
or	2.3	2.4	2.8	2.9	3.0	3.4	3.6	4.3	4.9
ab	18.2	17.6	19.1	19.9	18.0	19.0	21.2	22.1	23.8
an	27.6	21.6	24.1	25.3	23.9	25.8	25.3	23.1	22.0
di	18.1	17.4	20.2	21.8	22.7	23.9	21.4	18.7	16.7
hy	23.7	25.7	23.0	19.3	18.7	18.3	18.4	17.7	17.7
ol	—	8.0	2.2	—	6.2	—	—	—	—
mt	3.4*[2]	2.8	3.4*[2]	3.3*[2]	2.2	2.7	2.3	2.7	2.7
il	4.8	4.1	4.7	5.1	4.9	5.2	5.7	6.8	7.1
ap	0.6	0.5	0.6	0.6	0.6	0.6	0.7	0.9	1.0
M-value	57	67	61	54.5	61	54	51.5	45.5	41

1 = average composition of tholeiitic basalt (tholeiites and olivine tholeiites) for entire Hawaiian group (G.A. Macdonald and Katsura, 1964, p.124, table 9, analysis *8*).

2 = glassy tholeiite with 13.5 vol.% olivine phenocrysts recovered from depth of 2590 m, east rift zone of Kilauea (J.G. Moore, 1965, p.44, table 2, analysis *3*).

3 = glassy tholeiite with ~9 vol.% olivine phenocrysts and 1 vol.% each augite and plagioclase phenocrysts recovered from a depth of 1400 m, east rift zone of Kilauea (J.G. Moore, 1965, p.44, table 2, analysis *2*).

4 = glassy tholeiite with 2 vol.% olivine, 3 vol.% augite and 1 vol.% plagioclase phenocrysts recovered from a depth of 400 m, east rift zone of Kilauea (J.G. Moore, 1965, p.44, table 2, analysis *1*).

5 = most mafic glass found among lavas of the 1959 summit eruption of Kilauea Iki (Murata and D.H. Richter, 1966, p.24, table *8*, analysis *2*, and discussion on p.24).

6 = average composition of six Kilauea summit lavas, at end of major stage of olivine crystallization (Murata and D.H. Richter, 1966, p.14, table 3, analysis *7*, and discussion on p.23).

7 = average composition of glassy groundmass of three Kilauea summit lavas at beginning of plagioclase crystallization (Murata and D.H. Richter, 1966, p.15, table *4*, last column; and discussion on p.23).

8 = most differentiated glass in 1960 Kilauea summit lavas (Murata and D.H. Richter, 1966, p.15, table 4, column *4*; and discussion on p.23).

9 = most differentiated lava of 1955 Kilauea eruption (G.A. Macdonald, 1955, p.35, table 13, analysis *2*; see also discussion in Murata and D.H. Richter, 1966, p.23).

*¹ Small ($0.0x\%$) contents of CO_2, Cl and F also reported; their aggregate total does not exceed $0.0x\%$.

*² $Fe_2O_3/(Fe_2O_3 + FeO)$ reduced to 0.2.

fractionation on lava compositions at accreting plate margins, with particular reference to Iceland, the FAMOUS segment of the Mid-Atlantic Ridge, and the Troodos ophiolite complex.

Table 9.2 presents nine analyses of Hawaiian tholeiites. The first is an overall average of analyses of tholeiites and olivine tholeiites from the entire Hawaiian group (some may be accumulative rocks enriched in olivine crystals). The next three analyses are of very fresh porphyritic glassy tholeiites dredged from the submarine part of the east rift zone of Kilauea, in which increased amount of MgO clearly correlates with higher content of phenocrysts which are predominantly olivine. In contrast to these analyses of rocks which are, at least in part, accumulative or fractionated, analysis five represents the most mafic aphyric glassy rock (highest MgO content) of the 1959--1960 eruption of Kilauea, and thus represents an approach to a true parental liquid magma composition. Despite the high content of MgO and high M-value in this aphyric rock that indicate little high-level fractionation, the contents of TiO_2, K_2O and P_2O_5 are clearly higher than the range of values in MORB (see Table 9.1; Fig. 9.1). The remaining four analyses of Table 9.2 are of rocks believed to represent fractionation stages of this parental magma. Murata and D.H. Richter (1966) present an interesting detailed study, reproduced in Table 9.3, of the compositional control exercized by the fractionation first of olivine, then later joined by augite and plagioclase. It is inferred that from an MgO content of 10.0 down to 7.3 wt.% (or even 6.8 wt.% according to a comparable study by T.L. Wright, 1971), olivine fractionation essentially controls the compositions of the successively derived tholeiitic liquids. Note how fractionation of a total of ~36.5% of crystals results in a change in SiO_2 content from 49.2 to 51.2 wt.% only, i.e. the resultant product is still basaltic. Note also that augite precedes plagioclase as a phenocryst phase in these lavas from the Kilauea summit eruption where the parental magma, while still tholeiitic, may trend to a slightly more alkalic and/or water-rich composition than on the subaerial and submarine flanks of the volcano.

The situation in Iceland is apparently considerably more complex. First, compared to Hawaii, there is a higher proportion of more fractionated rock such as for example in the Thingmuli series (Table 9.4, analyses $1-6$, indicating, like Kilauea, relatively high initial TiO_2, K_2O and P_2O_5). Note, in passing, how M-values give a much greater separation than SiO_2 contents of early stages of fractionation. Furthermore, there appears to be a significant pattern of areal control over the compositions of basaltic rocks erupted within post-glacial and probable late Pleistocene time (Jakobsson, 1972) (see Fig. 9.3 and Table 9.4, analyses $7-12$).

Following the line of the Mid-Atlantic Ridge northwards onto land, the olivine tholeiites of the Reykjanes peninsula are apparently progressively enriched in TiO_2, K_2O and P_2O_5 relatively to MORB. In a more central part of Iceland, situated significantly just east of the ridge line, in an area

TABLE 9.3

Control of compositions (in wt.%) of recent Kilauea summit lavas by fractionation of olivine, augite and plagioclase (after Murata and D.H. Richter, 1966)

Stage of differentiation	Percentage of			Major minerals previously crystallized
	MgO	SiO$_2$	crystals removed	
Parental magma	10.00	49.20	0	none
End of major olivine crystallization	7.34	50.10	6.4	olivine
Start of main feldspar crystallization	6.75	50.45	14.1	pyroxene, olivine
Most differentiated glass in early-1960 lavas	5.74	50.69	30.3	plagioclase, pyroxene, olivine
Most differentiated 1955 lava	5.12	51.24	36.5	plagioclase, pyroxene, olivine

of voluminous volcanism comprising what has been termed the "Kverkfjöll hotspot" (Sigvaldason et al., 1974), two groups of tholeiite, one just saturated (Veidivötn) and one oversaturated (Askja-Mývatn), show high TiO$_2$, K$_2$O and P$_2$O$_5$ contents along with low M-values and high Na/Ca. On the basis of this kind of major-element data, these latter rock groups look like candidates for a Schilling—O'Hara debate all over again. However, to the south of this hotspot the character of the basic rocks changes via varieties transitional to alkali basalt (Torfajökull) to alkali basalts (Westman Islands) of decisively nepheline-normative composition, even without any corrections of Fe$_2$O$_3$ of the average analysis from its fairly high analysed value of 3.39 wt.%.

Alkali basalts also occur at Snaefellsnes in the extreme west of Iceland. One interpretation of the data is that superimposed on and partly mixing with the north—south ocean ridge volcanism there is a pattern of extraordinarily copious tholeiitic volcanism centered significantly to the east of the ridge axis (the Kverkfjöll hotspot). This hotspot centre is flanked by a discontinuous ring of eruption of transalkaline basalt and an outer ring of alkali basalt, thus reproducing in space the observed time sequence in the Hawaiian Islands where, in contrast to a relatively static situation in eastern Iceland, lithosphere is fairly scampering over underlying mesosphere (Fig. 9.5).

Just to emphasize the magnitude of these phenomena, it is worth noting that the Hawaiian archipelago is the highest and largest single mountain chain on Earth (if elevation above base level is considered), and a substantial fraction of all subaerial lava in the world at the present time is erupted within the confines of Iceland. These facts cannot be disregarded in any petrogenetic and tectonic synthesis.

Compared with MORB the following significant differences emerge: (1) OIT have significantly higher values of $^{87}Sr/^{86}Sr$ as noted above, pre-

TABLE 9.4

Analyses and CIPW norms of representative volcanic rocks from Iceland

	1	*2*	*3*	*4*	*5*	*6*	*7*	*8*	*9*	*10*	*11*	*12*
SiO_2	47.44	49.31	54.22	61.68	69.60	73.94	48.01	49.51	50.26	47.00	46.83	46.88
Al_2O_3	14.17	12.67	13.01	14.35	12.05	12.30	14.09	13.97	13.52	13.84	16.12	15.53
Fe_2O_3	4.37	4.75	4.41	3.95	0.85	1.14	2.69	2.15	3.52	2.74	3.39	2.30
FeO	7.97	10.31	8.38	4.31	1.95	0.92	10.06	11.37	11.87	12.68	8.69	9.42
MnO	0.20	0.26	0.25	0.20	0.08	0.04	0.21	0.21	0.24	0.21	0.19	0.22
MgO	7.25	4.69	3.29	1.04	0.23	0.05	8.29	5.97	5.58	5.96	7.47	7.51
CaO	11.11	9.10	7.08	4.29	1.45	0.96	11.77	10.76	9.18	9.74	10.26	11.07
Na_2O	2.46	3.01	3.55	4.38	4.60	4.74	2.17	2.71	2.48	2.98	3.27	2.90
K_2O	0.33	0.56	1.17	2.11	2.90	3.44	0.29	0.42	0.56	0.65	0.60	0.71
TiO_2	2.19	3.13	2.73	0.98	0.27	0.25	1.87	2.39	2.47	3.71	2.38	2.41
P_2O_5	0.27	0.57	0.82	0.34	0.06	0.04	0.19	0.23	0.23	0.33	0.32	0.44
H_2O^+	1.25	0.68	0.60	1.44	4.65	1.94	}0.28	}0.42	}0.31	}0.38	}0.30	}0.28
H_2O^-	0.81	0.76	0.40	0.87	1.15	0.22						
Total	99.82	99.80	99.91	99.94	99.84	99.98	99.92	100.11	100.22	100.22	99.82	99.67
Q	—	2.9	8.7	18.5	29.7	32.2	—	—	3.2	—	—	—
or	2.0	3.4	7.0	12.8	18.2	20.8	1.7	2.5	3.3	3.9	3.6	4.2
ab	21.3	25.9	30.4	38.0	41.4	41.0	18.4	23.0	21.0	25.3	24.3	22.9
an	27.3	19.8	16.3	13.6	3.9	2.2	28.0	24.8	24.1	22.5	27.7	27.4
ne	—	—	—	—	—	—	—	—	—	—	1.9	1.0
di	22.4	18.7	11.7	5.0	2.9	1.1	24.0	22.6	16.6	19.7	17.5	20.4
hy	13.0	17.5	15.0	3.7	1.9	0.5	12.6	17.8	22.0	7.1	—	—
ol	5.4	—	—	—	—	—	7.5	1.1	—	9.8	16.2	15.1
mt	3.6[1]	4.4[1]	3.7[1]	5.9	1.3	1.7	3.7[1]	3.1	4.5[1]	4.0	3.5[1]	3.4
il	4.3	6.1	5.3	1.9	0.6	0.5	3.6	4.6	4.7	7.1	4.6	4.6
ap	0.6	1.4	1.9	0.8	0.2	0.1	0.4	0.5	0.5	0.8	0.8	1.0
M-value	52	37	32.5	19	13	4	54	44	40	41	53	53.5

1 = average of three olivine tholeiites from Thingmuli (Carmichael, 1964, p.439, table 2, analyses *1—3*).

2 = average of seven tholeiites from Thingmuli (Carmichael, 1964, p.439, table 2, analyses *4—10*).

3 = average of three basaltic icelandites from Thingmuli (Carmichael, 1964, p.441, table 3, analyses *11—13*).

4 = average of four icelandites from Thingmuli (Carmichael, 1964, p.442, table 4, analyses *14—16*).

5 = average of two acid rocks (SiO_2 less than 70 wt.%) from Thingmuli (Carmichael, 1964, p.443, table 5, analyses *18* and *19*).

6 = average of five acid rocks (SiO_2 greater than 70 wt.%) from Thingmuli (Carmichael, 1964, p.443, table 5, analyses *20—24*).

7 = average of twelve olivine tholeiites from the Reykjanes peninsula (Jakobsson, 1972, p.367, table 1, analysis I*²).

8 = average of eleven "saturated tholeiites" from Veidivötn district (Jakobsson, 1972, p.367, table 1, analysis II*²).

9 = average of twelve quartz tholeiites from Askya and Mývatn districts (Jakobsson, 1972, p.367, table 1, analysis III*²).

10 = average of nine "transitional alkali basalts" from Torfajökull (Jakobsson, 1972, p.367, table 1, analysis IV*²).

11 = average of eight alkali basalts from the Westman Islands (Jakobsson, 1972, p.367, table 1, analysis V*²).

12 = average of four alkali basalts from Snaefellsnes (Jakobsson, 1972, p.367, table 1, analysis VI*²).

*¹ $Fe_2O_3/(Fe_2O_3 + FeO)$ recalculated to 0.2.

*² Averages compiled from various sources, published and unpublished, acknowledged by S.P. Jakobsson (Jakobsson, 1972, p.367).

sumably a function of derivation from mantle material with a similarly slightly higher Rb/Sr ratio; (2) $^{207}Pb/^{206}Pb$ lead ratios similarly are consistently higher in OIT; (3) OIT have a different REE "signature" (Fig. 9.6), in that they show appreciable enrichment in light REE, presumably a function of derivation in part from greater depths than MORB, conveniently shown by a higher La/Sm ratio (lanthanum, atomic number 57, is the lightest REE, and samarium, atomic number 62, is the sixth lightest). The La/Sm ratio in fact shows the greatest difference of any of the parameters derived from the composition of MORB and OIT; and (4) TiO_2, K_2O and P_2O_5 are similarly higher in OIT.

However, debate still continues on such possibilities as (1) the extent of the possible lateral flow of PHMP material in the asthenosphere and, where a plume underlies a spreading ridge system, the possible extent of lateral

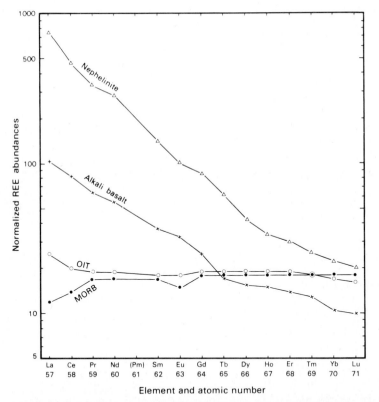

Fig. 9.6. Representative normalized REE abundances in MORB (*solid circles*), OIT (*open circles*), alkali basalt (*crosses*) and nephelinite (*triangles*) (generalized from data in Gast, 1968; Kay and Gast, 1973; Schilling, 1975). Note that this is diagrammatic only and that individual rock analyses of each grouping are found to lie in a broad spectrum of compositions.

plume flow below the axis of the ridge, a flow which may be partly or wholly blocked at the positions of major transform faults; (2) the extent of possible mixing of magmas derived from the PHMP and the DLVL; and (3) the possible generation of primordial alkali basalt magma within a plume at depth and addition of it to tholeiitic magma generated at shallower levels in the plume system. Thus, within the boundaries of the hypothesis, concepts are now being explored in slightly different terms to the relatively simple model first presented by Schilling.

Further suggestive evidence comes from the geochemistry and petrology of dredged basalts from the Bouvet triple junction, South Atlantic, (Dickey et al., 1977) which are more fractionated than typical oceanic tholeiites. Fractionation, however, does not fully account for their chemical characteristics, and it appears that they were derived from special source materials, contaminated perhaps by a mantle plume. The authors also caution that alternative concepts to rising plumes such as anomalously deep melting in the mantle are worthy of consideration.

In this context see also discussions on the possible roles of fractionation and plume origin in producing the relatively iron-rich basalts of the Juan de Fuca Ridge (Barr and Chase, 1974), and the proposed "hotspot" origin of the Kodiak—Bowie seamount chain apparently extended by the J. Tuzo Wilson knolls marking the site of a young submarine volcano (Chase, 1977).

Recently, suggestive evidence of a more complex pattern of mantle convection associated with magma generation than individual plumes has come a study of the Easter volcanic line (Bonatti et al., 1977). On this elongate line some 5000 km long, recent (less than 2 Ma old) activity has occurred in at least five areas. Rather than individual hotspots the authors account for this phenomenon by proposing not merely a line of lithospheric weakness but a mantle "hot line" of a type that may be related to major mantle convection cells which could develop below fast moving plates with axes parallel to the plate motion (F.M. Richter, 1973).

9.4. ALKALI BASALT SERIES AND BASANITES

The evolution of the term "olivine basalt" via "olivine alkali basalt" to "alkali basalt" has already been traced. In principle, this igneous rock series and its mafic progenitor, alkali basalt, has now been recognized for several decades. Its name is based on a compositional attribute and not on a precisely defined tectonic setting as with MORB and OIT. This is indeed reasonable as alkali basalt occurs in diverse tectonic settings. For example, rocks of the alkali basalt series build many of the smaller oceanic islands away from the mid ocean ridges, though they apparently form no more than ~1% to 2% of the bulk of the Hawaiian islands where they have been intensively studied. They also occur abundantly in some continental provinces associated with doming and/or rifting, notably the Neogene to present East African Rift Valley province and the Tertiary rocks of eastern Australia. Alkali basalts are also found to a lesser extent in areas of convergent plate margins such as New Zealand though they are not universally present in such an environment. Schwarzer and J.J.W. Rogers (1974), in a useful compilation, list no less than 39 examples of differentiated alkali basalt series: 23 in an oceanic island setting, 11 in a continental setting, and 5 in an island-arc setting. Schwarzer and Rogers also record a further 26 examples of alkali basalt alone without differentiated series members in the proportion of 8 oceanic to 15 continental and 3 island arc.

Schwarzer and Rogers tend to emphasize an overall chemical similarity in alkali basalts from these different environments. Although the quoted oceanic and continental examples are found in hotspots plausibly related to a common denominator in inferred plume activity, the island-arc examples, however, are unlikely to have originated in a similar way. Schwarzer and Rogers' analysis is based on major-element chemistry alone, always a dangerous procedure with mafic rocks in which trace-element signatures may have great petrogenetic significance. Of the quoted examples, some are relatively rich in potassium, while a few contain modal feldspathoid (and are thus basanites). Some exception can thus possibly be taken to the rather broad grouping attempted by Schwarzer and Rogers. Nevertheless it is clear that alkali basalts are characterized by significantly lower silica activities than those of tholeiites. The most evident mineralogical result of this lower silica activity is that olivine is the ubiquitous ferromagnesian mineral (never a ferromagnesian pyroxene) in rocks belonging to alkali basalt series, commonly present in two generations as phenocrysts and as a groundmass mineral. A summary of the numerous petrographical differences between tholeiite and alkali basalt, incorporating those of R.W. White (1966, pp. 252–255) based on Hawaiian rocks, is presented in Table 9.5. Numerous as these petrographic distinctions and clues may appear, in practice many are difficult to apply because of the very fine-grained or even glassy nature of individual samples, the presence of some transitional varieties, and also on account of some of the criteria being negative or of a relative nature. The affinity of a particular specimen with respect to these two important mafic groups may therefore remain in doubt pending chemical analysis.

Bulk chemical analysis has become more speedy and much more accessible in recent years, and has therefore come to play an increasingly important role in comparative work on fresh volcanic rock series. To a first approximation the main chemical difference between tholeiite and alkali basalt is that the distinctly lower silica activity of the latter results in the appearance of nepheline in the CIPW norms of alkali basalts unlike tholeiites which are all hypersthene-normative. Rocks with *modal* nepheline, however, are termed basanite (or tephrite if olivine-free), not alkali basalt. In crystalline rocks it has been empircally observed that modal nepheline appears

TABLE 9.5

Summary of petrographical differences between tholeiite and alkali basalts including differences observed by R.W. White (1966) on basaltic rocks from Hawaii

Tholeiites	Alkali basalts
(1) Ultramafic xenoliths very rare (more in the nature of glomeroporphyritic olivine phenocrysts when present)	ultramafic xenoliths fairly common, dunite and wehrlite predominating (mineralogy indicates equilibration at high pressures)
(2) Infrequent large olivine phenocrysts	common medium-sized olivine phenocrysts
(3) Olivine phenocrysts not commonly zoned	zoning of olivine phenocrysts often marked, with more highly birefringent, iron-rich rims often altered to iddingsite
(4) Olivine may show Bowen—Anderson reaction producing orthopyroxene rim	no orthopyroxene overgrowths on olivine
(5) Orthopyroxene phenocrysts may occur	orthopyroxene absent
(6) Plagioclase phenocrysts often accompany mafic phenocrysts (phenocryst sequence is generally olivine—plagioclase—augite)	plagioclase phenocrysts relatively less common (phenocryst sequence is generally olivine—augite—plagioclase)
(7) Associated accumulative rocks are oceanites (rich in olivine phenocrysts)	associated accumulative rocks are often ankaramites (rich in olivine + augite phenocrysts)
(8) If there are phenocrysts of augite, it is a common augite of a pale-brown colour	calcic augite phenocrysts are commonly titaniferous; rims especially may show the typical purplish colour
(9) Groundmass usually relatively fine-grained; intergranular textures common	groundmass is commonly relatively coarser (for a flow of comparable dimensions), and may range in texture from intergranular to subophitic, even ophitic
(10) No groundmass olivine	olivine present as groundmass constituent (to exclusion of any ferromagnesian pyroxene)
(11) Groundmass pyroxene (if it can be determined) is variable and can comprise either (1) pigeonite, rarely hypersthene, together with a subcalcic augite or augite, or (2) one species, a subcalcic augite	one species of groundmass pyroxene, a titansalite

TABLE 9.5 (*continued*)

Tholeiites	Alkali basalts
(12) No alkali-feldspar (except as component of glassy oversaturated mesostasis) or analcite	interstitial alkali feldspar may be recognized with difficulty; less commonly analcite
(13) Frequent intersertal texture with a mesostasis of glass often altered to chlorophaeite	interstitial glass absent or sparse
(14) Biotite generally absent	sporadic accessory biotite

when a content of ~5 wt.% normative nepheline is surpassed, so that this value serves to divide alkali basalt from basanite, and a very fine-grained or glassy rock with more than 5 wt.% normative nepheline could thus be termed "basanitoid" (G.A. Macdonald and Katsura, 1964, p.86). This incidentally is a good example of a term of petrographic origin being given a reasonable redefinition in chemical terms, a procedure frequently indulged in at present often without changing or at least defining the term, so that the reader is commonly not alerted to the redefinition*.

A further complication is that some rocks transitional between tholeiite and alkali basalt may be just hypersthene-normative but are apparently of alkali basalt petrographic affinity (Poldervaart, 1964), as they and mem-

---

*There is a difficulty involved in this apparently simple and straightforward concept of bracketing alkali basalts as containing between 0 and 5 wt.% normative nepheline. The problem arises where a rock may have been subjected to late-stage or post-crystallization oxidation producing high amounts of Fe_2O_3 relative to FeO. The calculation of the norm applies all Fe_2O_3 plus an equal molar amount of FeO to the formation of normative magnetite, which, of course, contains no silica at all, so that a rock analysis containing a high ferric to ferrous oxide ratio of secondary origin will result in fortuitously increased amounts of silica being available for other normative minerals, and hence the indication of a misleading magmatic affinity. Clearly, it could be desirable to make some correction to the raw analysis before calculating the norm and employing it for comparative purposes. A correction is frequently made and can take several forms, viz. to reduce Fe_2O_3 to some constant value, or again to some constant proportion of total iron oxide present, or yet again to employ more sophisticated corrections relating Fe_2O_3 content to other chemical parameters such as alkali content or TiO_2 content (to take account of the fact that "true" igneous Fe_2O_3 contents and ratios apparently vary within different groups of mafic rocks). As no consensus of treatment has emerged (see discussion in C.J. Hughes and Hussey, 1976), this correction factor (or the lack of any correction) has a considerable nuisance value when comparison of data is made. A plea for the adoption of some reasonable median ratio (make Fe_2O_3 equal to 0.2 total iron oxide by weight) is made by Hughes and Hussey, but as "true" ratios are lower than this in MORB and komatiites and apparently tend to be higher in more alkaline mafic rocks this procedure can be criticized; nevertheless, the adoption of it or some similar agreed convention would reduce the present confusion considerably (see also C.K. Brooks, 1976).

bers of their associated series show no olivine reaction relationship with ferromagnesian pyroxene.

The empirical division of alkali basalt (and more alkaline mafic rocks generally) from tholeiite (and "subalkaline" rocks generally) by use of a Harker variation diagram in which total alkalis, (Na_2O + K_2O), are plotted against silica, has already been mentioned (see Fig. 8.2). It is probably the most widely used major-element variation diagram in use for discrimination purposes at the present date, particularly for fresh volcanic mafic rocks. A difficulty with the use of this diagram for older rocks is that degradation involving the partial or complete replacement of pyrogenetic plagioclase by albite is apparently accompanied by the metasomatic addition of soda to the rock. Spilitized tholeiites, for example, commonly plot in the alkali basalt field of this diagram for this reason (see Chapter 14).

A further chemical difference between tholeiite and alkali basalt pertains to the composition of augite phenocrysts (if present). Augite phenocrysts of tholeiite are hypersthene-normative whereas the more salic augite phenocrysts of alkali basalt are nepheline-normative due to solid-solution components expressible as jadeite ($NaAlSi_2O_6$) and Tschermak's molecule ($CaAl_2SiO_6$). These silica-undersaturated components (stoichiometrically equivalent to albite and anorthite, respectively, less one molecule of silica each) reflect the crystallization in alkali basalts of clinopyroxene from melts of lower silica activity than those of tholeiites.

Fractionation within the alkali basalt series preserves a just undersaturated to saturated series of rocks known as hawaiite, mugearite, benmoreite, and trachyte. The first three names are specific to the alkali basalt series, a good example of a necessary new series nomenclature cutting across a traditional petrographic nomenclature of the type developed in Chapter 4. Average analyses of these rocks from Hawaii, where they have been intensively studied, are given in Table 9.6, taken from G.A. Macdonald (1968)*. Petrographically, hawaiite is a basic rock with a somewhat low colour index for a basalt, and characterized by plagioclase of andesine composition (for this reason it has confusingly been termed andesite in some older accounts). Mugearite, though still basaltic in aspect, has a still lower colour index, ~20—25, and typically contains plagioclase microphenocrysts of potassium-rich oligoclase—andesine composition grading to "lime anorthoclase" in the groundmass (Muir and Tilley, 1961). Due to the relative abundance and alignment of tabular feldspar microlites many mugearites are fissile in outcrop and difficult to collect. Both hawaiite and mugearite may contain

---

*Note again that an igneous rock series is essentially compositionally defined and does not necessarily have any areal restriction as with the term province. An illustration of this is the appropriate utilization of names from the Hebrides where an alkali basalt series is also present (Mugeary is a village in Skye, Ben More a mountain in Mull) for the classic alkali basalt series in Hawaii.

TABLE 9.6

Chemical compositions and CIPW norms of averaged members of Hawaiian alkali basalt series

	1	2	3	4	5	6
SiO_2	44.1	45.4	47.9	51.6	57.1	61.7
Al_2O_3	12.1	14.7	15.9	16.9	17.6	18.0
Fe_2O_3	3.2	4.1	4.9	4.2	4.8	3.3
FeO	9.6	9.2	7.6	6.1	3.0	1.5
MnO	0.2	0.2	0.2	0.2	0.2	0.2
MgO	13.0	7.8	4.8	3.3	1.6	0.4
CaO	11.5	10.5	8.0	6.1	3.5	1.2
Na_2O	1.9	3.0	4.2	5.4	5.9	7.4
K_2O	0.7	1.0	1.5	2.1	2.8	4.2
TiO_2	2.7	3.0	3.4	2.4	1.2	0.5
P_2O_5	0.3	0.4	0.7	1.1	0.7	0.2
Total	99.3	99.3	99.1	99.4	98.4	98.6
Q	—	—	—	—	3.4	—
C	—	—	—	—	0.2	—
or	4.2	6.0	9.0	12.5	16.8	25.2
ab	10.8	18.7	31.9	42.4	50.8	63.6
an	22.6	23.9	20.3	15.8	13.1	3.6
ne	2.9	3.8	2.2	2.0	—	—
di	26.5	21.2	12.6	6.3	—	1.0
hy	—	—	—	—	5.8	0.1
ol	23.4	16.0	12.3	10.9	—	1.5
mt	3.7[*1]	3.9[*1]	3.6[*1]	3.0[*1]	6.0[*2]	3.7[*2]
il	5.2	5.8	6.5	4.6	2.3	1.0
ap	0.7	0.9	1.7	2.6	1.7	0.5
M-value	65	52	41.5	37	27.5	13

1 = average of nine ankaramites; 2 = average of 35 alkali olivine basalts; 3 = average of 62 hawaiites; 4 = average of 23 mugearites; 5 = average of five benmoreites; 6 = average of five trachytes (all analyses taken from G.A. Macdonald, 1968, p.502, table 8).

[*1] $Fe_2O_3/(Fe_2O_3 + FeO)$ recalculated to 0.2.
[*2] Fe_2O_3 and FeO recalculated to contain equal weights of Fe.

olivine which, in line with the progressively more sodic compositions of the plagioclase, is found to be progressively more iron-rich in these more differentiated members of the series. Benmoreite is not well-defined mineralogically, but trachytes are thoroughly felsic, usually with a silica content over 60 wt.%. In practice, silica content is often used to group analysed rocks between demarcated limits within the series; for example, Duncan (1975), working on volcanic rocks from Mount Etna, utilized the following arbitrary subdivision of an alkali basalt series:

	SiO$_2$ content (wt.%)
Hawaiites	47—52
Basic mugearites	52—55
Mugearites	55—58
Benmoreites	58—62

G.A. Macdonald (1968, pp. 516 and 517) shows how successive members of the alkali basalt series in Hawaii could plausibly be derived by the subtraction of compositions of appropriate amounts of observed phenocryst phases from the composition of a parental alkali basalt magma and successively derived liquids. A mechanism of crystal fractionation thus seems proved, or at least acceptable, for this series of closely associated differentiated rocks.

The ultimate product of fractional crystallization in alkali basalt series may not be trachyte but may include small quantities of rocks that are either distinctly oversaturated (e.g., pantellerite and comendite as in Terceira, Azores, (Self and Gunn, 1976), or undersaturated (e.g., phonolite as in St. Helena) (I. Baker, 1969). These compositions, of course, correspond more closely than does trachyte to the compositions of the two minima in the system SiO$_2$—Ne—Ks, compositions that would indeed be expected to result from extreme fractionation. A peralkaline mineralogy commonly found in both reflects the "plagioclase effect" (Bowen, 1945) whereby prolonged fractionation of plagioclase (not only more lime-rich but also more alumina-rich than the crystallizing liquid, at least in the simple system An—Ab) may result eventually in a peralkaline condition of the melt. The existence of both undersaturated and oversaturated end-products associated with different alkali basalt series suggests that, unless other possible influences were at work such as gaseous transfer of alkalis, the alkali basalt series could be further subdivided presumably on the basis of small initial differences of silica activity reflected in the ultimate path of fractional crystallization. Major-element compositions of two oceanic alkali basalt series illustrating the above two contrasted possibilities are presented in Tables 9.7 and 9.8, namely an Azores series ending with comendite, and the St. Helena series ending with phonolite.

The analyses in Table 9.7 are selected from the description by Self and Gunn (1976) of the island of Terceira, Azores archipelago. They comprise rocks collected from the Fissure Zone and Santa Barbara volcano, the two most recently active volcanic areas of the island. The quoted analyses had been recast water-free to 100 wt.% with small rounding-off deviations in some of their totals. Total iron was reported as Fe$_2$O$_3$, and in accompanying normative calculations the proportions of iron oxides have been appropriately recalculated. The alkali basalts and hawaiite prove to be nepheline-normative and the remainder hypersthene-normative; the pantelleritic trachyte and comendite are acmite- and quartz-normative.

TABLE 9.7

Selected analyses and CIPW norms of young volcanic rocks of the island of Terceira, Azores; an alkali basalt—trachyte—comendite series

	1	2	3	4	5	6	7
SiO_2	46.76	46.61	48.11	54.15	57.36	64.83	67.43
Al_2O_3	13.99	13.84	16.13	15.62	16.71	13.98	13.88
Fe_2O_3	12.17*[1]	13.20*[1]	12.03*[1]	10.37*[1]	8.61*[1]	6.05*[1]	5.25*[1]
MnO	0.18	0.21	0.18	0.20	0.16	0.27	0.21
MgO	7.39	5.91	4.06	2.77	2.71	0.41	0.13
CaO	11.71	11.10	10.14	6.40	5.48	0.98	0.71
Na_2O	3.37	3.76	3.97	4.97	5.17	8.01	7.13
K_2O	0.87	1.18	1.22	2.19	2.38	5.00	4.91
TiO_2	2.90	3.42	3.56	2.40	1.43	0.44	0.34
P_2O_5	0.66	0.76	0.59	0.91	—	0.04	—
Total	100.00	99.99	99.99	99.98	100.01	100.01	99.99
Q	—	—	—	3.9	4.3	5.5	10.7
or	5.2	7.1	7.3	13.0	14.1	29.6	29.1
ab	20.8	21.8	29.1	42.3	43.9	44.2	44.2
an	20.7	17.6	22.8	13.9	15.4	—	—
ne	4.3	5.6	2.6	—	—	—	—
ns	—	—	—	—	—	3.2	1.8
ac	—	—	—	—	—	8.8	7.6
di	27.3	27.1	19.9	9.5	9.4	4.0	3.1
hy	—	—	—	3.2	3.8	1.7	2.9
ol	11.3	9.0	6.9	—	—	—	—
mt	3.3*[2]	3.6*[2]	3.2*[2]	7.6*[3]	6.3*[3]	—*[3]	—*[3]
il	5.6	6.6	6.8	4.6	2.7	0.8	0.7
ap	1.6	1.8	1.4	2.1	n.d.	0.1	n.d.
M-value	54.2	46.6	39.7	34.1	37.9	11.3	4.5

1 = alkali olivine basalt, Fissure Zone; analyst B.M. Gunn (Self and Gunn, 1976, appendix, p.310, analysis T-0059).

2 = alkali olivine basalt, Fissure Zone; analyst B.M. Gunn (Self and Gunn, 1976, appendix, p.310, analysis T-0050).

3 = hawaiite, Santa Barbara volcano; analyst B.M. Gunn (Self and Gunn, 1976, appendix, p.311, analysis T-0066).

4 = aphyric mugearite, Santa Barbara volcano; analyst B.M. Gunn (Self and Gunn, 1976, appendix, p.311, analysis T-0072).

5 = benmoreite, Santa Barbara volcano; analyst B.M. Gunn. (Self and Gunn, 1976, appendix, p.311, analysis T-2067).

6 = pantelleritic trachyte, Santa Barbara volcano; analyst B.M. Gunn. (Self and Gunn, 1976, appendix, p.311, analysis T-0084).

7 = comendite obsidian, Santa Barbara volcano; analyst B.M. Gunn (Self and Gunn, 1976, appendix, p.311, analysis T-2068).

n.d. = not determined.

*[1] Total iron reported as Fe_2O_3.

*[2] $Fe_2O_3/(Fe_2O_3 + FeO)$ recalculated to 0.2.

*[3] Fe_2O_3 and FeO recalculated to contain equal weights of Fe.

Table 9.8 presents averaged analyses of six petrographic groupings of rocks, ranging in composition from alkali basalt through trachyte to phonolite, taken from the account by I. Baker (1969) of the petrology of the island of St. Helena. The rocks are nepheline-normative throughout and the phonolite is acmite-normative. Baker concludes that whereas processes of crystal fractionation led to the formation of the series from an alkali basalt to trachyte, the chemistry of the phonolites may reflect in addition some late-stage volatile transfer and migration of alkali-rich fluids. Note that M-values, even among the most basic representatives of the St. Helena series, are low suggesting the operation of strong fractionation; i.e. petrologically, as well as topographically, we may be witnessing just the top of a large pyramid of fractionated rock.

For a discussion of the significance of trace-element contents in alkali basalts, particularly in contrast to those in tholeiites, one turns to the landmark presentation by Gast (1968) who states:

"Most of the discussions of the problem of parent magmas and genetic relations between magma types are in terms of bulk compositions (major elements) and phase equilibria relations. It is thus pertinent to establish in more detail the restrictions that may arise from dispersed element geochemistry.

Even a cursory review of the trace element concentrations in volcanic rocks indicates that they must carry very significant information bearing on the genetic relations of various liquids. For example, the abundance of the elements, Rb, Cs and Ba varies by more than two orders of magnitude in the class of rocks designated as basalt. The different trace element abundances of the saturated and undersaturated volcanic rocks are in fact as characteristic of these magma types as their mineralogy and major element chemistry."

Crude median values of some significant trace-element contents of large groups of tholeiite and alkali basalt are as follows (Prinz, 1967):

	Trace-element contents (ppm)					
	Cr	Ni	Ba	Rb	Sr	Zr
Tholeiite	140	75	200	5	400	100
Alkali basalt	142	80	400	30	700	125

It can be seen that whereas Cr and Ni contents are comparable, suggesting that neither group has undergone appreciably more early fractionation of olivine than the other, the contents of incompatible elements, notably Rb, are consistently higher in the alkali basalt group. The relative abundances of the REE in alkali basalt is very different to those in MORB tholeiite and shows an accentuation of the trend observable in OIT, that is to say a relative enrichment in the LREE (see Fig. 9.6).

$(^{87}Sr/^{86}Sr)^0$-values are consistently higher in oceanic island alkali basalts (0.7030—0.7045; Peterman and Hedge, 1971) than in MORB and on the

TABLE 9.8

Average chemical compositions and CIPW norms of members of an alkali basalt—phonolite series, Saint Helena Island, South Atlantic Ocean

	1	2	3	4	5	6
SiO_2	45.68	47.63	55.52	59.00	60.81	60.28
Al_2O_3	16.04	16.37	17.46	17.76	18.24	19.08
Fe_2O_3	2.50	4.46	2.83	3.11	2.05	2.22
FeO	9.61	8.08	5.93	3.93	2.78	1.44
MnO	0.22	0.26	0.24	0.26	0.21	0.20
MgO	5.72	4.06	1.52	0.33	0.22	0.07
CaO	9.06	7.12	3.50	2.20	1.82	0.95
Na_2O	3.72	4.98	6.24	7.50	7.24	8.94
K_2O	1.41	1.85	3.01	3.97	4.68	5.02
TiO_2	3.39	2.86	0.80	0.31	0.15	0.05
P_2O_5	0.59	0.87	0.44	0.11	0.05	0.04
H_2O^+	1.82	1.24	1.98	1.16	1.45	1.54
H_2O^-	0.28	0.24	0.40	0.31	0.24	0.18
Total	100.04	100.02	99.87	99.95	99.94	100.01
or	8.5	11.1	18.2	23.8	28.2	30.2
ab	23.6	30.5	51.8	53.9	54.8	45.3
an	23.4	17.1	11.0	3.1	3.5	—
ne	4.6	6.7	1.3	5.7	4.1	14.1
ac	—	—	—	—	—	4.9
di	15.3	10.9	3.3	6.3	4.7	4.0
hy	—	—	—	—	—	—
ol	13.1	12.5	7.6	1.7	1.3	0.5
mt	3.6*	3.6*	4.2	4.6	3.0	0.8
il	6.6	5.5	1.6	0.6	0.3	0.1
ap	1.4	2.1	1.1	0.3	0.1	0.1
M-value	46	37	24	8	7.5	3

1 = average of four analyses of alkali basalt, Saint Helena Island (I. Baker, 1969, p.1290, table 2, analyses 3—6).
2 = average of three analyses of "trachybasalt", Saint Helena Island (I. Baker, 1969, p.1290, table 2, analyses 7—9).
3 = average of three analyses of "trachyandesite", Saint Helena Island (I. Baker, 1969, p.1291, table 2, analyses 10—12).
4 = average of three analyses of trachyte, Saint Helena Island (I. Baker, 1969, p.1291, table 2, analyses 13, 14 and 16).
5 = average of three analyses of "phonolitic trachyte", Saint Helena Island (I. Baker, 1969, p.1291, table 2, analyses 15, 17 and 18).
6 = average of two analyses of phonolite, Saint Helena Island (I. Baker, 1969, p.1291, table 2, analyses 19 and 20).

*$Fe_2O_3/(Fe_2O_3 + FeO)$ recalculated to 0.2.

whole higher than in OIT (see p.276). The difference between the $^{87}Sr/^{86}Sr$ ratios in MORB and OIT has been attributed to a different mantle source; the same reason may well apply to oceanic alkali basalts but another factor must be considered. Radiogenic ^{87}Sr in the mantle is an extremely incompatible element, or better said an incompatible isotope, because it would likely be a violent misfit in the lattice of any minor mantle mineral (phlogopite?) or on crystal boundary surfaces that accommodated K and hence Rb in the first place. Hence, in common with incompatible elements and for the same reason, it would be strongly partitioned into the liquid phase during partial melting. Small degrees of partial melting therefore might result in a perceptibly higher $^{87}Sr/^{86}Sr$ ratio in the first liquids produced, relative to liquids produced by higher degrees of melting.

Following Gast, it seems that available data on incompatible trace-element contents of alkali basalt are consistent with the hypothesis that the primary magmas that gave rise to alkali basalts were derived by lower degrees of partial melting of accepted upper-mantle mineralogies than those which gave rise to tholeiites, possibly accompanied by more pronounced zone-refining processes (see Chapter 13). Thus the trace-element data support the distinct nature of tholeiite and alkali basalt stems. The REE pattern suggests in addition that a further difference is the operation of high-pressure eclogite fractionation on primordial alkali basalt magma, to a lesser degree in the case of primordial OIT and not at all in the case of primordial MORB.

In contrast, G.A. Macdonald (1968, p.573) suggests that the relatively small amounts of alkali basalt in Hawaii could be the result of pyroxene fractionation from an olivine-rich tholeiitic primordial magma at depth. However, this conclusion is based only on major-element data, and apparently ignores considerations based on trace elements.

An interesting example of an alkali basalt series and more alkaline basanites, showing typical traits of each, is found on Grande Comore, the youngest island of the Comores archipelago (Strong, 1972). There the alkali basalts of Kartala volcano form a typical fractionation series that can plausibly be related to the subtraction of olivine and augite of observed phenocryst composition (see Fig. 8.7) from a parental magma logically assumed to be equivalent to the most magnesium-rich aphyric analysis available (and thus excluding possibly accumulative rocks as candidates for parental magma). By contrast, the basanites of the contiguous La Grille volcano, a shield volcano with relatively shallower slopes suggesting extrusion of hot fluid magma with surface irregularities resulting from some 120 pyroclastic cones, are petrographically and chemically a more difficult group to rationalize. They contain abundant ultramafic nodules and are highly porphyritic in olivine that may be in part xenocrystal (i.e. derived by the physical disintegration of nodules), in part skeletal (suggesting quench crystallization of inherently magnesium-rich magma), and in part euhedral (raising the possibility of accumulation affecting the bulk chemistry of the rock). Chemically, they do not form a coherent series that can be related to high-level fractionation as in the case of Kartala lavas, but were apparently erupted rapidly from depths in discrete small magma batches that apparently reflect varying degrees of high-pressure fractionation (see Strong, 1972, for discussion).

In contrast to La Grille which lacks evolved members of a basanite series,

TABLE 9.9

Selected analyses and CIPW norms from Las Canadas volcanoes, Tenerife; a basanite—phonolite series

	1	2	3	4
SiO_2	42.64	48.00	55.51	58.57
Al_2O_3	13.85	17.77	17.04	18.12
Fe_2O_3	4.61	2.43	2.49	2.01
FeO	7.90	6.20	4.07	2.44
MnO	0.20	0.20	0.23	0.19
MgO	8.63	4.42	2.25	0.98
CaO	11.86	8.48	2.75	0.88
Na_2O	3.96	5.48	8.13	10.18
K_2O	0.91	2.36	4.45	5.27
TiO_2	4.30	2.76	1.93	0.83
P_2O_5	0.81	0.84	0.40	0.07
H_2O^+	0.38	0.31	1.06	0.47
H_2O^-	0.15	0.34	0.17	0.12
Total	100.20	99.59	100.48	100.13
or	5.4	14.1	26.5	31.1
ab	8.1	21.5	33.3	35.6
an	17.4	17.1	—	—
ne	13.9	13.7	16.3	15.6
ns	—	—	—	3.5
ac	—	—	5.3	6.0
di	29.4	16.3	9.1	3.4
hy	—	—	—	—
ol	12.1	7.5	3.9	3.1
mt	3.6*	2.5*	1.1	—
il	8.2	5.3	3.7	1.6
ap	1.9	2.0	0.9	0.2
M-value	55.7	47.8	38.0	28.2

1 = aphyric basanite flow, Las Canadas, Tenerife, (Ridley, 1970, p.146, table 4, analysis 3).
2 = trachybasanite ropy lava, Las Canadas, Tenerife, (Ridley, 1970, p.148, table 4, analysis 28).
3 = plagioclase phonolite flow, Las Canadas, Tenerife, (Ridley, 1970, p.146, table 4, analysis 10).
4 = phonolite flow, Las Canadas, Tenerife, (Ridley, 1970, p.151, table 4, analysis 46).

*$Fe_2O_3/(Fe_2O_3 + FeO)$ recalculated to 0.2.

a well documented example of a basanite to phonolite series is provided by the Pliocene—Pleistocene volcanic rocks of Tenerife, the largest of the Canary Islands with an area of over 2000 km^2, notable for a large proportion of intermediate and felsic rocks. Following Ridley (1970) the rock types, comprising both porphyritic and aphyric varieties, can be conveniently grouped into four: basanite, trachybasalt, plagioclase phonolite, and phonolite (Table 9.9). However, a continuum is present and together with the clustering of the phonolites, some mildly peralkaline, around a composition corresponding to the undersaturated minimum in the SiO_2—Ne—Ks system points strongly to an origin for the series by fractionation of a parental basanite magma.

An unusually potassium-rich alkali basalt series is found on Tristan, the largest island, some 100 km^2 in area, of a tiny cluster of islands known as the Tristan da Cunha group, situated 500 km east of the Mid-Atlantic Ridge. Tristan exposes the youngest volcanic rocks of the archipelago and has the overall form of a simple cone, with parasitic vents and domes, 2060 m above sea level and thus ~5800 m above base level. Following the well-documented eruption of October 1961 and evacuation of inhabitants, the work of a Royal Society of London expedition in early 1962 has provided a wealth of detailed information (P.E. Baker et al., 1964). The rocks of Tristan comprise a series ranging from accumulative ankaramites and alkali basalts through intermediate varieties to phonolites. All rocks are nepheline-normative (although felsic members are alluded to as trachyte in the report), but nepheline is virtually absent. Leucite, however, occurs as interstitial crystals and subhedral groundmass crystals in many of the group of trachybasalts, which are thus technically basanites on account of the presence of modal feldspathoid. The leucite appears to be an expression of a relatively high K/Na ratio (Table 9.10). This Tristan series is thus characterized by a lower alkalinity than the Tenerife rocks but has a higher K/Na ratio. It is instructive to plot total alkalis and $K_2O/(K_2O + Na_2O)$ ratios against M-values for these and other alkali basalt series.

9.5. HIGHLY ALKALINE ROCKS OF OCEANIC ISLANDS

An example of highly alkaline rocks in an oceanic island environment is provided by the Honolulu series in the island of Oahu, Hawaiian archipelago, (see Fig. 9.4). Rocks belonging to the Honolulu series are very young and are separated in time by several million years from the main island-building stage of tholeiitic volcanic eruption in Oahu. The Honolulu series comprise the products of some forty vents scattered over an area of ~400 km^2 in southeastern Oahu. Eruption has usually been explosive, typically forming small cones, ultramafic nodules are common, and in some vents the magma is present only as spatter. In these circumstances one cannot speak of a magma

TABLE 9.10

Chemical compositions and CIPW norms of representative rocks from Tristan Island, Tristan da Cunha archipelago; a highly potassic alkali basalt series

	1	2	3	4	5
SiO_2	42.78	45.98	49.44	54.57	60.23
Al_2O_3	14.27	17.05	18.46	19.47	20.15
Fe_2O_3	5.89	3.70	2.87	2.83	2.20
FeO	8.55	7.12	5.37	2.87	0.91
MnO	0.17	0.17	0.16	0.18	0.14
MgO	6.76	4.61	3.31	1.50	0.49
CaO	12.01	10.20	7.47	5.65	2.15
Na_2O	2.79	3.99	4.95	5.82	6.58
K_2O	2.06	3.02	3.68	4.85	6.23
TiO_2	4.14	3.41	3.19	1.75	0.80
P_2O_5	0.58	0.75	1.10	0.51	0.11
Total	100.0[1]	100.0[1]	100.0[1]	100.0[1]	100.0[1]
or	12.2	17.9	21.8	28.7	36.8
ab	3.2	9.5	22.8	28.8	42.0
an	20.4	19.7	17.3	12.7	7.1
ne	11.1	13.2	10.4	11.1	7.5
di	29.1	21.4	10.4	10.1	2.3
hy	—	—	—	—	—
ol	10.6	7.0	6.4	2.5	0.3
mt	4.1[2]	3.1[2]	2.4[2]	1.6[2]	2.3[3]
il	7.9	6.5	6.1	3.3	1.5
ap	1.4	1.8	2.6	1.2	0.3
M-value	46.2	43.6	42.1	32.3	22.4

1 = olivine basalt, Tristan (P.E. Baker et al., 1964, table 6, analysis 6).
2 = trachybasalt, Tristan (P.E. Baker et al., 1964, average of thirteen analyses of trachybasalt with SiO_2 less than 48 wt.% from table 6).
3 = trachybasalt, Tristan (P.E. Baker et al., 1964, average of three analyses of trachybasalt with SiO_2 greater than 48 wt.% from table 6).
4 = trachyandesite, Tristan (P.E. Baker et al., average of nine analyses from table 6).
5 = trachyte, Tristan (P.E. Baker et al., average of five analyses from table 6).

[1] Analyses recast water-free to 100 wt.%.
[2] $Fe_2O_3/(Fe_2O_3 + FeO)$ recalculated to 0.2.
[3] Fe_2O_3 and FeO recalculated to contain equal weights of Fe.

series as the eruptive rock is of restricted mafic compositions. Typical rock types, contrasting in their high alkalinities with any of the earlier shield-forming lavas, include nephelinite, ankaratrite and melilite nephelinite (see representative analyses in Table 9.11). Their M-values, 61—66, are consistently at the high end of the range of M-values for non-accumulative mafic rocks, suggesting little or no fractionation since equilibration of magma with mantle material. This supposition is strengthened by the ubiquitous occur-

TABLE 9.11

Representative analyses and CIPW norms of highly alkalic Hawaiian lavas

	1	2	3		1	2	3
SiO_2	39.7	39.4	36.6	or	—	3.2	—
Al_2O_3	11.4	10.2	10.8	ab	—	—	—
Fe_2O_3	5.3	6.5	5.7	an	10.8	12.5	8.7
FeO	8.2	7.0	8.9	ne	17.8	12.7	20.1
MnO	0.2	0.1	0.1	lc	5.7	3.2	5.0
MgO	12.1	14.1	12.6	di	34.5	35.1	23.3
CaO	12.8	12.3	13.6	hy	—	—	—
Na_2O	3.8	2.7	4.1	ol	15.8	15.4	25.7
K_2O	1.2	1.2	1.0	mt	7.9	9.7	8.8
TiO_2	2.8	3.3	2.8	il	5.4	6.4	5.7
P_2O_5	0.9	0.8	1.1	ap	2.1	1.9	2.7
Total	98.4	97.6	97.3	M-value	62	66	61

1 = average of 10 nephelinites; *2* = representative ankaratrite (olivine nephelinite); *3* = average of 7 melilite nephelinites (all analyses taken from G.A. Macdonald, 1968, p.502, table 8).

rence of a suite of interesting ultramafic nodules that by reason of their size, typically up to 10 cm, more rarely up to 30 cm, and relatively high density, are generally believed to have been rapidly elutriated by the magma from depths of up to 100 km as indicated by their mineralogy (but see Sparks et al., 1977a, for a dissenting conclusion on speed of ascent, based on field studies of Hawaiian and Etnean lavas that show that these lavas behave, at least on eruption, as Bingham liquids with high yield strengths).

It is appropriate here to digress and consider the subject of these enigmatic **ultramafic nodules**, also called *ultrabasic xenoliths*, the presence of which is highly characteristic of highly alkaline mafic rocks. A full descriptive account of the Honolulu series nodules is given by R.W. White (1966) who concludes that although inclusions and enclosing mafic rocks appear to be genetically related, more than one origin is possible for them. Ultramafic nodules are absent from tholeiites (apart from glomeroporphyritic olivine crystals), rare in alkali basalts, but increasingly common in progressively more alkaline mafic rocks. Mineralogically, the comprise variable proportions of highly-magnesian olivine, orthopyroxene and clinopyroxene, commonly together with one of the aluminous minerals spinel or garnet. Textures range from coarse isotropic fabrics to rocks showing effects of strain with polygonization and recrystallization to fine-grained aggregates. A few contain quenched glassy material (Pike and Schwarzman, 1977). Typically they show complicated exsolution phenomena such as the exsolution of spinel from aluminous pyroxenes. Spectacular exsolution of garnet + orthopyroxene from clinopyroxene and garnet + clinopyroxene from orthopyroxene has occurred in nodules predominantly of websterite and garnet websterite composition at Salt Lake Crater, Oahu, (Beeson and Jackson, 1970).

It is possible to envisage several possible origins for the nodules:

(1) Fractionation products of primary magmas at depth (i.e. during ascent of primary magma on its way to becoming parental magma at crustal levels).

(2) "Restites" produced from original mantle material (i.e. the nodules may represent the complementary solid fraction to the liquid magma produced during an episode of partial melting).

(3) "Accidental" inclusions of mantle material, itself possibly spanning a compositional range from primary mantle to one depleted by previous partial melting episodes, and thus unlike cases (1) and (2) not directly genetically related to the host magma.

Criteria are difficult to establish as a highly-magnesian ultramafic mineralogy would be anticipated in all three possibilities and textures have been very much reconstituted. A convergence in composition plus appropriate chemistry could be held to indicate primary mantle material. P.G. Harris et al. (1972) show that some ultramafic nodules may indeed be rich enough in incompatible elements to represent original mantle material. Their geochemical survey strongly indicated that, if this were the case, regional differences in composition must exist in the upper mantle (for further discussion of this subject see Section 10.2 on kimberlites and associated rocks and also pp. 446—449). Progressive depletion in alkalis and incompatible elements generally should chemically characterize mantle restites. Fractionation products should generally be characterized by few crystallizing mineral phases or even one mineral phase, conceivably by layering or even cumulus textures [though by analogy with recrystallization textures within large crustal layered intrusions, for example, in the Stillwater Complex (Jackson, 1961), distinctive cumulus textures may not be expected to survive appreciable residence times at mantle temperatures]. Some alkali basalts, in particular, contain nodules that exhibit what appear to be a series of fractionation products appropriate to different depths ranging in mineralogy from familiar low-pressure plagioclase-bearing assemblages to higher-pressure parageneses. Nodules containing two-pyroxene-bearing assemblages in inferred fractionation products lend support to models of pyroxene fractionation having been operative at some stage in the history of alkali basalts. Of particular interest is a possible criterion of origin resulting from the probable behaviour of strontium isotopes during the process of magma formation. Fractionation products have equilibrated with liquid and their mineral phases should have identical initial $^{87}Sr/^{86}Sr$ ratios; there is accumulating evidence however that isotope equilibration has not occurred between mantle phases despite long residence times at high temperatures. Strontium is ~100 times more abundant in mantle clinopyroxene than in olivine or orthopyroxene whereas rubidium (although not abundant in any) has a more uniform distribution, so that the the latter two minerals acquire appreciably higher $^{87}Sr/^{86}Sr$ ratios during mantle residence (P.G. Harris et al., 1972).

The most highly alkalic, oceanic-island rocks known are those of the Fernando de Noronha archipelago and Trinidade Island. These rocks together with others from the oceanic island environment are well described in Carmichael et al. (1974, ch. 8). See also the review by P.E. Baker (1973) of the petrology of the various islands of the South Atlantic, revealing a spectrum from transitional to increasingly alkaline series.

In comparison with MORB, OIT and alkali basalts, these more alkaline rocks of oceanic islands differ in being highly variable in composition and very much less voluminous in amount. Chemically, their contents of Ti and P, and other incompatible elements are high, they show strong enrichment in the LREE, and have relatively high, somewhat capricious, $^{87}Sr/^{86}Sr$ initial ratios.

In some of the larger oceanic island archipelagoes, highly alkaline rocks appear to succeed the eruption of OIT and alkali basalt, although some smaller oceanic islands situated away from ocean ridges appear to be built solely of relatively alkaline magma types. It is prudent to remember, however, that exposure in oceanic islands, at best representing only the tip of a much more massive submarine pedestal of volcanic rock, is often limited to a thin veneer of surface flows plus possible sections of small stratigraphical extent in caldera walls and sea cliffs, and that extrapolation from this meagre evidence may be dangerous.

In line with a plume model and previous précis observations on petrogenesis, it would seem that this spectrum of highly alkaline mafic rocks originates by low degrees of partial melting at depths where equilibration with garnet has occurred. This might take place at the edge of the upwelling plume where the lithospheric plate has passed over the more central axial part of the plume in which relatively copious melting has resulted in voluminous OIT, or conceivably within a "weak" plume in which only low degrees of melting have occurred without formation of OIT.

In the case of the islands of Fernando Po, Principe, São Tomé and Annobon, however, which lie on a direct continuation of the "Cameroun Line" (see, e.g., Fitton and D.J. Hughes, 1977), a lineamant reflecting a fundamental lithospheric weakness may be a more likely cause of volcanic activity of this type than plumes.

Chapter 10

IGNEOUS ROCKS OF CONTINENTAL AREAS

10.1. INTRODUCTION

The first observation to be made is a negative one but nevertheless significant: great expanses of cratonic areas and relatively younger fold belt terrain, practically whole continents in fact, are *devoid* of igneous activity at the present time, apart from the supra-Benioff-zone igneous activity around some continental margins. Furthermore, within the geologically recent past, igneous activity in continental areas, with the notable exception of parts of the East African rift valley system, has been neither voluminous nor continuous over prolonged time periods. However, what igneous rocks erupted in continental areas may lack in quantity they make up for in a variety (and presumably in an unusual cause) that the following account attempts to portray.

A more genetic basis for this chapter could perhaps have been that of "mid-plate" or "intra-plate" igneous activity (see, e.g., Gass et al., 1978) as opposed to igneous activity either at diverging or at converging plate margins. This apparently neat three-fold division is belied, however, as we have seen in Chapter 9, by the concentration of inferred plumes at, or near, diverging plate boundaries, notably in Iceland. Also, in continental areas we see a significantly wider spectrum of igneous rocks than those formed by mid-plate igneous activity in areas of oceanic crust. Hence an overall grouping into oceanic, continental, and supra-Benioff-zone igneous activity has been adopted in Chapters 9, 10 and 11, respectively.

10.2. KIMBERLITES AND RELATED ROCKS

Kimberlites and associated rocks are widely, albeit sparsely, distributed in continental, mainly cratonic, areas, the youngest crustal host probably being of Palaeozoic age in Borneo. They typically occur in very narrow dykes, a metre or less in width, that have been proved in some places where depth of dissection permits observation, to underlie the well-known diatremes ("pipes") that form only at relatively shallow crustal levels within 2—3 km from the surface. Extrusive kimberlite is virtually unknown; see Reid et al. (1975) for an account of a kimberlite-like lava from Tanzania with an unusual texture of curious oblate spheroids of olivine in a calcite-rich matrix. Notable concentrations of kimberlite occur in Siberia and southern Africa.

A translation has been made of the work of Frantsesson (1970) on the Siberian occurrences. A detailed study of Lesotho kimberlites is provided by articles in the book of that name edited by Nixon (1973). An important conference and accompanying field trip to kimberlite localities was held in Cape Town in 1973, and the many excellent papers presented there, many relating to southern Africa, have been published as volume 9, 1975, of *Physics and Chemistry of the Earth*. A useful introductory review of kimberlites is that by Cox (1978a).

Kimberlites differ among themselves most conspicuously in the relative amounts of phlogopite phenocrysts relative to ubiquitous (commonly serpentinized) forsteritic olivine phenocrysts. Groundmass mineralogy comprises a second generation of phlogopite and serpentinized olivine together with serpentine, calcite, apatite, ilmenite and perovskite, the presence of the latter indicating a very low silica activity. Kimberlites with a high proportion of phlogopite in the groundmass have been called "micaceous kimberlites", those with less phlogopite "basaltic kimberlite". Because of their explosive and inferred rapid emplacement (necessarily so in the case of diamondiferous kimberlites or else the diamond would have inverted to carbon), there is generally no time during kimberlite emplacement for differentiation to have operated to produce an igneous rock series, particularly in the usual sense of the term as related to crystal fractionation. A rare and notable exception of kimberlite differentiated in situ is provided by the occurrence of three kimberlite sills near Benfontein, South Africa, (Dawson and Hawthorne, 1973). In these bodies megacryst crystals of olivine and phlogopite, together with smaller crystals of spinel, perovskite, and apatite, have demonstrably accumulated to form layered rocks at the base of the sill; much of the intercumulus material is carbonate.

This carbonate is a clue to a vital component of kimberlitic magmas. The "carbonatization" and "serpentinization" that characterize the petrography of kimberlite are now realized to be not entirely the result of later alteration but at least in part a function of originally high CO_2 and H_2O contents in the kimberlitic magma. For example, primary platy quench crystals of carbonate of magmatic crystallization have been recognized, and in addition recent experimental work has shown surprisingly high possible solubilities of CO_2 in kimberlite-like melts at high pressures (Wyllie and Huang, 1976). Numerous examples of carbonate ocelli in kimberlite, presumably formed by unmixing of carbonate-rich liquid from a kimberlitic melt as it ascends to regions of lower confining pressures have now been described (e.g., D.W. Hawkins, 1976). Janse (1969) plausibly attributed dykes of carbonate rock, previously vaguely thought of as having been formed by hydrothermal alteration processes, to the intrusion of bodies of carbonatite penecontemporaneously with emplacement of kimberlitic rocks in the Gross Brukkaros complex, South West Africa/Namibia. Whereas most kimberlites are fairly low in CaO which presumably has been fractionated into a highly fluid carbonatite phase and thus lost to the magmatic silicate system in many instances, another possibility is the partial loss of CO_2 and CaO, etc., from a kimberlitic magma at some intermediate pressure, resulting in the crystallization of such calcium-bearing phases as melilite and monticellite (Janse, 1971), from a melt still of very low silica activity but retaining relatively more CaO than most kimberlites. The formation of a suite of melilite basalts, monticellite-bearing rocks, and alnöites sometimes found in association with some kimberlitic provinces has been attributed to such a mechanism. To the extent of this inferred differentiation related to a partial or complete fractionation and escape of a carbonatite liquid phase, one could speak of a "kimberlite series".

Given the habit and behaviour of kimberlite intrusions, determination of their chemical composition is clearly going to present problems in sampling and interpretation. With some reservations on this score some chemical analyses are presented in Table 10.1. The highly ultrabasic character is evident. High Cr and Ni presumably reflect a lack of early crystal fractionation of chromite and magnesium-rich silicates from a primitive mantle melt. A remarkably high tenor of incompatible elements suggests long continued processes of zone-refining during a long residence and equilibration of partial melt with mantle material*.

Initial $^{87}Sr/^{86}Sr$ ratios of kimberlite, which might be anticipated to be uniformly low in a mantle-derived melt transported rapidly to the surface without intimate contamination with crustal material (as opposed to the large-scale incorporation of xenoliths in diatremes), are variable and surprisingly high, commonly in the vicinity of 0.710—0.711. In this parameter, kimberlites and other highly potassium-rich series of continental areas contrast with the spectrum of values of oceanic series, ranging from a low of ~0.7025 in MORB through 0.7035 in OIT and values in the range of 0.703—0.705 in alkali basalts and up to ~0.706 in the most potassium-rich oceanic alkaline series (Peterman and Hedge, 1971). As discussed on p.305, this could be due to the preferential incorporation of extremely incompatible radiogenic ^{87}Sr into primordial kimberlitic melt during low degrees of partial melting accompanied by high degrees of zone-refining processes.

Just as extreme economic interest focuses on the diamonds that (some) kimberlites contain, so scientific interest focuses on the ubiquitous inclusions occurring in kimberlites as they provide us with available hand-specimen material of mantle below ophiolite levels. Some inclusions are clearly accidental inclusions of crust including interesting parageneses and compositions appropriate to lower crust, such as, for example, scapolite-bearing granulites (Goldsmith, 1976), and providing indications of temper-

---

*Contrast the peridotitic komatiites (see Chapter 12), also ultrabasic in composition, in which similarly high M-values but *low* incompatible-element contents strongly suggest a *high proportion* of partial melt of mantle — a different set of conditions to equilibrating with mantle at high pressures and inferred high P_{H_2O} and P_{CO_2}.

TABLE 10.1

Chemical composition of kimberlites: (a) major elements and (b) selected trace elements

	(a) Major elements (wt.% oxides)		(b) selected trace elements (ppm)	
	1	2		3
SiO_2	35.2	31.1	Cr	1,440
Al_2O_3	4.4	4.9	Co	77
FeO*	9.8	10.5	Ni	1,140
MnO	0.11	0.10	Rb	21
MgO	27.9	23.9	Sr	445
CaO	7.6	10.6	Ba	740
Na_2O	0.32	0.31	Y	46
K_2O	0.98	2.1	Zr	445
TiO_2	2.32	2.03	Nb	240
P_2O_5	0.76	0.66		
H_2O^+	7.4	5.9		
CO_2	3.3	7.1		
Total	100.05	99.20		

1 = average basaltic kimberlite (Dawson, 1972, p.302, table 1, analysis *2*); *2* = average micaceous kimberlite (Dawson, 1972, p.302, table 1, analysis *3*); *3* = average selected trace-element contents of fourteen Basutoland kimberlites (Dawson, 1967, p.273, table 8.6, analysis *6*).

*Total iron as FeO.

ature regime there (N.W. Rogers, 1977); see also a review by Dawson (1977). The inclusions of inferred mantle provenance can be grouped into two major broad categories: eclogitic and ultramafic, of which the eclogitic are found in highly variable but overall minor proportions.

Eclogites are of basic composition and could represent (once subducted?) basic rocks that have been metamorphosed under mantle conditions. Conceivably, some eclogites showing mineral banding could be cumulates formed during crystallization of primary magmas in the mantle. It has even been suggested that diamond crystals, which usually occur scattered in the kimberlite matrix but which, when present in inclusions, are almost invariably found in the eclogitic rather than in the ultramafic inclusions, may once have been cumulus crystals.

The ultramafic inclusions contain variable proportions of magnesian olivine, pyroxenes, and garnet. An observed convergence in composition towards garnet lherzolite, together with compositional considerations including the condition that mantle should be capable of generating basic magmas, led P.G. Harris et al. (1967) to conclude that a garnet lherzolite, (composition 67 vol.% olivine, 12 vol.% orthopyroxene, 11 vol.% clinopyroxene and 10 vol.% garnet) could be representative of average upper mantle. Later work, however, has tended to suggest that simple models of uniform upper mantle (convenient as this would be as a starting point for modeling igneous petrogenesis) are not tenable [see P.G. Harris et al. (1972) and discussion in Chapter 13, Section 13.2].

Detailed analysis of the minerals from these kinds of inclusions and the equating of these compositions with experimental data coupled with theoretical considerations show that eclogitic and ultramafic inclusions have equilibrated at high pressures and temperatures (Meyer, 1977). Parameters such as the Al_2O_3 content and Ca/Mg ratio of both clinopyroxenes and orthopyroxenes have been particularly useful in this regard (see Wyllie, 1967, p.396; Boyd, 1973; B.J. Wood and Banno, 1973; B.J. Wood, 1974). Assuming that kimberlitic magma has incorporated accidental inclusions of mantle from various depths during its ascent, and that mineralogical equilibrium in the inclusions has been frozen, in theory at least it would be possible from a study of pyroxene compositions to reconstruct the geothermal gradient for the upper mantle in the region of the kimberlite at the time of its eruption. The garnet lherzolite nodules exhibit two broadly contrasted textures: (1) "sheared" to varying degrees ranging from kink-bands in olivine to mylonitic textures with eyes of orthopyroxene and garnet only; and (2) coarsely granular with grain size of ~5 mm. The sheared nodules appear to have equilibrated at pressures equivalent to depths approaching 200 km and have a more primitive, more "fertile" composition, richer in K, Na, Ti and P, and incompatible trace elements than the granular nodules which appear to have equilibrated at lower pressures equivalent to shallower depths around 150 km. The eclogite nodules prove to have equilibrated at still lower pressures.

Sobolev (1972) has drawn attention to the occurrence in some Siberian kimberlites of yet another class of rare inclusions of deep-seated provenance mnemonically termed "grospydites" from their paragenesis of grossularite-rich garnet, (clino)pyroxene, and kyanite (syn. diclase), together with rutile and corundum. The bulk chemistry of these inclusions, rich in Ca, Al and Ti, is completely distinct from the more common mafic and ultramafic inclusions noted above, and supports the postulate of the occurrence of discrete bodies of refractory Ca—Al—Ti-rich material in the (deeper) mantle (D.L. Anderson, 1975). In view of this considerable evidence from kimberlite inclusions, it would now seem that anyone who can sanguinely envisage compositionally homogeneous mantle in a petrogenetic scheme is making a dangerous oversimplification.

One question on which opinion is sharply divided at the time of writing is whether kimberlite eruptions are related to (a) a steady-state mantle phenomenon, or (b) mantle plume activity. The first possibility can be broadly stated as follows. Seismic work has identified a low-velocity layer forming a semi-continuous shell around the Earth with a fairly well-defined lower boundary at 250 km and an upper boundary at roughly 60 km under oceanic crust and 100—200 km under continental crust (but even non-existent under some cratonic areas). This low-velocity layer is equated with the asthenosphere, the independently determined level of isostatic compensation, and is bounded by moving relatively-rigid lithosphere plates above and static (?) mantle (mesosphere) below. From observation of attenuation and reduced velocities of compressional P waves traversing the low-velocity layer, it seems probable that the latter contains small amounts, $\sim 1\%$, of interstitial melt (D.L. Anderson, 1970). This melt, representing the first volatile-enriched partial melt of mantle and existing in small amounts below inferred dry-liquidus temperatures (see, e.g., Mysen and Boettcher, 1975) would be increasingly magnesium-rich at depth and would be expected to be extremely enriched in incompatible elements; it could well approximate to a kimberlite melt in composition. Deep fracturing of lithosphere could tap pockets of this at intervals, thus producing kimberlite eruptions in a spasmodic devolatilizing process unrelated to plate boundaries or plumes. The second possibility envisages (1) that rising geotherms above a plume result in uprise and/or increase of small amounts of localized partial melt and that kimberlite eruptions therefore likely signal a rising plume (H.W. Green and Gueguen, 1974), or alternatively (2) that kimberlites may represent the eruption of left-over pockets of magma following a cycle of plume-generated igneous activity (Cox, 1972, p.324). The high incidence of kimberlites in southern Africa, many of them of Cretaceous age, is not inconsistent with this latter view. Siberian kimberlites occur over a wide area but apparently significantly to the north of the Aldan upwarp and Baikal rift system. Indeed, not all kimberlites appear to fit easily into an "incipient plume" or "plume left-over" model.

10.3. CARBONATITES

As well as occurring as ocelli in kimberlite and ultrabasic lamprophyres and veins in kimberlite complexes, discrete carbonatite masses of larger dimensions typically occur in epizonal ring dykes, cone sheets, dykes and diatremes, and may form lavas and tuffs. A useful bibliography and annotated compilation (Gittins, 1966) showed ~ 320 then known carbonate localities outcropping over an aggregate area of no more than ~ 500 km^2. These localities include notable concentrations in, and near, rift valley systems and fragmented continental margins such as the East African rift valley system, the Baikal rift and Aldan upwarp, the Rhine graben, the much older inferred St. Lawrence rift valley system, and Brazilian coastal localities.

Accumulative field and experimental evidence strongly indicates that carbonatites originate: first by the immiscible separation of a CO_2- and H_2O-rich fluid, rich also in alkali and alkaline earth elements and virtually devoid of silica, from certain highly alkaline magmas including kimberlites, ultrabasic lamprophyres and mafic magmas of nephelinite and related compositions; then subsequently by the crystallization of this fluid as carbonate accompanied by the further loss of some components, particularly K_2O and Na_2O, in highly reactive hydrothermal solutions. Thus carbonatite does not constitute a series in itself, but rather is a possible adjunct to other alkaline series. These, in their turn, represent only the crystallized products from the silicate melt fraction which originated from a primary magma containing significant CO_2 and H_2O. Thus, considered in conjunction, carbonatites and highly alkaline rocks illustrate to a high degree the fact that a crystallized igneous rock may reflect a complex history of magmatic evolution, and correlation of crystallized igneous rock compositions with liquid magma compositions is inherently a dangerous procedure, other factors such as loss or accumulation of primocrysts and subsequent alteration apart.

Notwithstanding this lack of series status, it is convenient to consider features of carbonatite at this point (see also Chapter 4, Section 4.4.9). Incidentally, it may be noted that although carbonatites are typically associated with highly alkaline magmas of continental regions, a carbonatite has been recorded from Cape Verde Islands, associated with highly alkaline magma in an oceanic setting.

An excellent historical survey of the evolution of views on carbonatite is presented in a definitive publication by Tuttle and Gittins (1966, pp. xi—xix), from which it will be seen that the conclusion on the origins of carbonatite presented above has not been reached without difficulty. A highlight of this evolution is the contrast between, on the one hand, objective field observation coupled with logical deduction by W.C. Brögger at Fen, by H. von Eckermann at Alnö (numerous publications over several decades beginning in 1928), by F. Dixey and associates in Nyasaland (now Malawi) and, on the other hand, hypotheses such as limestone assimilation (R.A. Daly) and the impossibility of carbonate magma (N.L. Bowen) that have proved either only partly correct or even downright misleading. Neglected by textbooks of the period, truth fought a losing battle with prestige for many decades until 1960. In that year J.B. Dawson, not without personal risk, witnessed erupting carbonatite magma at the world's only active carbonatite volcano, Oldoinyo Lengai, Tanganyika, and Wyllie and Tuttle (1960) demonstrated an unsuspected eutectic in the system $CaO-CO_2-H_2O$, ~65:19:16, respectively, in weight.% proportions, at remarkably low temperatures, 640°—685°C, and over an appreciable pressure range up to at least 4 kbar. As calcite stoichiometrically contains 56 mol.% CaO and 44 mol.% CO_2, a eutectic liquid of the above composition would thus not only be undersaturated with respect to silica but indeed if it were to crystallize calcite the remaining liquid would be drastically oversaturated with respect to CaO. Subsequent work has both demonstrated comparable eutectics with Na_2O, K_2O, and MgO in the place of CaO, and the fact that even in the presence of excess silica very little silica enters the inferred base-rich carbonatite-like liquid. In the meantime, one of the early successes of strontium-isotope work was the demonstration that carbonatite is not related to limestone as the latter, like seawater, has a $^{87}Sr/^{86}Sr$ ratio close to 0.709 whereas carbonatites have a more variable but generally lower initial $^{87}Sr/^{86}Sr$ ratio. Further experimental work has shown that only undersaturated silicate magmas would be expected to give rise to carbonatite liquids. Crystallization in the simple system nepheline—calcite—H_2O in equilibrium with vapour, for example, can yield successively: (1) nepheline; (2) melilite + nepheline; (3) hydroxyhaüyne + melilite; (4) cancrinite + melilite; and (5) calcite + cancrinite + melilite; late-stage liquid fractions are rich in $CaCO_3$, whereas the vapour phase is enriched in Na (Watkinson and Wyllie, 1971). An ingenious study of fluid inclusions in apatites from some East African carbonatites and ijolites revealed that (Rankin and Le Bas, 1974):

"Carbonate-rich and silicate-rich melts can coexist as immiscible fractions in naturally-occurring ijolite magmas"

The Oldoinyo Lengai carbonatite lavas proved to be rich in alkalis, particularly Na_2O (Dawson, 1966). In contrast, epizone carbonatite intrusions are poor in alkalis and are usually composed of calcite, more rarely of ankeritic varieties or dolomite, possibly with minor amounts of siderite, rhodocrosite and rare-earth carbonate (composite intrusions may exhibit an incomplete sequence along this trend). Most of the alkalis that the intrusive parental carbonatitic liquid contained apparently entered the country rock thus causing ubiquitous intense fenitization. Complementary igneous rocks associated with large carbonatite complexes include the distinctive urtite—ijolite—melteigite—jacupirangite suite, the approximate plutonic equivalents of nephelinite to melanephelinite. In this respect ngurumanite, a nepheline-pyroxene rock with "vug-like patches of primary calcite, analcite and zeolites", is a hypabyssal, highly alkaline rock that appears to have retained some of its pristine CO_2 and other volatile content (Saggerson and Williams, 1964).

A distinctive suite of abundant trace elements, presumably first concentrated during the genesis of primary alkalic mafic magmas and then strongly partitioned into carbonatitic liquid, includes Ba, Sr, Nb, P, REE, Zr, F and S. Concentrations of these elements give rise here and there to workable deposits of barite, pyrochlore, phosphate, monazite, fluorite, etc. No less than 170 distinct mineral species, many of them containing relatively rare elements, have been identified in carbonatites. This is in sharp contrast to the very limited number of minerals in silicate igneous rocks, in which diadochy permits the incorporation of most trace elements into the lattices of common rock-forming minerals.

In petrogenetic terms, the highly diverse and interesting group of rocks known as carbonatites must thus be considered as an adjunct to highly alkaline igneous rock series wherever they may occur. The chemical evolution of carbonatites reflects a history of partitioning between silicate and carbonatite liquids followed by a drastic loss of bases in solution (highly reactive alkalis first) from the carbonatite liquid on final crystallization.

10.4. HIGHLY POTASSIUM-RICH SERIES

These are a very sparse and highly variable group of rocks, many of them so rare as to be petrological curiosities, apparently genetically linked in some way by extraordinarily high contents of incompatible elements and high initial $^{87}Sr/^{86}Sr$ ratios. Nevertheless, they exist and their presence should be explained in any philosophy of igneous rocks. Because of their variability more attention is given here to descriptions of individual localities than in the case of most other igneous rock series. Although nothing quite so potassium-rich occurs in areas of oceanic crust, not all the potassium-rich types occur in solely continental areas; some are found in a possible supra-Benioff-zone situation, admittedly a long way behind the associated trench system (active or fossil). There may be a reason for this similarity based on a common factor of great depths of generation. We shall begin, following our theme of continental igneous activity, with two areas that can be plausibly linked to plume-generated centres, beginning first with the most inherently potassium-rich group of all igneous rocks, the lamproites of Western Australia.

10.4.1. Lamproites of Western Australia

Though of no great volume, this group of rocks has attracted attention because of an unusual mineralogy reflecting high contents of Mg, K, Ti, P, Zr and Ba, and a high K/Na ratio and low Al_2O_3 content.

Occurrences (Wade and Prider, 1940; Prider, 1960) are situated near the Fitzroy River, ~300 km inland from Broome, Western Australia, at approximately 125°E, 18°30'S. A cluster of fifteen small plugs, diatremes, flows, craters, cryptovolcanic structures and dykes were found there within a 25-km radius, and a further four occurrences were found within an 80-km radius. The majority cut Permian sediments and show a crude alignment in NNE and WNW directions, reflecting fault directions and structural trends in underlying crystalline basement rocks. Although these features were regarded as Tertiary in age on the 1971 tectonic map of Australia, a Rb—Sr isochron, complicated because of the high common strontium content of the rocks, has yielded a Jurassic age of 145 ± 10 Ma.

The largest of several plugs forms a topographical feature ~60 m high and exposes a core of grey to grey-green igneous rock, ~50 m in diameter. Around all the igneous occurrences the country rock is noticeably hardened and discoloured, and is locally brecciated and cemented by igneous rock. Here and there intrusive breccias are found. Conspicuous masses and veins of chalcedony and crystalline quartz are also found sporadically around the occurrences. The presence of these together with discoloured and brecciated country rock has indicated several small diatremes in which no igneous rock occurs; one showing, for example, of volcanic breccia no more than 5 m across was found to be situated at the junction of two faults. Several radial dykes were noted, some of them vesicular. Crater-like structures, maximum diameter 1.5 km, expose breccias and flow-banded and vesicular lava, often highly weathered. The most imposing remnant is of a lava flow or flows (sills?) forming a topographic feature, 160 m long and up to 90 m in height. Narrow dykes of lamproite have also been intersected in boreholes in the area. Crude but generous calculations indicate that probably no more than 1 km^3 of eruptive rock was emplaced, a far cry indeed from the volumes of major igneous rock series.

The igneous rocks have been variously regarded as belonging to the leucite basalt, or lamprophyre, or kimberlite families; the name *"lamproite"*, defined as a volcanic lamprophyre with mica phenocrysts, has been applied to them. Abundant, usually small, phenocrysts of leucite (or pseudomorphs after leucite in some of the hypabassal rocks) characterize all rocks; in addition, dark-mineral phenocrysts are conspicuously magnesium-rich and comprise one or more of the following: nontronite pseudomorphs after olivine (rarely present in a few rocks and in small amounts), phlogopite (a pleochroic titaniferous variety), diopside, and magnophorite (an amphibole of magnesian kataphorite composition). Phenocryst content gives rise to the following rock names: fitzroyite (phlogopite and leucite), cedricite (diopside and leucite), mamilite (magnophorite and leucite), wyomingite (phlogopite, diopside and leucite) and woldigite (phlogopite, diopside, magnophorite and leucite). Prominent accessory minerals are priderite (essentially a potassium titanate, resembling rutile in its optical properties), apatite, and rarer wadeite (a potassium—zirconium silicate). The groundmass is difficult to determine but has been reported to contain serpentine and zeolite. A zonation has been claimed in some hypabyssal rocks from outer zones of fitzroyite to inner zones of cedricite or mamilite, and a (generalized) sequence seems to be brecciation followed by the extrusion of fitzroyite lavas, then the intrusion of cedricite and mamilite plugs, and finally woldigite dykes. The order of phenocryst crystallization deduced from this and from textures (these are panidiomorphic lamprophyric but sometimes one phenocryst species may be included in another) is olivine, phlogopite, diopside, leucite, priderite, wadeite and magnophorite. As all rocks are porphyritic and many may have undergone fractionation, only one analysis [out of twelve reported in Prider (1960)], that of a fitzroyite, is reproduced here (Table 10.2). Within an overall range of silica contents from 44 to 54 wt.% this fitzroyite has the lowest, in addition to having lowest alkalis

TABLE 10.2

Chemical compositions (in wt.%) of members of various highly potassium-rich rock series

	1	2	3	4	5	6	7	8	9	10
SiO_2	44.02	43.7	35.00	37.95	46.4	45.53	43.56	50.23	55.14	46.74
Al_2O_3	6.30	10.0	7.69	7.98	15.7	8.5	7.85	10.15	10.35	12.21
Fe_2O_3	5.98	4.0	7.12	5.84	5.3	2.93	5.57	3.65	3.27	0.95
FeO	2.01	7.0	4.76	5.01	3.4	4.31	0.85	1.21	0.62	7.20
MnO	0.06	0.21	0.26	0.22	0.15	0.11	0.15	0.09	0.06	0.15
MgO	11.98	11.2	12.37	14.27	6.0	14.86	11.03	7.48	6.41	7.87
CaO	4.61	13.8	16.02	12.31	11.6	9.06	11.89	6.12	3.45	11.67
Na_2O	0.28	1.89	1.33	0.95	1.6	1.5	0.74	1.29	1.21	1.30
K_2O	6.59	2.90	3.54	6.33	6.6	3.6	7.19	10.48	11.77	5.80
TiO_2	6.57	3.41	4.84	4.56	0.94	1.57	2.31	2.30	2.58	0.77
P_2O_5	1.55	0.52	0.97	0.75	0.58	1.82	1.50	1.81	1.40	0.96
H_2O^+	3.83	0.68	2.82	1.90	0.94	4.1	2.89	2.34	1.23	2.08
H_2O^-	3.42	—	1.32	0.84	—	1.13	2.09	1.09	0.61	0.40
CO_2	nil	—	1.53	0.59	—	0.6	—	—	0.20	0.91
BaO	2.55	—	—	—	0.17	0.41	0.66	0.61	0.52	0.56
SrO	0.07	—	0.22	0.28	0.19	0.22	0.40	0.32	0.26	0.25
ZrO_2	0.28	—	—	—	—	0.12	0.27	0.25	0.27	—
Cr_2O_3	0.09	—	—	—	—	—	0.04	0.06	0.04	—
SO_3	0.12	—	—	—	—	—	0.52	0.35	0.40	—
S	—	—	—	—	—	—	—	—	—	0.14
Li_2O	—	—	0.21	0.22	—	—	—	—	—	—
Total	100.31	99.50*	100.00	100.00	99.6	100.3	99.51	99.83	99.79	99.91

1 = fitzroyite, middle flow, Mt. North, Western Australia (Prider 1960, p.79, table 1, analysis II).

2 = selected parental composition of leucitite series, Bufumbria, Birunga volcanic field (A.K. Ferguson and Cundari, 1975, p.32, table 2, analysis 7).

3 = average of seven analyses of katungite, Toro—Ankole volcanic field (K. Bell and Powell, 1969, table 3, analysis *32*).

4 = average of six analyses of mafurite, Toro—Ankole volcanic field (K. Bell and Powell, 1969, table 3, analysis *31*).

5 = average of four analyses of leucitite, "high-K series", Roccamonfina volcano, Roman province, Italy, (Appleton, 1972, p.429, table 1, analysis *1*).

6 = jumillite, Jumilla, southern Spain, (Borley, 1967, p.365, table 1, analysis *1*).

7 = madupite, Pilot Butte, Rock Springs, Leucite Hills, Wyoming, (Carmichael, 1967b, p.50, table 12, analysis LH*16*).

8 = wyomingite, margin of dyke, Boars Tusk, Leucite Hills, Wyoming, (Carmichael, 1967b, p.50, table 12, analysis LH*1*).

9 = orendite, North Table Mountain, Leucite Hills, Wyoming, (Carmichael, 1967b, p.50, table 12, analysis LH*12*).

10 = leucitite from chilled margin of Shonkin Sag sill (average of two analyses, one of lower chilled margin, one of upper chilled margin) (Nash and Wilkinson, 1970, p.244, table 2, average of analyses SS*2* and SS*7*).

*Includes "others" 0.19 wt.%.

and highest magnesia, and apparently the earliest phenocryst assemblage. For these reasons this fitzroyite might come nearest to parental composition although some degree of fractionation and accumulation of primocrysts can not be ruled out. However, all analyses have an unmistakable family resemblance in that K_2O, K/Na, TiO_2, P_2O_5, BaO and H_2O are all very high. Perhaps surprisingly, normative calculations show that all rocks are on the borderline of silica saturation and rough calculations show that the groundmass must be quite siliceous. An extremely peralkaline condition is reflected by the occurrence of potassium metasilicate in all norms (the simplified normative calculation scheme presented in Chapter 4 does not allow for this extremely rare possibility for which references given there should be consulted). Kimberlitic affinities are indicated by high abundances of Ti, Zr, P, Ba, Sr, V, F and H_2O, but the lamproites as a group have distinctly lower MgO, and higher K_2O and Al_2O_3 than kimberlite. Neither is there a record of their containing any of the typically kimberlitic suite of inclusions. Furthermore, the lamproites were not so gas-rich as kimberlites which almost invariably reach near-surface levels as fluidized systems whereas the lamproites were in part extruded as lavas. Apart from one analysis of a "carbonated woldigite" there is no mention of CO_2 activity associated with the lamproites either directly or indirectly. Interestingly, there are persistent rumours of the occurrence of diamonds in the Fitzroy basin, where no actual kimberlite has yet been recorded.

Initial $^{87}Sr/^{86}Sr$ ratios of the lamproites range from 0.7125 to 0.7215 (Powell and K. Bell, 1970), the highest ever recorded for any mafic rock type, the high end of the range coincidentally about as high as estimated present-day $^{87}Sr/^{86}Sr$ ratios of average crustal material.

A model of olivine fractionation from a potassium-rich peridotitic magma in depth has been proposed to account for the chemical compositions of these lamproites, but the extremely high concentrations of incompatible elements strongly suggest long continued zone-refining processes between an incipient melt phase and mantle.

Whatever the petrogenesis may be, small quantities of magma of comparable and extreme composition were erupted at, or near, the surface in late Jurassic time within an area of $\sim 2 \cdot 10^4$ km^2, most within an area of 2000 km^2. The crust had been stable cratonic in character for ~ 1750 Ma. Crustal instability associated presumably with the inception of the breakup of Gondwanaland led to basins filled with Permo-Triassic sediments developing in Western Australia. Sedimentation continued on the site of the present littoral in Jurassic and early Cretaceous time accompanied by Upper Jurassic alkali basalt and Lower Cretaceous tholeiite. In late Cretaceous times, a clastic wedge thickening seawards presumably correlates with the development of a new ocean basin as crustal separation became complete.

10.4.2. Highly alkaline rocks of the western East African rift valley

The greatest concentration in the world of highly potassium-rich lavas, including numerous occurrences of the very rare mineral kalsilite, are found in two young volcanic fields, Birunga and Toro—Ankole, situated near the equator in the western East African rift valley, near the central part of a domal structure including the unique (non-volcanic) horst of Ruwenzori (Fig. 10.1).

The Birunga field comprises eight major volcanic peaks, heights between 2975 and 4510 m, including two active volcanoes, one of them the well-known Nyiragongo, and over 400 subsidiary cones all within a 70 km length of the rift valley between Lake Kivu and Lake Edward.

In the Toro—Ankole field, comprising several scattered areas situated north of Lake Edward and ~150 km NNE of the Birunga field, there are numerous explosion craters and small cones but no large volcanoes, and indeed lava is much less common than pyroclastic rock in this field.

There is considerable variation in the volcanic rock types, many of them of uncommon mineralogies, and a consequent profusion of rock names. Broadly speaking, whereas both suites are undersaturated and highly alkaline, nepheline is always present in Birunga rocks, reflecting a Na_2O content comparable to the (high) K_2O contents; however, the Toro—Ankole rocks have K_2O/Na_2O ratios around 3—5 and are virtually nepheline-free, and in addition are more highly magnesian reflected in more abundant olivine.

The Birunga rocks comprise abundant nepheline- and leucite-bearing "leucite-basanites" and "nepheline leucitites", and two low-pressure fractionation series ending with phonolite and trachyte (via leucite tephrite) have been recognized (A.K. Ferguson and Cundari, 1975).

In the Toro—Ankole field, feldspar is virtually absent and porphyritic feldspathoid minerals comprise leucite, kalsilite, and potassium-rich nepheline. Main rock types, distinguished on the basis of their major porphyritic constituents, are as follows (see K. Bell and Powell, 1969, p.539): "ugandite", with olivine, augite and leucite phenocrysts; "mafurite", with olivine, augite and kalsilite phenocrysts; "katungite" with olivine and melilite phenocrysts. The rocks contain abundant opaque oxide, perovskite, and commonly also biotite, in a glassy matrix. There is little discernible trend of low-pressure fractionation, although some compositional variation may well have been effected in these conspicuously porphyritic rocks by the fractionation at some depth of proportions of such phases as olivine, biotite and augite that are locally very abundant in xenoliths (the "OBP group" of Holmes and Harwood, 1932). Carbonatite lavas and tuffs are also well known in the Toro—Ankole field, the silicate lavas of which are clearly of a more extreme and primitive composition than the Birunga lavas.

Some analyses illustrating the compositions of the major rock types are given in Table 10.2. An ultrabasic character is evident throughout, along with extraordinarily high contents of the incompatible major elements Ti and P, and of the incompatible trace elements Rb, Ba, Sr and Zr (Higazy, 1954). Initial $^{87}Sr/^{86}Sr$ range from moderately to distinctly high (0.7036—0.7111), averaging 0.705 for melilite- and nepheline-bearing varieties, and 0.707 for feldspar-bearing varieties (K. Bell and Powell, 1969).

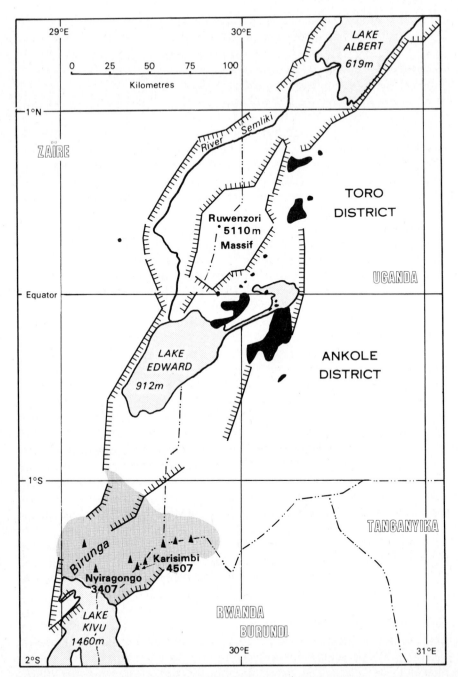

Fig. 10.1. Sketch map of a 400 km length of the western East African rift valley straddling equator and showing the positions of eight major volcanic peaks in the Birunga volcanic field (*shaded grey*) and centres of eruption (in *black*) in the Toro—Ankole field (after A. Holmes, 1965, pp. 1070 and 1076, figs. 780 and 784).

The three groups of highly potassium-rich rocks so far considered are by no means as comparable as some have claimed. On the contrary, they exhibit highly variable mineralogies reflecting variation in such compositional parameters as K_2O/Na_2O ratios from 6 in Western Australian examples to ~1 in Birunga, and an amazing span in silica activity from those with normative silica to those carrying melilite and kalsilite. However, they do share a pattern of extreme enrichment in incompatible elements and a high initial $^{87}Sr/^{86}Sr$ ratio. The fitzroyites of Western Australia and the Toro- -Ankole ultrabasic rocks in particular seem to epitomize the trend already noted in the relatively more alkaline of the oceanic lavas towards explosive eruption in small quantities of batches of magma of erratic and extreme compositions.

Yet another common feature is the geotectonic environment of eruption; Veevers and Cotterill (1976), for example, have drawn attention to the detailed parallelism in terms of time and space between the evolution of the East African rift valley system and Western Australia.

10.4.3. Other examples of highly potassium-rich rocks

It must be admitted, however, that not all occurrences of highly potassium-rich rocks in continental areas can be fitted into this kind of model. The highly potassium-rich rocks of the Roman province (see, e.g., Appleton, 1972) and southern Spain (Borley, 1967) are of Plio-Pleistocene age and occur in areas of plate convergence, and thus quite likely overlie former Benioff zones. The highly potassium-rich rocks of the Leucite Hills, Wyoming, (Carmichael, 1967b, with references to earlier classic work) and of Montana (E.S. Larsen, 1940), including the Highwood Mountains with the famous differentiated Shonkin Sag sill of leucitite and the capped dyke of monticellite-bearing peridotite of kimberlitic affinity, are even more difficult to characterize. They also may reflect igneous activity over a fossil Benioff zone at great depth, or they could perhaps reflect left-over igneous activity associated with overridden plumes along the former site of the East Pacific Rise.

The analyses in Table 10.2 reveal the great ranges in K_2O content and K_2O/Na_2O ratio, a generally very high TiO_2 content contrasting with low TiO_2 in the unequivocally supra-Benioff-zone examples, and extraordinarily high contents throughout of incompatible elements, notably Ba. In silica content the rocks range from extreme ultrabasic compositions (35 wt.% SiO_2 in katungite) through undersaturated basic rocks such as madupite to oversaturated wyomingite and orendite, the latter of intermediate silica content.

In short, as well as an inherent and characteristic chemical variability, highly potassium-rich rocks in continental areas obstinately refuse to be fitted into one sole geotectonic environment. To anticipate further discussion in Chapter 13, they probably represent the final erupted produced from small quantities of primordial magmas that equilibrated with mantle at depth as incipient melts or "leftovers" after partial-melting episodes. Their high K_2O activities may quite simply be a function of depth of genesis (cf. Marsh and Carmichael, 1974). Readers might also refer to Edgar and Condliffe (1978) for an ingenious model involving small amounts of partial melting of a deep peridotitic mantle source containing both H_2O and CO_2 followed by fractionation and resorption of phlogopite.

10.5. VARIOUS ALKALINE SERIES AS EXEMPLIFIED BY IGNEOUS ACTIVITY
IN THE KENYA DOME

This area has been chosen as the next example in the exposition of volcanic igneous activity in continental areas as it has witnessed the eruption within the latter half of Cenozoic time of not only voluminous lavas (several hundred thousands of cubic kilometres) but also a wide and interesting spectrum of igneous rock series ranging from highly alkaline through basanite and alkali basalt to transalkaline basalt series.

The Kenya dome and associated rift valleys form part of the East African rift valley system that extends both southwards into further domes and rift valley systems in Tanzania and Malawi and northwards similarly into Ethiopia (Fig. 10.2). It is important to realize that despite the apparent north—south elongation of the rift valley system the integral units that must be considered for petrogenetic purposes are the domes for, despite complexities in detail, the volumes and compositional sequences of erupted igneous rocks seem to show a pattern in space and time with respect to identifiable domal upwarpings of different ages. The rift valleys themselves, ~50 km wide and ~1 km deep, so conspicuous a topographic feature of this part of the world, are nevertheless small in scale when compared with the domal structures that they transect. The domes have diameters of around 1000 km and maximum uplift centrally up to 2 km. Commonly three, less commonly four or two, rift valley arms radiate outwards from a more or less central position in each mature dome. Although clearly a reflection of tension, the actual rate of crustal extension across East African rift valleys can generally be shown to have very low upper limits, and many have existed over long periods of time with no sign of plate separation. Along other former rift valleys, however, plate separation has occurred, notably within the relatively recent past along the Gulf of Aden and the Red Sea. It seems therefore that in East Africa, we are witnessing a distinctive igneous activity which may (or may not) be accompanied by plate separation.

A strong, albeit speculative, case can be put forward for regarding the domes plus their associated igneous rocks (hotspots) as the surface expression of developing plumes over a period of time extending from the present back to the time of the breakup of Gondwanaland. Some of the con-

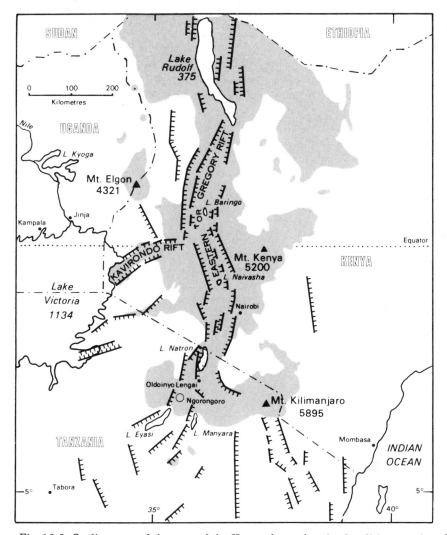

Fig. 10.2. Outline map of the area of the Kenya dome showing localities mentioned in the text. Areas underlain by Tertiary and Quaternary volcanic rocks are *shaded* in *grey*. Note the contrast between the chain of elongate lakes — Rudolf, Baringo, Naivasha, Natron, Eyasi, Manyara — lying within the rift valley system, and the large but shallow Lake Victoria occupying a shallow depression on the surface between the Kenya and Kivu domes (based on compilations by L.A.J. Williams, 1969, p.63, fig. 1; and by King, 1970, fig. 1, facing p.264).

cepts which have led to this hypothesis were developed in the following papers: an attempt to see order in terms of space and time in the magmatic affinities of rift valley volcanic activity (P.G. Harris, 1969); a review of Karroo volcanism and the breakup of Gondwanaland (Cox, 1970); the

evolution of the Afro-Arabian dome (Gass, 1970), these last two papers
dealing with the old and young end, respectively, of the East African igneous
cycle; later accounts of the Afar triangle (Tazieff, 1970; Tazieff et al.,
1972), which must surely be one of the most critical, not to say fascinating,
areas in terms of understanding the present regime of world tectonic/igneous
activity; and recognition of a pattern of plume-generated triple junctions,
leading either to plate separation or to "failed arms" (*aulacogens*), (Burke
and Dewey, 1973).

The structural evolution of the Kenya dome has been documented by
B.H. Baker and Wohlenberg (1971) following a broad survey of the whole
East African rift valley system by King (1970). Based on studies of the shape
and height of well-preserved Miocene and late-Pliocene erosion surfaces, it
is clear that epeirogeny began in early Miocene time in central Kenya with
local upwarping and faulting, followed by broad domal uplift of 300 m in
the late Miocene, and a major uplift of 1400 m during late Pliocene to mid-
Pleistocene time. Contours of the aggregate amount of uplift show that the
area affected forms a dome ~1000 km across with maximum uplift of
~1700 m in the middle of the dome on the equator at 36°E. From this
central position three rift valley arms radiate: two major arms aligned just
east of north and just east of south form the Gregory rift valley, superficially very close to a north—south continuity; a third rift valley, the
Kavirondo rift valley, extends in a WSW direction (See Fig. 10.2).

The first extensive rift faults of the Kenya dome developed early in the
Pliocene at the culmination of the Miocene uplift phase, and were followed
by further rift-faulting in late Pliocene time and local rejuvenation until
mid-Pleistocene time. The net result has been the formation of the complex
graben termed rift valleys. Their widths vary from 60—70 km (Gregory),
to 30—40 km (Kavirondo). Present-day differences in altitude between
marginal plateaux and rift valley floors range up to 2000 m, although the
relatively greater thicknesses of penecontemporaneous volcanic rocks within
the rift valleys suggests that the aggregate throw of bounding faults may
reach 3000—4000 m in places.

Igneous rocks are clearly spatially associated with domal centres; in the
Kenya dome, for example, voluminous lavas extend for nearly 1000 km
along the Gregory rift valley and have a maximum outcrop width of 500 km
across the domal structure in an east—west direction. Furthermore, the age
of eruption and composition of lavas seem to follow a broadly consistent
sequence within each dome.

Volcanism within the area of the Kenya dome began in early Miocene
time along the line of the Kavirondo rift just west of the present centre of
the dome as defined both by maximum uplift and intersection of the three
rift valley arms. Large central volcanoes were formed and are typically composed of nephelinite and melanephelinite lavas, and pyroclastic rocks with
minor basanite, basalt, phonolite and trachyte. Several of these now
much-dissected volcanoes display carbonatite cores, fenitized basement

and intrusive bodies of ijolite and other rock types characteristically found in the carbonatite association (Le Bas, 1977). In later Miocene time, within the central parts of the dome, extraordinarily voluminous phonolites erupted in vast floods apparently from fissures and largely filled the incipient Gregory rift and spread far across its shoulders. These Kenya phonolites are said to exceed the total volume of phonolite lavas found elsewhere in the world by several orders of magnitude (see Goles, 1976). Nephelinites, basanites, and basalts are intercalated with the phonolites, particularly in the lower parts of the phonolite pile.

Tracing the continuing pattern of eruptive activity in the Kenya dome now becomes increasingly complicated as not only do younger volcanic rocks in part cover up and obscure the older ones but also different magma types came to be erupted in different parts of the dome. For example, in mid-Pliocene time there appear to have been voluminous eruptions of nephelinites and phonolites to the west and south. These were followed in late Pliocene time by massive eruptions of trachyte, many of them ignimbrites, in the central part of the dome. Large central volcanoes, e.g., Mount Kenya and Mount Kilimanjaro, built largely of alkali basalts, trachytes and phonolites, were constructed in a flanking position to the east of the Gregory rift during the late Pliocene and early Pleistocene.

In the later Quaternary extending to the present day (at which we see as it were a snapshot — a still — in the changing scene produced by evolution, be it tectonic, igneous, or of life) there develops a strong tendency towards axial volcanism within the Gregory rift, and basalt and trachyte with small volumes of pantelleritic trachyte (the first appearance of silica-oversaturated rocks) occur near the axis of the dome (see B.H. Baker et al., 1977). These contrast with occurrences of alkali basalt up to 200 km away to the east and small quantities of nephelinite, phonolite and carbonatite at the southern extremity of the dome, notably the above-mentioned carbonatite volcano Oldoinyo Lengai, situated ~350 km away from the centre of the dome.

The chemical distinction of this rich spectrum of young and fresh igneous rocks series (Table 10.3) can be well shown by the familiar alkalis vs. silica Harker variation diagram (Fig. 10.3). Note that none of the selected representative analyses of Table 10.3 lies within the subalkaline field, i.e. below the alkali basalt—tholeiite divide of Macdonald and Katsura.

The foregoing is a highly simplified account of a complex volcanic province that has been described in more detail, for example, by L.A.J. Williams (1969, 1972) and King and Chapman (1972). However, some degree of generalization is a necessary prelude before attempting to discern order at what at first sight may seem bewildering perplexity. One of the first to attempt this was P.G. Harris (1969) who both saw a recurrent pattern in the composition of rift valley volcanic rocks and provided a possible explanation for this pattern which fits remarkably well with plume models that were more fully developed later.

TABLE 10.3

Chemical compositions and CIPW norms of selected rocks from the Kenya dome

	1	2	3	4	5	6	7	8	9
SiO_2 (wt.%)	43.90	50.85	55.22	46.5	54.06	47.68	47.64	59.24	63.65
Al_2O_3	14.38	19.69	21.09	16.1	16.39	15.55	14.20	15.92	14.12
Fe_2O_3	4.58	3.01	2.02	3.5	3.87	2.87	3.49	1.67	2.01
FeO	7.68	3.17	2.19	7.5	6.38	8.62	10.47	5.02	6.03
MnO	0.23	0.20	0.26	0.2	0.27	0.20	0.24	0.20	0.27
MgO	6.10	1.96	0.48	7.4	1.88	6.73	5.41	2.27	0.04
CaO	10.48	5.86	2.21	10.9	4.97	12.69	10.86	4.77	1.31
Na_2O	4.55	6.16	9.57	3.1	5.58	2.64	2.82	5.20	6.34
K_2O	1.56	3.69	4.56	1.1	2.84	0.65	1.23	4.10	5.22
TiO_2	2.79	1.51	0.77	2.3	1.45	2.02	3.07	1.31	0.94
P_2O_5	0.54	0.44	0.06	0.4	0.39	0.35	0.58	0.28	0.07
H_2O^+	2.42	1.81	1.41	—	1.37				
H_2O^-	0.35	1.18	0.23	—	0.61				
CO_2	—	0.37	—	—	—				
Zr (ppm)						88	184	387	764
Total	99.56	99.90	100.07	99.0	100.06	100.00*[1]	100.00*[1]	100.00*[1]	100.00*[1]
Q	—	—	—	—	—	—	—	0.7	4.4
or	9.6	22.6	27.4	6.6	17.1	3.8	7.3	24.2	30.9
ab	10.6	26.9	33.8	19.3	46.9	20.9	23.9	44.0	43.6
an	14.7	15.8	1.1	27.1	11.5	28.7	22.5	8.0	—
ne	15.9	14.7	26.2	3.9	0.7	0.8	—	—	—
ns	—	—	—	—	—	—	—	—	0.8
ac	—	—	—	—	—	—	—	—	5.8
di	29.1	9.5	5.6	20.3	9.3	26.2	22.9	11.5	5.4
wo	—	—	1.2	—	—	—	—	—	—
hy	—	—	—	—	—	—	3.3	6.0	7.3
ol	9.8	4.6	—	14.3	5.1	11.6	8.8	—	—
mt	3.6*[2]	1.8*[2]	3.0	3.2*[2]	5.7	3.3*[2]	4.0*[2]	2.4	—
il	5.5	3.0	1.5	4.4	2.8	3.8	5.9	2.5	1.8
ap	1.3	1.1	0.1	0.9	0.9	0.8	1.4	0.7	0.2
M-value	47.5	36.5	16.7	54.9	24.9	51.3	41.4	37.6	0.9

1 = analcite basanite, Saimo, Kamasia range ("Group I", Miocene) (King and Chapman, 1972, p.202, table 1, analysis *1/690v*).

2 = analcite mugearite, Samburu basalts, Laikipia ("Group I", Miocene) (King and Chapman, 1972, p.202, table 1, analysis *10/95*).

3 = plateau phonolite, Kamasia range ("Group II", late Miocene to early Pliocene) (King and Chapman, 1972, p.202, table 1, analysis *2/235*).

4 = average of 8 alkali basalts, Kamasia range ("Groups III and IV", Pliocene) (King and Chapman, 1972, p.203, table 2, analysis A).

5 = mugearite, Kamsoror ("Group III", Pliocene) (King and Chapman, 1972, p.203, table 2, analysis *1/125*).

6—9: 6 = basalt, analysis KLR-*32*; 7 = ferrobasalt, analysis KLR-*38*; 8 = benmoreite, analysis KLR-*18*; 9 = trachyte, analysis KLR-*54*. Analyses 6 to 9 are representatives taken from 21 analyses (each including concentrations of 20 trace elements) of a transitional basalt series of Pleistocene age from the Gregory rift, Kenya, (after B.H. Baker et al., 1977, pp. 310—312, table 1). The authors state that the basalts may contain minor amounts of normative nepheline or hypersthene, and that the trachytes are mildly but progressively peralkaline and contain minor modal and normative quartz. The authors arrange their analyses in order of increasing Zr content, an effective index of differentiation in this series.

*[1] Recast water-free to 100 wt.%.
*[2] $Fe_2O_3/(Fe_2O_3 + FeO)$ reduced to 0.2.

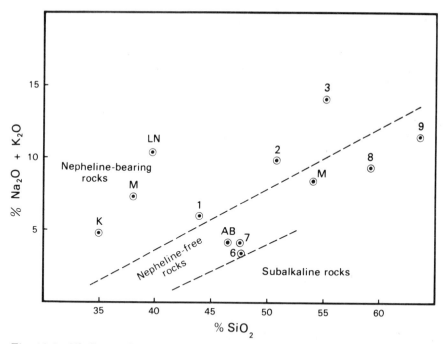

Fig. 10.3. Alkalis vs. silica diagram for selected rift valley rocks. The *lower dashed line* is the boundary dividing Hawaiian tholeiites below from alkaline rocks above (after G.A. Macdonald and Katsura, 1964). The *upper dashed line* divides a field of nepheline-bearing rocks above from nepheline-free alkali basalts below, based on some 90 analyses of volcanic rocks from northern Tanganyika (Saggerson and L.A.J. Williams, 1964, p.75, fig. 11). The rocks plotted here are an average katungite (*K*), an average mafurite (*M*), and a leucite-nephelinite (*LN*), all from Table 10.2; and an analcite basanite (*1*), an analcite mugearite (*2*), a plateau phonolite (*3*), an average alkali basalt (*AB*), a mugearite (*M*), and four representative rocks from a transitional basalt series (*6—9*), all from Table 10.3.

Harris began by contrasting the chemistry of MORB with that of more alkaline mafic rocks found on ridge flanks and within oceanic basins and, with respect to what we now perceive, made the highly prescient comment:

"To date, no-one seems to have discussed whether there is any systematic change in basalt type or heat flow *along* [present writer's italics] the length of the ocean ridge systems."

Harris then did precisely this for a rift valley environment by tracing the pattern of composition of geologically recent lavas along the length of the East African rift valley system. He noted oceanic tholeiite in the Red Sea, the mildly alkalic lavas with comenditic fractionation products of Aden (see Section 10.6), the apparently more alkalic nature of lavas flanking the Ethiopian rift valley compared to those within it (see also Mohr, 1971), the pattern we have just noted in the Kenya dome, and so on. The dimension of time was then added by tracing also the sequential changes in volcanic activity.

Harris concluded that volcanism is likely to be concentrated at areas of intersecting rifts (dome centres) and is likely to be less alkalic there than in areas farther away from the centre whether axially along the rift valley or transverse to it; with the passage of time the composition of lavas erupted at the dome centre becomes less alkalic in nature, a change that also occurs later in areas radially outwards from the centre. Harris proceeded to plausibly link magma type with depth of generation, the less alkalic at progressively shallower depths, and to link these inferred different levels of magma generation most probably to changing geothermal gradients in the upper mantle over long periods of time, of the order of tens of millions of years. The lowest geothermal gradient would thus correlate with the most alkaline products (kimberlite, high-potassium mafic rocks, nephelinites, etc.) and the highest geothermal gradients with tholeiitic rocks (not represented in the Kenya dome). In other words, part of the sequence ultrabasic alkaline rocks—basanite—alkali basalts—transalkaline basalt—tholeiite can be recognized (a) in a progressively younging succession in any one dome centre; and (b) in a more or less radial arrangement around a dome centre at any one time, the more alkalic rocks farthest away from the centre.

Consideration of the data previously presented shows that the distribution of the spectrum of eruptive rocks occurring in the Kenya dome fits this pattern well, and the various eruptive rocks could have been derived from primary magmas generated in a rising plume.

The Bouger gravity profile* of the Kenya dome shows a broad negative anomaly across the dome with a smaller but sharper positive anomaly over the position of the rift valley itself (B.H. Baker and Wohlenberg, 1971). After making further allowance for the strong influence exerted by the presence of some thousands of metres of less dense volcanics on the flanks and on the floor of the rift valleys, refined residual values can be fitted to a model such as in Fig. 10.4. A broad body of relatively low-density upper mantle (the relatively hot top of a plume?) underlying the dome is overlain by relatively dense rock within the crust below the rift valley (mafic igneous rocks representing the complementary fraction to the voluminous fraction-

---

*A correction known as the *free-air correction* is applied to compare gravity readings that may have been taken at different heights above sea level, as the value of gravity decreases regularly with distance from the centre of the Earth. A further correction, the *Bouguer correction* is applied, in the opposite sense to the free-air correction, to correct for the gravitational attraction of rock (assumed to be a flat slab) between sea level and the spot height at which the raw gravity observation was made. Refinements in this process are possible to allow for the gravitational effects of near-surface rocks of known specific gravities. The *Bouguer anomaly* is a residual value obtained after making both these corrections and comparing the resultant value with sea-level gravity on the spheroid at the same latitude. The Bouguer anomaly therefore is a first-order attempt to compensate for variations in gravity due to topographic and near-surface effects, and it is thus an indicator of more fundamental differences in gravity, attributable to underlying crustal and upper-mantle material.

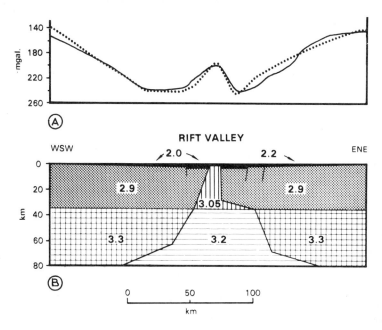

Fig. 10.4. Bouguer gravity profile across the Gregory rift in a central part of the Kenya dome. The *continuous line* is the observed Bouger gravity anomaly and the *dotted line* is the computed anomaly for the crustal model shown below (after B.H. Baker and Wohlenberg, 1971, p.541, fig. 7).

ated phonolites and trachytes on surface?). To the limited extent of this physical emplacement of igneous rock in the crust some "plate separation" may have occurred, but surface exposures of basement rocks place a maximum upper limit to this of only 10 km since Miocene time.

10.6. IGNEOUS ACTIVITY IN THE AFRO-ARABIAN DOME ASSOCIATED WITH PLATE SEPARATION

The igneous activity and the tectonic evolution of the Afro-Arabian dome have become the subject of extreme interest and intensive research in recent years. Like the Kenya dome, a synthesis seems to have emerged.

Compared with the Kenya dome the Afro-Arabian dome (Fig. 10.5) is older and plate separation is actively progressing across two of its rift valley arms, the Gulf of Aden and the Red Sea. A focus of interest is the incipient separation at the centre of the dome in the fascinating area of the Afar triangle (Tazieff, 1970) and the nearby Ardoukôba rift forming the Gulf of Tadjoura (Needham et al., 1976). Igneous activity at the present day, in large part submarine, is producing rocks overwhelmingly of the tholeiite series plus small amounts of rocks of the transalkaline igneous rock series, but

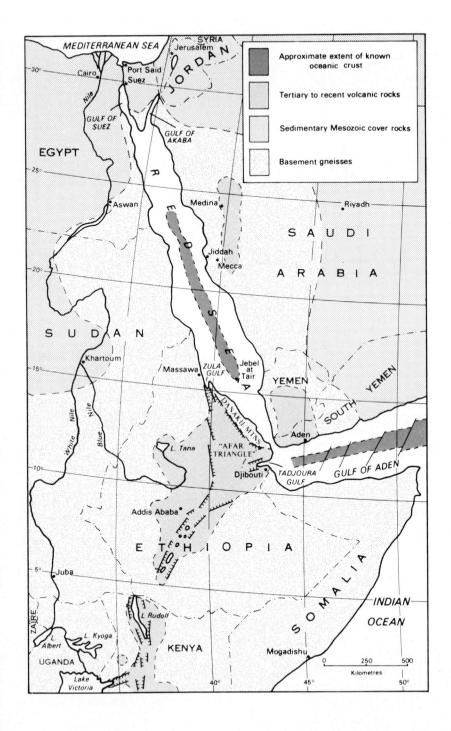

relatively more alkalic rocks are also present in the geological record of the area. Taken in conjunction with the igneous rocks of the Kenya dome, the igneous rocks of the Afro-Arabian dome thus complete a spectrum of igneous rock series within continental areas from highly alkaline to tholeiitic (with such rocks as the potassic series of Birunga and Toro—Ankole, together with kimberlite and carbonatite, constituting the rare, highly variable, and sporadically distributed alkaline ultrabasic end of the spectrum).

From a review by Gass (1970), it is clear that uplift of the Afro-Arabian dome began in early Tertiary time and continued until the end of the Eocene, and that rift valley formation took place within Miocene time when plate separation was also initiated. Since that time ~200 km width of new oceanic crust has been generated in the wider part of the Gulf of Aden and ~50 km width of new oceanic crust in the southern Red Sea. It is possible that other contiguous domes centred on the Red Sea and (originally) off the Gulf of Aden may have contributed to this tectonic evolution.

On a regional scale within the Afro-Arabian dome, it appears that the observed variety of magmatic products can be broadly correlated with style of crustal deformation (Gass, 1970, p.287): (1) alkali basalts preceding and concurrent with vertical uplift of the dome; (2) transalkaline basalts (lavas intermediate in composition between alkali basalt and tholeiite and that characteristically produce peralkaline rhyolites on fractionation) in zones of crustal attenuation; and (3) tholeiitic basalts in areas of new oceanic crust. There are chemical indications that these tholeiites are not of MORB but of OIT affinity (Schilling, 1973a), attributed as we have seen in the oceanic areas to the influence of an underlying plume.

The pre-rifting alkali basalts are voluminous; they cover ~750,000 km^2 in Ethiopia and ~30,000 km^2 in southwest Arabia. They are known as the "Trap Series" and appear to be anaiogous structurally (but not quite petrographically as we shall see) to flood basalts such as those of the Karroo, Deccan and Paraná. The Trap Series thins radially outwards in all directions from a position approximating to the centre of the dome where a lava sequence 3.5 km thick has been recorded. A lower member of thick aphyric flows (erupted from fissures?) is overlain by an upper member of more varied porphyritic and differentiated flows (erupted from central volcanoes?).

The transalkaline basalts and their relatively abundant differentiated affiliated lavas erupted from central volcanoes in areas of attenuated (and apparently at the time attenuating) sialic crust and constitute by volume only some 1/500 of that of the Trap Series. They have been best docu-

---

Fig. 10.5. Sketch map showing outline of major geological features and localities mentioned in the text in connection with discussion of the igneous rocks of the Afro-Arabian dome. No attempt has been made in this sketch to portray the complexity of "half-graben" faulting within the area of the Afar triangle (based on maps by Gass, 1970, p.286, fig.1; Tazieff, 1970, pp. 34 and 35; and A.A.G.S., 1963).

TABLE 10.4

Average chemical compositions and CIPW norms of basalts of the Trap Series of Ethiopia and Yemen, the attenuated zones, and tholeiitic associations of the Gulf of Aden and Red Sea areas, all within the Afro-Arabian dome (after Gass, 1977, table 2, facing p.290, with references to source of data)

	1	1a*	2	2a*	3	3a	4
SiO_2	46.60	46.70	47.47	47.75	47.30	47.56	47.46
TiO_2	2.15	2.15	2.61	2.62	2.00	2.01	2.59
Al_2O_3	16.20	16.23	15.98	16.07	14.10	14.17	16.26
Fe_2O_3	3.80	1.50	7.07	1.50	6.70	1.50	5.23
FeO	8.45	10.54	5.38	10.46	7.30	12.04	6.72
MnO	0.23	0.23	0.28	0.28	0.20	0.20	0.19
MgO	6.65	6.66	3.29	3.31	5.95	5.98	6.72
CaO	9.60	9.62	9.43	9.48	10.80	10.86	10.08
Na_2O	2.85	2.86	3.84	3.86	2.50	2.51	3.29
K_2O	1.25	1.25	1.67	1.68	1.30	1.31	0.84
H_2O	1.80	1.80	2.03	2.04	1.45	1.46	—
P_2O_5	0.44	0.44	0.45	0.45	0.40	0.40	0.61
Total	100.02	99.98	100.00	100.00	100.00	100.00	99.99
Q	—	—	—	—	0.60	—	—
or	7.39	7.39	9.87	9.93	7.69	7.74	4.97
ab	24.11	22.83	31.40	24.69	21.15	21.24	27.84
an	27.72	27.76	21.44	21.57	23.42	23.53	27.12
ne	—	0.74	0.59	4.32	—	—	—
di	13.83	14.10	17.69	19.02	22.08	23.09	15.05
hy	2.49	—	—	—	9.18	0.99	4.66
ol	12.06	18.09	0.74	10.25	—	15.03	6.45
mt	5.51	2.18	10.25	2.18	9.71	2.18	7.58
il	4.08	4.08	4.95	4.97	3.79	3.82	4.92
ap	1.04	1.04	1.06	1.06	0.94	0.94	1.44
M-value	49.7	49.7	33.0	33.0	44.1	44.1	50.8

Trap Series: 1 = Ethiopia, lower unit, average of sixteen analyses; *2* = Yemen, average of six analyses.
Attenuated zones: 3 = Ethiopia, Aden series, average of 25 analyses; *4* = Jebel Khariz, Aden Volcanic Series, south Arabia, average of five analyses; *5* = Central chain of Afar depression, average of nine analyses.

*Columns with the suffix a are the preceding analysis after the Fe_2O_3 content has been recalculated to 1.5 wt.%.

mented in the Aden area — the "Aden Volcanic Series" — but are found also in the Afar triangle and within the Ethiopian rift valley. Whereas the Aden and Afar transalkaline series rocks are Neogene in age, the Ethiopian rift valley volcanoes, constituting the world's most abundant peralkaline silicic rock province, are Pliocene to Recent in age (Gibson, 1975).

*	5	5a*	6	6a*	7	7a*	
7.64	49.46	49.58	48.93	48.88	50.04	50.12	SiO_2
2.60	2.31	2.32	1.09	1.09	1.93	1.93	TiO_2
6.32	12.81	12.84	16.60	16.59	16.56	16.58	Al_2O_3
1.50	4.14	1.50	2.09	1.50	3.31	1.50	Fe_2O_3
0.13	8.77	11.19	6.83	7.35	7.49	9.13	FeO
0.19	0.17	0.17	0.24	0.24	0.19	0.19	MnO
6.75	6.32	6.34	7.38	7.37	5.19	5.20	MgO
0.12	10.97	11.00	12.67	12.66	11.33	11.34	CaO
3.30	3.01	3.02	2.42	2.42	3.00	3.00	Na_2O
0.84	0.71	0.71	0.19	0.19	0.36	0.36	K_2O
—	0.98	0.98	1.56	1.56	0.65	0.65	H_2O
0.61	0.35	0.35	0.15	0.15	—	—	P_2O_5
0.00	100.00	100.00	100.15	100.00	100.05	100.00	Total
—	0.19	—	—	—	—	—	Q
4.97	4.20	4.20	1.12	1.12	2.13	2.13	or
6.23	25.47	25.55	20.48	20.48	25.38	25.38	ab
7.24	19.35	19.39	33.88	33.85	30.66	30.72	an
0.92	—	—	—	—	—	—	ne
5.59	26.67	27.22	22.69	22.77	20.83	21.12	di
—	11.95	7.37	10.06	8.86	10.95	11.30	hy
5.53	—	7.90	4.92	6.78	—	2.86	ol
2.18	6.00	2.18	3.03	2.18	4.80	2.18	mt
4.94	4.39	4.41	2.07	2.07	3.67	3.67	il
4.44	0.83	0.83	0.35	0.35	n.d.	n.d.	ap
0.8	47.3	47.3	59.7	59.7	46.7	46.7	M-value

Tholeiitic associations of the Gulf of Aden and Red Sea: 6 = Gulf of Aden, average of eight aphyric basalts; 7 = Jebel at Tair, volcanic island within the Red Sea median trough, average of six analyses.

Tholeiitic rocks characterize, as in the mid-ocean ridges, areas of active plate separation. They have been found in the Gulf of Aden, the median trough of the Red Sea, and on the island of Jebel at Tair situated near the southern end of the Red Sea. In addition, tholeiitic rocks are apparently being erupted at the present time within central parts of the Afar triangle

TABLE 10.5

Chemical compositions and CIPW norms of selected rocks from the Aden volcanic suite

	1	2	3	4	5	6	7
SiO_2	46.80	46.98	54.11	59.08	63.74	65.61	69.54
Al_2O_3	13.71	12.88	15.34	14.70	14.93	14.17	12.77
Fe_2O_3	5.10	10.85	6.45	5.05	2.72	6.24	4.03
FeO	7.35	3.98	4.40	3.25	3.72	0.35	0.99
MnO	0.17	0.27	0.17	0.17	0.23	0.10	0.17
MgO	7.75	3.42	2.87	1.43	0.79	0.53	0.13
CaO	10.28	9.23	6.18	4.05	2.62	2.15	0.62
Na_2O	2.54	3.42	4.71	5.37	5.82	5.18	5.48
K_2O	0.87	1.34	3.46	2.89	3.43	3.67	4.96
TiO_2	3.00	3.80	1.50	1.48	0.94	0.53	0.33
P_2O_5	0.42	0.62	0.67	0.53	0.22	0.16	0.04
H_2O^+	1.38	1.64	0.40	0.55	0.19	0.43	0.13
H_2O^-	0.36	0.41	0.82	1.16	0.47	0.65	0.44
CO_2	0.08	0.86	—	—	0.09	pres.	—
F	—	—	0.11	0.18	—	—	0.26
Cl	—	—	—	—	—	—	0.07
Less O for F, Cl	—	—	0.05	0.08	—	—	0.13
Total	99.81	99.70	100.14	99.81	99.91	99.97	99.83
Q	—	—	3.3	9.0	10.6	16.9	18.2
or	5.3	8.3	14.7	17.4	20.4	22.0	29.6
ab	22.0	30.1	40.3	46.4	49.7	44.6	38.5
an	24.0	16.5	13.6	7.6	4.5	4.7	—
ac	—	—	—	—	—	—	7.4
di	20.7	22.7	10.5	7.6	6.0	4.3	2.5
hy	8.2	5.3	4.8	1.5	2.4	1.3	3.0
ol	9.5	3.9	—	—	—	—	—
mt	3.6*[1]	4.2*[1]	8.3*[2]	6.4*[2]	4.0	4.9*[2]	0.1*[2]
il	5.8	7.5	2.9	2.9	1.8	1.0	0.6
ap	1.0	1.5	1.6	1.3	0.5	0.4	0.1
M-value	53.3	30.3	33.0	24.2	18.0	13.5	4.6

1 = basalt lava, Amen Khal parasitic centre (Cox et al., 1970, p.442, table 3, analysis A.*803*).

2 = hawaiite lava, Main Cone Series, south Aden (Cox et al., 1970, p.442, table 3, analysis A.*761*).

3 = mugearite intrusion into Tawahi Series, southeast of Hospital, Ma'alla (Cox et al., 1970, p.442, table 3, analysis A.*874*).

4 = trachyandesite lava, overflow from Shamsan Caldera Series, 700 m west of Ras Taih, southeast Aden (Cox et al., 1970, p.442, table 3, analysis A.*849*).

5 = trachyandesite lava, Shamsan Caldera Series, Signal Station (Cox et al., 1970, p.442, table 3, analysis A.*783*).

6 = trachyte sheet (lava?) within the agglomerates of the Tawahi Series 700 m east of Amen Khal (Cox et al., 1970, p.442, table 3, analysis A.*862*).

7 = rhyolite lava, Main Cone Series, Sira Island, east Aden (Cox et al., 1970, p.442, table 3, A.*843*).

pres. = present.

*[1] $Fe_2O_3/(Fe_2O_3 + FeO)$ recalculated to 0.2.

*[2] Fe_2O_3 and FeO recalculated to contain equal weights of Fe.

(Barberi et al., 1972). Estimates of the volume of tholeiitic rocks depend on estimates of the area underlain by new oceanic crust which are difficult to make precisely because of the thick blankets of sedimentary material on the margins of the Gulf of Aden and the Red Sea that obscure the underlying crustal structure. A very crude estimate would be $1.25 \cdot 10^5$ km^2 of new oceanic crust which, assuming an average oceanic crustal thickness, would imply $\sim 8 \cdot 10^5$ km^3 of tholeiitic rock.

Following Gass (1970, table 2) average analyses of the three mafic lava types are presented in Table 10.4. Many of the analyses contain high Fe_2O_3 relative to FeO, presumably as a result of weathering. For comparative purposes a uniform and fairly drastic reduction of the Fe_2O_3 contents to 1.50 wt.% was made by Gass for all analyses and the derived norms given for both sets of data. It is very instructive to study the effect of this correction (so clearly displayed by Gass and by contrast so commonly glossed over in many petrological accounts) on such critical normative parameters as normative hypersthene and nepheline contents. Some minor reservations on the magnitude of the correction apart, it does appear that the Trap Series basalts are just nepheline-normative and that the oceanic basalts by contrast have a typical strongly hypersthene-normative tholeiitic character. Of the basalts of attenuated zones of alleged transalkaline type which are believed to be parental to peralkaline silicic fractionation products, one from Ethiopia is just hypersthene-normative, a second from the Aden Volcanic Series is just nepheline-normative, and a third from the central part of the Afar depression is strongly hypersthene-normative (and thus more akin to the tholeiite category?). However, more recent analyses of the Aden Volcanic Series coupled with petrography (Cox et al., 1970) do seem to indicate a well-documented fractionation sequence of parental hypersthene-normative "transalkaline basalt" via trachybasalt and trachyte to comendite* (Table 10.5).

To account for the change in composition of erupted lavas in both time and space Gass (1970, p.298, fig. 2) suggested a model of partial melting within a "rising lithothermal system". In this model, partial melts at relatively deep levels produce alkali basalts, and as the lithothermal system rises partial melting at increasingly shallower levels results in the production of magmas of increasing silica activity. The model of Gass parallels the "mushroom" model (Fig. 10.6) of the distribution and amount of partial melting within a rising plume developed by Oxburgh and Turcotte (1968). Some of the chemical peculiarities of the Afar tholeiites match those of Iceland (Schilling, 1973a), so that if one accepts a plume hypothesis in Iceland then

---

*Reduction of high raw Fe_2O_3 contents of mafic rocks in this Aden series to generally accepted levels would in fact convert one or two analyses to nepheline-normative (see discussion in Cox et al., 1970, pp. 449 and 450). Nevertheless, on an alkalis vs. silica variation diagram (Cox et al., 1970, p.437, fig. 3) the series clearly lies in an intermediate position between typical tholeiites and alkali basalts.

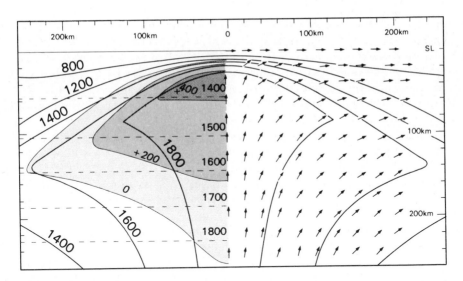

Fig. 10.6. Detail at top of an ascending plume in a model developed by Oxburgh and Turcotte (1968, pp. 2643—2661) and based on their fig. 11 (Oxburgh and Turcotte, 1968, p.2653). Isotherms in the plume head are shown as *heavy lines* (values in degrees Celsius). Liquidus temperatures for olivine tholeiite are indicated by *horizontal dashed lines*. The contour line marked *0* is the locus of intersection of identical values of plume temperature and olivine tholeiite liquidus temperatures, and is hence an envelope of partial melting within the plume. Two other contours indicate where the predicted plume temperature exceeds olivine tholeiite liquidus temperatures by 200° and 400°C, respectively, thus connoting inferred higher proportions of mantle melting within these smaller, more central parts of the plume head. The *small arrows* on the right-hand side of the diagram are predicted velocity vectors.

a plume attribution of the whole cycle of igneous activity in the Afro-Arabian dome does seem plausible, and indeed for a depth of 300—400 km seismic velocities in the mantle below Afar are anomalously low.

Assuming for the moment that the plume hypothesis is broadly applicable it is legitimate to enquire into the patent differences between the Afro-Arabian and Kenya domes (and indeed other domes). The very large amounts of flood basalts erupted during the early stages of the development of the Afro-Arabian dome are paralleled in other continents but are absent from the Kenya dome. They indicate much higher amounts of melt reaching the surface per unit of time presumably produced by a larger plume or one that was hotter (as the result of near-adiabatic ascent of mantle from deeper levels?) or by one that was rising faster. In any of these possible circumstances producing a "strong plume", it can easily be envisaged that eventual plate separation would more easily occur than above a weak plume. Certainly, it would seem reasonable that some massive event within the upper mantle is necessary to initiate plate separation in old, thick, continental lithosphere. Weak plumes might also be ineffectual in penetrating thick cratonic lithosphere; magma generation in them might be limited to small amounts at relatively great depths -- the set of conditions that fits the highly potassium-rich series; or again, in a weak plume, magmas once generated, although of less extreme compositions than the highly potassium-rich ones, may undergo considerable fractionation during penetration of thick lithosphere eventually producing large proportions of fractionated rock types at surface, such as the voluminous phonolites of the Kenya dome*.

Although these variations on a plume model may appear disingenuously elastic and thus philosophically unacceptable to some petrologists, at least a self-consistent pattern of igneous and tectonic evolution is beginning to emerge that can be plausibly explained by a plume model. It is for this reason that the presentation of igneous activity in cratonic areas largely in terms of a plume hypothesis has been adopted here. It remains to be seen, however, how well petrogenetic theory concerning magma compositions may fit what predictive constraints there may be in plume models.

10.7. CONTINENTAL FLOOD BASALTS MAINLY OF THOLEIITE COMPOSITION

In their tectonic setting and in their great volumes erupted within geologically short time spans, flood basalts and associated diabase sills form a recognizable group that has its closest analogue among rocks we have so far considered in the alkali basalt Trap Series of Ethiopia and Yemen. However, the majority of flood basalts differ in, if anything, even greater volumes and, more generally, in an overall tholeiitic affinity. Spectacular terraced scenery formed by the erosion of numerous superimposed flat-lying flows is the rule. Well-known examples include those in Table 10.6. On account of their huge scale any digestible description demands generalization, always a dangerous enterprise in geology, and in connection with these flood basalts in particular one has the uneasy feeling of significant data being overlooked. The following thumbnail sketches are of the main outlines of some provinces plus some details of interest.

10.7.1. Columbia River basalts

Waters (1961) documented stratigraphic and lithologic variation based on 28 stratigraphical sections and mapping of selected areas. Basalts belonging to a lower division known as the Picture Gorge type are characterized by a few per cent of modal olivine. Olivine is present both as phenocrysts

---

*There is evidence that some trachytes erupted in continental areas have been produced by fractionation at high pressures: the presence of ultramafic inclusions in some Nigerian trachytes (J.B. Wright, 1969b) strongly indicates their elutriation in a trachytic host magma from mantle depths; the high M-values of some Australian hawaiites, mugearites and benmoreites (D.H. Green et al., 1974) similarly indicates an origin by fractionation under mantle conditions (rather than by fractionation at crustal levels that has probably produced superficially comparable igneous rock series in some oceanic islands, for example).

TABLE 10.6

Comparative data on ages and volumes of well-known flood basalt provinces and related rocks

Locality	Age	Remarks
Snake River Plain basalts	Quaternary	$0.5 \cdot 10^5$ km² extent
Iceland*	Neogene—Recent	10^5 km² extent, forming whole crustal thickness up to a maximum of 16 km
Columbia River plateau basalts	mainly Miocene	$0.13 \cdot 10^5$ km² extent; average thickness 1000 m; pre-existing relief of 1500 m buried in places
Ethiopia and Yemen Trap Series	early Tertiary	$8 \cdot 10^5$ km² extent; maximum thickness of 3500 m in central part of dome
Blosseville Coast basalts, east Greenland	early Tertiary	$0.54 \cdot 10^5$ km² extent, probably once $1.5 \cdot 10^5$ km²; average thickness ~3000 m
Deccan Traps	late Cretaceous to to Paleocene	present outcrop more than $5 \cdot 10^5$ km², once 10^6 or even $2 \cdot 10^6$ km²; thickness commonly up to 2000 m, locally 3000 m
Paraná plateau basalts	Jurassic to early Cretaceous	present outcrop $0.75 \cdot 10^6$ km², once $1.2 \cdot 10^6$ km²; average thickness 1000 m, locally in excess of 1500 m
Karroo basalts	Jurassic	present outcrop $0.15 \cdot 10^6$ km², once 10^6 km²; thickness 400—1500 m, locally 9000 m in Lebombo monocline; many associated sills
Tholeiitic lavas and minor intrusions of Nova Scotia, New Jersey, Carolinas and Morocco	Triassic	much smaller in exposed area than most, but extension known to underlie off-shore Mesozoic—Tertiary sedimentary sequence
Northern Australia	Cambrian	probable original outcrop area of $4 \cdot 10^5$ km²; thickness up to 1000 m
Keweenawan lavas	~1100 Ma ago	outcrop area at least 10^5 km²; thickness up to 5000 m
Oceanic crust*	produced in last 200 Ma; in part consumed	$3 \cdot 10^8$ km²; thickness normally close to 6500 m

*Not continental flood basalt, of course, but inserted for purposes of comparison.

(commonly in association with augite—plagioclase phenocrysts and thus suggesting a history of significant cotectic crystallization), and also in the matrices of some specimens which thus have petrographic affinity with alkali basalt. Silica contents are commonly in the range of 47—50 wt.% and K_2O content is around 0.6 wt.%. A younger division known as the Yakima type rests in places with distinct unconformity on the Picture Gorge-type basalts. Yakima-type basalts only rarely carry olivine, have silica contents generally in the range 53—55 wt.% (and are thus strictly intermediate rather than basic rocks), and contain around 1.5 wt.% K_2O. The Yakima-type basalts characteristically carry a high proportion of glassy mesostasis in an exaggerated form of intersertal texture. Unlike the Picture Gorge type they are frequently aphyric or microphyric. Nevertheless their relatively high differentiation indices and relatively low M-values would suggest an even longer continued history of fractionation than that affecting Picture Gorge types, if fractionation were indeed the cause of the chemical differences between the two types. Average analyses of the two types are included in Table 10.7. This broad two-fold division has been refined but not essentially changed by later work (T.L. Wright et al., 1973). Initial $^{87}Sr/^{86}Sr$ ratios are rather uniform (0.7035—0.7039) in the Picture Gorge type but higher and more variable (0.7045—0.7080) in the basalts of Yakima type (McDougall, 1976).

10.7.2. Keweenawan lavas

Work on the Proterozoic Portage Lake Lava Series of the Keweenaw peninsula, northern Michigan, (Jolly and R.E. Smith, 1972) has documented the effects of degradation consequent on burial metamorphism. Degradation has involved localized metamorphic differentiation, particularly along flow tops and other permeable channelways with drastic effects on major-element chemistry (see also Chapter 14). Nevertheless a broad distinction, based on the petrography and chemistry of relatively less altered flow rocks, into rela-aphyric to slightly porphyritic olivine tholeiite containing phenocrysts fine-grained to slightly porphyritic olivine tholeiite containing phenocrysts of olivine (rarely more than 10 vol.%), plagioclase and augite. Most of the remaining flows are tholeiites with plagioclase and augite phenocrysts. Average analyses* are included in Table 10.7. Although apparently defining a broadly similar series to the Columbia River basalts, the Keweenawan analyses contain much higher Fe_2O_3/FeO ratios and H_2O contents attributable to some degree of post-crystallization alteration, despite only the freshest available material being selected for analysis. Note the marked

---

*Note how recasting of the Keweenawan analyses water-free, as has been done for several other analyses in Table 10.7, would result in an upward revision in weight per cent SiO_2 of between 1.5 and 2%, a significant amount when comparisons are made.

TABLE 10.7

Chemical compositions of various flood basalts and diabases

	Columbia River basalt		Keweenawan basalts		Karroo basalts		
	1	2	3	4	5	6	7
SiO_2	49.5	53.8	47.7	50.0	49.7	52.1	51.8
Al_2O_3	15.5	13.9	15.1	13.9	8.9	14.1	14.8
Fe_2O_3	3.8	2.6	6.4	8.7	2.0	3.7	3.9
FeO	7.8	9.2	5.5	5.3	10.5	8.4	7.3
MnO	0.2	0.2	0.2	0.2	0.2	0.2	0.2
MgO	6.2	4.1	8.2	4.4	15.3	5.5	7.1
CaO	10.2	7.9	7.6	5.4	7.7	8.7	10.6
Na_2O	2.8	3.0	2.9	3.9	1.6	2.6	2.4
K_2O	0.6	1.5	0.8	1.6	1.6	1.7	0.7
TiO_2	1.6	2.0	1.5	2.7	2.7	2.6	1.1
P_2O_5	0.3	0.4	0.2	0.4	0.4	0.4	0.1
H_2O^+	}1.6	}1.2	3.7	3.3	—	—	—
H_2O^-			—	—	—	—	—
CO_2	—	—	0.2	0.3	—	—	—
Total	100.1	99.8	100.0	100.1	100.6	100.0	100.0

1 = average Picture Gorge type, Columbia River Group (Waters, 1961, p.592, table 2, analysis *8*).

2 = average Yakima type, Columbia River Group (Waters, 1961, p.593, table 3, analysis *8*).

3 = olivine tholeiite average of six analyses from Portage Lake lava series, Keweenawan peninsula (Jolly and R.E. Smith, 1972, p.276, table 1, average of analyses *1—6*).

4 = tholeiite, average of three analyses from Portage Lake lava series, Keweenawan peninsula (Jolly and R.E. Smith, 1972, p.276, table 1, average of analyses *7—9*).

5 = average of 47 Nuanetsi olivine-rich basalts, Karroo System (Cox, 1972, p.317, table 2, analysis C).

6 = average of 14 Nuanetsi basalts with MgO below 8 wt.%, Karroo System (Cox, 1972, p.317, table 2, analysis B).

7 = average of 21 Lesotho basalts, Karroo System (Cox, 1972, p.317, table 2, analysis A).

*Total iron as FeO.

sympathetic variation, correlatable with an inferred history of fractionation, of K_2O, TiO_2 and P_2O_5 with silica content. Broad statements, of a type all too common in petrological writing, to the effect that a rock type such as flood basalt is "rich" in K_2O, etc., could obviously be more precisely expressed by reference to the position of the rock in question to its appropriate series, as defined by such parameters as differentiation index, MgO content, or M-value, or in an appropriate variation diagram. Only in this way may vital intrinsic differences in magma type, perhaps related as we have seen to subtle mantle inhomogeneities, become apparent.

an basalts		Tasmanian diabase	Ferrar diabases			Blosseville Coast basalt	
	9	10	11	12	13	14	
1	52.65	53.4	50.40	53.75	55.65	48.88	SiO_2
5	14.42	15.4	15.51	14.23	13.95	13.68	Al_2O_3
2	2.76	0.8	0.99	2.23	3.24	—	Fe_2O_3
3	10.02	8.4	7.83	7.61	7.38	13.43*	FeO
3	0.14	0.1	0.17	0.18	0.17	0.20	MnO
5	4.98	6.7	10.60	6.64	4.50	6.57	MgO
7	8.96	11.1	10.87	10.60	8.51	11.29	CaO
3	3.01	1.7	1.42	1.83	2.50	2.35	Na_2O
1	1.08	1.0	0.37	0.81	1.45	0.26	K_2O
5	1.70	0.6	0.44	0.70	1.03	2.57	TiO_2
5	0.28	0.1	0.08	0.18	0.23	0.28	P_2O_5
	—	0.7	1.21	0.64	0.98	—	H_2O^+
	—	—	0.34	0.67	0.69	—	H_2O^-
	—	—	—	—	—	—	CO_2
	100.0	100.0	100.23	100.07	100.28	99.51	Total

8 = average basalt of Lower Deccan Traps (Sukheswala and Poldervaart, 1958, p.1487, table 3, analysis 4).

9 = average basalt of Upper Deccan Traps (Sukheswala and Poldervaart, 1958, p.1487, table 3, analysis 6).

10 = average Tasmanian chilled dolerite (McDougall, 1962, p.295, table 8, analysis 2).

11 = olivine tholeiite, chilled margin, Ferrar Dolerite Group, Antarctica (Gunn, 1966, p.908, table 2, analysis 1).

12 = hypersthene tholeiite, average of five chilled margins, Ferrar Dolerite Group, Antarctica (Gunn, 1966, p.908, table 2, average of analyses 2—6).

13 = pigeonite tholeiite, average of four chilled margins, Ferrar Dolerite Group, Antarctica (Gunn, 1966, p.908, table 2, average of analyses 7—10).

14 = average Blosseville Coast plateau basalt (C.K. Brooks et al., p.286, table 2, analysis 1).

10.7.3. Karroo volcanic cycle

Predominantly mafic volcanic rocks of Jurassic age are widespread in southern Africa (Cox et al., 1967; Cox, 1970, 1972), where although only $\sim 1.4 \cdot 10^5$ km^2 of lava outcrop assignable to late Karroo time remains it seems probable that as much as $2 \cdot 10^6$ km^2 of southern Africa was once affected by eruptive activity at this time. Some of the evidence for this comes from the distribution of numerous comagmatic sills in Karroo strata, some thick and fractionated. The structure and petrography of these sills, all tholeiitic, have been reviewed by F. Walker and Poldervaart (1949). The great bulk of Karroo volcanism was apparently confined to a short period of 20 Ma in the early Mesozoic. In more inland parts of the province the lavas are generally preserved as high plateau outliers, in broad basins, and in the Zambezi rift valley. In eastern Karroo outcrops, where lava sequences are thick, the structure is dominated by deeper troughs such as the east—west Nuanetsi syncline apparently developed during volcanism which abuts the meridionally-trending Lebombo monocline containing an apparent thickness of 9000 m of lavas in its limb and letting down the eastern side of the structure ~9—13 km below a coastal veneer of younger sediments (Fig. 10.7). Crustal extension is indicated by a structure common to many of the basins wherein a series of parallel normal faults, all heading in the same direction, separate elongate blocks tilted in the opposing sense in such a way that the effect of the tilt balances the vertical throw of the faults so that the nett effect is only a horizontal extension. This structure has been termed "factory roof structure" or "half-grabens"; similar faulting within the floor of the Afar depression has been called "ratchet faulting". A greater degree of stretching of the crust leading to its eventual rupture is indicated by the Lebombo monocline. Comparable structures to this include the East Greenland crustal flexure, the Panvel flexure of northwest India, and a similar monoclinal flexure along the northern margin of the Gulf of Aden, all of them along inferred lines of crustal separation in Tertiary time. The actual locations of the various troughs, synclinal basins, rift valleys, crustal flexures and breaks associated with Karroo volcanism appear to reflect in part the development of a pattern of dome—rift valley systems modified in part by pre-existing lines of weakness around basement cratons and along older orogenic belts.

Petrographically and compositionally the Karroo lavas are complex. Most are tholeiites originating from presumed fissure eruptions as central complexes are rare. The upper parts of the Lebombo and Nuanetsi successions contain voluminous interbedded rhyolite constituting about a third of the total lava pile. Their acid rocks yield initial $^{87}Sr/^{86}Sr$ ratios of 0.7042 ± 0.0005 and 0.7081 ± 0.0008, respectively (Manton, 1968). Manton suggested a mantle origin for the rhyolite but Cox (1972) has given reasons for preferring an origin by crustal anatexis, a not unlikely possibility in view of the accession of such vast amounts of molten basic rock to continental crustal levels. Manton also reported initial $^{87}Sr/^{86}Sr$ ratios of $0.7042—0.7125$ for the mafic rocks, a large span including surprisingly high values for tholeiitic rocks, and moreover suggestively encompassing the rhyolite values. Leaving aside this controversy over the associated acid rocks, the mafic rocks themselves are of considerable interest. Many are quartz-normative and have a sparsely feldsparphyric appearance. At the intersection of the Nuanetsi trough and the Lebombo monocline the very thick lava succession consists of a few basal flows of nephelinite overlain by olivine-rich tholeiitic varieties

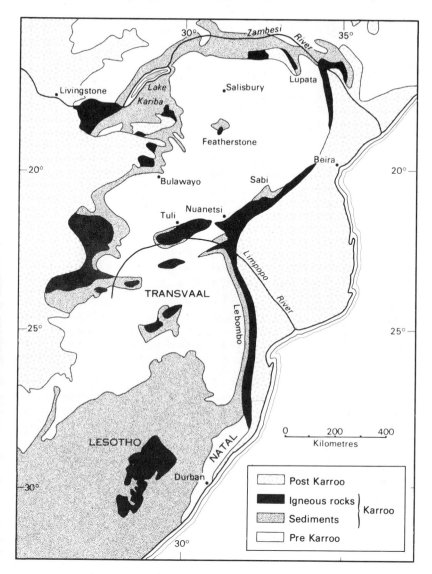

Fig. 10.7. Geological sketch map of southeastern Africa showing distribution of rocks of the Karroo System and localities mentioned in the text (after Cox, 1970, p.212, fig. 1).

which are, in turn, overlain by a thick succession of more typical tholeiites. These tholeiites, at this apparent focal point of voluminous volcanism, are characterized by overall more potassium-rich compositions than those in more peripheral areas to the west and south.

Following Cox (1972), it is tempting to think of this Nuanetsi-Lebombo node as a centre overlying a vigorous plume that, after initial eruption of small quantities of alkalic mafic rocks, produced large amounts of primitive Mg-rich magma (the "culmination phase") by voluminous partial melting with relatively little fractionation en route to the surface, followed by more normal tholeiite (the "steady-state period"), accompanied by downsagging of the crust attendant on the vast bulk of erupted lavas.

Average analyses of olivine-rich Nuanetsi basalt with MgO below 8 wt.%, and of Lesotho basalts (from much further south) are included in Table 10.7, taken from Cox (1972, p.317). The relative enrichment of mafic lavas in the Nuanetsi area in K_2O, TiO_2 and P_2O_5, despite an otherwise highly "primitive" nature, i.e. with high Mg, Cr and Ni (Jamieson and Clarke, 1970), is well shown by these analyses (see also p.353).

10.7.4. Deccan Traps

The Deccan Traps of northwestern India, constituting from their present and inferred former outcrop area possibly the most voluminous flood basalt province known, are in the main flat-lying and give rise to monotonous terraced scenery inland. Nearer the coast, however, tilting has produced dips of up to 20° or more and is accompanied by normal faulting and the Panvel flexure akin to the Lebombo and other monoclines. Eruption was predominantly basaltic with a very low explosion index. Individual flows are commonly thin, averaging ~5 m and seldom exceeding 15 m. Although some dykes have been recorded there is an apparent deficiency of obvious feeders for the great pile of lavas, and it is presumed, therefore, that extrusion took place by fissure eruptions as a result of which younger flows covered up much of the evidence of the mode of formation of the earlier ones. A few central complexes are known but they cannot possibly be regarded as source areas for the whole of the immense volcanic pile. They are distinguished by the local occurrence of sub-volcanic intrusions, pyroclastic beds exhibiting relatively abrupt changes in thickness, and the presence both of accumulative basaltic types such as oceanite on the one hand and of fractionated rocks including rhyolites on the other. Rocks of intermediate compositions are more rarely reported. In the Girnar igneous complex initial $^{87}Sr/^{86}Sr$ ratios in gabbros, etc., range from 0.7051 to 0.7080, not unusually high for the continental flood basalt association, but initial ratios of spatially associated silicic porphyries cluster around 0.7275, thus strongly suggesting an origin by crustal anatexis for them (Paul et al., 1977).

To a first approximation however, petrographical and chemical uniformity is a feature of the Deccan basalt sequence. In hand specimen, the basalts are generally amygdaloidal and are either aphyric or sparsely porphyritic. Phenocrysts generally comprise plagioclase and augite although olivine sometimes is found. The groundmass is typically intersertal in texture and contains no olivine. There is a suggestion of a difference in composition between a lower and upper grouping of the basalts of which average analyses are included in

Table 10.7. However, it should be noted that in arriving at these average figures some judicious exclusion of older suspect analyses had to be made, and also that analyses of some relatively fractionated varieties may have been included (Sukheswala and Poldervaart, 1958). All common flow rocks are quartz-normative.

In addition to this comprehensible picture of voluminous tholeiitic volcanism, Ghose (1976, with references to earlier work) has recorded basanites, nephelinites and carbonatites of comparable age broadly aligned in two major rift zones in western India. Although younger than the earliest Deccan tholeiites of central India, these strongly alkalic rocks predate other Deccan tholeiites in western India and thus belong to the same overall igneous cycle.

10.7.5. Tasmanian and Ferrar diabases

Intrusive tholeiitic diabases of Jurassic age occur on a massive scale, comparable to that of flood basalt eruption, in Tasmania and Antarctica. The predominance of shallow intrusives rather than extrusive rock reflects merely the presence of thick flat-lying sedimentary sequences that favour sill development at the time of eruption (see pp. 60 and 61) and probable considerable post-depositional erosion of lava, some of which is in fact known to have accompanied the Antarctic sills.

The Tasmanian diabases, emplaced ~165 Ma ago, comprise sills that commonly reach or exceed 300 m in thickness, oblique transgressive inclined sills and dykes, all intruding gently-dipping Permian—Triassic sediments over a preserved area of $1.5 \cdot 10^4$ km^2. In places the sills display spectacular differentiation in situ, of a type akin to that of the Palisades diabase sill (see Chapter 7), into a series of differentiated rocks ranging to fayalite granophyre (McDougall, 1962). The parental magma of the intrusions, as indicated by the composition and petrography of chilled margin material, was quartz-tholeiite commonly containing microphenocrysts of orthopyroxene on intrusion.

The Ferrar Dolerite Group, first described by H.T. Ferrar in 1907, comprises sills, inclined transgressive sheets, dykes and lavas, emplaced in mid-Jurassic times between 167 and 147 Ma ago, into a 2-km thick sequence of block-faulted but essentially still flat-lying deltaic sediments that overlie a Palaeozoic peneplain. One massive sill has been emplaced along the actual horizon of the unconformity. The mafic rocks crop out over a distance of 1600 km in Victoria Land and may well extend across the continent in the Trans-Antarctic Mountains for a total distance of 4700 km, and apparently include the gigantic Dufek layered intrusion. The aggregate thickness of preserved mafic rocks in any section generally ranges up to 1 km. Sills predominate, ranging in individual thickness from 100 to 500 m, and like the Tasmanian examples they commonly show marked differentiation in situ (Gunn, 1966). Chilled marginal material shows a grouping into three distinct tholeiitic types: an olivine tholeiite containing olivine micropheno-crysts and slightly olivine-normative; a hypersthene diabase containing hypersthene, augite, and plagioclase microphenocrysts; and a pigeonite diabase containing pigeonite, augite, and plagioclast microphenocrysts. Average major-element analyses are included in Table 10.7. These chilled marginal rocks may well be "parental" to the individual differentiated sill rocks that they enclose, but their chemistry and phenocryst assemblages very strongly suggest that the last two, if not all three, types are derived by fractionation of a true parental magma to the series that may approximate to the olivine tholeiite type in view of the latter's relatively higher MgO content (this dual use of the term parental has been mentioned before in Chapter 8, p.254).

Initial $^{87}Sr/^{86}Sr$-values of both the Tasmanian and Ferrar diabases lie in the narrow range 0.711—0.712, coincidentally comparable to the median value of kimberlites. This is a puzzlingly high figure for voluminous mafic rock, possibly explicable in terms of crustal assimilation, or alternatively, or perhaps more probably in view of their gross chemistry, by derivation from a mantle source of an unusually high Rb/Sr ratio (Compston et al., 1968). These values are at the high end of the variable and high range of values of initial $^{87}Sr/^{86}Sr$ ratios in flood basalts in part already noted, viz. 0.7035—0.7039 (Picture Gorge), 0.7045—0.7080 (Yakima), 0.7042—0.7125 (Karroo), 0.7051—0.7080 (Deccan), 0.7057 (Paraná) and 0.708—0.712 (Snake River; see Leeman and Vitaliano, 1976).

10.7.6. Blosseville Coast basalts of east Greenland

Basalts of the Blosseville coast between Scoresby Sund, latitude 71°30′ N, and Kangerdlugssuaq, latitude 69°N, have a present-day outcrop area of $5.4 \cdot 10^4$ km^2. Inclusion of further basalt localities to the north and south and a probable western extension beneath the Greenland ice sheet would raise this figure to $17.5 \cdot 10^4$ km^2, with an average thickness of around 3 km (C.K. Brooks et al., 1976). Eruption was confined to a rather short period, not exceeding 3 Ma, within early Tertiary time. In these respects the Blos-seville Coast basalts are typical flood basalts. The bulk of the sampled lavas appear to have a very uniform tholeiitic composition (see Table 10.7). Quantitatively minor varieties include picritic lavas with up to 35 wt.% modal olivine among the earliest lavas, and more evolved types in some of the younger lavas.

Although most of the flood basalts as exemplified above are tholeiitic in character, even a cursory examination of their major-element composition reveals, however, that they are far from uniform. In some sequences a pro-gression from older, relatively more alkaline, types to younger, less alkaline types can be discerned (e.g., Columbia River basalts). Some variation within the predominant tholeiites can be accounted for, notably in the case of the Ferrar diabases, if different analyses represent magmas belonging to a series plausibly linked by processes of fractional crystallization. Then to meaningfully compare one series with another, one should compare inferred parental magma compositions of the series, or, if this is not practical owing to the apparently evolved nature of much flood basalt material, compare compositions at an equivalent stage in series evolution, i.e. at some comparable SiO_2 content, MgO content, differentiation index, or M-value. No matter what method is adopted one is left with fundamental chemical differences between and within provinces, notably in contents of K_2O, TiO_2 and P_2O_5, among major elements. The Tasmanian and Ferrar diabases, for example, are much poorer in these elements than the Nuanetsi lavas of the Karroo which are conspicuously rich in K_2O, etc., even in rocks of highly magnesian compositions. Again, within the Karroo province itself, the Lesotho basalts, situated ~ 800 km south of the Nuanetsi—Lebombo focus, are, relative to the Nuanetsi lavas, poor in K_2O, etc. These differences extend to, and indeed are enhanced by, differences in trace element compositions as shown by the compilation in Table 10.8.

To a first approximation, low nickel contents should reveal the extent of early olivine fractionation beginning with mafic magmas, which, assuming that they had attained equilibrium with mantle of an assumed average composition and using partition coefficients indicated by experimental work, may initially have contained around 240—400 ppm Ni. This value would be halved even by the fractionation of 6—12 wt.% forsteritic olivine or increased in a like proportion by the accumulation of olivine crystals (H. Sato, 1977). High chromium content should similarly attest relatively unfractionated mafic magmas. Against this crude index of early fractionation as measured by Ni and Cr contents it can be seen that there is a wide variation in the tenor of such typical incompatible elements as Rb, Sr, Ba and Zr. Whereas the Ferrar magma types could be plausibly linked by fractionation processes, there is a strong hint that within a group such as the Karroo we may be facing a similar situation as the contrast between MORB and OIT, i.e. essentially differing mantle source rocks including relatively primitive, but fertile, mantle.

In this context, Jamieson and Clarke (1970) documented and commented upon the range of potassium and associated elements in tholeiitic basalts, many of them of the flood basalt association, and concluded that

"the relationship between K_2O and MgO in the average analyses indicates that the range of K_2O content cannot be solely the result of near-surface differentiation"

and also that

"crustal contamination is considered not to be a generally significant factor in the development of continental tholeiites."

TABLE 10.8

Average concentrations in ppm of selected trace elements from representative flood basalts

	Columbia River basalts		Karroo basalts			Ferrar diabases			Blosseville Coast basalt
	1	2	3	4	5	6	7	8	9
Ni	44	19	763	170	73	249	85	53	79
Cr	79	38	875	280	317	352	142	59	148
Rb	11	33	—	—	—	12	30	50	2.5
Sr	241	280	1,200	706	190	100	126	138	287
Ba	440	460	925	630	256	157	232	376	68
Zr	160	200	285	196	85	53	83	157	151

1 = average composition of samples from type section of Picture Gorge basalts (McDougall, 1976, p.780, tables 1 and 2).
2 = average composition of samples from Lower and Middle Yakima basalts (McDougall, 1976, pp.781 and 782, tables 3 and 4).
3 = average composition of samples of olivine-rich basalts (MgO > 8 wt.%), Nuanetsi (Cox et al., 1967, p.1462, table 3, analysis D_3).
4 = average composition of samples of basalts with MgO between 5 and 8 wt.%, Nuanetsi (Cox et al., 1967, p.1462, table 3, analysis D_2).
5 = average composition of samples of basalts from Lesotho (formerly Basutoland) (Cox et al., 1967, p.1462, table 3, analysis F).
6 = average composition of olivine dolerite type, Ferrar Dolerites (Gunn, 1966, p.907, table 1).
7 = average composition of hypersthene dolerite type, Ferrar Dolerites (Gunn, 1966, p.907, table 1).
8 = average composition of pigeonite dolerite type, Ferrar Dolerites (Gunn, 1966, p.907, table 1).
9 = average composition of Blosseville Coast plateau basalt (C.K. Brooks et al., 1976, p.286, table 2, analysis 1).

The Karroo, Paraná, Ferrar, Deccan and Blosseville occurrences are all associated with Phanerozoic plate separation and possible plume activity. The tectonic setting of the much older Proterozoic Keweenawan lava pile is unclear (see Chapter 12). The Columbia River basalts have been interpreted in terms of eruption in a "back-arc" extensional environment (see Chapter 11) albeit on continental crust (McDougall, 1976). The nearby Snake River Plain basalts have been interpreted as the product of a plume situated at a former East Pacific Rise spreading plate boundary now overridden by continental crust of the American plate. Flood basalts should therefore be thought of as a polygenetic association comprising different igneous rock series* with further subdivision probably to come in the future.

---

*Richly illustrative of this kind of complexity even within one field is the survey of the geochemistry of the Newer Basalts of Victoria and South Australia by Irving and D.H. Green (1976).

Megatectonic considerations apart, continental flood basalts have in fact provided a rich field for exploring possible significant inhomogeneities in the upper mantle; see, for example, the discussion by C. Brooks et al. (1976) of the role of ancient lithosphere in young continental volcanism.

10.8. INTRUSIVE ROCKS OF ANOROGENIC CONTINENTAL TERRAIN

In considering igneous rock series and their possible relationship to concepts arising from the new global tectonics it is, of course, appropriate to focus attention on young lava suites. Not only is the probable relationship of such young igneous rocks to accompanying plate movements and accompanying inferred mantle processes more easy to ascertain but also young and correspondingly fresh lavas lend themselves to meaningful chemical analysis and determination of liquid lines of descent. However, we have already turned aside from an exclusive consideration of extrusive rocks to include evidence from the ophiolite association with respect to MORB and have also mentioned in passing the intrusive sill equivalents found in many flood basalt provinces. Intrusive rocks of epizonal affinities, and thus belonging to the "volcanic association", are relatively conspicuous in areas of older continental igneous activity and thus deserve mention here. For reasons given in Chapter 3, pp. 69 and 70, whereas mafic magma is commonly erupted at the surface (although, given the right conditions of young flat sedimentary cover, sill intrusion may be quantitatively important) the more felsic magmas commonly do not reach the surface but crystallize within the crust. Thus in older provinces, where lavas may have been largely or even completely eroded away, study of the exposed intrusive rocks of the suite may give a violently exaggerated picture of the relative abundance of felsic rocks in the igneous rock series in question, even to the virtual exclusion of mafic members. It should be remembered also that major mafic intrusions, where they might occur, are likely to be cumulitic and thus analyses of their constituent rock types will not be equatable to magma compositions. For these reasons, it is clearly intrinsically more difficult to reconstruct a full liquid line of descent from work in intrusive terrain.

The several provinces selected here — Oslo, northern Nigeria, Afar, White Mountains and Monteregian, the British Tertiary, Gardar — include, however, numerous features of petrological interest and significance and thus cannot be omitted from an account of continental igneous activity although overall considerations of space only permit abbreviated accounts. All of them contain fine examples of high-level epizonal ring dykes and associated cauldron-subsidence intrusions.

10.8.1. The Oslo region

A definitive account of this famous area by Oftedahl (1960) includes references to his own previous work and to the classic work of W.C. Brögger and subsequent work by O. Holtedahl and T.F.W. Barth. Abundant epizonal plutons intrude and preserve downfaulted blocks of a lava series within a graben structure some 250 km long and ~30—60 km wide, reminiscent in these dimensions of rift valleys. The intrusive rocks, which are broadly co-magmatic with the extrusive rocks and are penecontemporaneous with them, were emplaced and crystallized over a short time span around 276 Ma ago (Heier and Compston, 1969). The geology is shown diagrammatically in Fig. 10.8: note the circular outcrop patterns of intrusive rocks, displayed best by the relatively younger ring intrusions.

The lavas overlie a thin series of fossiliferous Lower Permian fresh-water sediments with interbedded tuffs. In contrast to an original inferred volume in excess of 6000 km^3, the lavas now have actual outcrop areas as follows:

	(km^2)	(%)
Basalt and trachybasalt	220	15
Rhomb-porphyry	1,160	80
Trachyte, rhyolite, tuff, explosion breccia	80	5

The basalts are of alkali basalt affinity with fairly high potassium content. The "rhomb-porphyries" are just-saturated intermediate rocks, silica contents in the mid-50's, that carry conspicuous phenocrysts of potassium-rich plagioclase of a sodic andesine to oligoclase range in composition. These plagioclases typically have as crystal faces only (110), (1$\bar{1}$0), and ($\bar{2}$01), thus accounting for their rhombic shape in cross-section, and crystals are sometimes mantled by alkali feldspar which is the sole feldspar phase of the groundmass. The rhomb-porphyries would appear to represent an equivalent fractionation stage to mugearite in the more common, less potassium-rich alkali basalt series. The rhomb-porphyries were extruded as voluminous flows presumably from fissure eruptions in two periods separated by a period of development of explosive centres and cauldron-subsidence intrusions. One flow unit (RP$_1$) can be recognized in separated outcrops over an area of 6000 km^2. It seldom has a thickness less than 100 m, and thus apparently had a volume of at least 600 km^3. The manner of eruption invites a comparison with the even more voluminous phonolite flows of the Kenya dome.

Outcrop areas of the intrusive rocks are as follows:

Fig. 10.8. Geological sketch map of the Oslo region [after a compilation by Petersen (1978, p.332, fig. 1), following earlier, therein quoted classic work].

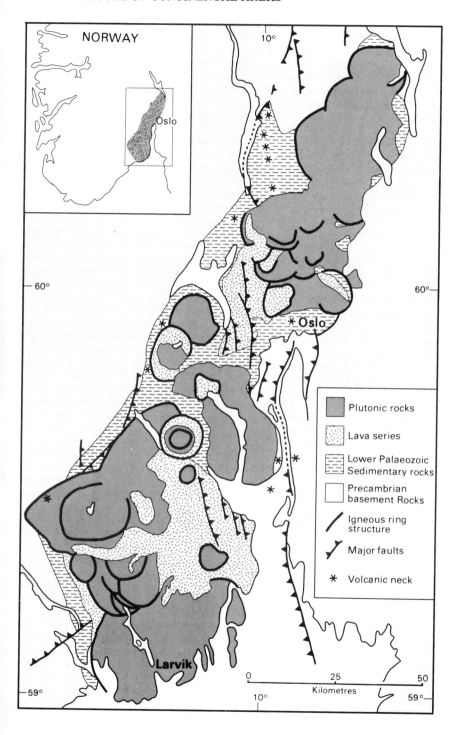

NORWAY

Oslo

10°

60°

60°

*Oslo

Plutonic rocks

Lava series

Lower Palaeozoic
Sedimentary rocks

Precambrian
basement Rocks

Igneous ring
structure

Major faults

Volcanic neck

Larvik

0   25   50
Kilometres

10°

59°

59°

	(km²)	(%)
Essexite	15	0.3
Larvikite and other monzonites	1,906	37.8
Nepheline monzonite and nepheline syenite	65	1.3
Syenite and nordmarkite	1,400	27.7
Ekerite and other peralkaline granites	821	16.3
Biotite granite	840	16.6

The essexites, the only basic component of the intrusive rocks, are found only as volcanic necks. Gravity measurements, however, show large positive anomalies within the graben that may well be due to the presence of large masses of basic rock within the crust. The felsic intrusives occur in classic epizone structures. The larvikites, compositionally the plutonic equivalents of the rhomb porphyries, contain a schillerized feldspar that proves to be an oligoclase—orthoclase cryptoperthite. Decisively undersaturated rocks are rare. Peralkaline syenites grading to more quartz-rich peralkaline rocks are abundant. This broad series of rocks has been regarded as a fractionation series in relation to which the (non-peralkaline) biotite granites appear to be extraneous. Surprisingly however, the biotite granites plot on the same Rb/Sr isochron as the other rocks, sharing a common age and fairly low initial ratio of 0.7041 (Heier and Compston, 1969) that would thus suggest a common mantle source.

In view of the relatively small volume and restricted spectrum of series composition of the Oslo rocks, an exact comparison with rift valley volcanism does not really appear very convincing in spite of a rift-valley-sized graben structure.

10.8.2. The Younger Granites of northern Nigeria

Some forty complexes of Jurassic age are grouped within a rectangular area ~400 × 150 km in northern Nigeria and display spectacular shallow epizone-type intrusions (Jacobson et al., 1958). The rock types are predominantly granitic in composition, and texturally range from coarse-grained granitic through porphyritic to intrusive rhyolite porphyries, the latter often showing autobrecciation and associated with tuffisite veining. Some lavas are also found, mainly in central downfaulted blocks. Compositionally the rocks comprise, in percentages of a total outcrop area of 6700 km², biotite granites (56 vol.%), undifferentiated rhyolites (19 vol.%), riebeckite granites (12 vol.%), fayalite-amphibole granites (8 vol.%), intermediate and basic rocks (5 vol.%). Although peralkaline rocks are relatively subordinate, the province is said to contain the greatest amount of riebeckite granite in the world.

Fig. 10.9. Geological sketch map showing the distribution of the peralkaline and other associated ring complexes of the Niger—Nigeria province, and the younger intrusive and volcanic rocks along the Chad—Cameroun line (based on Bowden and D.C. Turner, 1974, p.331, fig. 1; and A.A.G.S., 1963).

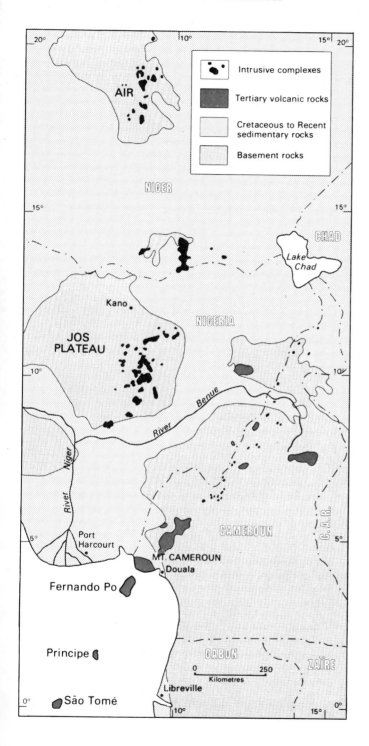

Fluorite is a common accessory mineral in most of the granites. Cassiterite is commonly found in greisened patches within the biotite granites and is also, together with columbite, a frequent accessory mineral constituent. Plagioclase of the granites is rarely more calcic than An_{15}, and hedenbergite and fayalite are common dark minerals. This mineralogy reflects the fact that the granites are consistently low in Ca and Mg, and that Na/Ca and Fe/Mg ratios are high. As well as high contents of F, Sn and Nb (consistently associated with a high Nb/Ta ratio) there are ubiquitous high contents of Y and REE. Zr is also high particularly in the peralkaline granites, and Hf/Zr ratios are very high. B is very low and tourmaline has never been recorded in any of the Younger Granites or their associated tin deposits. This suggests a different petrogenesis to that of the typical tin-mineralized granites of mesozone affinity where tourmaline is generally conspicuous. The data are in fact compatible with strong fractionation, although the repetition and alternation in individual complexes of riebeckite granites, some of which are strongly peralkaline and contain normative sodium metasilicate in addition to normative acmite, together with peraluminous corundum-normative biotite granites is not easily accounted for. An origin by crustal fusion is, however, preferred by Bowden and D.C. Turner (1974). This finds support in a surprisingly high $^{87}Sr/^{86}Sr$ initial ratio of 0.7212 ± 0.0040, yielded by a straight-line whole-rock isochron on two peralkaline and four non-peralkaline rocks of one complex.

The Younger Granites of northern Nigeria constitute one of three comparable provinces situated at widely spaced intervals more or less along the ninth meridian (see review by R. Black and Girod, 1970; Fig. 10.9). However, ages range widely from ~430 Ma in northern Niger (the Aïr province) to 340—290 Ma in southern Niger and to around 114 Ma for the Younger Granites of northern Nigeria. It has been suggested (Bowden et al., 1976) that this apparent progression may chart the slow drift of the African plate northwards over a relatively fixed sub-lithospheric heat source, i.e. a continental hotspot over a possible mantle plume. The relatively small volume of igneous rocks produced should be noted however, especially when compared with those of the Kenya and Afro-Arabian domes. Also significant is the irregular age pattern of igneous activity in the sub-parallel Cameroun volcanic line of Chad—Cameroun—Fernando Po—São Tomé (see Fitton and D.J. Hughes, 1977). Perhaps therefore we are witnessing in these provinces the reactivation of small amounts of igneous activity along fundamental lines of lithosphere weaknesses.

10.8.3. The White Mountain plutonic series of New Hampshire

Rocks of this grouping outcrop in thirty or so centres in a 250 km long NNW-trending belt. The term "series" has been traditionally applied to these rocks but does not carry a modern compositional connotation. The linear trend is apparently continued seawards to the southeast by the New England seamount chain and to the west-northwest by the Monteregian Hills (Fig. 10.10). Recent age work (Foland and Faul, 1977; and a short review by Zartman, 1977) has decisively indicated, however, that the ages of the intrusive centres do not show any regular progression along the belt, thus eliminating any simple migrating hotspot hypothesis (Fig. 10.11). No graben structures or doming accompany the emplacement of the igneous rocks, which show a surprisingly long time span of ages ranging from ~235 to 100 Ma with apparent concentrations of magmatic activity at 180 and 120 Ma ago.

The total outcrop area of plutonic rocks is ~1100 km² and that of extrusive rocks (preserved only as downfaulted blocks in classic cauldron-subsidence complexes) is ~75 km². The lavas are said to comprise mainly alkali basalts, intermediate rocks, trachytes and rhyolites. The intrusive rocks, including numerous minor intrusions, show a comparable succession from basic to acid in each complex with granitic members greatly predominating.

The "Conway Granite", a fairly coarse, near-ideal hypersolvus granite, is a prominent late-state intrusive facies in about half of the individual complexes, although the absolute age of this rock type varies considerably from complex

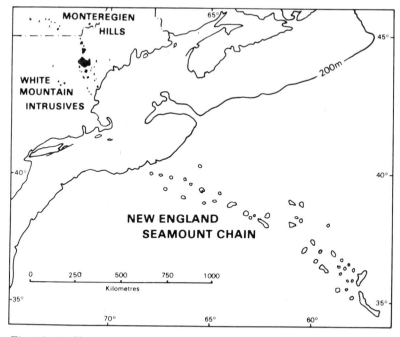

Fig. 10.10. Sketch map showing the location of the White Mountain intrusive centres in relation to the Monteregian Hills and the New England seamount chain (after Foland and Paul, 1977, p.891, fig. 2).

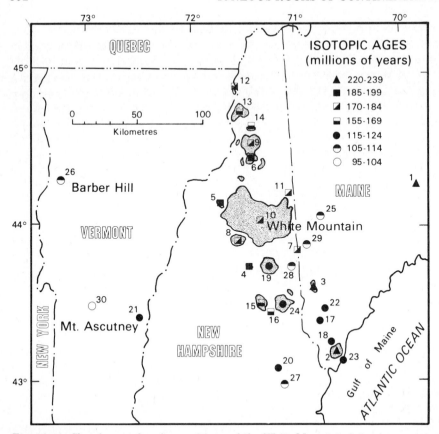

Fig. 10.11. Sketch map showing positions of the White Mountain intrusive centres and their ages (after Zartman, 1977, p.259, fig. 1).

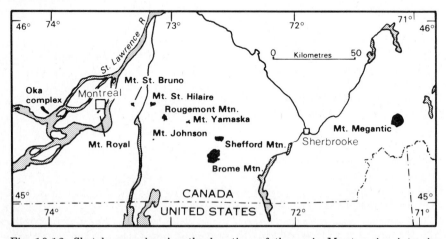

Fig. 10.12. Sketch map showing the location of the main Monteregian intrusive complexes (after Gold, 1967, p.289, fig. 9.1).

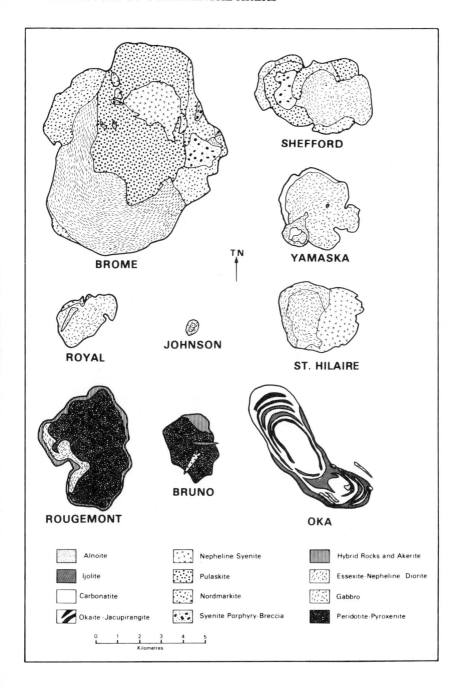

Fig. 10.13. Geology of the nine westernmost Monteregian intrusive complexes (after A.R. Philpotts, 1969, plate 1).

to complex. Overall proportions of intrusive rock types based on outcrop areas are as follows:

	(wt.%)
Gabbro, norite, diorite and quartz diorite	0.4
Monzonite	1.4
Feldspathoidal syenite	0.01
Syenite	19.7
Granite	78.3

Undersaturated rocks are thus exceedingly rare, and only a minority of the acid rocks are peralkaline. Some of the rare basic intrusive members carry orthopyroxene and are thus of a subalkaline affinity. Initial $^{87}Sr/^{86}Sr$ ratio is 0.7060 ± 0.0007, a value possibly just consistent with a mantle origin, but at least equally likely to be the result of some crustal contamination.

The outcrop belt of the White Mountain series intrusives is apparently continued to the WNW by the Monteregian Hills, which comprise ten centres outcropping over a linear distance of ~150 km (Fig. 10.12). These centres expose intrusive rocks ranging in age from ~125 to 95 Ma (Gold, 1967). In contrast to the White Mountain rocks, the Monteregian rocks are small in bulk and are decisively alkaline in character. They comprise peridotites, alkali gabbros, some of which are nepheline-bearing, syenogabbros, pulaskites, strongly alkaline syenites, nepheline- and sodalite syenites, carbonatites (Fig. 10.13), together with some nearby alnöite dykes and diatremes. The Monteregian Hills, the White Mountain series occurrences, and the New England seamount chain constitute a continuous belt ~2000 km long, along which, at least in the observable land area, igneous eruption has recurred over an appreciable time span. Rather than appeal to a migrating hotspot, it seems more reasonable to envisage sporadic emplacement of igneous rocks along this inferred lithospheric fracture zone, perhaps related to, or reactivated by, transform faulting during the formation of the North Atlantic. D.K. Bailey (1977) draws attention to the reactivation of alkalic igneous activity in the Monteregian area in no less than four distinct episodes, namely 1000—820, 565, 450 and around 110 Ma. The length of this time span lends further support to a control of continental magmatism by lines of lithospheric weakness rather than by transient plumes.

10.8.4. The Tertiary volcanic province of the Inner Hebrides, Scotland

An overview of the geology of this area (Fig. 10.14), which has remained classic for over a century, can be obtained from Richey et al. (1961) and Stewart (1965). Radiometric age dating has confirmed the early Tertiary age indicated by fossil plants in sediments interbedded with the lavas, and

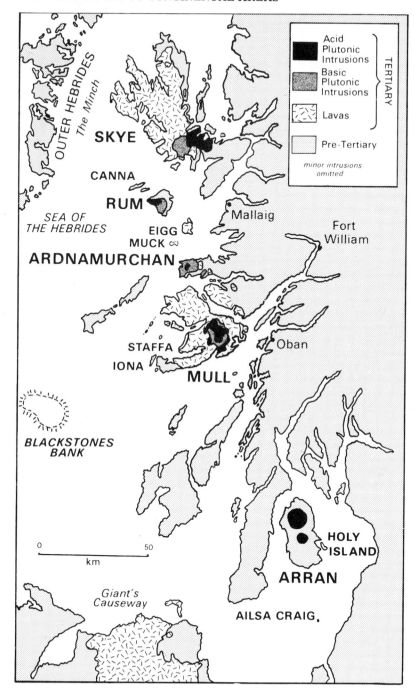

Fig. 10.14. Geological sketch map of the islands of the Inner Hebrides, Scotland, and adjacent areas showing positions of the Tertiary central complexes (after Richey et al., 1961, p.2, fig. 1; and McQuillin et al., 1975, p.180, fig. 1).

has furthermore shown that the time span of igneous activity was quite short within the time range of 60—52 Ma, with a much shorter time span for individual complexes (references in J.D. Bell, 1976). The province is younger than those described above, and in keeping with this the proportion of extrusive to intrusive rocks is higher:

	Outcrop area (km^2)
Lavas (overwhelmingly basaltic)	2,000
Intrusive rocks	
basic and ultrabasic	290
acid (mainly granite and granophyre)	500

The intrusive rocks, with the exception of some minor intrusions, are confined to the central complexes of Skye, Rum, Ardnamurchan, Mull, and Arran, where there is abundant evidence that they were the site of volcanoes. To this list should be added: (1) the entirely submarine Blackstone complex where the observation of a gravity high followed by a dredging evidence strongly indicates the existence of another complex (McQuillin et al., 1975); (2) the much eroded small islands of the St. Kilda group situated some 100 km to the west of the Outer Hebrides; and (3) three complexes in Northern Ireland. With the exception of the "off-axis" St. Kilda complex the other nine complexes are aligned more or less north—south over a distance of nearly 400 km.

A relatively early group of "plateau lavas", believed to have been erupted from fissures, is of an overall alkali basalt affinity, within which in Skye two slightly divergent sub-series have been recognized (R.N. Thompson et al., 1972). A later group of "central magma type" lavas are tholeiitic. Intermediate rocks are rare although an interesting study of cone sheets from the Ardnamurchan complex (Holland and G.M. Brown, 1972) revealed evolved tholeiite series compositions spanning the intermediate range of silica content. The acid rocks show widely differing initial $^{87}Sr/^{86}Sr$ ratios (Moorbath and J.D. Bell, 1965) ranging from values comparable with the basic rocks to distinctly high values suggesting an origin by crustal anatexis. A few minor intrusions of peralkaline compositions are known such as the riebeckite trachyte of Holy Island and the riebeckite microgranite of Ailsa Craig. The spectrum of igneous rock series thus ranges from alkali basalt through presumably transitional types to tholeiite, a spectrum associated in its later stages with active sea-floor spreading in the Afro-Arabian dome, for example. Although the age of these Scottish igneous rocks is suggestively coincident with the age of opening of the North Atlantic between Greenland and Scotland, evidence of crustal extension in the Hebrides is limited to the small aggregate thickness of a regional dyke swarm (maximum ~2 km) and to large-scale half-graben-type faulting in the Minch separating the Inner Hebrides from the non-volcanic Outer Hebrides, with sialic crust, however, continuously underlying the entire area.

10.8.5. The Gardar province

The Gardar province, one of the most notable alkaline igneous provinces in the world, has been recently reviewed by Upton (1974) and by Emeleus and Upton (1976). Within the time span of 1330—1140 Ma there was a history of complex faulting, subaerial sedimentation, lava extrusion, and the emplacement of about ten major intrusive complexes in a fairly compact area in southern Greenland, ~200 km in an east—west direction by ~70 km in a north—south direction. Total outcrop area of igneous rocks is ~1400 km² exposing predominantly intrusive rocks.

The lavas comprise mainly alkali basalts and hawaiites, although trachytes and comendites together with rare monchiquite and carbonatite flows have been recorded. Hypabyssal rocks are numerous and comprise mainly diabases (often containing apparently cognate plagioclase megacrysts) together with trachydolerites, phonolites, trachytes, quartz-trachytes, and comendites, and rare ultrabasic mica-rich lamprophyric dykes. The hypabyssal rocks thus exhibit a parallel range in composition to the preserved lavas.

The major intrusive centres comprise mainly felsic epizonal plutons, generally either granitic or foyaitic, and commonly peralkaline. Many of them show exceptionally well-developed layering indicative of a cumulate origin (J. Ferguson and Pulvertaft, 1963). In the Ilimaussuaq intrusion, possibly the most well known, there is an association of both undersaturated and oversaturated peralkaline rocks. Small bodies of carbonatite, including the famous siderite—cryolite deposit of Ivigtut, have also been recorded. Allowing for the universal predominance of felsic rocks in the geological plutonic record, these intrusives thus exhibit a parallel spectrum of compositions to the lavas and hypabyssal rocks.

10.9. CONCLUSIONS

From the above accounts it seems that although a plume model may be compatible with the tectonic and igneous evolution of the East African domes of different ages, difficulties are encountered in attempting to explain all continental igneous activity in terms solely of plumes. To illustrate our present attempts at understanding the range of igneous activity considered in this chapter, it is fitting to quote from the summary conclusions of a recent review by Turcotte and Oxburgh (1978):

"It does not appear that either the plume hypothesis or the propagating fracture hypothesis can easily explain all aspects of intra-plate volcanism. The principal advantages of the plume hypothesis are:

1. Ascending convection in a plume explains the production of magma by pressure-release melting (but deep plumes give too much melting).

2. The fixed plume hypothesis predicts the general trends of volcanic chains, and in

a number of cases, the rates of propagation.

3. Crustal doming is a natural consequence of the flow processes associated with plumes.

The principal advantages of the propagating fracture hypothesis are:

1. It explains the widespread occurrence of intra-plate volcanism, particularly the continued volcanism along volcanic lineaments.

2. It explains the common association of extensional tectonics with intra-plate volcanism.

3. It does not require anomalous asthenosphere beneath volcanic areas.

4. It provides an explanation for the migration of magmas through the lithosphere by corrosion fracturing.

With further refinement either hypothesis may be able to explain intra-plate volcanism or both hypotheses might be valid: the similarity of the products of intra-plate volcanism, however, suggests one process rather than two, or possibly an entirely new hypothesis is required."

Chapter 11

IGNEOUS ROCKS ABOVE BENIOFF SEISMIC ZONES

11.1. INTRODUCTION

The most conspicuous volcanic features on the Earth today are the chains of stratovolcanoes arranged in narrow elongate zones overlying Benioff zones. Their morphologies and flanking clastic debris reflect processes of rapid growth and rapid subaerial erosion, in some cases modified by late-stage caldera formation and the eruption of widespread ash flows and airfall tuffs. Clearly this is a tectonic setting quite distinct from those of plate separation and intraplate igneous activity that have been discussed in Chapters 9 and 10. The commonest rock extruded subaerially in this setting is andesite (indeed andesite is the commonest igneous rock type *seen* to erupt in the world, but is paradoxically rare in the geological record for reasons discussed in Chapter 2) and this observation may lull one into dangerous overgeneralizations. The fact is that supra-Benioff zone (SBZ) igneous activity is quite extraordinarily varied and demands correspondingly detailed discussion. One reason for this, of course, is that an understanding of present-day SBZ igneous activity may enable one in turn to confidently identify corresponding rocks in the geological record and thus by analogy to identify former convergent plate boundaries, an important megatectonic parameter, particularly when viewed in comparison with the relative impermanence of oceanic crustal rocks.

We shall begin therefore by examining the very young and apparently uncomplicated Scotia arc, which is composed mainly of rocks belonging to the island-arc tholeiite series. Then we shall examine a more complex sequence of island-arc tholeiite, calc-alkaline, and shoshonite series in Fiji where the crust has been thickened to average continental dimensions by inferred continued SBZ volcanism. From this follows an examination of polarity, that is to say the *transverse* variation in chemical composition encompassing the above spectrum of series that is shown in whole or in part by eruptive rocks *across* most arcs. At this stage, it is appropriate to pause and review the various attempts at classification and terminologies that have been inflicted upon SBZ rocks. An important additional complication is *back-arc spreading* that can apparently occur in both an oceanic and continental crustal environment, illustrated by the Lau Basin and the Basin and Range province, respectively. A consideration of the North Island of New Zealand affords a glimpse of the great complications caused by the important extra dimension of *time* involving relatively rapid changes in plate boundary regime and position; these complications can be resolved when dealing with geologically youthful rocks but could well remain indecipherable when occurring in the older geological record. Next, mention must be made of the apparent, not easily explicable, *lateral* variation in composition *along* the arc of the Lesser Antilles, including some low-potassium alkali basalts. The relatively great preponderance of acid rocks in the Andes and lack of back-arc spreading there may reflect a fundamental difference in plate motion relative to underlying mesospheric mantle, yet another possible complication. Yet another dimension to the study of SBZ igneous activity is provided by the varied *plutonic* rocks exposed in the older geological record and apparently originating in this tectonic setting.

Attempts to relate SBZ igneous activity to megatectonics illustrate to a high degree, on the one hand, an intuitive feeling widely held by igneous petrologists that such a synthesis is now possible but, on the other hand, some practical difficulties and apparent anomalies. In such a situation, rather common in igneous petrology at the present day, there will always be some who stress the broad similarities and there will be those who point to the perplexing anomalies, which of course are often the source of advances in scientific knowledge. Certainly at the time of writing no consensus on petrogenesis has been reached, but let us at least attempt to get some facts straight first, trusting that these are sufficiently representative and accurate to be of service. Most questions of petrogenesis are left until Chapter 13.

11.2. VARIATION AMONG YOUNG SBZ VOLCANIC ROCKS

11.2.1. The Scotia arc: an island-arc tholeiite series

The visible expression of this structure is the archipelago known as the South Sandwich Islands (Fig. 11.1). These islands form a slightly curved chain, some 450 km long, comprising eight active volcanic centres that emerge above sea level to form islands, groups of islands, and in one instance an eruptive seamount transiently above sea level. The volcanic centres are spaced out at intervals of ~60 km and the largest island group has a land area of only 200 km². The islands are in some cases simple volcanic cones rising up to 1500 m above sea level; parasitic cones are also common. The explosion index is ~40. The curvature of the chain (the French term, "guirlande d'îles" is evocative) is convex to the east, and ~160 km away, on this eastern or outer* side, there is an elongate deep or trench where depths up to 8000 m

---

*The terms *inner* and *outer* are in common use in island-arc terminology referring, for example, in the case of Japan to the inner Japanese Sea plus continent side on the one hand, and to the outer oceanic side on the other hand. In the context of the Scotia arc occurring entirely within oceanic crust, these terms relate to the convexity of the arc and more fundamentally to the "inward" dipping direction of the underlying Benioff zone (the Tonga—Kermadec trench—arc system is virtually straight).

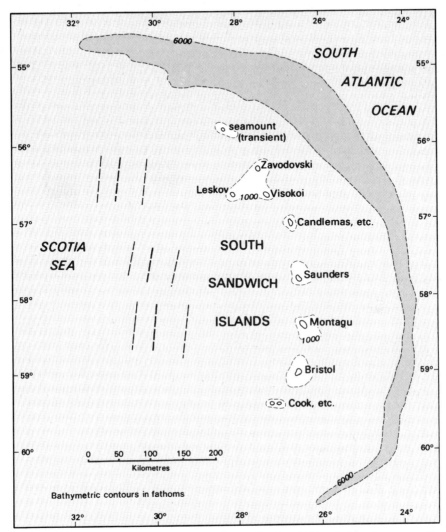

Fig. 11.1. The volcanic island arc of the South Sandwich Islands (also known as the Scotia arc) with their curious admixture of British and Russian names. The 6000-m depth contour effectively delineates the associated trench, and the 1000-m depth contour delineates the substantial submarine volcanic pedestals underlying the peaks that rise above sea level. The position of a spreading centre is indicated by *thick dashed lines* and that of the 1-Ma sea-floor isochron is shown by *thin dashed lines* (compiled from data in P.E. Baker, 1968, pp. 189—206; Barker and Griffiths, 1972, pp. 151—183).

have been recorded. Each island (or island group) represents merely the tip of a substantial pedestal of volcanic material built up from the level of the ocean floor. Between the volcanic centres sea depths are moderate, in places over 2000 m. A ninth island, Leskov, is situated ~50 km to the west of the curved axis of the chain, i.e. on the inner concave side furthest away from the trench; its volcanic products are significantly different from the others.

Shallow-focus earthquakes occur between the trench and the island-arc and intermediate-focus earthquakes* occur beneath the arc and for some distance to the west on the inner concave side where the deepest focus recorded to date is 170 km; the ensemble of foci define a Benioff zone dipping westwards under the arc.

The presence of a central trough in a bathymetric rough zone flanked by young oceanic crust as revealed by its magnetic stripe pattern indicates active sea-floor spreading along a north—south axis, ~240 km west of the island arc (Barker, 1972). A small plate, bounded by a trench to the east, a ridge to the west, and transform faults to the north and south, is growing and, like the Roaring Forties and the Flying Dutchman, moving implacably eastwards relative to the American and Antarctic continents. The *total* spreading rate appears to be ~7—9 cm/a, i.e. rather fast, and is asymmetric favouring the eastern, island-arc side (spreading rates are usually quoted as half the total spreading rate which is commonly observed, and usually assumed, to be symmetrical). Like many east-facing arcs the Scotia arc is thus associated with *back-arc spreading.*

The observational record of volcanism in historical times is incomplete as the islands have been relatively little visited but all have every appearance of contemporaneous volcanic activity such as the presence of uneroded land forms, fumarolic activity, ash deposits on ice, etc. In keeping with this overall youthful aspect two potassium—argon ages on what look like some of the oldest exposed volcanic products have yielded dates of 0.7 and 4.0 Ma, respectively. Nevertheless, a recent synthesis (Barker and Griffiths, 1972) indicates that the arc may be a little older than the 5—7.5 Ma previously suggested.

The predominant rock types are basalt and basaltic andesite; estimated frequencies, taken from the useful account by P.E. Baker (1968), are as follows:

	Frequency %
Basalt and basaltic andesite (<54 wt.% SiO_2)	67
Andesite (54—63 wt.% SiO_2)	29
Dacite (63—70 wt.% SiO_2)	3
Rhyolite (>70 wt.% SiO_2)	1

A striking feature of the petrography of the lavas is the rarity or lack of phenocrysts resulting in sparsely porphyritic or aphyric lavas; even the dacites are dark, commonly aphyric rocks. This, of course, contrasts with the

---

*Shallow-focus earthquakes are those which originate within the crust and uppermost mantle down to depths of 70 km; intermediate-focus earthquakes originate in the mantle at depths between 70 and 300 km; deep-focus earthquakes originate at still greater depths down to a maximum recorded depth of ~750 km. Intermediate- and deep-focus earthquakes are known only from Benioff zones.

characteristically porphyritic habit and appearance of many island-arc rocks of calc-alkaline affinity (see p.376). The commonest rock type is an aphyric basaltic andesite containing plagioclase, augite, and some orthopyroxene (although some of the basaltic andesites have a sub-calcic augite as the groundmass pyroxene). Phenocrysts where they are present comprise plagioclase, olivine and augite. Of these, plagioclase is the most abundant and commonly exhibits oscillatory zoning in the bytownite—labradorite range. Some of the plagioclase phenocrysts are, however, as calcic as anorthite. Average chemical compositions are reproduced in Table 11.1.

The basalt members of the series have similarities to ocean ridge tholeiites such as low alkalis, low K_2O/Na_2O ratios, and low Fe_2O_3/FeO ratios among major elements and low Rb, Sr and Ba contents among trace elements. Furthermore, when analyses of the Scotia series are plotted on an *AFM* diagram they show a characteristically tholeiitic iron-enrichment trend in rocks of intermediate composition (Fig. 11.2), admittedly not as marked as in some tholeiites but distinct from rocks belonging to the calc-alkaline series that are characteristic of subaerial eruption in mature island arcs. Using a Harker variation diagram of total alkalis vs. silica the Scotia rocks plot well into the sub-alkaline field. These Scotia arc rocks thus provide a fairly clear-cut

TABLE 11.1

Representative chemical compositions of volcanic rocks from the South Sandwich Islands (Scotia arc) recast to 100 wt.% on a water-free basis

	1	2	3	4
SiO_2	51.2	60.0	65.4	73.7
Al_2O_3	18.1	16.2	13.8	13.7
Fe_2O_3	2.7	2.3	1.3	0.7
MnO	0.2	0.2	0.2	0.1
FeO	7.4	5.5	7.1	2.2
MgO	6.2	2.6	1.4	0.5
CaO	11.0	6.9	4.7	3.3
Na_2O	2.0	3.3	4.0	4.7
K_2O	0.3	0.8	1.1	0.7
TiO_2	0.8	0.9	0.8	0.3
P_2O_5	0.1	0.2	0.2	0.1

1 = average of ten basalts from South Sandwich Islands (P.E. Baker, 1968, p.199, table 2, analysis 2).

2 = average of six andesites from South Sandwich Islands (P.E. Baker, 1968, p.199, table 2, analysis 4).

3 = average of two dacites from South Sandwich Islands (P.E. Baker, 1968, p.199, table 2, analysis 6).

4 = rhyolite pumice, Protector Shoal, South Sandwich Islands (P.E. Baker, 1968, p.199, table 2, analysis 8).

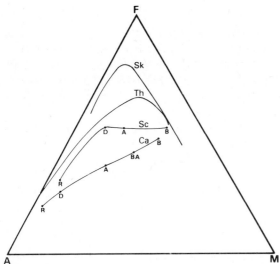

Fig. 11.2. *AFM* plot of calculated Skaergaard liquid compositions (*Sk*), Thingmuli lava compositions (*Th*), Scotia arc lava compositions (*Sc*), and average Cascades lava compositions (*Ca*). $A = (Na_2O + K_2O)$; $F = (FeO + Fe_2O_3)$; $M = MgO$ (all in weight per cent). $B =$ average basalt, BA = average basaltic andesite, A = average andesite, D = average dacite, and R = average rhyolite (after I.S.E. Carmichael, 1964, p.448, fig. 3; P.E. Baker, 1968, p.199, table 2).

example of an *island-arc tholeiite* (IAT) series, also known as low-K tholeiites (LKT), although a few of the Scotia analyses plot just within the field of the high-alumina basalt series, a proposed subdivision of the subalkaline field (see p.384). Note that the overall analogy chemically to oceanic tholeiite is so great that to postulate any fundamentally different origin for the parental basic member of the series would be unreasonable.

Petrographically the rocks of Leskov Island, geographically anomalous in that it is situated ~50 km farther from the trench than the other islands of the archipelago, are distinct and are more akin to calc-alkaline rocks in that they include typical porphyritic two-pyroxene andesite. As well therefore as affording an example of island-arc tholeiite series rocks in a young island arc, the Scotia arc also gives us a hint of *transverse* variation in rock type, a feature that is much more in evidence in other island arcs, e.g., Japan and Sunda (see p.380).

The abundant basaltic andesites and intermediate rocks in the Scotia arc series imply (given a presumed parental basaltic magma) that some appreciable fractionation has occurred, as indeed do the low Ni and Cr contents of 14 and 34 ppm, respectively, of rocks comprising even the "average basalt". The exponential decrease in observed abundance of more highly fractionated products with higher SiO_2 contents is in line with theoretical predictions based on a model of fractional crystallization. The relatively minute amount of acid rocks (only one example of a rock of rhyolitic composition is so far known) contrasts sharply, for example, with the situation in Iceland suggesting that there may have been relatively less opportunity for *high-level* crystal fractionation processes in the Sandwich Islands. Alternatively, the relatively abundant Iceland rhyolites may owe their existence to some other cause than fractionation such as refusion within a relatively thick basaltic crust in an area of high heat flow.

Similar low-potassium tholeiitic series are known from the leading edge (nearest the trench) of older and more complex SBZ volcanic belts. Two examples with analytical data include a basalt to rhyolite series of the Talesea peninsula of New Guinea reviewed by Carmichael et al. (1974, pp. 536—538), and the easternmost Nasu—Kuril belt of low-alkali tholeiites and associated intermediate rocks of Hokkaido, the northernmost major island of the Japanese archipelago, well documented by the pioneering work of Katsui (1961, with references to earlier work).

11.2.2. Fiji — a product of mature island-arc igneous activity comprising rocks of the island-arc tholeiite, calc-alkaline, and shoshonite series

Whereas the dimensions, bathymetry, available ages, volume and composition of eruptive products, and crustal thickness of the Scotia arc are all entirely compatible with it being a young arc of relatively simple magmatic derivation, the Fijian Islands have clearly had a longer and more complex history. The geochemistry of the rocks of Viti Levu, largest island of the archipelago, centered at $18°S, 178°E$, dimensions $\sim 130 \times 80$ km, has been the subject of a study by Gill (1970). Crustal thickness is $\sim 25—30$ km but there is no evidence to indicate that this represents a fragment of previous continental crust; on the contrary, the island is apparently built of the accumulated products of island-arc igneous activity, extending over a period of time from late Eocene to Pliocene, roughly 30 Ma, a history considerably longer than that of the Scotia arc to date. Though now quite separated from the active Kermadec—Tonga and New Hebrides trench—arc systems to the east and west, respectively, by recent back-arc spreading and transform faulting and no longer actively volcanic, and indeed no longer underlain by a Benioff seismic zone, igneous and stratigraphical similarities to other circum-Pacific island arcs leave little room to doubt that the construction of the Fiji archipelago was effected in an island-arc environment.

The exposed rocks, comprising volcanic rocks, some comagmatic intrusive rocks and volcanogenic sediments had been divided into seven groups by conventional mapping on the basis of unconformities and stratigraphical relationships. They have been regrouped by Gill on a geochemical basis into three and recognized as belonging to the following three igneous rock series:

(1) Island-arc tholeiite series (mainly submarine; the oldest).
(2) Calc-alkaline series (subaerial).
(3) Shoshonite series (subaerial; the youngest).

Although present outcrops of the three groupings show a spatial distribution, in Viti Levu there is also a clear temporal sequence. Intermediate rocks of the calc-alkaline series predominate in the subaerial record and basic rocks of the island-arc tholeiite series predominate in the submarine record. Comparable series are recognizable in many other SBZ provinces.

11.2.3. Comparison of petrographical and chemical features of rocks belonging to the island-arc tholeiite series, calc-alkaline, and shoshonite series

Obvious differences between these series, in Fiji and elsewhere, are that there is an upward sequence from submarine to subaerial volcanic activity and that there are greater proportions of intermediate and acid volcanic rocks in the calc-alkaline and shoshonite series than in the island-arc tholeiite suite. A crude estimate of the volumetric proportions of the three series in mature island arcs has been made by Jakeš and White (1971): as much as 85% may belong to the island-arc tholeiite series, some 12.5% to calc-alkaline series, and only 2.5% to the shoshonite series. These proportions can be further subdivided on the basis of the observed average frequency of different members within series as represented in Table 11.2.

In distinction to the island-arc tholeiite series, the calc-alkaline series (here used broadly synonymously with the term "*orogenic andesite series*" and including "*high-alumina basalt*" as its basic member) comprises relatively large amounts of intermediate rocks. There are usually conspicuously porphyritic with phenocrysts of plagioclase and both pyroxenes (together with hornblende in relatively more alkali-rich varieties) set in groundmasses with distinctive textures (see Chapter 5). Plagioclase phenocrysts are commonly conspicuously oscillatory-zoned and corroded, features they share with plagioclase phenocrysts of intermediate members of the island-arc tholeiite series. Plagioclase phenocrysts are, however, generally more abundant in calc-alkaline series rocks. Groundmass ferromagnesian pyroxene, particularly in intermediate members of the calc-alkaline series, is more likely to be hypersthene than the pigeonite or sub-calcic augite of island-arc tholeiite series rocks.

TABLE 11.2

Relative abundances of members of the island-arc tholeiite, calc-alkaline, and shoshonite series in mature island arcs

Series	"Basalt" member, <53 wt.% SiO_2	"Andesite" member, $53-62$ wt.% SiO_2	"Dacite" member, >62 wt.% SiO_2
Tholeiitic	42.5	29.7	12.7
Calc-alkaline	1.6	7.0	4.0
Shoshonitic	1.3	1.0	0.2

The shoshonite series is characterized by predominant basic to intermediate varieties containing labradorite and augite phenocrysts. If the matrix is not too glassy, plagioclase microlites can be seen to be mantled by sanidine; leucite may occur in small amounts, although its presence is not part of the essential definition of shoshonite. Relatively more primitive, though not necessarily accumulative, members of the shoshonite series with higher contents of magnesia carry olivine phenocrysts and are then known as absarokite. Relatively more evolved members, lacking in olivine, and possessing a lower colour index than shoshonite are known as banakites and resemble latite.

In major-element chemistry the most obvious distinction between the three series is an increasingly high alkali content from the island-arc tholeiite series through the calc-alkaline series to the shoshonite series, K_2O showing proportionately the greater increase. K_2O content is in fact the main major-element discriminant between the series. Al_2O_3 tends to be high in basic and intermediate members of the calc-alkaline series. TiO_2 is noticeably and characteristically low throughout, nearly always less than 1.2 wt.%, and commonly less than ~ 1.0 wt.% even in mafic members of the three series (even although incompatible elements in general are higher than, for example, in MORB where TiO_2 is similarly low). Some representative analyses are given in Table 11.3. The "average andesite" of R.A. Daly and the "average Cenozoic andesite" of F. Chayes are widely quoted but overgeneralized figures that probably give too little weight to common and relatively basic two-pyroxene andesites and too much weight to relatively evolved and/or alkalic andesitic rocks. Also included in Table 11.3 are the most basic representatives of analyses quoted by Kuno of tholeiite, high-alumina basalt, and alkali basalt types from a SBZ setting in Japan. These can be compared with representative SBZ tholeiitic basalt, high-alumina basalt, and shoshonite of Jakeš and White (1971).

In comparing trace-element contents of the different SBZ series, as indeed with major-element content, we also face the familiar problem of disentangling variations *within* a series from intrinsic differences *between* different series. A laudable attempt to cope with this by comparing trace-element contents at comparable SiO_2 levels within the three series, using data from many sources, is a compilation by Jakeš and White (1972, p.37) reproduced in part in Table 11.4. From these data it is apparent that the tenor of Ni, Cr and V is comparable in mafic members of all three series and shows a marked depletion with increasing SiO_2 content in each; Ni and Cr are low in mafic members when compared with other mafic rocks. Incompatible elements such as the large alkali and the alkaline earth elements, Rb, Sr and Ba, are markedly enriched towards the shoshonite end of the spectrum of series as are the LREE both absolutely and relatively to HREE. The island-arc tholeiite series, in fact, shows a flat chondrite-normalized REE distribution whereas the calc-alkaline and shoshonite series show progressively greater amounts of LREE enrichment.

TABLE 11.3

Chemical compositions and CIPW norms of average andesites and of representative mafic members of SBZ volcanic series

	1	2	3	4	5	6	7	8
SiO_2	59.59	58.17	51.25	49.51	47.95	51.57	50.59	53.74
Al_2O_3	17.31	17.26	14.73	18.19	16.46	15.91	16.29	15.84
Fe_2O_3	3.33	3.07	3.82	2.89	4.40	2.74	3.66	3.25
FeO	3.13	4.17	10.22	7.66	5.86	7.04	5.08	4.85
MnO	0.18	—	0.28	0.28	0.21	0.17	0.17	0.11
MgO	2.75	3.23	5.47	7.07	8.99	6.73	8.96	6.36
CaO	5.80	6.93	11.73	9.83	10.46	11.74	9.50	7.90
Na_2O	3.58	3.21	1.85	2.49	2.72	2.41	2.89	2.38
K_2O	2.04	1.61	0.26	0.48	1.09	0.44	1.07	2.57
TiO_2	0.77	0.80	0.81	0.64	1.09	0.80	1.05	1.05
P_2O_5	0.26	0.20	0.13	0.17	0.41	0.11	0.21	0.54
H_2O^+	}1.26	}1.24	0.11	0.45	0.37	}0.45	}0.81	}1.09
H_2O^-			0.02	0.27	0.28			
Total	100.00	99.89	100.68	99.93	100.29	100.11	100.28	99.68
Q	14.8	13.8	5.5	—	—	2.3	—	5.4
or	12.2	9.6	1.5	2.9	6.5	2.6	6.4	15.4
ab	30.7	27.5	15.6	21.2	23.1	20.5	24.6	20.4
an	25.5	28.3	31.0	37.3	29.6	31.4	28.5	25.3
di	1.6	4.2	21.6	8.7	15.8	21.4	14.0	8.8
hy	8.3	10.0	17.5	22.1	2.9	16.1	14.2	16.6
ol	—	—	—	1.8	12.7	—	4.5	—
mt	4.9	4.5	5.5	4.2	6.4	4.0	5.3	4.8
il	1.5	1.5	1.5	1.3	2.1	1.5	2.0	2.0
ap	0.6	0.5	0.3	0.4	1.0	0.3	0.5	1.3
M-value	44	45	41	54	61	55	65	59

1 = average andesite (Daly, 1933, p.16, table 1, analysis 49).
2 = average of 1775 Cenozoic andesites (Chayes, 1969, p.2, table 1).
3 = aphyric basalt, representative of island-arc tholeiite (Kuno, 1960, p.125, table 1, analysis 2; with source reference).
4 = olivine basalt, representative of high-alumina basalt (Kuno, 1960, p.125, table 1, analysis 4; with source reference).
5 = olivine basalt, representative of (SBZ) alkali basalt (Kuno, 1960, p.125, table 1, analysis 8; with source reference).
6 = representative island-arc tholeiite (Jakeš and A.J.R. White, 1971, p.226, table 2).
7 = representative high-alumina basalt (Jakeš and A.J.R. White, 1971, p.226, table 2).
8 = representative shoshonite (Jakeš and A.J.R. White, 1971, p.226, table 2).

Note how a study of the Fiji province reveals that although a majority of the subaerial volcanic rocks are calc-alkaline andesites (and rock types thus likely to be witnessed in contemporaneous SBZ volcanic activity) most of the volcanic pile is composed of basic and intermediate rocks of the island-arc tholeiite series. Furthermore, subaerial andesites are erupted at the present

TABLE 11.4

Selected trace-element abundances of volcanic rocks from island arcs (following Jakeš and A.J.R. White, 1972, p.33, table 2B)

	Island-arc tholeiites			Calc-alkaline association			Shoshonite association		
	basalt	andesite	dacite	basalt	andesite	dacite	basalt	andesite	dacite
SiO_2 (wt.%)	52	58	63	52	58	63	—	—	—
Rb (ppm)	5	6	15	10	30	45	75	100	120
Ba (ppm)	75	100	175	115	270	520	1,000	850	900
Sr (ppm)	200	220	90	330	385	460	700	850	850
K/Rb	1,000	890	870	340	430	380	200	200	200
La (ppm)	1.1	2.4	5.5	9.6	11.9	14	14	18	—
Yb (ppm)	1.4	2.4	2.7	2.7	1.9	1.4	2.1	1.2	—
La/Yb	0.8	1.0	2.0	3.6	6.3	10	6.7	15	—
Th (ppm)	0.5	0.31	1.6	1.1	2.2	1.7	2.0	2.8	—
U (ppm)	0.15	0.34	0.85	0.2	0.7	0.6	1.0	1.3	—
Ni (ppm)	30	20	1	25	18	5	20	—	—
V (ppm)	270	175	19	255	175	68	200	—	—
Cr (ppm)	50	15	4	40	25	13	30	—	—

day from volcanoes with extremely high explosion indices (commonly 95, even as high as 99) and thus liable to be in large part fragmentary on eruption. This fact, coupled with their exposed position on the flanks of rapidly-eroding stratovolcanoes, means that, in contrast to their conspicuous abundance in visible eruptions, andesite flows are relatively rare rocks in the geological record. Note, for example, the abundance of pillowed basaltic flows in a well-documented fossil island-arc sequence (Kean and Strong, 1975).

11.2.4. Polarity in areas of SBZ igneous activity

What is meant by polarity in this context is an observed subtle change in the compositions of eruptive rocks transverse to the direction of the trench. This, of course, is most obvious in areas of active SBZ volcanism but may also prove to be a significant tectonic indicator in the fossil igneous record. What is found is that nearest the trench the erupted rocks at the volcanic front have a low potassium content whereas those farther away have a progressively higher potassium content. Along with this variation in K content, the most obvious variant in major-element chemistry, there is an accompanying systematic variation in content of other elements, particularly trace elements, and to some extent in the composition of evolved members of the different series. Of course, *within* any igneous rock series there is a large increase in K_2O content and large variations in trace-element contents in more evolved members so that effective methods must be devised to distinguish differences in K_2O content, etc., *between* different igneous rock series. The three series already mentioned, the island-arc tholeiite, the calc-alkaline, the shoshonite, that we have already noted in a *temporal* sequence in Viti Levu, Fiji, can in fact be shown to have erupted concurrently in a *spatial* sequence across some Benioff-zone systems, although not all active arcs by any means exhibit this wide spectrum of compositional variation at one time.

Historically it is interesting to note, with respect to what we now conceive as the new global tectonics, the curiously prescient observations of du Toit (1937, p.4) who recorded amongst much other data the more alkalic compositions of volcanic rocks away from what he termed the leading edge (i.e. the west coast) of South America. Dutch investigators such as W.P. de Roever, summarized in van Bemmelen (1949), had similarly documented a distinct and apparently systematic change in composition, northwards across the Sunda arc in Sumatra, Java and islands of the Banda Sea, from normal and voluminous calc-alkaline rocks to highly potassic rocks such as the leucite basanites of the island of Batu Tara, situated ~50 km north of the main axis of volcanic activity. Follow up work on this interesting discovery was unfortunately disrupted by the events of World War II. In a series of papers in the late 1950's Katsui (1961, with references to his previous work) documented consistent chemical differences across the active SBZ volcanic zone in Hokkaido, the northernmost island of Japan. Katsui ascribed recognisably different igneous rock series to different parental magmas, namely "alkali-poor tholeiite", "weakly alkali tholeiite" and "alkali olivine basalt", and noted their distribution pattern in parallel curved zones at increasing distances from the associated trench system. However, the best-known demonstration of polarity in the Japanese and other SBZ systems has probably been the work of Kuno (1959, 1960, 1966). Reference will be made in the section dealing with classification (p.384) to some of the concepts arising from Kuno's work.

Hatherton and Dickinson (1969, with reference to previous work) drew attention to the existence in several SBZ belts of a sharp **volcanic front** situated ~150 km inwards from the trench or, probably more significantly, where the vertical distance, h, to the underlying Benioff zone attains ~80—100 km. No volcanic activity occurs between the trench and the volcanic front, and no less than ~75% of SBZ volcanism occurs within a relatively narrow zone, ~50 km wide inwards from the volcanic front. They dealt with the phenomenon of polarity by comparing the K_2O content of various igneous rock series at similar SiO_2 contents (e.g., their K_{55}-value equals the K_2O content at a point in a series where SiO_2 equals 55 wt.%). Plotting of K_{55}-values against h for different SBZ provinces shows a fairly consistent pattern of steady increase of K_{55} with h as indicated in Fig. 11.3. The more accurately the position of intermediate- and deep-focus earthquakes are determined the more closely the actual Benioff seismic zone appears to correlate with a

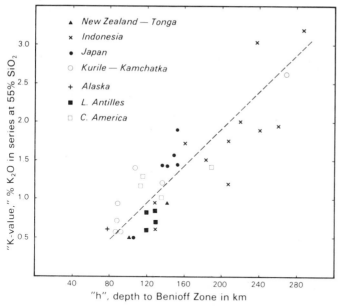

Fig. 11.3. Variation in K_2O content of andesite series rocks (as expressed by their K_{55}-values) with depth to Benioff zone (after Hatherton and Dickinson, 1969, p.5308, fig. 6).

narrow planar zone situated at the top of a thick mass of high $Q^{(*)}$, now believed to represent the top of a downgoing slab of cold lithosphere. This observed correlation between K_{55} and h suggests that a significant contribution to magma generation occurs at or close to this zone. Marsh and Carmichael (1974), for example, attempt to explain progressively increasing amounts of K_2O quite simply by the higher calculated K_2O activities of andesitic melts equilibrating at progressively greater depths with a high-pressure eclogitic assemblage containing sanidine, originally oceanic crust.

11.2.5. Review of terminology of SBZ igneous rock series

This has had a tangled history involving both muddled and changing mineralogical and chemical definitions and petrogenetic assumptions that may have not won general acceptance during the test of time. One consequence unfortunately is the persistence of overlapping terms and definitions. Analytical investigation has been on the whole curiously loath to disentangle

---

*Q is a dimensionless quantity expressing seismic measure of anelasticity within the mantle, derived from the observed damping of seismic waves. The reciprocal of Q is essentially a measure of the fraction of elastic energy dissipated per cycle. Q is roughly proportional to viscosity, and for a given mantle depth a relatively high Q-value correlates with relatively cold rock.

groundmass from whole-rock analyses so as to establish liquid lines of descent of these abundantly porphyritic rocks, implying the possibility, indeed probability, of accumulation or loss of a proportion of primocryst material in the finally erupted product (see, however, Ewart, 1976, for a survey of modern orogenic lavas and his conclusions that the majority of them have been modified by crystal fractionation processes dominated by plagioclase). Petrographical descriptions are non-uniform and often abbreviated, commonly referring only to the nature of mafic phenocryst phases, and assuming an implied and unspecified (but generally large) amount of plagioclase phenocrysts.

One valiant attempt at a simplified compositional classification based on silica and K_2O contents (S.R. Taylor, 1969), reproduced in graphical form in Fig. 11.4, was designed to include

"rocks typically associated with andesites in orogenic areas."

Note in this the prevalent tendency to redefine rock names not in terms of mineralogy but of chemical composition. Not all workers might agree on 53 wt.% SiO_2 content as a suitable division between basalt and andesite in view of the long established use of 52 wt.% as a division between basic and intermediate rocks; others would use the term "basaltic andesite" for SBZ rocks of andesitic aspect containing between 52 and 56 wt.% SiO_2. Another possible shortcoming for more general application is the neglect of basic rock types other than high alumina basalt. The classification, however, recognizes the great variance of K_2O in SBZ igneous rock series and anticipates in its ordinates later diagrams of the type of Fig. 11.6 (see p.385).

An apparent mineralogical difference was used by Kuno (1959) to separate a *pigeonitic series* from a *hypersthenic series* among SBZ rocks. The terms pigeonite and hypersthene here refer to the compositions of the pyroxene of

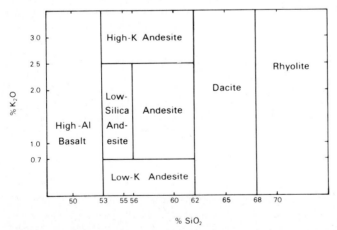

Fig. 11.4. Classification of rocks "typically associated with andesites in orogenic areas" (after S.R. Taylor, 1969, p.45).

the *groundmass* of basic members of the proposed series. This distinction may be very difficult indeed to make in fine-grained rocks, necessitating unusually high magnification objective lenses plus the use of corresponding condenser lenses in order to obtain diagnostic figures on small grains; the distinction is, of course, impossible to make in rocks with a glassy matrix or in rocks that have been degraded. Broadly speaking, the pigeonitic series of Kuno corresponds with the island-arc tholeiite series and the hypersthenic series corresponds with the subaerially more common orogenic andesite association.

The recognition, or rather the assumption, of the parental role of basic magma in SBZ igneous rock series has led, as in other igneous rock series, to the use of an inferred parental basic magma for a series nomenclature, as for example an island-arc tholeiite series, a high-alumina basalt series, and a shoshonite series in the Fijiian rocks discussed on p.375.

High-alumina basalt was originally defined by Kuno (1960) as typically erupted in an SBZ environment and lying compositionally between tholeiite and alkali basalt. A perceived high alumina content was defined graphically by plotting Al_2O_3 vs. alkali content within certain specified ranges of silica content. Kuno also quoted average analyses of the three types and in addition listed petrographical criteria. These three overlapping criteria are, however, difficult to apply consistently and unambiguously, particularly when the rocks have been degraded. Many petrologists would agree that the high-alumina basalt series includes the common "calc-alkaline" suite of "orogenic andesites".

The term *calc-alkaline* itself has been used in several different senses. Many users of this common term, if subjected to the embarrassment of Socratic questioning, would not be able to supply a strict definition, and even if they did, it might not satisfy others. Holmes (1920) recorded the usage of the confusingly similar term *calc-alkali* for rocks in which the dominant minerals are feldspars, hornblende and/or augite, and in which feldspathoids and soda inosilicates are absent. This definition of calc-alkali rocks represents a very broad grouping indeed, including all tholeiites and the great majority of orogenic andesites, and in practice is replaced by the term "subalkaline" in current usage. Peacock (1931) defined yet another term *calc-alkalic* to refer to an igneous rock series in which the content by weight of ($Na_2O + K_2O$) comes to equal the content by weight of CaO at a silica content between 56 and 61 wt.% (see reference to alkali-lime index in Chapter 8, p.260). Many orogenic andesite suites are calc-alkalic on this definition; some, however, have an alkali-lime index greater than 61 and are thus technically *calcic*, and a few have alkali-lime indices between 51 and 56 and are thus *alkali-calcic* according to Peacock's definitions (see McBirney, 1969, for examples). There has undoubtedly been a strong tendency to apply the term calc-alkaline to the intermediate members of what we now perceive as different SBZ series, as in Katsui (1961); see also comments in

G.M. Brown et al. (1977), concerning the petrographic similarities among intermediate members of different SBZ series.

Most petrologists today, however, equate a calc-alkaline series simply with subalkaline rocks occupying an intermediate position in a compositional spectrum of SBZ series lying between island-arc tholeiite series (syn. low-K arc tholeiite) on the one hand, and various alkali basalt series (including shoshonites) on the other hand.

A chemical distinction can be made on the basis of the familiar alkalis vs. silica variation diagram. Using this method specifically for SBZ igneous rock series, Kuno (1969) effected a three-fold division into an alkali basalt series, a hypersthenic (= high-alumina basalt series), and a pigeonitic (= island-arc tholeiite) series, the latter two by a subdivision of the subalkaline field (Fig. 11.5). It is worth noting in passing that the position of Kuno's line dividing SBZ alkali basalts on the one hand from the subalkaline high-alumina basalts and island-arc tholeiites on the other hand is not quite coincident, although very nearly so, with the comparable line drawn by G.A. Macdonald and Katsura (1964) to divide Hawaiian alkali basalt from tholeiite, and also extends to higher silica contents. An *AFM* diagram is also helpful in dis-

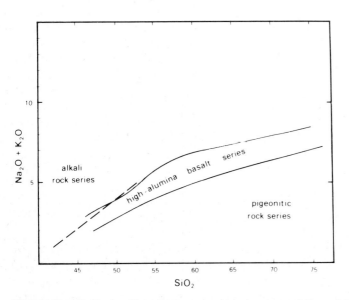

Fig. 11.5. Alkalis vs. silica diagram showing a proposed three-fold subdivision of SBZ vol-canic rocks into "pigeonitic series", "high-alumina basalt series" and "alkali rock series". The terms are those used by Kuno, (1969, p.14, fig. 1), in introducing this diagram. In practice, the term "pigeonitic series" is closely synonymous with the terms "island-arc tholeiite series" and "low-K tholeiite series"; similarly, the term "high-alumina basalt series" is apparently synonymous with Kuno's former term "hypersthenic series". The *dashed line*, inserted for purposes of comparison, is that used by G.A. Macdonald and Katsura (1964) to separate subalkaline Hawaiian rocks with tholeiitic mineralogies from those of alkaline mineralogies.

tinguishing island-arc tholeiite series rocks from high-alumina basalt series rocks as the former show a degree of iron enrichment in intermediate members which is characteristically lacking in the latter (see Fig. 11.2). Irvine and Baragar (1971) find that a plot of Al_2O_3 against normative plagioclase composition is also helpful for distinguishing what they term calc-alkaline series from tholeiitic ones.

Present practice in characterizing SBZ series also recognizes the relatively great variance in K_2O contents between series and quotes either K_{55}- and K_{60}-values or uses quite simply a K_2O vs. SiO_2 plot in which various series may be compared as in Fig. 11.6. Note in this diagram an apparent three-fold grouping into: (1) a low-K tholeiite series (IAT); (2) an intermediate K content series equating with typical subaerial andesitic series; and (3) more erratic high-K series (shoshonites). This three-fold grouping seems to be reinforced rather than blurred by more data plots (see Gill, 1970, p.195).

Not all SBZ alkali basalt series are as potassium rich as shoshonites, and the term shoshonite series, employed, for example, by Jakeš and White (1971) for the most potassium-rich rocks of the Viti Levu spectrum, has itself not been immune from criticism. Joplin (1968), in an informative review, shows that rocks that apparently fit petrographical definitions of shoshonite occur not only in well-defined SBZ environments, such as Indonesia, but also in the completely anorogenic environment of the East African rift valley.

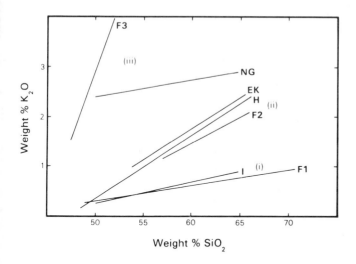

Fig. 11.6. Plot of K_2O content against SiO_2 content for various orogenic series: "first period", Fiji, *F1*; Izu Islands, *I*; "second period", Fiji, *F2*; Mount Hood, Cascades, *H*; eastern Kamchatka, *EK*; shoshonite association of New Guinea Highlands, *NG*; "third period", Fiji, *F3* (data selected from Wise, 1969, pp.969—1006; Gill, 1970, p.195, fig. 8; Jakeš and A.J.R. White, 1972, p.30, fig. 1).

(*i*) represents typical island-arc tholeiite series; (*ii*) represents typical subaerial andesite series; and (*iii*) represents inherently variable, more alkaline series including shoshonite series rocks.

The terms absarokite, shoshonite and banakite were all originally employed in the Yellowstone Park area* to describe Upper Tertiary rocks erupted a considerable distance away from the position of a former inferred Benioff zone. Nicholls and Carmichael (1969) provide an interesting commentary on the petrography, chemistry and nomenclature of these rocks in their type area. They demonstrate that far from belonging to an alkaline magma series these rocks are tholeiitic in character in possessing residual siliceous glass. Following Willmott (1972), current usage of a chemical definition of shoshonite, however, would be essentially as follows: alkaline rocks with K_2O/Na_2O ratios near or greater than 1.0 compared with a ratio of ~ 0.5 for alkali basalt, and with a low TiO_2 content. I.E. Smith (1972) documents high-potassium shoshonitic rocks of Miocene age in southeastern Papua, a time that coincided with initiation of a period of block faulting and vertical movements there. At the present time there does not seem to be sufficient evidence to justify adding on a tectonic rider to any definition of shoshonite.

The difficulties in connection with the nomenclature of SBZ volcanic rocks alluded to in this section continue to be of nuisance value and this ground has to be continually re-covered in research papers (see, e.g., a discussion in Johnson et al., 1978, pp. 97—102).

11.2.6. Back-arc spreading

In several localities of the Atlantic and Pacific Oceans, remote from the major ocean ridge systems, there are areas of very young oceanic crust known as *back-arc basins* (syn. *"marginal basins"*), that lie on the inner side of active volcanic arcs. One of them situated to the west of the Scotia arc has already been alluded to; a second underlies the Caribbean Sea, west of the volcanic arc of the Lesser Antilles. Pacific examples comprise the western part of the Bering Sea behind the western part of the Aleutian arc, the seas of Okhotsk and Japan behind the Kamchatka—Kuril and Japanese arc systems, part of the Philippine Sea behind the Izu—Bonin and Marianas arcs, and the Lau Basin west of the Tonga—Kermadec arc. Considered as a group (see Moberly, 1972) they clearly define a distinctive tectonic environment in which igneous rock is currently being erupted.

The account by Karig (1970) of the Lau Basin was one of the earliest to summarize such distinctive and otherwise anomalous features as high heat flow, thin or nonexistent sediment cover, very rough topography of linear ridges and troughs with relief ranging up to 1000 m, lack of a coherent pattern of magnetic striping, and low seismic velocities in the underlying mantle, and to attribute these features to previously unsuspected spreading in an environment distinct from that of mid-ocean ridges. The Lau Basin

---

*Readers are reminded of the excellent etymological sources of Holmes (1920) and Johanssen (1931—1939).

seems to be a particularly clear-cut example of a back-arc basin as it is situated above a currently active Benioff zone and contains very young oceanic crust, much of it less than 5 Ma in age. It thus seems to be a youthful feature that has split and divided a former arc into two — a western, at present non-volcanic ridge extending southwards from the Fijiian archipelago as the submarine Lau—Colville ridge and an eastern ridge, the presently active Tonga island arc.

The mechanism of spreading in this back-arc environment is imperfectly understood. Moberly (1972) has shown that the downgoing lithospheric slab may sink at an angle greater than the Benioff seismic zone, i.e. it may in part move by "falling" in addition to "faulting" with respect to overlying mantle. If this is true some compensatory mantle upflow must occur above the downgoing slab and this in turn could initiate magma production. J.T. Wilson (1973) also emphasizes that the *relative movements* of converging plates *with respect to underlying mesospheric mantle* may well play a vital role in determining whether back-arc spreading occurs, in sharp contrast to an Andean situation (see pp. 403—405). In a different connection, recognition of crust formed in marginal basins may also prove to be a vital link in the elucidation of the parentage and genesis of allochthonous ophiolite klippen, which invariably yield ages of formation only slightly older than that of their obduction.

Samples of the Lau Basin basalts commonly lack large phenocrysts and contain microphenocrysts of plagioclase and skeletal olivine in glassy, variolitic, or subophitic to ophitic textured groundmass, and show a limited range in composition. Their major-element contents are apparently indistinguishable from MORB. The average major-element composition of eleven unaltered basalt samples from the Lau Basin (J.W. Hawkins, 1976, p.290, table 4, analysis A) is as follows in weight per cent: SiO_2, 48.8; Al_2O_3, 16.4; Fe_2O_3, 2.0; FeO, 6.9; MgO, 8.6; CaO, 12.6; Na_2O, 2.4; K_2O, 0.18; TiO_2, 1.2; and P_2O_5, 0.08. Trace-element abundances (Ni, 160; Cr, 390; Sr, 200; and Rb <1 ppm) also resemble MORB values. Quoted $^{87}Sr/^{86}Sr$ ratios, however, range from 0.7020 to 0.7051, a range that both straddles that of average MORB and extends to considerably higher values.

Independent work on this classic back-arc spreading area by Gill (1976) suggests, however, the:

"Lau Basin basalts are transitional between ocean-floor and island-arc tholeiites, sharing with the latter their higher Rb, Ba, light REE and ^{87}Sr contents, and lower Ti, Zr and Hf contents."

Gill included, however, in his sampling mafic rocks from localities other than the central part of the Lau Basin, some of which could therefore conceivably be transitional in space and time with the preceding island-arc-type eruptive activity. Gill acknowledges that:

"further studies are necessary for meaningful comparisons".

11.2.7. The North Island of New Zealand — a cautionary tale

Not only does apparent back-arc spreading extend southwards from the Lau Basin into this area of crust of continental thickness with attendant voluminous eruption of acid rocks, but also a study of this area reveals several other complications in a simple model of SBZ volcanism making it worthy of detailed consideration.

The green and mountainous islands of New Zealand rise unexpectedly from the southern ocean on the present boundary between the Pacific and Australian plates between latitudes of 34°30' and 47°S[*]. Although one or two orders smaller than landmasses commonly regarded as continents, together they constitute the largest island mass not connected by shallow seas to a nearby continent. They have a surface area of $2.6 \cdot 10^5$ km^2 and are underlain by continental crust with thickness in excess of 30 km. Moreover, there are contiguous areas of shallow seas, including the Chatham Rise and Campbell Plateau to the east and southeast and the Lord Howe Rise to the northwest, underlain by crust generally 15—25 km thick and locally of average continental crustal thickness, that in the aggregate are several times larger in area.

The lateral dying out southwards of the Tonga—Kermadec Benioff zone underneath the North Island of New Zealand at a latitude of ∼40°N is well documented by seismic data on the foci of intermediate- and deep-focus earthquakes. Beneath the northern and central parts of the island a high proportion of foci lie between 150 and 250 km, and foci as deep as 350 km have been recorded. Curiously, and in contrast to many other island arcs, a very small proportion of the foci lie between 35 and 100 km. Collectively the foci define well a Benioff zone dipping west at ∼60° below the northern two-thirds of the North Island. In sharp contrast, however, south of approximately latitude 40°N intermediate- and deep-focus earthquakes have not been recorded, but numerous shallow earthquakes reflect movement along the northern splayed end of the famous Alpine fault, best developed in the South Island, a tear fault with lateral movement of at least 500 km in Cenozoic time that forms part of the boundary between the Pacific and Australian plates. Within the North Island active volcanism is confined to the northern and central areas above the Benioff zone.

"Basement" to the North Island volcanic zone, wherever it is not covered by lavas, ash flows, ashes, lahars and volcanically derived sediments of Pliocene and Quaternary age, is seen to be largely composed of Mesozoic sediments in which greywackes figure prominently. The total thickness of

---

*The earlier name of Aotearoa, bestowed on the islands by intrepid Maori mariners centuries before they were visited by pakeha (whites), literally means "land of the long white cloud" and refers to the common presence of cloud formed where the prevailing westerlies encounter this barrier to their passage.

these sediments is unknown but very great, perhaps 15 km. Lateral variation in them is consistent with a volcanic source area to the west. In the Coromandel peninsula and parts of the Auckland peninsula there is a later suite of andesitic and related volcanic rocks of calc-alkaline types, high-level intrusions and flysch-like volcanogenic sediments, all of early Miocene age. Among features of considerable interest to students of possible "fossil" island arcs in this Miocene assemblage are widespread alteration, polymetallic mineralisation, a pyrophyllite deposit, ophiolites, submarine lahars and slumping on a large scale, and the preservation of casts of an abundant fauna dominated by soft-bodied, sediment-eating organisms in penecontemporaneous flysch-like sediments. These may well therefore represent the remains of a volcanic arc trending in a different direction to the present plate boundary, and thus accounting not only for the apparently aberrant northwest-trending peninsulas but also for part of the submarine continuation of thick crust in that direction.

The Taupo volcanic zone of the North Island (see map, Fig. 11.7) occupies, and its ignimbrite members overlap the edges of, a structurally complex graben that seismic evidence indicates has undergone maximum aggregate subsidence of at least 3000 m. Numerous parallel normal faults, some still active, trend northeastwards and are in part coincident with the northeastern most branched splays of the Alpine Fault of the South Island. The recent faulting has in places resulted in the formation of more evident but smaller young graben within the overall major structure.

Although exact limits are difficult to determine because of the mantling by young volcanic material, the graben extends from the middle of the North Island into the Bay of Plenty and onward into the Lau Basin. The length along which active volcanism occurs including White Island is ~250 km, and the width of the structure ~50 km. The southwestern extremity of the graben is coincident with, and hidden by, the andesitic stratovolcanoes Tongariro, Ngauruhoe and Ruapehu, that overlie the southernmost extension of the underlying Benioff zone. The graben contains four youthful centres of rhyolitic volcanism, defined by high-level rhyolitic intrusions and calderas. The eruptive sequence began with andesite (first recorded in the Lower Pleistocene), continued with copius ignimbrites (first recorded in the Middle Pleistocene) accompanied by the formation of calderas and other volcano—tectonic depressions, followed by the eruption of more ignimbrite, pumice, ash, rhyolite domes and minor basaltic fissure eruptions that continue to the present day. Superimposed on this broad chronological time scale there appears to be a diachronous shift southwestwards of volcanism.

The evolution and morphology of the major young andesite volcanoes Tongariro, Ngauruhoe and Ruapehu has already been traced (pp. 36 and 37). Although some of the earlier flows of Tongariro contain hornblende phenocrysts, the majority of the andesites in the area and all the recent flows are pyroxene andesites containing varying proportions of plagioclase, hypersthene

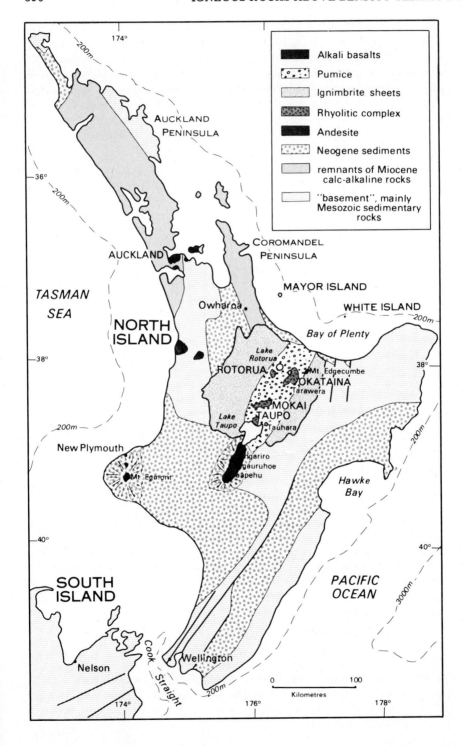

TABLE 11.5

Chemical compositions of representative New Zealand andesites

	Mount Egmont				Tongariro group			
	1	*2*	*3*	*4*	*5*	*6*	*7*	*8*
SiO_2	51.75	54.50	57.18	53.56	53.7	56.94	59.2	59.46
Al_2O_3	17.94	17.95	18.24	19.08	15.1	16.73	17.2	16.30
Fe_2O_3	4.16	4.25	3.49	4.16	2.5	1.35	3.1	1.26
FeO	4.72	3.83	2.89	3.52	6.65	6.72	3.3	5.26
MnO	0.17	0.17	0.16	0.18	0.16	0.14	0.05	0.15
MgO	4.24	3.37	2.55	3.92	7.6	4.72	3.7	4.46
CaO	9.37	8.36	7.09	8.31	10.5	8.05	6.5	7.03
Na_2O	3.43	3.61	3.95	3.21	2.4	2.80	3.3	3.14
K_2O	1.77	1.67	2.05	2.65	0.6	1.33	0.9	1.59
TiO_2	0.93	0.93	0.73	1.02	0.71	0.81	0.52	0.71
P_2O_5	n.d.	n.d.	n.d.	n.d.	0.12	0.17	0.09	0.18
H_2O^+	n.d.	n.d.	n.d.	n.d.	0.20	0.21	n.d.	0.25
H_2O^-	n.d.	n.d.	n.d.	n.d.	0.08	0.06	n.d.	0.11
Total	98.48	98.64	98.33	99.61	100.32	100.03	97.86	99.90

1 = average of four "basalts" (51—53 wt.% SiO_2) from analyses of Mt. Egmont andesites (Gow, 1968, p.181, table 3, analyses *3, 5, 6* and *11*).
2 = average of three "low-silica andesites" (53—56 wt.% SiO_2) from analyses of Mt. Egmont andesites (Gow, 1968, p.181, table 3, analyses *4, 8* and *10*).
3 = average of five "andesites" (56—60 wt.% SiO_2) from analyses of Mt. Egmont andesites (Gow, 1968, p.181, table 3, analyses *1, 2, 7, 9* and *12*).
4 = average of two "high-K andesites" (>2.5 wt.% K_2O) from analyses of Mt. Egmont andesites (Gow, 1968, p.181, table 2, analyses C and E).
5 = olivine andesite, Tongariro (Ewart, 1965a, p.90, table B-B, analysis *3*).
6 = average of two plagioclase pyroxene andesites, Ngauruhoe (Ewart, 1965a, p.90, table B-B, analyses *6* and *7*).
7 = hornblende andesite, Ruapehu (Ewart, 1965a, p.90, table B-B, analysis *8*).
8 = average of three plagioclase andesites, Tongariro and Ruapehu (Ewart, 1965a, p.90, table B-B, analyses *1, 2* and *4*).
n.d. = not determined.

and augite phenocrysts, with minor olivine in some. Uniaxial pigeonite has been noted as a groundmass phase and also mantling hypersthene phenocrysts. Chemically, they form a fairly compact group with SiO_2 content ranging from 53.5 to 60 wt.% (Ewart, 1965a). Analyses have been averaged and grouped in Table 11.5. One of the original analyses with 11.29 wt.% MgO and 12.72 wt.% Al_2O_3 has been omitted from this regrouping as its composi-

Fig. 11.7. Diagrammatic sketch map of the geology of the North Island of New Zealand showing localities mentioned in the text, and in particular the distribution of the Quaternary igneous rocks of the Taupo volcanic zone extending from Ruapehu to White Island (based on D.S.I.R., 1972).

tion suggests enrichment in accumulative pyroxene. To what degree crystal accumulation processes may have affected the bulk chemistry of the other analysed rocks is, of course, a problem common to the study of most andesites. Other occurrences of andesite are known from farther north in the central volcanic zone, e.g., Mount Edgecumbe; they are petrographically and chemically comparable to the Tongariro-group andesites and predate overlying rhyolitic rocks.

Standing alone, ~125 km to the west of the Tongariro group of volcanoes, rising from sea level to a summit of 2518 m, is Mount Egmont, the "Fuji" of New Zealand. Mount Egmont is a majestic stratovolcano, capped by ice, mantled by forest, and surrounded by an apron of laharic ring plain. Slightly to the north is the basal wreck of presumably a once comparable volcano, and to the north, again close to New Plymouth, are the Sugar Loaf Islands composed of dykes, plugs and volcanic necks, representing the erosional remnants of a third, yet older volcano now almost completely reduced to sea level. Although there are records of possible volcanic activity in the 19th century, Mount Egmont is currently dormant. The lavas of Mount Egmont comprise a fairly compact group, SiO_2 content of 51—59 wt.%, predominantly hornblende andesites as opposed to the predominantly pyroxene-bearing andesites of the Tongariro group. The Egmont andesites do not contain orthopyroxene but the more basic varieties contain up to 7 vol.% olivine, so that the petrographic varieties comprise olivine-augite andesite, augite andesite and augite-hornblende andesite. Sampling of pebbles collected from detritus derived from the growing volcano provided a wider spectrum than sampling from present outcrops of the volcano which is, of course, clothed only by its most recent eruptive products (Gow, 1968). Of the 22 available analyses, following the terminology of Taylor alluded to above, 4 would be high-alumina basalt, 9 low-silica andesite, 7 andesite, and 2 high-K andesite, and this grouping is a useful basis for averaging and presenting the analyses (see Table 11.5). Although somewhat erratic they are clearly higher in K_2O than the Tongariro andesites of the central volcanic zone, and have a K_{55}-value of ~2% as opposed to a K_{55}-value of just under 1% for the Tongariro andesites. This higher K_{55}-value correlates with the deeper position of the Benioff zone under Mount Egmont in the normal transverse polarity sense.

North of the Tongariro group of prominent andesitic stratovolcanoes, the Taupo volcanic zone presents a relatively much more subdued relief that belies considerable vertical movements along faults and reflects a complex and voluminous recent eruptive history mainly of rhyolitic rocks. Four extrusive centres, Okataina, Rotorua, Mokai and Taupo, have been recognized on the basis of the distribution of pyroclastic rocks, rhyolite domes, and most prominently by caldera structures, all except Mokai occupied by lakes. Of these the largest, Lake Taupo, affords magnificent views of the stratovolcanoes to the south; the most perfectly circular lake, expressive of the form of a young caldera, is Rotorua. The extrusion of rhyolite ash flows in batholithic volumes ($2 \cdot 10^4$ km^3 over an area of $2.5 \cdot 10^4$ km^2) preceded the emplacement of much smaller volumes of rhyolite domes in resurgent calderas. In this classic province of Pleistocene ignimbrites, one is impressed by the relative paucity of glassy ignimbrites and the predominance of soft ashy varieties. Much of the material is undergoing erosion and would appear destined to be preserved in the geological record as epiclastic material. Several individual ignimbrites had volumes up to 300 km³, one possibly 600 km³. A very detailed study of the lateral variation and cooling history within one ignimbrite unit has been made by Briggs (1976a, b). On an historical note, the first appreciation of the mode of origin of ignimbrite is generally credited to Marshall (1935) in connection with a rock type described as "lenticulite" erupted during earlier Miocene volcanism in the lower Coromandel peninsula. This lenticulite, which can still be readily examined in a quarry at Owharoa, is a remarkable eutaxitic rock that contains parallel black glassy discs of flattened pumice with flame ends in a gray, lithoidal matrix of shards and crystals; the structure of the rock clearly called for a novel origin distinct from that of a lava flow. Rocks of the numerous rhyolite domes comprise glassy, spherulitic and felsitic varieties, and exhibit autobrecciation and flow banding. These minor rhyolite masses are often closely clustered — in the Tarawera complex for example, situated at the southeastern edge of the Okataina caldera structure, there are 11 rhyolite domes, three of them associated with lava flows, and one plug. Rocks of coarser-grained texture such as granophyre and porphyritic granophyre with euhedral quartz and plagioclase phenocrysts in a granophyric matrix are known but occur only as fragments in reworked beds of pyroclastic material. Presumably they represent samples of material that crystallized at a slightly deeper level in the crust and that were brought to the surface by explosive activity*. Much of the central part of the Taupo volcanic zone is mantled by recent deposits of rhyolitic ash and pumice. Although these airfall beds frequently show reworking, individual beds can commonly be identified, measured and isopached; they provide useful datum horizons and can be related to recent eruptions in several centres. Where interbedded with fossiliferous sediments near the coast, age dating by using the spontaneous fission tracks in shards has enabled an absolute time scale of the Pleistocene to be constructed (Seward, 1974).

Andesite and dacite have been dredged from several of the seamounts that occur within an area ~40 km wide and up to 70 km offshore contiguous to and extending seawards the Taupo volcanic zone. The volcanically active White Island, ~2 km in diameter, is situated in this same area ~50 km offshore. It consists of a crater within walls rising to ~150 m. In the floor of this crater there is intensive fumarole activity. An entire working party of twelve miners exploiting the sulphur deposits was overwhelmed there by

---

*This kind of depth relationship between felsite and granophyre can in fact be seen in the eroded early Tertiary volcano of Rum, Scotland (C.J. Hughes, 1960).

a laharic flow in 1912. The extrusive rock is a dacitic andesite, SiO_2 content 62 wt.%, which appears to fall within an apparently continuous compositional spectrum embracing the various andesites of Tongariro and the more acidic rocks of the Taupo volcanic zone.

Petrographically, typical ignimbrites such as the Te Whaiti and Matahina ignimbrites (Ewart, 1965b; Ewart and Healy, 1965) contain phenocrysts of quartz, plagioclase, magnetite, ubiquitous small amounts of hypersthene and hornblende, and on the average less than 2 vol.% of lithic fragments. Rhyolites of the eleven domes of the Tarawera complex (Cole, 1970) contain phenocrysts of quartz, plagioclase (analysed range between An_{41} and An_{28}, optically determined zoning An_{53} to An_{16}), ubiquitous hypersthene and magnetite, together with hornblende and/or biotite in some. Dacites of five domes of Tauhara volcano, part of the Taupo complex (Léwis, 1968b), SiO_2 range 65—68 wt.%, contain phenocrysts of quartz, plagioclase (oscillatory and normal zoning in range An_{55}—An_{30}), ubiquitous hypersthene, together with augite, hornblende, magnetite, and perhaps surprisingly 1—3 wt.% of olivine (Fo_{83}) rimmed by granular orthopyroxene.

Representative analyses of these dacitic to rhyolitic rocks are given in Table 11.6. Lime content remains fairly high, even in the rhyolitic glasses of highest silica content; total FeO is low throughout. It seems that the acid rocks of the Taupo volcanic zone form a petrographically and chemically coherent suite of typically calc-alkaline character.

Recent eruptions of alkali basalt have created prominent landforms in the environs of Auckland, comprising small shield volcanoes, scoria cones, and maar-shaped features up to 1 km across, with relatively low rims of pyroclastic material formed by phreatic explosive activity in low-lying waterlogged ground*. Petrographically, the basic rocks are all conspicuously olivine-bearing with olivine contents in the range of 9—22 vol.%, and comprise olivine alkali basalt, augite-rich picrite basalt, and nepheline basanite, the latter found only as ejected blocks in some scoria mounds. Chemically, all the rocks are nepheline-normative (although the basalts have no modal nepheline) and have the low SiO_2, high TiO_2, high alkalis, and high Na_2O/K_2O of typically oceanic alkali basalts and associates such as those of Hawaii, from which

---

*Something of a hazard is posed by eruptive activity of this type. The last dated eruption occurred ~760 years ago, and the time span of the eruptive activity of this field is from ~60,000 years ago to the present day. It would be optimistic to assume that volcanic activity conveniently ceased in the Auckland area at the time of its intensive settlement by humans who now number some 600,000. In an interesting paper (Searle, 1964) the volcanic risk, taking into account such factors as the probable periodicity, size of built-up area at risk, area of devastation in one phreatic eruption, has been calculated at ~2% per century. Assuming that volcanic risk is a euphemism for death, this figure over a lifespan is coincidentally very close to the risk of being killed by vehicles in the more highly developed countries (~1.7% of deaths in North America are caused by motor accidents and ~20% of the population suffer serious injury in them). If one risk is acceptable so presumably is the other. Amen.

TABLE 11.6

Representative analyses of dacite and rhyolite rocks of the Taupo volcanic zone, North Island, New Zealand

	1	2	3	4	5	6
SiO_2	62.15	64.6	66.4	70.60	73.85	74.22
Al_2O_3	14.32	16.0	15.3	13.23	13.55	13.27
Fe_2O_3	1.64	4.1	2.2	0.75	1.25	0.88
FeO	4.33	0.85	1.7	1.39	0.60	0.92
MnO	0.15	0.06	0.06	0.11	0.05	0.05
MgO	3.97	3.0	2.0	0.30	0.30	0.28
CaO	5.89	5.1	4.3	1.56	1.53	1.59
Na_2O	2.49	3.20	3.51	4.58	3.71	4.24
K_2O	2.37	1.84	2.44	2.86	3.60	3.18
TiO_2	0.71	0.5	0.35	0.20	0.23	0.28
P_2O_5	0.12	0.14	0.11	0.04	0.05	0.05
H_2O^+	1.21	0.9	1.6	3.26	0.59	0.80
H_2O^-	0.29	0.2	0.2	0.70	0.38	0.23
Total	99.65	100.49	100.17	99.58	99.69	99.99

1 = dacitic andesite, White Island, Bay of Plenty (Ewart, 1965a, p.90, table B-B, analysis 12).
2 = dacite, Tauhara complex (Lewis, 1968b, p.672, table 6, analysis 2).
3 = dacite, Tauhara complex (Lewis, 1968b, p. 672, table 6, analysis 5).
4 = average composition of recent Taupo pumice deposits (Ewart and Stipp, 1968, p.704, table 1, analysis 3).
5 = average composition of Quaternary ignimbrites (Ewart and Stipp, 1968, p. 704, table 1, analysis 2).
6 = average composition of rhyolite lavas and domes (Ewart and Stipp, 1968, p.704, table 1, analysis 1).

they are claimed to be indistinguishable (Searle, 1960). An average analysis is given in Table 11.7.

Mayor Island is a recently extinct volcano, diameter 4 km with a broad pedestal 14 km across at the 130-fathom contour, situated in the Bay of Plenty some 90 km west of White Island, more or less half way between White Island and the Auckland district, and thus some distance to the west of the seaward extension of the Taupo volcanic zone. Its special interest lies in the peralkaline composition of its exposed rocks which are predominantly partly devitrified glassy porphyritic pantelleritic lavas and pumice (Ewart et al., 1968). The rocks contain phenocrysts of quartz, anorthoclase ($\sim Ab_{60}$), aegirine-hedenbergite, and cossyrite in a matrix of alkali feldspar, quartz, aegirine and sometimes other sodic dark minerals. An overall average analysis is given in Table 11.7 together with average Taupo rhyolite for purposes of comparison. The pantellerite contains some 30% normative quartz and several per cent each of normative acmite and normative sodium metasilicate. Compared to the Taupo rhyolite, the pantellerites have: (1) virtually the same Si,

TABLE 11.7

(a) Chemical compositions (wt.%) and CIPW norms (wt.%) of Auckland alkali basalt and Mayor Island pantellerite compared to Taupo volcanic zone rhyolite; and (b) selected trace-element contents (ppm) of Mayor Island pantellerite and Taupo volcanic zone rhyolite

	(a) Auckland alkali basalt and Mayor Island pantellerite compared to Taupo volcanic zone rhyolite					(b) Mayor Island pantellerite and Taupo volcanic zone rhyolite	
	1	2	3	4		5	6
SiO_2	44.8	53.10	73.4	74.22	Rb	143	108
Al_2O_3	13.86	15.85	9.05	13.27	Cs	5.5	3.3
Fe_2O_3	2.91	4.3	2.35	0.88	K/Rb	249	249
FeO	9.63	6.3	3.80	0.92	Sr	2	125
MnO	0.17	0.22	0.19	0.05	Ba	21	870
MgO	11.07	2.79	<0.01	0.28	Sc	0.8	4.7
CaO	10.16	7.14	0.21	1.59	Total REE	488	123
Na_2O	3.19	4.6	6.20	4.24	Eu	1.5	1.0
K_2O	1.09	2.09	4.25	3.18	Th	20.2	11.3
TiO_2	1.96	2.04	0.15	0.28	U	6.0	2.5
P_2O_5	0.55	0.63	0.01	0.05	Nb	72	5.6
H_2O^+	0.73	0.6	0.15	0.80	Sn	6.5	1.3
H_2O^-	—	0.7	—	0.23	Mo	18.3	2.6
CO_2	—	—	0.21	—	Zr	1,250	160
S	—	—	0.02	—	Hf	19.5	4.5
Total	100.0	100.36	99.99	99.99			
Q	—	1.6	32.0	33.8			
c	—	—	—	0.1			
or	6.5	12.5	25.2	19.0			
ab	12.7	39.3	23.0	36.3			
an	20.4	16.6	—	7.6			
ne	7.8	—	—	—			
ns	—	—	5.1	—			
ac	—	—	6.8	—			
di	21.6	12.3	0.9	—			
hy	—	6.1	6.7	1.3			
ol	22.3	—	—	—			
mt	3.7*	6.3	—	1.3			
il	3.8	3.9	0.3	0.5			
ap	1.3	1.5	0.02	0.1			
M-value	61.4	32.4	0.3	22.1			

1 = average composition of Auckland alkali basalt (Searle, 1960, p.25, table 1).

2 = trachybasalt, Mayor Island (Ewart et al., 1968, p.124, table 4, analysis 6).

3 = pantellerite, Mayor Island (Ewart et al., 1968, p.124, table 4, analysis 1).

4 = average rhyolite of Taupo volcanic zone (Ewart and Stipp, 1968, p.704, table 1, analysis 1).

5 = average content of selected trace elements in Mayor Island pantellerites (after Ewart et al., 1968, pp. 126 and 127, table 5).

6 = average content of selected trace elements in Taupo volcanic zone rhyolites (after Ewart, 1968, pp. 126 and 127, table 5).

*$Fe_2O_3/(Fe_2O_3 + FeO)$ recalculated to 0.2.

K, Rb and Cs contents, and K/Rb ratio; (2) somewhat higher contents of Na, Fe, Mn and REE, and highly charged cations such as Th, U, Hf, Nb, Sn and Mo; (3) approximately one order higher Zr; (4) somewhat lower Al and Eu (relative to normalized chondritic abundances of other REE); (5) approximately one order lower Ca and Sc; (6) one to two orders lower Ba; and (7) approximately two orders lower Mg and Sr. This typically peralkaline chemistry has been attributed to extreme fractionation first of relatively magnesium-rich dark-mineral silicates, followed by plagioclase, and then by anorthoclase (of a more potassium-rich composition than the magma). A possible clue to the igneous parentage of the pantellerites is the presence on Mayor Island of inclusions of porphyritic trachybasalt, containing phenocrysts of labradorite, olivine, titanaugite and alkali feldspar in a matrix of augite, magnetite, andesine and alkali feldspar, and taken to indicate a mildly alkaline alkali basalt parent. Ewart et al. acknowledge that some additional differentiation process such as volatile transfer may be involved, and that post-eruptive devitrification may have also resulted in some modification of the chemistry of the lavas.

Therefore, within a radius of less than 150 km, recent examples of calc-alkaline rocks can be found that are predominantly rhyolitic but include a series from andesite through dacite to rhyolite, distinctly higher potassium andesites, oceanic type alkali basalts, and peralkaline rhyolite. If we were to include volcanic rocks erupted within Neogene time, a time span that would probably not be easily resolvable by isotopic dating methods in Palaeozoic and older rocks, all the above rock types and more can easily be found within the yet smaller radius of 60 km. It is salutary to be aware of these complications in SBZ volcanism.

11.2.8. Basin and Range province

The Basin and Range province in its most classical development in Nevada, southernmost Idaho, western Utah, southeastern California and southern Arizona is very distinctive, characterized by numerous short parallel mountain ranges aligned approximately north—south and rising precipitously to heights of 1000 m or more above encircling piedmont slopes. This topography is the reflection of tilted blocks lying between numerous young normal faults, which may in places cut and tilt slightly older tilted blocks and faults in a complex manner (Proffet, 1977). Thornbury (1965, p.471) emphasizes: (1) the great width (1100 km) of the province at the Mexican border; (2) the transitional nature of the junction with the Colorado Plateau to the east; (3) that the Sierra Nevada and Wasatch range, apparently bounding a width of 800 km of typical Basin and Range province to the west and east, respectively, at higher latitudes, could themselves be regarded as extra large fault blocks in the same overall structural context; and (4) that faulting of a comparable nature and trend extends northwards into Oregon. Indeed, following Gilluly (1963), Christiansen and Lipman (1972, p.255, fig. 4) show a still wider distribution of basin—range faulting extending, for example, northward into Idaho and Montana and including structures to the east of the Colorado Plateau such as the Rio Grande "rift valley".

Within the Nevada area at least it is known that the continental crust is thin, only ~20 km in places, that heat flow is high, that within the underlying upper mantle density is low and seismic waves are attenuated (i.e. the uppermost mantle here has the properties of the low-velocity layer), and that the aggregate lateral extension on the numerous normal faults is very great and attains the order of at least 90 km.

Around the early 1970's several workers (McKee and Silberman, 1970; Scholz et al., 1971; Christiansen and Lipman, 1972) recognized a unifying theme in the midst of complexity: andesitic volcanism (accompanied in some areas by present-day exposed quartz-diorite to adamellite intrusive rocks), associated in its last stages with prolonged and voluminous ignimbrite eruption, was apparently succeeded everywhere by the initiation of normal faulting and the later eruption of a bimodal suite of basalt and rhyolite. It was also recognized that these changes were diachronous, occurring first in central Nevada and apparently younging peripherally. This observed sequence of igneous activity coupled with what is known of the megatectonics at the present day plus reasonable inferences on the recent past has led to persuasive interpretations of the history of the area as follows:

(1) Widespread SBZ andesitic eruptive activity occurred during early Tertiary times; K_2O—depth plots suggest that here may have been two shallowly-dipping subduction zones, the more easterly underlying the alkali-calcic andesites of such areas as the San Juan Mountains that are situated a surprising distance inland from the western coast of North America; the Colorado Plateau could conceivably represent the continental equivalent of the observed volcanically inactive trench—arc gap in contemporary trench—arc systems.

(2) Subduction along the western margin of the North American plate ceased at varying times within middle Tertiary times, related to the timing of the juxtaposition of segments of the East Pacific Rise ridge system bounded by transform faults with the American plate margin, coupled with a possible major change in relative plate motions in the early Miocene; eruptive activity continued above the previously subducted lithosphere, and showed a comparable diachronous change to voluminous ignimbrite eruption (Christiansen and Lipman, 1972, p.259, fig. 5).

(3) After a varying interval of time, prominent ensialic back-arc spreading began, notably above the position of the former inferred western subduction zone, accompanied by the eruption of basalt with some rhyolite. Whether this is some kind of self-triggering tensile mechanism acting on a belt previously "softened" by the generation and passage of andesitic magmas, or a

hotspot reflecting the physical overriding by the American plate of an up-welling mantle plume previously underlying part of the East Pacific Rise, or merely reflects a major change in relative plate motions independently inferred to have commenced in early Neogene times is still hotly debated.

Whatever may be the fundamental cause or causes (a combination of the above is not impossible), this ensialic back-arc spreading constitutes a distinctive sub-regime of SBZ igneous activity, paralleled in the recent geological record in the Province of Yunnan of China and to a discernible extent in the North Island of New Zealand (see p.389). The several igneous components require characterization, viz. the nature of the ignimbrites associated with the waning of andesitic volcanism, the type of basalt that was later erupted, and the nature of the rhyolitic rocks associated with these later basalts.

The ignimbrites of Nevada covered $\sim 1.5 \cdot 10^5$ km^2 to an aggregate thickness exceeding 1000 m in places. Individual ignimbrites average 60 m in thickness; some have been traced for distances in excess of 150 km. Analysis of ignimbrite is notoriously unreliable due to the effect of differentiation produced by the selective winnowing of shards from crystals during eruption, to alkali transfer accompanying primary devitrification, and to inclusion of variable amounts of lithic fragments ranging downwards in size to ones only seen easily under the microscope. Following E.F. Cook (1965), the crystal content of the Nevada ignimbrites ranges from 5 to 65 vol.%; phases present include nearly ubiquitous quartz, sanidine, plagioclase and biotite, and rare hornblende, augite and hypersthene — a typically calc-alkaline petrography.

The later basaltic volcanism comprises fields of: (1) basalt; (2) differentiated alkali basalt series, and (3) bimodal associations of basalt and high-silica rhyolite (Christiansen and Lipman, 1972). The basaltic fields comprise mainly tholeiites and minor alkali basalts; they extend into Oregon and include the flood basalts of the Snake River plains, where a possible plume derivation is debatable (see, e.g., R.N. Thompson, 1975). The differentiated alkali basalt series include some peralkaline rhyolites, presumably fractionation products of a parental transalkaline magma composition. The rhyolitic ignimbrites which interdigitate with basalt flows in Yellowstone, in contrast to both the peralkaline rhyolites and the earlier calc-alkaline ignimbrites, are high in silica and contain fayalite and hedenbergite phenocrysts that are suggestive of a history of strong fractionation of a tholeiitic parental magma. The association of these various basaltic and other rock types is clearly a complex topic within the apparent context of ensialic back-arc spreading.

11.2.9. Lateral variation in SBZ *parental magmas including alkali basalt as shown by rocks of the Lesser Antilles*

Grenada, the southernmost major island of the volcanic island arc of the Lesser Antilles (Fig. 11.8), reveals an unusual suite of basanitoids and alkali basalts, basaltic andesites, andesites and dacites, together with nodules of basic and ultramafic cumulate rocks. As Cawthorn et al. (1973) report:

"There is apparently a complete chemical and petrographic gradation between these various rock types"

such that one may justifiably regard them as an igneous rock series.

"This unusual sequence, in which the liquids migrate through the low-pressure thermal divide of olivine—clinopyroxene—plagioclase from critically undersaturated compositions, is interpreted from experimental and chemical data as the result of crystallization of amphibole under a hydrous pressure of several kilobars."

The abundance of amphibole in some of the nodules also lends support to this conclusion. The silica contents of four analysed amphiboles from ejected blocks in the neighbouring island of St. Vincent cluster closely around 41 wt.%; fractionation of material of this composition therefore could result in silica enrichment of residual melts, such as would be demanded of a fractionation model to produce the abundant andesites and more acid members of orogenic andesite series.

This surprising result from Grenada has been put in a broader context by a review of the geochemistry of 1518 rocks, mainly younger than Miocene in age, from the fifteen larger islands of the whole 700 km length of the Lesser Antilles volcanic arc (G.M. Brown et al., 1977). A subduction zone dipping westwards at $\sim 40°$ underlies all these active or recently active volcanic islands at approximately the same vertical distance, ~ 100 km, the minimum distance at which volcanic activity usually occurs and defines the volcanic front. Although the whole Lesser Antilles assemblage of volcanic rocks is characterized by an appropriately low abundance of potassium, there is a pronounced *lateral* variation along the arc in mafic members from alkalic mafic rocks in the south to tholeiitic mafic rocks in the north. The associated andesites and dacites of the southern alkalic suites, including Grenada, can not be distinguished petrographically from the andesites and dacites of the other apparently different magmatic lineages further north, although a distinction is evident in trace-element abundances, particularly those of Zr, Ni and Cr. As the authors point out, variation *along* an arc thus adds a new dimension in SBZ volcanism to the now well-established transverse polarity *across* the arc. The authors appeal to differences in mantle composition, specifically partial melting of garnet lherzolite mantle in a relatively "fertile" zone of upper mantle below the southern sector of the arc to produce the more alkalic parental magmas erupted there.

For comparative purposes Brown et al. use a classification based on silica

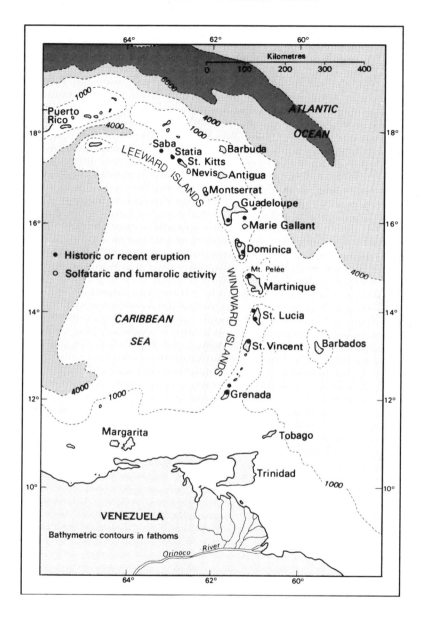

Fig. 11.8. Locality map of the island arc of the Lesser Antilles (comprising the Leeward Islands and Windward Islands) and adjacent areas. The volcanic arc is well delineated by the examples of historic and recent volcanic activity. Barbados has no history of recent volcanism. Note that typical trench bathymetry is here masked by sediment infill from the Orinoco, but a prominent axis of negative gravity anomaly runs through Barbados and parallels the volcanic arc (after Robson and Tomblin, 1966, pp. vii and ix).

TABLE 11.8

Average compositions (weight per cent major oxides and ppm selected trace elements) and CIPW norms at comparable SiO_2 contents of three basalts and three andesites from Grenada, Dominica, and St. Kitts, Lesser Antilles

	Basalts			Andesites		
	Grenada 1	Dominica 2	St. Kitts 3	Grenada 4	Dominica 5	St. Kitts 6
SiO_2	51.02	50.90	51.13	58.84	59.07	59.15
Al_2O_3	17.73	19.83	19.77	17.54	17.60	17.57
Fe_2O_3	2.24	2.51	2.32	1.70	1.94	1.91
FeO	6.22	6.97	6.43	4.71	5.38	5.30
MnO	0.16	0.19	0.18	0.14	0.16	0.18
MgO	6.23	3.74	3.92	2.94	2.51	2.60
CaO	10.44	11.10	11.09	7.39	7.67	7.66
Na_2O	3.24	2.50	3.00	3.85	3.03	3.52
K_2O	0.94	0.49	0.36	1.51	1.19	0.62
TiO_2	0.86	0.85	0.98	0.65	0.71	0.70
P_2O_5	0.18	0.12	0.10	0.18	0.12	0.11
Total	99.26	99.20	99.28	99.45	99.38	99.32
Rb	23	16	5	43	46	13
Ba	398	105	126	546	220	174
Sr	738	310	290	877	287	284
Zr	96	44	65	169	102	96
Ni	129	36	11	66	38	3
V	215	n.d.	235	140	n.d.	110
Cr	285	109	17	152	133	17
Q	—	3.9	2.3	10.1	15.2	14.6
or	5.6	2.9	2.1	9.0	7.1	3.7
ab	27.6	21.3	25.6	32.8	25.8	30.0
an	31.3	41.8	39.7	26.3	31.1	30.5
ne	—	—	—	—	—	—
di	16.0	10.7	12.4	7.8	5.3	5.8
hy	8.3	13.8	12.4	10.0	11.1	11.0
ol	5.9	—	—	—	—	—
mt	3.3	3.7	3.4	2.5	2.8	2.8
il	1.7	1.6	1.9	1.2	1.4	1.3
ap	0.4	0.3	0.2	0.4	0.3	0.3
M-value	57	41.5	44.5	45	38	39

1 = average basalt from Grenada (G.M. Brown et al., 1977, p.791, table 6, analysis *1*).
2 = average basalt from Dominica (G.M. Brown et al., 1977, p. 791, table 6, analysis *2*).
3 = average basalt from St. Kitts (G.M. Brown et al., 1977, p.791, table 6, analysis *3*).
4 = average andesite from Grenada (G.M. Brown et al., 1977, p.791, table 6, analysis *4*).
5 = average andesite from Dominica (G.M. Brown et al., 1977, p.791, table 6, analysis *5*).
6 = average andesite from St. Kitts (G.M. Brown et al., 1977, p.791, table 6, analysis *6*).

content into picrite (<48 wt.%), basalt (48—53 wt.%), andesite (53—62 wt.%), dacite (62—68 wt.%), and rhyodacite (68—72 wt.%). On this basis, outcrop areas in the whole Lesser Antilles amount to 4% picrite, 13% basalt, 42% andesite, 39% dacite, and 2% rhyodacite. The exposed volcanic rocks are thus predominantly of intermediate compositions. The authors present abundant analytical data from Grenada, Dominica and St. Kitts (islands representative of the southern, central and northern sectors of the arc, respectively) conveniently averaged and arranged in 2-wt.% silica intervals (G.M. Brown et al., 1977, p.790, tables 2, 3 and 4), together with average compositions of basalt and andesite from these three islands (G.M. Brown et al., 1977, p. 791, table 6), reproduced here as Table 11.8. The syndrome in the Grenada basalts of high Ni and Cr, suggestive of a primitive unfractionated composition, coupled, however, with relatively high contents of incompatible elements such as Rb, Ba, Sr and Zr, is one we have met before in discussing the possible influence of relatively fertile mantle related to plumes on the generation of flood basalts. Average $^{87}Sr/^{86}Sr$ is 0.7048 from the southern islands, and 0.7038 from the central and northern islands, a difference which in its amount and consistency is by analogy with the data from Iceland, etc., not incompatible with the proposed petrogenetic mechanism of different mantle source rocks.

A total alkalis vs. silica variation diagram for the three series (G.M. Brown et al., 1977, p.792, fig. 3) clearly shows the Grenada series crossing from the alkaline field to the subalkaline field. The distinction between the Dominica and St. Kitts series that both lie entirely in the subalkaline field is not pronounced; neither is there any difference between them on a *AFM* plot (G.M. Brown et al., 1977, p.792, fig. 6). K_2O is lowest in the St. Kitts series (representative of the northern sector) which is of typical island-arc tholeiite affinity. The characterization by the authors of the central sector as "calc-alkalic", in contrast to the alkalic southern series and tholeiitic northern series, is not at all sharply drawn vis-à-vis the northern series. Neither do the authors bestow on this central series the name of a parental mafic magma as they do for the other two. In view of the authors' demonstration of the petrographic similarity of the predominant intermediate members of all three groupings, it is clear how the term calc-alkaline has come to be identified with intermediate SBZ volcanic rocks belonging to various lineages.

11.2.10. Volcanism of the central Andes: voluminous andesite and rhyolite with no basic rocks

The Andes form a continuous mountain chain ~6500 km in length clearly situated over an east-dipping Benioff zone. In detail, however, the Andes are complex and can be subdivided along their length into a series of tectonic segments (Sillitoe, 1974) with differing patterns and ages of eruptive activity and associated ore deposits. There is evidence that the transverse boundaries of these distinct segments may be fundamentally related to discontinuities in the underlying Benioff zone.

Late Cenozoic volcanic activity is mainly confined to three of these segments (Pichler and Zeil, 1969): (1) a northern zone in Colombia and Ecuador between latitudes 5°N and 2°S; (2) a central zone close to the frontier between northern Chile and Argentina but extending also into southern Peru and Bolivia between latitudes 15° and 27°S; and (3) a southern zone in central and southern Chile between latitudes 33° and 44°S.

The central zone considered here has an extent of ~1100 km in length by ~100—200 km in width. It witnessed two late-Cenozoic volcanic episodes: an older series of rhyolitic to rhyodacitic ignimbrites (the "Rhyolite Formation") erupted 12—2 Ma ago, and a younger series of andesites (the "Andesite Formation") erupted from stratovolcanoes in Pleistocene and Holocene times. The number of young, mostly now extinct, stratovolcanoes is ~600 and represents the greatest concentration of them on Earth. The belt of young andesitic rocks is no more than 30—50 km in width localized around the elongate chain of volcanoes. The rhyolitic rocks spread out over a larger extent due to their mode of eruption and cover an area of $1.5 \cdot 10^5$ km^2. No basaltic rocks whatsoever have been found associated with these young volcanic rocks.

Rocks of the Andesite Formation show an appreciable range of silica contents, ~52.5—64.5 wt.% SiO_2, and have been named quartz-bearing andesite, latite—andesite, quartz-bearing latite—andesite, and quartz-bearing latite (Pichler and Zeil, 1969). Phenocrysts comprise plagioclase, commonly labradorite, exhibiting reverse and oscillatory zoning and abundant glass inclusions, and generally two pyroxenes, hypersthene and augite. Some rocks contain olivine, mainly corroded cores, and some may contain hornblende and/or biotite. A generally potassium-rich affinity is claimed by Pichler and Zeil, and inspection of their analytical data shows the K_{55}-value to be fairly high, ~1.6 wt.% K_2O.

The ignimbrites of the Rhyolite Formation are peraluminous. They have trace-element contents that appear to be serial with those of the andesites, although Cu contents of some of the rhyolites are surprisingly high, and Zr also is high in the rhyolites.

Pichler and Zeil favour a primary andesite magma produced by partial fusion of lower crustal material. James et al. (1976), however, on the basis mainly of strontium-isotope data, conclude that parental magma was produced by partial melting of (or equilibration with) a thick lithosphere; the lack of basic members of the series could be explained by prolonged fractionation during ascent of parental magma through thick and hence relatively cold lithosphere.

The central Andes demonstrate that SBZ volcanism is by no means necessarily a steady-state phenomenon but rather episodic and paroxysmal even on a geologic time scale, as shown by the segmentation of zones of current volcanic activity and by the ages of the volcanic products in the area described.

Some writers make a facile two-fold division into "Andean" and "island-arc" types of SBZ igneous activity, related to subduction beneath continental or oceanic crust, respectively. However, the igneous products vary not only between but also within each environment, and the more subtle distinction made by J.T. Wilson (1973), based on relative lithosphere—mesosphere velocities, may prove to be significant.

11.3. PLUTONIC ROCKS APPARENTLY FORMED DURING SBZ ERUPTIVE ACTIVITY

In establishing igneous rock series and relating them to new global tectonic regimes, attention naturally focuses on very young, preferably recent, lavas: not only can they yield quenched liquidus compositions and therefore a liquid line of descent, but also because of their youthfulness they are more likely to be unaltered, and also they can be easily tied in precisely to tectonic regimes at their time of eruption. Intrusive rocks, to be exposed, are necessarily older and thus their precise relation to inferred plate motions, plumes, etc., at their time of formation may remain in some doubt (although unroofed Neogene, even Pleistocene, high-level intrusions are known from Afar, Mount Kenya, and Bougainville Island, for example). However, just as intrusive rocks form an essential part of an ophiolite suite, so too are recognizably distinct groupings of intrusive rocks abundant in what may legitimately be inferred to have been a SBZ environment at the time of their formation. The reader is warned however, that a high degree of contention surrounds the petrogenesis of many of the rocks to be described below.

11.3.1. The low-potassium island-arc plutonic complex of Tanzawa

One of the few plutonic complexes in a SBZ setting that seem to be directly comparable with observed extrusive SBZ rocks is the Tanzawa complex situated near the eastern margin of Honshu, the main island of the Japanese archipelago. The plutonic rocks (Ishizaka and Yanagi, 1977) are of Miocene age and outcrop in the south Fossa Magna region which structurally forms the northern part of the Izu—Bonin arc. Petrographically the rocks comprise olivine-, pyroxene- and hornblende-gabbros, diorites, tonalites and leucocratic trondheimite. $(^{87}Sr/^{86}Sr)^0$ values lie between 0.7030 and 0.7040, values which are indistinguishable from those of recent volcanic rocks of the Izu— Bonin arc. In keeping with the absence of K-feldspar, K_2O contents are very low (K_{55}-value from inspection of analytical data is ~ 0.3 wt.%). The rocks, along with a somewhat comparable series in Viti Levu, Fiji (Gill, 1970), can reasonably be considered therefore as a plutonic equivalent to an island-arc tholeiite series.

11.3.2. Ultramafic rocks of Alaskan pipe type

Some thirty-odd, mineralogically comparable, ultramafic pipe-like bodies of early Cretaceous age are found in a 40 km wide belt, ~550 km long between latitudes 54°30' and 59°30'N in southeastern Alaska (H.P. Taylor, 1967). The ultramafic pipes for the most part are emplaced within and intensely metamorphose to amphibolite and granulite facies a similarly elongate complex of gabbros, norites and diorites of overall subalkaline affinity. Individual pipes very greatly in size, the eight largest ranging from 5—30 km² in outcrop area.

In sharp mineralogical contrast to the igneous country rocks, the ultramafic rocks are composed of olivine, chromite, calcic augite, magnetite and hornblende, and contain virtually no hypersthene or plagioclase. Rock types, which are repeated with remarkable similarity in the different pipes, comprise dunite, peridotite, olivine clinopyroxenite, hornblende clinopyroxenite and hornblende-magnetite clinopyroxenite. A zoning pattern, frequently incomplete or discontinuous, can be established from dunite centrally to hornblende clinopyroxenites on the outer parts of the pipes. It is noteworthy that a belt containing similar ultramafic pipes of Devonian age in the Urals (H.P. Taylor, 1967, p.99):

"has so many features in common with those of southeastern Alaska that most of the descriptions are interchangeable."

Perhaps equally surprising, in view of this parallelism, is the apparent lack of other described examples of such rocks elsewhere.

Mineral parageneses and compositions, the Fo contents of olivines for example, suggest that the dunites are the highest-temperature products; the pipes thus present the apparent anomaly of having crystallized from the centre outwards. This feature plus the presence of apparently intrusive contacts led Taylor to propose the successive injection of magmas of differing compositions beginning with pyroxenite magmas and concluding with dunite magmas.

The detailed and well-illustrated work of Irvine (1974) on the Duke Island Complex, one of the larger well-exposed and accessible pipes, suggests a different mode of origin in that there is compelling evidence in the exposed rock sequences for the following sequential order of mineral crystallization from one parental magma type: olivine with minor chromite, followed by a probably prolonged period of cotectic crystallization of olivine and clinopyroxene, followed finally by clinopyroxene and magnetite. Most of the rocks are cumulates (although the cumulate textures are modified in places by recrystallization) exhibiting such syn-depositional features as layering, gravity layering, slumping, cross-bedding, "intraformational" brecciation, and incorporation to a marked degree of brecciated cumulate blocks in younger cumulate sequences, indicative presumably of violent current action*. Hornblende becomes an increasingly prominent intercumulus phase with evolution in the series, and also in part replaces original clinopyroxene, but was probably never a cumulus mineral itself, so that the name hornblende should not figure in a cumulate nomenclature for the rock even although modal hornblende may become prominent.

Although direct evidence (e.g., composition of chilled marginal material) is lacking, Irvine (1974, pp. 141—143) makes the vital distinction between *cumulate* composition and corresponding *magma* composition and shows that appeal to various ultramafic liquids is an improbable, indeed an untenable, proposition. Irvine stresses the difficulties of estimating a magma composition from a cumulate sequence alone, and proceeds to make some logical deductions. In the parental magma, activity of silica was apparently low, TiO_2 was low as indicated by the low titania contents of augites, and Al_2O_3 was not necessarily low despite the absence of plagioclase as the hornblendes and some of the augites are rich in alumina. A high P_{H_2O} and high f_{O_2} can be inferred from the abundance of hornblende and the relatively early appearance of magnetite as a cumulus phase. The conclusion is reached that the parental magma was a critically undersaturated alkaline ultrabasic magma, primitive in the sense of being rich in MgO, and possibly akin to a nearby suite of extrusive "potassium-rich ankaramites" of Jurassic or early Cretaceous age with a phenocryst assemblage of olivine, augite and magnetite (Irvine, 1974, pp. 143, 144 and 176). The present relative positions of masses of dunite, olivine clinopyroxenite, etc., although in part due to being part of a complexly cross-bedded and faulted cumulate sequence, is attributed by Irvine to:

"the diapiric re-emplacement of rudely stratiform sequences"

either by continued rise of magma from depth or by tectonic compression.

Murray (1972) doubts the existence of a parental ultrabasic alkaline magma, and has suggested that the parental magma was akin to olivine-rich tholeiite and that the Alaskan pipes represent quite simply the feeder pipes of andesitic volcanoes. The evidence reviewed by Murray in support of this interpretation includes the following: (1) the probably non-fortuitous spatial association of the elongate belt of pipes with abundant subalkaline gabbroic and noritic rocks and also with a belt of extrusive rocks of early Cretaceous age comprising in the main basaltic andesites; (2) experimental work has shown that under sufficiently high P_{H_2O}, liquids of olivine tholeiite compositions can yield the following sequence of crystallizing phases: olivine, clinopyroxene, hornblende, all earlier than plagioclase; (3) the significant lack from the Alaskan pipe rocks of biotite, which is found in experimental work

---

*These "igneous" sedimentary features are curiously reminiscent of the poorly-sorted, massively cross-bedded sequences of epiclastic hyalotuff and pillow breccia material flanking Moberg ridges in Iceland, originating in an inferred similar environment of violent current activity.

to be a crystallization product of alkali basalt compositions under similarly high P_{H_2O}; (4) the presence in one Alaskan pipe, the Hope Complex, of the crystallization sequence olivine—orthopyroxene—clinopyroxene with hornblende and plagioclase appearing together at a later stage, a sequence also relatable to experimental work on tholeiitic melts under an appreciable but lower P_{H_2O} than the foregoing; (5) the overall trend in clinopyroxene compositions of the Alaskan pipe rocks is not significantly different from those crystallizing from tholeiitic magmas under differing values of P_{H_2O}, and the clinopyroxenes specifically lack the characteristically high sodium contents of clinopyroxenes crystallized from alkaline mafic magmas.

Whatever may be the final outcome of these arguments as to exact magmatic affinity of these Alaskan pipe rocks, it can not be denied that their locale of formation was one of long-standing subduction. One might legitimately ask why comparable rock types are not more common in dissected terrain of former andesitic volcanism. One reason for this apparent scarcity may lie in that experimental work indicates a high pressure and correspondingly considerable depth of formation, so that only relatively rarely might these rock types be exposed.

11.3.3. Coastal Batholith of Peru

The following account of this batholith, magnificently exposed in an area of mountainous desert relief, is based on several recent papers by Pitcher and colleagues: Cobbing and Pitcher (1972), Pitcher (1974, 1978), Myers (1975), Bussell et al. (1976) and Cobbing et al. (1977). The batholith outcrops in an elongate belt parallel to the present trench, coast line, and Andean cordillera. It has dimensions of ~1600 km in length by 50 km in width, and is probably up to 15 km thick. It is composed of some 800 separate mappable plutons and innumerable minor intrusions. The intrusive rocks have ages ranging from 30 to 100 Ma, and are spatially coincident with two groups of volcanic rocks of overlapping ages with the plutonic rocks: (1) the older Casma Group comprises submarine andesitic lavas, some pillowed, and volcaniclastic sediments including abundant chert, and is assigned on paleontological evidence to the Albian (~100 Ma); (2) the younger Calipuy Group of early Tertiary age lies unconformably upon gently folded Casma Group rocks, and comprises subaerial andesites, dacites, rhyolites including some ignimbrites, and derived sediments. Field relations leave little doubt that the plutonic rocks of the Coastal Batholith represent the intrusive equivalents, at least in part, of these extrusive rocks, the whole assemblage having originated above a Benioff zone.

The plutonic rocks comprise 16 vol.% gabbro and diorite, 58 vol.% tonalite and granodiorite, 25.5 vol.% adamellite, and 0.5 vol.% granite, based on outcrop areas in the central segment of the Coastal Batholith. The method of emplacement is in part by apparent diapiric ascent and forceful intrusion, but predominantly by cauldron subsidence; indeed typically epizone structures are wonderfully displayed in great detail in terrain with vertical relief ranging up to 4500 m [see sections in Myers (1975), in particular, and accompanying descriptions of shallow-seated fluidization effects].

Three contiguous elongate segments along the outcrop of the Coastal Batholith with lengths of ~200, 400 and 900 km, respectively, exhibit recognizably distinct groups of plutonic rocks ("superunits"), associated type and intensity of mineralization, etc., even though one superunit may have been emplaced within a similar time span and span a similar range in silica content as a superunit in another segment. These broad differences may in turn be related to discontinuities between different segments of underlying subduction zone at the time of emplacement in a similar manner to the recognizable discontinuities associated with contemporary volcanic regimes (see pp. 403 and 404). Within the spectrum of intrusive rocks forming a superunit, a small number of distinctive rock types, "units", can be recognized that comprise the bulk of the total in any one segment. Furthermore, within different complexes in any one segment an identical order of intrusion of these units can be recognized although their relative abundances vary somewhat, and even although their absolute age ranges may differ. Some plutons may show internal differentiation in place, but an overall time sequence from relatively basic to relatively acid units in each complex is clear, over a time span of ~10 Ma.

With respect to magmatic affinity, the close spatial association within the complexes of a significant proportion of basic rocks of relatively early age should be noted. The basic rocks are largely hypersthene gabbros ± olivine with much magmatic amphibole and some are cumulitic (see Mullan and Bussell, 1977, for a review). This would suggest an origin for the several series by differentiation (fractionation modified by large-scale assimilation? — see discussion of Sierra Nevada intrusives in Chapter 7) of batches of parental basic magma each over a 10-Ma time span. The overall chemistry of the rocks of the Coastal Batholith shows that they belong to the "I-type" of Chappell and A.J.R. White (1974), in that they have low $Al_2O_3/(Na_2O + K_2O + CaO)$ molecular ratios less than 1.1, very little (less than 1 wt.%) or no normative corundum, and initial $^{87}Sr/^{86}Sr$ ratios in the range 0.704—0.706. These features invite correlation with mantle-derived liquids progressively modified by assimilation of lower crust. By contrast, the granitic rocks of the Neogene Cordillera Blanca (situated inland of the relatively older Coastal Batholith) have a restricted high range of silica contents, are muscovite-bearing, are corundum-normative with $Al_2O_3/(Na_2O + K_2O + CaO)$ ratios greater than 1.1, and have higher initial $^{87}Sr/^{86}Sr$ ratios: these are the "S-type" granitic rocks of Chappell and White, derived presumably from the crust itself, with little or no input of mantle-derived material. Nevertheless, one does wonder if these young rocks exposed only to shallow depths could represent merely the end-product of the same kind of differentiation sequence as that indicated by the Coastal Batholith rocks.

11.3.4. Porphyry coppers

The term "porphyry copper" is so well established and accepted that a justification of its use is scarcely necessary. A definition, flexible enough to include all examples, however is difficult. To an economic geologist the connotation is a large deposit containing disseminated values of copper and/ or molybdenum, occurring in and around a body of granitic rock. In petrological terms, the connotation is of a high-level calc-alkaline intrusion of intermediate to acid composition displaying a high degree of deuteric alteration.

The distribution of known porphyry copper deposits shows a high correlation with the landward or SBZ side of circum-Pacific destructive plate margins. Others are known in a comparable SBZ situation from the Alpine—Himalayan belt. Sillitoe (1972) records some 110 from the western Americas, some 20 from the Southwest Pacific, a further 20 from the Alpine—Himalayan belt, and only two occurrences outside of these post-Palaeozoic orogenic belts, namely in Uzbekistan and Kazakhstan, U.S.S.R., where any relationship to possible older destructive plate margins in not established. The correlation of known porphyry coppers with an inferred SBZ environment at their time of formation is thus nigh perfect. Indeed Sutulov (1974) goes further and claims that no porphyry coppers are older than Triassic in age, and that all originated in a SBZ environment.

An invaluable summary and synthesis, based on a selection of porphyry copper deposits, is that given by Lowell and Guilbert (1970), from which a distinctive pattern of alteration and associated mineralization emerges as a unifying theme. A "typical" porphyry copper deposit would be centred on a quartz-monzonite or adamellite porphyry stock with an oval cross-section, in plan $\sim 1 \times 2$ km. Concentric zones of alteration, commonly in the form of vertical cylindrical rings include: (1) a central "potassium-rich" zone within the granitic rock where K-feldspar and biotite accompanied by anhydrite are stable alteration phases occurring both as replacement and veinlet fillings of pyrogenetic minerals; (2) a "phyllic" zone, occurring outside the above and separated from it by a transition zone only some tens of metres wide, characterized by extensive sericite; (3) an outermost "propylitic" zone in which both granitic and country rock are degraded to greenschist facies with extensive development of greenish minerals such as chlorite, epidote and actinolite, with an indefinite outer envelope of alteration. Mineralization in the inner potassium-rich zone consists of disseminated sulphides (0.3 wt.% Cu; pyrite/ chalcopyrite = 1:2); in the outer potassium-rich zone there is an ore zone proper, its richness and extent often controlled by supergene enrichment, containing both disseminated ore and veinlets (more than 0.5 wt.% Cu; pyrite/chalcopyrite = 1:1); this is followed outward in turn by the occurrence of abundant pyrite (6—25 wt.%) with low copper values, and largely coincident with the phyllic zone. These zones, although centred on the igneous body are not exactly related to igneous contacts; on an average $\sim 70\%$ of ore is found in the igneous host rock and 30% in the country rock, but it should be noted that this figure of 70% includes a range of from 10 to 100%. Farther out still from the mineral zones listed in the above paradigm, there may be relatively wide veins containing small shoots of copper values plus gold and silver values.

Metallic mineralization thus appears to be consistently related to a pattern of alteration, the whole forming an envelope partly within and partly without a particular intrusion. The form of the mineralized zone commonly approximates to that of a vertical annular cylinder, though variants such as stubby cylindrical and inverted flatly conical also occur. The temperature at the centre of alteration where K-feldspar and biotite are stable must have been high and could only have been attained at such a high crustal level during the period of cooling of the intrusion itself. This essential continuum of magmatic, deuteric, and hydrothermal processes in space and time has been indicated, for example, by Robertson (1962). There has been a marked telescoping of normal mineral zones into small distances across porphyry—country rock contacts, and it appears likely from isotopic studies that circulating groundwater participated in the alteration process (see also Henley and McNabb, 1978).

The igneous intrusions on which the porphyry copper deposits centre range from diorite to adamellite in composition, including monzonite, quartz-monzonite and granodiorite. The magma type seems to belong to a high-potassium andesite series. Rocks more mafic than diorite are typically absent and rocks more acid than adamellite are typically confined to minor intrusions. Some intrusions, in addition to the prominent deuteric and hydrothermal alteration effects, are internally differentiated and zoned. Many intrusions are part of composite complexes with the order of emplacement generally in order of increasing silica content. The rocks have the typically homophanous structure of passively emplaced high-level magmatic plutons. Texture may be granitic in places but is frequently porphyritic. In some areas the intrusions are seen to be associated with penecontemporaneous volcanic rocks of a comparable range in composition. The detailed evidence for shallow depth of emplacement of porphyry copper stocks is compelling. Many of the intrusions penetrated to shallow depths where subterranean vesiculation took place as shown by drusy structures and the common occurrence of subterranean bodies of explosion breccia, some of them diatremes and some of them apparently blind. Vesiculation commonly occurred after a proportion of the intrusive body had crystallized thus increasing the volatile content of the remaining liquid fraction and resulting in the explosive shattering in places of the congealed shell of the intrusion. Volatile loss from vesiculating magma chambers was not complete as indicated by the typically fine-grained or microcrystalline texture of the groundmass of porphyritic rocks rather than a glassy or felsitic texture. The remarkably young age (1.11—1.24 Ma) of an unroofed Pleistocene porphyry at Ok Tedi in Papua

New Guinea (Page and McDougall, 1972) attests extreme shallowness of emplacement, even allowing for the effects of possibly rapid erosion in that area. Comparison of many deposits and the pattern of variation that they display with depth leads to the conclusion that porphyry copper deposits originated within 3000 m of surface, thus accounting for the pronounced telescoping of conventional mineral zones owing to a high temperature gradient having been set up in this near-surface environment between hot high-level magma and cold country rock.

Several quieries might legitimately be asked about this apparently unique porphyry copper syndrome. In the first place, how were such large concentrations of copper and/or molybdenum generated? Very conservative estimates of the tonnage of metal (occurring both in ore-grade rock and below ore grade), expressed as a fraction of the tonnage of igneous rock present extended down to 3000 m, are several times the average abundance of the metals in comparable igneous rocks. As porphyry copper deposits of comparable grade are found emplaced in a variety of country rocks and in areas of different crustal thickness and type, an origin by crustal contamination would appear unlikely. Sillitoe (1972) suggests that calc-alkaline magmas may on occasion incorporate relatively large amounts of Cu and other metals from rocks within a descending slab of older oceanic crust. However, if this were the case, logically one might expect to find similarly anomalously high metal values in some of the consanguinous lavas of andesitic to dacitic compositions, but such high values have not been recorded. Another explanation could follow from the work of G.C. Kennedy (1955) on the equilibrium distribution of H_2O within a body of magma. Particularly at shallow depths, a high relative concentration of H_2O would be effected in the upper parts of a vertical body of magma. If this escaped by diffusion and vesiculation processes the equilibrium would tend to be maintained by the accession of more H_2O rising through the magma chamber as long as it remained molten or partly molten, or conceivably crystalline but hot. This H_2O could carry with it halides and metals in solution and thus effect a leaching and redeposition of metals within a larger magma system than that exposed to view by erosion within upper-crustal levels.

Why do not more intrusions of similar compositions and environment show, if not economic mineralisation, at least comparable alteration processes? Why are there not more porphyry copper deposits in the pre-Mesozoic geological record? Convincing answers to these questions have so far not been forthcoming.

11.3.5. Contrasting styles of batholith emplacement and genesis

The complex dialogue on the interpretation of the petrography, chemistry, internal differentiation and isotope contents of the Sierra Nevada batholith and other SBZ plutonic rocks has already been outlined in Chapter 7. The data seem to demand contributions both of mantle-derived material and of crustal material to produce this plutonic igneous rock series, i.e. an origin probably comparable to that of the young batholiths of Peru.

The Coast Range batholith of British Columbia on the other hand has been interpreted (also see Chapter 7) in terms of remobilization of sialic crustal material, notably in the synthesis presented by Hutchison (1970). If this is true, the scale is impressive — $\sim 12.5 \cdot 10^4$ km^2 of batholithic rocks along a north—south elongate zone some 1500 km long, and presumably in an SBZ situation at time of generation. Unlike the Coastal Batholith of Peru and the Southern California and Sierra Nevada batholiths, gabbro is very rare indeed and migmatites are abundant in the Coast Range batholith.

In summary, we see that not only is the petrogenesis of the variable group of volcanic products erupted in a SBZ situation hotly debated, but also the plutonic rocks are similarly diverse and not amenable to facile interpretations. This is perhaps a fitting note to conclude these chapters on the challenging and not yet completed task of documenting recognizable young igneous rock series in their inferred global tectonic settings.

Chapter 12

IGNEOUS ROCKS OF THE PRECAMBRIAN

12.1. INTRODUCTION

Following Chapters 8—11 it would appear that there *is* an essential connection between igneous activity and both the present global tectonic regime of plate movements and a necessarily accompanying though less well understood mantle convection. One topical question is how far back before Phanerozoic time a comparable megatectonic regime was operative. Although many classes of Precambrian igneous rocks are comparable to those in the recent geological record, some are notably different. The interpretation of igneous activity in Precambrian time thus presents a distinct set of challenging and perplexing problems, that in turn are inextricably related to crustal evolution and megatectonic concepts. A recent account of Archaean and Proterozoic rocks is included in *The Evolving Continents* by Windley (1977). Some items of specific interest to the igneous petrologist are singled out for attention in this chapter.

12.2. EXTRATERRESTRIAL IGNEOUS ROCKS

12.2.1. Meteorites

From the fascinating and complex subject of the study of meteorites several items of interest emerge for the igneous petrologist. First, they reveal the timing of the end of a period of thermonuclear synthesis within our solar system, the accretion of its matter into bodies of various sizes, and an indication of a quickly ensuing core—mantle fractionation within some asteroid-sized bodies that subsequently disintegrated. As well as giving us this practical starting point for planetary accretion, meteorites also afford us glimpses of inferred primordial planetary matter including parameters, important to the igneous petrologist, such as absolute and relative abundance of REE, and initial isotopic ratios of Pb and Sr. A useful source of detailed information on meteorites, richly indicative of their complexity, is Wasson (1974).

One group of meteorites, the *achondrites*, accounts for ~14% of observed "falls" and includes, in order of abundance, stony meteorites, iron meteorites (the most common of all meteorite "finds"), and stony-iron meteorites. Stony meteorites consist of coarsely crystalline mafic and ultramafic rocks. Iron meteorites (syn. siderites) are composed mainly of Ni—Fe alloys showing coarse Widmanstätten textures interpreted as the product of exsolution consequent on slow cooling over a period of 10—100 Ma within a parent body ~100—500 km in diameter (or at equivalent depths within a larger body). Other phases in iron meteorites include sulphides, notably troilite, FeS, and very small amounts of phosphides and carbides. Stony-iron meteorites contain varying proportions of silicate and metal phases. The achondritic group is generally regarded as having formed from the break up of cooled asteroids of appreciable size that had undergone core—mantle fractionation (incomplete in the case of the stony-iron meteorites) of a similar type to that inferred for the Earth. This group of meteorites yields a straight line Rb—Sr isochron, age of ~4.5 Ga, and an initial $^{87}Sr/^{86}Sr$ ratio close to 0.699. An adjusted value for BABI (acronym for "basaltic achrondrite best initial") is 0.69897 ± 0.00003, and some meteorites of this group contain virtually no Rb so that their present $^{87}Sr/^{86}Sr$ ratio approximates to this very low value (see discussion in Faure, 1977, pp. 107—110). Similarly, very low primeval isotope ratios have been reported for lead in troilite from the Cañon Diablo iron meteorite: $^{206}Pb/^{204}Pb = 9.307 \pm 0.006$; $^{207}Pb/^{204}Pb = 10.294 \pm 0.006$; $^{208}Pb/^{204}Pb = 29.476 \pm 0.018$ (Tatsumoto et al., 1973).

The large group of *chondritic meteorites*, accounting for the remaining 86% of observed falls, is characterized in the main by chondritic textures reflecting the crystallization of magnesian olivine and orthopyroxene under conditions of extreme supercooling (see p.164). There are arguments to suggest that chondritic meteorites cannot be attributed solely to the cooling of condensates from an initial solar nebula; instead, they probably represent the quenching of liquid droplets produced by impact processes (see discussion in Wasson, 1974, pp. 198—200).

The sub-class of *carbonaceous meteorites* is altogether different. Although classified as chondrites (as distinct from the entirely achondritic stones, stony-irons and irons), and although some carbonaceous chondrites contain chondrules (C2 and C3 chondrites), some so-called carbonaceous chondrites, the C1 chondrites, actually lack chondrules and commonly have a breccia structure. The C1 chondrites are notable for their content of (abiogenic) hydrocarbons and other volatile constituents implying a low final temperature of accretion. One carbonaceous type C3 chondrite, the Allende meteorite, has the even slightly more primitive $^{87}Sr/^{86}Sr$ initial ratio of 0.69877 than BABI. The mineralogy of this meteorite, an observed fall in Pueblo Allende, Mexico, in 1969, has received unprecedented attention leading to the conclusion that it may contain a component of primitive high-temperature condensates (see, e.g., Haggerty, 1978, for a recent review).

It seems that the carbonaceous chondrites may come nearest to representing (C1) or containing (C2 and C3) an early condensate from primordial solar nebular material (see review by Grossman, 1975). It is the relative REE abundances in carbonaceous chondrites to which REE contents of igneous rocks are normalized (see, e.g., S.R. Taylor and Gorton, 1977).

One interesting interpretation of isotopic data indicates that the age of stony meteorites, and thus the timing of formation of at least one planetary body and subsequent mantle—core fractionation and cooling, followed the end of thermonuclear synthesis within our solar system in the remarkably short time interval of 120—290 Ma, depending on the limits of certain assumptions as to the duration of thermonuclear synthesis. This interpretation is based on the discovery in some stony meteorites of anomalous amounts of ^{129}Xe, the radiogenic product of ^{129}I, produced during thermonuclear synthesis, of which the relatively short half-life of 17 Ma means that it is now an "extinct" radioactive isotopic species in nature (J.H. Reynolds, 1960).

Having presented a simplistic model of meteorite origins, it is only proper to point out that the study of meteorites is enormously complex in detail, still with many unresolved conflicts of interpretation (see, e.g., reports on the Lunar and Planetary Science Conference in *Geotimes*, June 1978 issue). However, for our present purposes it is re-assuring to note that we do seem to have strong indications of a finite beginning in time for planetary evolution, igneous processes, and evolution of naturally occurring radioactive and radiogenic isotopes.

12.2.2. The Moon

No Earth scientist alive today can fail to be aware of the broad outlines of lunar geology. In a historical context, it is especially interesting to compare the two perceptive pre-Apollo books by Baldwin (*The Face of the Moon*, 1949; *The Measure of the Moon*, 1963) with later synoptic accounts, e.g., *Lunar Science: A Post-Apollo View* by S.R. Taylor (1975). A difficulty, lamented by Taylor, is the immense volume of recently published material, notably in annual proceedings of the Lunar Science Conferences (1970 onwards) published as supplements to *Geochimica et Cosmochimica Acta*. This difficulty incidentally was heroically tackled by Taylor in producing his book that combines condensation, information, scientific exactitude, and readability to a high degree. A later short review of igneous aspects of lunar evolution is by G.M. Brown (1977).

The ultimate origin of the Moon, some 2% of the Earth's volume and 1.2% of its mass, is still enigmatic. Whatever processes led to its formation as a planetary entity ~4.5 Ga ago apparently resulted in a notable overall enrichment in refractory elements such as Ti, Zr, Y, REE and Ca, and a notable depletion of volatile, siderophile, and chalcophile elements such as Na and K, Fe and Ni, and Cu, respectively, relative to chondritic and Earth abundances.

A very early igneous event, occurring shortly after the formation of the Moon, was the production of a crust over the whole surface of the Moon up to ~60 km in thickness, composed mainly of varieties of gabbro, norite and anorthosite, the so-called Lunar Highlands. This was achieved presumably by crystal fractionation processes, including upward floating of plagioclase crystals on a massive scale within large volumes of molten material of basic composition. Relatively minor amounts of extrusive basalt are known from the Lunar Highlands and reveal a compositional range from low-potassium varieties to small amounts greatly enriched in potassium, REE, phosphorus and other incompatible elements, the so-called KREEP basalts (Table 12.1).

This old crust has been intensively cratered by meteorite bombardment. Although the precise rate of decrease of meteorite flux inferred from crater densities remains somewhat debatable, any lunar surface older than 3.9 Ga has been effectively saturated by large overlapping meteorite impact craters of which many of the most evident are ∼50—100 km in diameter. In keeping with this decreasing flux most of the large maria basins are older than 3.7 Ga and none exceeds in size Mare Imbrium dated at 3.9 Ga. Tycho, however, a respectable 90 km in diameter, is a rare example of a relatively young crater (age estimated at ∼100 Ma).

Younger basaltic flows, known to range in age between 3.95 and 3.15 Ga, under a regolith some metres thick, cover about a quarter of the Moon's

TABLE 12.1

Representative analyses of lunar basalts

	1	2	3	4	5	6
SiO_2	46.6	49.0	44.2	48.8	40.5	49.5
Al_2O_3	18.8	17.9	8.48	9.30	8.7	15.9
Fe_2O_3	nil	nil	nil	nil	nil	3.0
FeO	9.7	7.9	22.5	18.6	19.0	8.0
MnO	n.d.	0.15	0.29	0.27	0.25	0.17
MgO	11.0	9.7	11.2	9.46	7.6	6.6
CaO	11.6	12.0	9.45	10.8	10.2	10.0
Na_2O	0.37	0.36	0.24	0.26	0.50	2.7
K_2O	0.12	0.73	0.03	0.03	0.29	1.0
TiO_2	1.25	0.97	2.26	1.46	11.8	1.9
P_2O_5	n.d.	0.22	0.06	0.03	0.18	0.33
Cr_2O_3	0.26	0.20	0.70	0.66	0.37	n.d.
S	n.d.	n.d.	0.05	0.03	n.d.	n.d.
H_2O	nil	nil	nil	nil	nil	0.9
Total	99.6	98.98	99.46	99.08	99.67	100.00

1 = low-K Fra Mauro basalt, Lunar Highlands (quoted by S.R. Taylor, 1975, p.234, table 5.8).
2 = Fra Mauro "KREEP" basalt, Lunar Highlands (average of six analyses) (S.R. Taylor, 1975, p.228, table 5.6).
3 = olivine basalt, Apollo 15, maria (S.R. Taylor, 1975, p.136, table 4.3).
4 = quartz basalt, Apollo 15, maria (S.R. Taylor, 1975, p.136, table 4.3).
5 = high-K basalt, Apollo 11, maria (S.R. Taylor, 1975, p.136, table 4.3.).
6 = average of analyses of terrestrial basalts (Manson, 1967, p. 226, table VI, analysis 1).
n.d. = not determined.

surface including much of the maria basins and thus accounting for their relatively smooth surfaces. In contrast to the only available brecciated and shocked samples of Lunar Highland plutonic and extrusive rocks many samples of these maria lavas, as old as or older than any known rock on Earth, are beautifully fresh and unmarred in appearance. Chemically comparable in some ways with terrestrial basalts (see Table 12.1), the lunar maria basalts are characterized by a complete lack of ferric iron, the presence of small amounts of native iron and iron sulphide in the groundmass, variable but generally high Ti, lack of hydroxyl-bearing minerals, and low alkali contents that result in plagioclase typically having a bytownite-to-anorthite compositional range. Augite is often subtly and irregularly zoned. A high content of the refractory elements Ti, Zr and Y is reflected in the presence of an unusual suite of accessory minerals including two first discovered in lunar basalts: armacolite, $(Fe,Mg)Ti_2O_5$, and tranquillityite, $Fe_8(Zr,Y)_2Ti_3Si_3O_{24}$. The REE, although abundant, show a flat chondrite-normalized pattern with the exception of a negative europium anomaly. This is held to reflect derivation of the basalt magmas from a source area depleted in the components of plagioclase below the anorthositic gabbro crust, itself containing abundant plagioclase complementarily enriched in europium. Some lunar basalts, when $\sim 95\%$ crystallized, show textural evidence of immiscible splitting into two liquid fractions, one represented by an iron-rich glass (SiO_2 46 wt.%, FeO 32 wt.%) and the other by a high-silica glass (SiO_2 75 wt.%) resembling a potassium-rich granite in composition. It was in part the discovery of immiscible relationships in these lunar basalts that revived interest in the possibility of immiscibility in terrestrial rocks (see Chapter 7).

At the present day the Moon has a rigid subsolidus "lithosphere" extending to a depth of 1000 km, and an interior "asthenosphere" extending from 1000 to 1740 km (centre), within which a partial melt seems to be present, and which may contain a differentiated core. Constraints on the amount or even existence of any metal core include the low overall density of the Moon (3.34 g/cm^3, compared with 5.52 g/cm^3 for the Earth) and a very low magnetic field. Contemporaneous igneous activity on the Moon is apparently restricted to lunar "transient phenomena", i.e. local degassing at some craters coupled with seismic activity.

12.3. ASTROBLEMES AND RELATED IGNEOUS ACTIVITY ON THE EARTH

As noted above, the Moon's surface bears testimony of a continuing meteorite bombardment decreasing both in flux and in the probability of large impacts with time. This same flux would be anticipated on Earth, and the extra terminal acceleration due to the Earth's greater gravity means that all but the smallest meteorites (which, of course, are subjected to a greater amount of frictional braking and volatilization due to the Earth's atmosphere) will impact on Earth with even greater velocity and consequent explosive effect. Heightened interest in lunar surface morphology in the late 1960's was reflected in the growing number of recognitions and descriptions of astroblemes on the Earth; contrast, for example, the 110 recognizable astrobleme structures listed in 1966 with the further 17 listed in 1969, recorded in the definitive publications of Freeberg (1966, 1969).

With reference to the clear evidence of saturation of the Moon's surface by large impact craters prior to 3.9 Ga, no evidence of a presumably parallel episode on Earth remains where the oldest known dated rocks are ~3.7—3.8 Ga. Even since that time a great many craters must have been obliterated by processes of erosion and major tectonic reworking of the Earth's crust that have no parallels on the Moon. Although meteorite flux has decreased markedly since early Precambrian times, most easily recognizable astroblemes on Earth are in fact late Phanerozoic in age, although some are older, more or less vestigial, circular structures preserved in hard-rock Precambrian terrain, notably, for example, Clearwater Lakes (Bostock, 1965; re-interpretation by McIntyre, 1968), Mistastin Lake (F.C. Taylor and Dence, 1969), and Brent (Lozej and Beales, 1975) in the Canadian shield (see also references to many others in annual indices of the *Canadian Journal of Earth Sciences*). As some 70% of the Earth's surface is covered by ocean, some catastrophic effects of large meteorites plunging into seawater might be anticipated in the stratigraphical record, but no one seems to have presented evidence to document this.

In general it has been shown (see, e.g., von Engelhardt and Stöffler, 1974) that at the moment of impact of a large meteorite two immensely powerful shock waves are propagated, one travelling back through the impacting meteorite, the second travelling into the Earth below the point of impact. As a result of these the meteorite itself and a hemispherical mass below the point of impact are volatilized. A more distant hemispherical shell within the Earth will be melted, and a yet more distant shell will be partly melted, brecciated, and shocked to varying degrees involving the momentary high-pressure—high-temperature effects associated with the rapid passage of the shock wave — *shock metamorphism* (French, 1968). The accompanying violent explosive effect forms an explosion crater with much brecciated material ejected laterally with low trajectories and much of the molten material ejected upwards in high trajectories. The accumulation of these products results in a cover around the meteorite crater of a widespread blanket of fragmental material (the Bunte Breccia of the Ries crater, for example) overlain by a more proximal ring of molten material partly mixed with fragmental material. The latter rock is known as *suevite* or suevite breccia, containing shocked fragments of country rock and pumice-like material (the latter often showing pronounced aerodynamic shapes) in varying proportions, the whole rock bearing a superficial resemblance to ignimbrite. These relationships and rock types are beautifully displayed by the Ries crater (diameter 24 km, age 14.8 Ma), the youngest large astrobleme on Earth, and once the subject incidentally of a classic debate as to an "internal" vs. "astrobleme" origin (Bucher, 1963; Dietz, 1963).

An analogous suevite deposit, the "Onaping Tuff", lies within the central portions of the *Sudbury Basin*, a much larger astrobleme, diameter 60 km, age 1700 Ma (see articles in Guy-Bray, 1972). The Sudbury Basin was the locale of a large, composite funnel-shaped intrusion comprising norites, quartz norites and granophyres developed from a parental quartz-rich norite emplaced after the impact (Naldrett et al., 1970). In the footwall of the composite intrusive body there are numerous, very large inwardly directed shatter cones and near the lower contact of the intrusive there are intrusive units that have solidified to produce sulphide bodies of considerable economic interest on account of their content of nickel and copper. At one time it was conjectured that the nickel content of the sulphides might in some way be related to the composition of an impacting iron meteorite. However, it now seems certain that intrusive silicate and sulphide magmas were generated entirely from internal sources within the Earth (see Naldrett and Kullerud, 1967), following both an instantaneous, residual temperature rise after passage of the shock wave and a gradual upward movement of mantle to achieve isostatic compensation after the gratuitous excavation of the astrobleme crater. This generation of mafic magma at some time after impact is echoed, of course, on the Moon where many extensive areas underlain by lunar basalts are obviously spatially related to large astroblemes; in this context see a discussion (Hulme, 1974) on the time interval required for internal recovery and remelting.

An astrobleme origin has been reviewed (Rhodes, 1975) for the *Bushveld* structure, site of the largest single igneous intrusion on Earth and believed by some to be the largest impact structure recognized on Earth. Multiple intrusions of tholeiitic magma were erupted resulting in numerous sills below the level of the massive Bushveld Complex itself, which was itself apparently formed by the thickening due to numerous rapid accessions of tholeiitic magma resulting in a very large magma chamber indeed. The resultant crystallized thick cumulate sequence follows a typically tholeiitic trend from orthopyroxene peridotites to highly fractionated ferrogabbros. The volume of magma produced, $\sim 2 \cdot 10^5$ km^3, well within the relatively short period of the cooling time of the intrusion as a whole (~ 1 Ma) represents a greater rate (though not necessarily a greater amount) of magma production than that of a flood basalt province, and indeed there does not appear to have been any plausible internal tectonic cause for magma production at this rate. Following Hulme, the depth of magma generation is related to crater diameter; the Sudbury astrobleme appears to have generated quartz tholeiite at relatively shallow mantle depths, the larger Bushveld structure tholeiite at greater depths.

To the extent of small superficial deposits of suevite and the much larger tholeiitic complexes accompanying a few larger astroblemes (Stillwater also ?), meteorite impact can thus be seen to have had a capricious effect on the Earth's igneous history.

12.4. ARCHAEAN IGNEOUS ROCKS

12.4.1. Introduction

Archaeozoic time refers to earliest Precambrian time extending from the beginning of the Earth's history to ~2.5 Ga ago. The oldest Archaean rocks yet known of this aeon have yielded ages approaching 3.8 Ga, and it is convenient therefore to distinguish a subdivision, proto-Archaean, to refer to (hypothetical) crustal rocks formed during the earliest times of all in the Earth's history, namely older than 3.8 Ga, of which we have no direct rock record.

A characteristic attribute of Archaean rocks is their occurrence in greenstone—granite terrain, and the above younger limit refers to the cessation of development of this distinctive igneous and tectonic style and consequent stabilization of the craton; the exact age of this apparently varies from ~2.8 to 2.4 Ga between different cratonic provinces. Archaean is thus a well-understood term but not precisely defined in terms of time, although it is very often used (technically incorrectly) as a time division. The upper boundary to the Archaeozoic aeon is taken by some workers to be 2300 Ma, defined chronostratigraphically by the appearance of widespread algal carbonate rocks.

Greenstone belts, composed of mafic and other volcanic rocks, ironstones, cherts and other sediments, reveal an apparently systematic development through the Archaean from relatively small and thin sequences forming arcuate outcrop patterns in older terrains to later, wide massive belts containing very thick sequences. In view of this pattern of overall regularity, greenstone belts are unlikely therefore to signal the sites of astrobleme-induced mafic eruption (see D.H. Green, 1972). The intervening "granite" areas are variable in their lithology and comprise both relatively high-level plutons and gneiss complexes, in places migmatitic, in upper amphibolite to granulite facies, inferred to have originated at lower crustal levels. In some areas this latter type of gneissic terrain may predominate to the point of virtual exclusion of recognizable greenstone belt material. Controversy has centered on the nature of the relationship between greenstone belts and granite, i.e. whether greenstone belts were everywhere laid down upon an older granitic basement, or, conversely, whether they were intruded by the surrounding granite. These two possibilities are not mutually exclusive given the possibility of remobilisation of basement [see, for example, the classic demonstration of mantled granite domes by Eskola (1949)]. More controversy centres on the relative proportions of freshly-derived mantle material and reworked basement in terrain constituting recognizably younger Archaean (and Proterozoic) metamorphic/intrusive events within gneissic areas.

Archaean rocks are exposed at the Earth's surface today in nearly a score of isolated cratons generally separated by Proterozoic and younger fold belts. The largest Archaean fragment is the Superior province of the Canadian shield extending into northern regions of the conterminous U.S.A.; other exposed fragments include the Slave province of the Northwest Territories; the North Atlantic craton now fragmented by recent plate separation into a part of Labrador, the pre-Ketilidian massif of southern Greenland, and parts of northwest Scotland; the oldest part of the Fennoscandian shield in northern Finland and adjoining areas; the Ukrainian and Aldan massifs of the U.S.S.R.: the Transvaal, Rhodesian and Tanzanian cratons of southern Africa; the Dharwar Complex of southern India and other small Archaean remnants to the north; the Yilgarn and Pilbara "blocks" of Western Australia; parts of the Brazilian and Guyanan shields, and a part of the Eastern Antarctica craton. All these Archaean cratons reveal many points of similarity in their development, which is, of course, essentially an igneous story, thus justifying considering them together [see Read and Watson (1975) for an informative factual overview]. The subject is not only of great intrinsic interest, but also of great topical interest — see, for example, the papers comprising *The Early History of the Earth* (Windley, 1976), and *Evolution of the Earth's Crust* (Tarling, 1978).

12.4.2. Oldest crustal rocks

At a time when the age of oldest greenstone belt rocks were gradually being established as just exceeding 3 Ga it came as a surprise (L.P. Black et al., 1971) to learn that some lithologically rather drab-looking grey tonalitic gneisses in west Greenland yielded significantly older isochron ages, and correspondingly primitive initial isotopic ratios, based both on Pb and Rb—Sr isotopic data, in the approximate range 3.6—4.0 Ga, later revised to the range 3700—3750 Ma* (Moorbath et al., 1972). Since this exciting discovery Archaean rocks of comparable antiquity have been proved in Labrador, Minnesota, Rhodesia and South Africa. The geological record in west Greenland remains, however, easily the best exposed and documented; the most recent definitive description of it is by Bridgwater et al. (1976) from which beautifully illustrated account the following abridgment has largely been made [see also a review of the development of the Archaean gneiss complex in the whole North Atlantic region by Bridgwater et al. (1978)].

---

*These later figures were based on Rb—Sr dating assuming a half-life for ^{87}Rb of 50 Ga. In view of the recent experimental determination (Davis et al., 1977) of this half-life to be 48.9 ± 0.4 Ga, the above Rb—Sr isochron ages should be reduced by ~2%. Note that in determining Rb—Sr isochron ages prior to 1977 most laboratories have assumed half-lives of *either* 50 Ga *or* 47 Ga — a 6% difference!

The oldest rocks so far identified in west Greenland are the *Isua* supracrustals outcropping in an arcuate belt ~15 km long and no more than 2 km across, probably a fragment of a once more extensive belt as suggested by the distribution of inclusions in surrounding gneissic granitic rocks. Rock types of the group include magnesian ultramafic rocks, metabasalts, metasediments, banded magnetite-quartz ironstone and metaconglomerates containing cobbles of acid volcanic rocks. Lithologically, the assemblage thus affords a comparison with younger greenstone belts. The ironstones have yielded a Pb—Pb whole-rock isochron age of 3760 ± 70 Ma. The presence of the clastic components requires that a yet older basement must have been in existence. The Isua supracrustals mantle a gneiss dome which has intrusive contacts with the supracrustal rocks. This particular gneiss has yielded a Rb—Sr whole-rock isochron age of 3700 ± 140 Ma, a date which is concordant within experimental error with that of the more widespread Amîtsoq gneisses.

The *Amîtsoq gneiss group* is the older of two major groups of quartzofeldspathic gneisses in the region, and is characterized by the abundance of tectonized and amphibolitized dyke intrusions within it, known as the *Ameralik dykes*. Amîtsoq gneisses have yielded consistent ages of ~3650 Ma (zircon concordia and Pb—Pb whole-rock isochron data) and in the range 3700—3750 Ma (Rb—Sr whole-rock isochrons, using the "long" half-life). Generally, the Amîtsoq gneisses are inhomogeneous and heavily reworked; locally more homogeneous bodies of grey tonalitic and granodioritic gneiss material prevail. The group is interpreted as a syn-tectonic, calc-alkaline, intrusive suite.

Although perhaps not all of the same age, the name *Malene supracrustals* is given to numerous infolded belts comprising amphibolites derived from ultrabasic and basic lavas in part pillowed, layered basic sills, and minor pelites, calc-silicate rocks and thin quartzites, that together form ~10—20% of the complex. These rocks have not been reliably dated.

Concordant layers of gabbros, leucogabbros and anorthosites are widespread in the Amîtsoq gneiss terrain, best studied in the *Fiskenaesset complex*, where igneous layering, mineralogies and cumulus textures have been remarkably well preserved in thick competent slabs that have remained in part untectonized.

A basic dyke swarm, now amphibolite, known as the *Ameralik dykes* cuts Amîtsoq gneisses as noted above. Some apparently comparable dykes cut Malene supracrustal rocks and rocks of the gabbro—anorthosite suite; in some areas, however, dykes cutting Amîtsoq gneisses do not continue into adjacent Malene supracrustal rocks. Notwithstanding this minor uncertainty as to age, these Ameralik dykes usefully serve to distinguish a younger quartzofeldspathic gneiss unit, the Nûk gneiss, which is not cut by these dykes, from the Amîtsoq gneisses.

The *Nûk gneisses* have intrusive relationships into the complexly tectonically-interlayered Amîtsoq gneisses, Malene supracrustals, the gabbro—anorthosite suite, and Ameralik dykes. Nûk gneisses from one area have yielded a whole-rock Rb—Sr isochron age of 3040 ± 50 Ma with an initial $^{87}Sr/^{86}Sr$ ratio of 0.7026 ± 0.0004. The Nûk gneisses include hornblende gabbros and ultramafic cumulate layers interlayered with greatly predominating biotite tonalite and biotite granodiorite. Like the Amîtsoq gneisses before them, the Nûk gneisses seem to constitute a syntectonic calc-alkaline suite.

A later group of *miscellaneous granitic rocks* comprises discrete bodies of granites, areas of granitization, and bodies of charnockitic granite. In age, most of this granitic activity intervenes between the emplacement of the Nûk gneisses and a younger *granulite facies metamorphism*, well dated at around 2800 Ma (it is possible that some of the charnockitic granites were emplaced under conditions of granulite facies metamorphism, although some yet younger granitic rocks postdate the granulite metamorphism).

The apparently still younger *Tartoq* supracrustal group, comprising mainly greenschists, were deposited on and tectonically juxtaposed with the earlier much more complexly deformed rock units. *Granitic rocks* intrude the margins of Tartoq greenschists.

In summary, the igneous elements of this magnificently exposed early Archaean crustal fragment comprise at the very least three episodes each of interdigitating mafic rock and granite eruption at the level exposed. Whereas a mantle derivation of the basic dykes, gabbro—anorthosite layer complexes, and various greenstones is unquestioned, the derivation of the granitic rocks has excited debate. Needless to say, all sialic crust must necessarily represent a fractionation product at some time of mantle material. However, there are arguments, based on REE distribution and inferred thermal gradients, to suggest that, despite their low initial $^{87}Sr/^{86}Sr$ ratios, a proximate lower-crustal granulite source for the calc-alkaline granitic material cannot be ruled out (see, e.g., the synthesis presented by Collerson and Fryer, 1978).

12.4.3. Greenstone belts

Within Archaean terrain, granitic gneisses and granites (sensu lato) account on the average for $\sim 70\%$ of outcrop area and greenstone belts the remaining 30%. Greenstone belts are thus more abundant overall than they are in the fragment of Archaean craton we have so far considered from west Greenland. Whereas attempts at interpretation of the genesis of the granitic rocks suggest a fundamental story of the process of cratonisation, the greenstone belts, mainly volcanic, partly sedimentary, are apparently remnants of superficial coverings. Nevertheless they contain many features of great interest to the igneous petrologist and their distinctive structural style must in some way be related to large-scale crustal processes in Archaean time. It is a generally held view that the apparently great differences between granite—greenstone terrain on the one hand and amphibolite to granulite facies gneiss terrain on the other hand reflect merely different depths of erosion of comparable Archaean crust. Certainly, it appears that there is no systematic age difference between them. The greenstone belts, while having many features in common, display a trend of increasing dimensions and complexity with time. We shall take as two type examples the Barberton greenstone belt, a compact, relatively old but quite large belt within the Transvaal craton, and the Abitibi belt, the largest and probably the youngest in the Superior province. In addition, some analogies with other areas will be made, principally in connection with the distinctive highly magnesian komatiites of the lower part of the Barberton sequence.

12.4.3.1. The Barberton greenstone belt. With an age of close to 3.4 Ga and underlying an area with maximum dimensions 120 × 40 km, the well-exposed rocks of the Barberton greenstone belt, situated in the Union of South Africa near its border with Swaziland, have been the subject of a very full presentation by M.J. Viljoen and R.P. Viljoen (1969). With the exception of localized cover rocks, the greenstone belt is bounded by granitic rocks, many of which occur as discrete diapiric plutons. The granitic rocks surround rather than penetrate the greenstone belt rocks, and no one granitic pluton is totally surrounded in outcrop by greenstones. The marginal piercement structures of the diapirs result in a succession of embayed concave contacts with the rocks of the belt itself (Fig. 12.1). The greenstone belt sequence has the enormous aggregate thickness of over 17 km. The greenstone belt rocks are ubiquitously steeply dipping and arranged in an overall synclinal structure like the lower hull of a ship except that the keel is not exposed to view. There are several smaller internal synclinal structures often separated by faults in the positions where anticlinal crests might have been anticipated. Axial cleavage is conspicuous by its absence, and vertical lineations suggest extension vertically downwards rather than horizontal compression. An overall northeast strike is locally strongly modified near contacts into conformity with the contact against encircling granitic rock. Metamorphic grade varies systematically from virtually nil centrally to greenschist facies marginally and locally to amphibolite facies in narrow zones near granite contacts and in slivers of greenstone belt rocks (usually ultramafic) included within the granite itself. The whole structural picture gives a strong impression of downward sinking and extension of the greenstone belt rocks relative to surrounding granitic gneiss and granitic diapirs.

The sequence of greenstone belt rocks with maximum thicknesses of the different groups is as follows:

(3) *Moodies Group* (uppermost), 3.7 km; mainly arenaceous, consisting of a basal conglomerate followed by thick, cyclic units of quartzite and shale.

(2) *Fig Tree Group*, 2.1 km; mainly pelitic, comprising turbidites, cherts, ferruginous cherts and tuffs of intermediate composition, and showing considerable lateral facies changes.

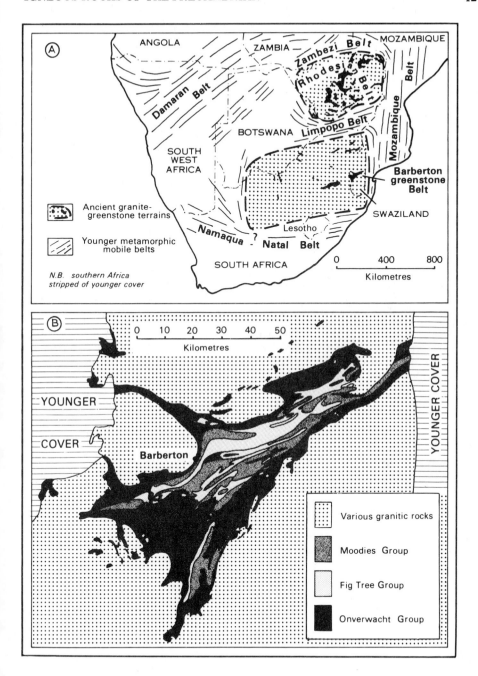

Fig. 12.1. A. Geological setting of the Barberton greenstone belt in relation to the Archaean terrain of the Kaapvaal and Rhodesian cratons and adjacent Proterozoic mobile belts.
B. Sketch map of the geology of the Barberton greenstone belt showing the overall relationship of the three groups referred to in the text.
(After M.J. Viljoen and R.P. Viljoen, 1969, pp. 10 and 11, figs. 1 and 2.)

(1) *Onverwacht Group*, subdivided into two:

(a) an upper "mafic to felsic unit", 7.7 km; this comprises repeated cycles of tholeiitic lavas passing upwards into felsic lavas with interbeds of tuffaceous sediments, tuffs and agglomerates. The felsic rocks are progressively more abundant in higher cycles of the succession. Some ultramafic sills occur in lower cycles.

(b) lower "ultramafic unit", 7.5 km; this comprises a komatiite suite ($\sim$60—70% of the unit) of ultramafic lavas, sills and pods (metaperidotites), metabasalts of high-Mg type, and tholeiitic basalts, and minor amounts of acidic rocks towards the top.

Within the lower unit of the Onverwacht sequence there are abundant fine-grained mafic rocks in part of ultrabasic chemistry and apparent ultramafic mineralogy. The very lowest part has everywhere been partly assimilated by granite, the grade of metamorphism is highest there, and the effects of carbonate replacement, silicification and serpentinisation upon the major-element compositions of these rocks needs careful consideration (M.J. Viljoen and R.P. Viljoen, 1969, pp. 29—55). Nevertheless, in thin section enough of the primary texture can be distinguished to say that many rocks, some of them obviously pillowed flows, some with distinctive quench textures (see p.163), contained extraordinarily high amounts of modal olivine. Chemical analyses show magnesia ranging from levels characteristic of common basaltic rocks up to as high as 30 wt.% MgO.

A family name, *komatiite*, was given to this newly recognized class of igneous rocks (M.J. Viljoen and R.P. Viljoen, 1969, pp. 55—85). Major-element chemical peculiarities of komatiite are overall high MgO accompanied by a very high MgO/FeO ratio, low alkalis and TiO_2, and an Al_2O_3/CaO ratio less than unity. Those with an MgO content in the range 10—20 wt.% were termed *basaltic komatiite*, and those with still higher-MgO *peridotitic komatiite*. The Viljoens recognized the great danger implicit in assuming that alteration was isochemical and they quoted (M.J. Viljoen and R.P. Viljoen, 1969, p.40) some alarming figures on the apparent diminution in contents of some major oxides, notably lime and alkali oxides, with increasing serpentinisation and content of H_2O. Notwithstanding this they felt able to quote the average chemical compositions of several komatiite lava types (see Table 12.2).

The Viljoens drew attention to the occurrence in numerous localities within the peridotitic komatiites of "crystalline quench textures" composed of large (up to 7 cm) bladed, skeletal, now serpentinised, olivine crystals set in a matrix containing a smaller but similarly bladed set of (now tremolitised) augite crystals.

Within the lower Onverwacht Group there are several types of layered and differentiated ultramafic bodies paralleling in their bulk compositions those of komatiite. The "Kaapmuiden" type sills, for example, attaining thicknesses in excess of 600 m, consist of thick lower dunitic and peridotitic zones passing upward into narrow zones of bronzitite and websterite terminated upward by a very narrow anorthositic norite—gabbro zone. Other sills of similar overall composition are characterized by prominent rhythmic layering in the lower ultramafic part of the sequence.

The upper "mafic to felsic unit" of the predominantly volcanic Onverwacht succession is mainly composed of tholeiites, locally pillowed in the lower part of the sequence with a considerable proportion of subaqueous pyroclastic material towards the top. Several cyclic repetitions of thick basaltic units containing some ultramafic pods at their base overlain by thinner acidic flows and pyroclastics can be distinguished. Some of the acid flow rocks are pillowed — an unusual feature. The proportion of acid rock within these cyclic units increases upwards in the pile and there is often a considerable lateral variation in the thickness of acid units. Chert beds commonly cap the acid rocks and at the Sheba Queen mine where underlying acid rocks are thick and probably indicate a vent area, the cherts host a stratiform auriferous pyrite deposit. Zircons from one of these acid rocks have yielded an age of 3360 Ma, the oldest yet recorded from Barberton. The acid rock types, now keratophyres, apparently comprised dacites, rhyodacites, and possibly rhyolites. Overall the felsic lavas comprise no more than 10% of the succession of this upper unit of the Onverwacht Group.

The greater proportion of material in greenstone belt terrain is "granite", often left a featureless pink on geological maps. In the Barberton region there appear to be at least three recognizable divisions within the granite terrain:

(1) Ancient tonalitic gneisses including some granulite facies rocks; these gneisses may in part represent older basement to the greenstone belt itself, although the gneisses at least in part appear to surround and pierce the margins of the greenstone belt. Unequivocal unconformable basement—cover relationships have not been seen in the Barberton granite—greenstone terrain, although an unconformity between a slightly younger greenstone belt and underlying basement gneiss has been recently demonstrated within the Rhodesian craton (Bickle et al., 1975).

(2) A later episode of emplacement at fairly high crustal levels of potassium-rich granites known as the "hood" granites; these are homogeneous in places but grade locally via zones of migmatite into the tonalitic gneisses.

(3) A distinctly younger set of medium- to coarse-grained, often conspicuously porphyritic, granite plutons with ages in the range 2.8—2.5 Ga.

The assemblage could be regarded as complementary, although over a relatively shorter time interval, to the events portrayed at lower crustal levels in west Greenland.

12.4.3.2. Komatiites from other areas. Although the Barberton greenstone belt has been lucidly described in considerable detail, disadvantages in interpretation include the incomplete preservation and metamorphism of the voluminous komatiites, its most interesting rock type, that occur adjacent to, and partly dismembered within remobilized and younger granitic rocks.

TABLE 12.2

Representative analyses of high-magnesium basalts including komatiites of the Barberton distric[t]

	1	2	3	4	5
SiO_2	45.94	47.49	50.02	53.29	53.7?
Al_2O_3	2.98	3.45	7.17	5.50	10.0?
Fe_2O_3	6.23	6.39	1.25	1.00	1.2?
FeO	4.80	5.86	8.53	9.06	9.9?
MnO	0.18	0.20	0.23	0.22	0.2?
MgO	33.79	26.92	21.53	15.56	10.3?
CaO	4.73	7.40	8.78	13.09	10.1
Na_2O	0.15	0.52	0.41	1.23	2.7?
K_2O	0.03	0.05	0.06	0.09	0.4
TiO_2	0.34	0.55	0.49	0.57	0.8
P_2O_5	n.d.	n.d.	n.d.	n.d.	n.d.
H_2O^+	—	—	—	—	—
H_2O^-	—	—	—	—	—
CO_2	—	—	—	—	—
LOI[*3]	—	—	—	—	—
NiO	n.d.	n.d.	n.d.	n.d.	n.d.
Cr_2O_3	n.d.	n.d.	n.d.	n.d.	n.d.
Total	99.17[*4]	98.83[*4]	98.47[*4]	99.61[*4]	99.6

1 = average "peridotitic komatiite", Komati Formation, Barberton greenstone belt (M.J. Viljoen and R.P. Viljoen, 1969, table V, facing p.80, analysis 2).

2 = average "peridotitic komatiite", Sandspruit Formation, Barberton greenstone belt (M.J. Viljoen and R.P. Viljoen, 1969, table V, facing p.80, analysis 1).

3 = average "basaltic komatiite, Geluk type", Barberton greenstone belt (M.J. Viljoen and R.P. Viljoen, 1969, table V, facing p.80, analysis 5).

4 = average "basaltic komatiite, Badplaas type", Barberton greenstone belt (M.J. Viljoen and R.P. Viljoen, 1969, table V, facing p.80, analysis 4).

5 = average "basaltic komatiite, Barberton type", Barberton greenstone belt (M.J. Viljoen and R.P. Viljoen, 1969, table V, facing p.80, analysis 3).

[*1] Total Fe as Fe_2O_3.
[*2] Total Fe as FeO.
[*3] Loss on ignition.
[*4] Recalculated water-free.
[*5] Recalculated to eliminate total loss on ignition (LOI).

A comparable suite of rocks has been described by D.A.C. Williams (1972) from the Mount Monger area of the Yilgarn Block, Western Australia. The succession recognized in the Mt. Monger area is ~11 km thick within which a cycle, comparable to the Onverwacht Group, containing tholeiites and komatiite-like rocks, unconformably overlies an earlier cycle ending with felsic volcanics and sediments. Williams finds difficulty in applying the term komatiite because of differing CaO/Al_2O_3 ratios, or even accepting it as a term because its definition was based on altered and probably metasomatised

6	7	8	9	10	11	
47.4	48.3	40.8	45.6	44.0	45.1	SiO_2
8.4	9.1	10.0	8.75	8.3	10.8	Al_2O_3
		2.94	13.55*1	2.2	2.4	Fe_2O_3
0.6*2	10.4*2	6.49		8.8	8.1	FeO
0.20	0.20	0.16	0.22	0.19	0.18	MnO
4.8	21.7	23.3	22.96	26.0	19.7	MgO
7.4	9.0	6.86	7.73	7.3	9.2	CaO
0.10	0.13	0.23	n.d.	0.90	1.04	Na_2O
0.02	0.03	0.07	n.d.	0.06	0.08	K_2O
0.34	0.43	0.25	0.48	0.58	0.76	TiO_2
0.04	0.07	0.02	n.d.	0.07	0.09	P_2O_5
—	—	5.91	—	0.97	2.14	H_2O^+
—	—	0.86	—	—	—	H_2O^-
—	—	0.28	—	—	—	CO_2
(6.4)	(5.4)	—	—	—	—	LOI*3
0.16	0.14	n.d.	n.d.	0.19	0.12	NiO
0.53	0.43	0.45	0.48	0.36	0.27	Ca_2O_3
100.0*5	100.0*5	98.62	99.77*5	99.9	100.0	

6 = spinifex-textured peridotite, Mt. Monger, Western Australia (D.A.C. Williams, 1972, p.175, table 1, analysis 3).

7 = representative high-Mg basalt (highest MgO content of five quoted analyses), Seabrook Hills, Mt. Monger district, Western Australia (D.A.C. Williams, 1972, p.175, table 1, analysis 5).

8 = spinifex-textured upper part of high-Mg basalt flow unit, Munro, Abitibi greenstone belt, Ontario (Pyke et al., 1973, p. 973, table 1, analysis 3; see also discussion on p.972).

9 = weighted average of five spinifex-textured rocks from the Dundonald sill, Abitibi greenstone belt, Ontario [Pyke et al., 1973, p. 973, table 1, analysis 8, see also discussion on p.972; and note detailed account of the Dundonald sill by Naldrett and G.D. Mason (1968)].

10 = average of twelve picritic basalts, Baffin Island (Clarke, 1970, p.205, table 1, analysis 1).

11 = average of four olivine basalts, Baffin Island (Clarke, 1970, p.205, table 1, analysis 3).

rocks. Representative analyses of mafic rocks from the Mt. Monger area are included in Table 12.2; those containing MgO in the approximate range of 8—24 wt.% are termed simply "high-Mg basalts"*. Thin-section examination shows that these rocks are less altered than the Barberton rocks, as indicated also by their lower H_2O contents. In many places they too are characterized

---

*See also Arndt and C. Brooks (1978) for discussion and references concerning the ongoing debate about definition of the term komatiite.

by comparable **remarkable** quench textures resulting in the bladed and skeletal growth of olivine and augite crystals; these textures have been called *spinifex textures* (after a genus of branched, spiky, desert grass indigenous to Western Australia). Among the high-magnesium basalts there are lenses of ultramafic rock ranging in dimensions from 60 to 600 m in thickness and from 0.12 to 10 km in length. They were apparently emplaced as a suspension of euhedral olivine crystals in liquid wherein the primocryst olivine crystals were in many instances concentrated centrally presumably by processes of flow differentiation. Some settling of primocrysts occurred although cooling was apparently rapid resulting in a second generation of olivine with skeletal quench habit. Williams quotes just over 33 wt.% MgO for the olivine-rich cores of some of these lenses, and estimates ~25 wt.% MgO for the parental liquid.

Naldrett and G.D. Mason (1968) described comparable layered ultramafic sill rocks from the Superior province with quench textures described by them as "bird-track" or "herring-bone". Although the curious textures had been noted previously in the field, this was apparently the first published recognition of them as primary quench textures* (for which the term spinifex is now generally accepted).

Perhaps the best exposures to date of highly magnesian flow rocks are those recorded and illustrated by Pyke et al. (1973, with an excellent brief historical introduction to komatiites) from Munro Township in the large Abitibi belt of the same Superior province. Some 60 flow units are exposed in a stratigraphic thickness of 125 m. Two analyses of spinifex-textured rocks believed to represent melt compositions are included in Table 12.2. Low alkali contents may reflect leaching during alteration (note high H_2O^+ contents); contents of TiO_2, believed to be stable under many conditions of low-grade alteration (see Chapter 14) are consistently low, and all these rocks appear to be of subalkaline affinity.

Analogies to these highly magnesian basaltic rocks are rare in younger rocks. Some primitive highly magnesian tholeiitic basalts of Tertiary age have been recorded from Baffin Bay, west Greenland (Clarke, 1970) with comparable high Cr (see Table 12.2). Olivine has a skeletal, but not spinifex habit in some of these rocks. Gale (1973) has recorded abundant magnesium-rich pillowed mafic flows in a lower Paleozoic ophiolite sequence in Newfoundland. MgO in these rocks averages just over 14 wt.% and TiO_2 has a characteristically low value of 0.16 wt.%. Other "basaltic komatiites" from ophiolite complexes are discussed by Sun and Nesbitt (1978). Other occurrences, including the high—Mg Nuanetsi flows (plume-generated like Baffin Bay?), are noted and discussed by Cox (1978b).

---

*Naldrett (1964) argued, from the existence of these primary quench textures, that liquid magmas of highly magnesian compositions must have been erupted.

12.4.3.3. The Abitibi greenstone belt. The Abitibi greenstone belt, age about 2.7 Ga, the largest in the Archaean Superior province with maximum dimensions roughly 750 × 200 km, has been lucidly described by Goodwin and Ridler (1970), from whose account this abridgement has in large part been made. In keeping with this great size the belt displays a more complex association of rocks than that of the Barberton and many other belts. Of a total area of $\sim 9.5 \cdot 10^4$ km^2, 45.6% is underlain by mafic volcanic rocks ("basalt-andesite"), 2.5% by mafic intrusions, 3.6% by felsic volcanic rocks, 16.0% by sediments and 32.3% by granitic rocks. Within overall mafic assemblages locally exceeding 12 km in apparent stratigraphic thickness a succession upwards from low-potassium tholeiites to high-alumina tholeiites can be discerned. Greenschist-grade metamorphism is ubiquitous but igneous features such as pillow lavas, hyaloclastic rocks, variolitic textures, etc., are commonly well preserved and attest subaqueous extrusion. Locally too, as noted above, spinifex-textured, high-magnesium flows occur together with layered ultramafic to mafic intrusions. Some dozen major anticlinal and synclinal structures roughly parallel the longer axis of the belt. Although significant north—south shortening across these structures seems to be indicated, the degree of structural complexity is not sufficient to obscure the existence of some eleven volcanic complexes. These are identified by concentrations of separate associations of intermediate to felsic volcanic rocks stratigraphically overlying the thick mafic sequence, but unlike the latter showing marked lateral variations in thickness, and commonly associated with epizone intrusions. Within the felsic rocks dacite predominates over rhyolite; trachyte occurs in one centre, Kirkland Lake. This area is also unusual in containing some basanite flows intercalated within a mafic flow sequence. The Abitibi greenstone belt contains several thousand known mineral occurrences and has been host to 150 producing mines. The economic mineralisation consists mainly of auriferous quartz veins and disseminated to massive sulphide deposits (Cu—Zn—Au—Ag), both strongly localized in the intermediate to acid volcanic complexes. Sedimentary rocks within the belt comprise: (1) sequences of volcanogenic sediments up to 3 km thick, often coarse-grained, polymictic, with distinctive chaotic structures, and of proximal derivation from nearby intermediate and felsic volcanic rocks, and (2) sequences of a more regularly bedded greywacke—flysch assemblage up to 2 km thick of a more distal facies. Iron formation rocks, widely distributed in oxide, sulphide and carbonate facies, afford important stratigraphical marker beds. The Abitibi belt is surrounded by foliated granitic rocks displaying apparently intrusive contact relations to the rocks of the belt, although some may represent older remobilized basement — the common greenstone—granite syndrome. Within the belt itself, entirely surrounded by volcanic and sedimentary rocks, there are at least seven major plutons, each 70—100 km in diameter, most situated along major anticlinal axes, the localisation and magnitude of negative gravity anomalies indicates that these plutons are deep-rooted. Preliminary age dating indicates that whereas some of these plutons are much younger than the belt, the ages of others may not be far removed from the time of formation of the belt. A crude bilateral symmetry was recognized on the basis of two major broad east-west-trending zones of mafic to felsic centres with associated clastic rocks and ironstones separated by an intervening median zone underlain by uniform tholeiitic basalts, fine-grained clastic rocks, and some major granitic batholiths (in contrast to any polarised island-arc system). The belt was attributed by Goodwin and Ridler to "downsinking of thin supple Archaean crust" or to "spreading apart of Archaean crustal blocks".

12.4.3.4. Structural setting of Archaean igneous activity. Greater concentrations of radioactive elements in Archaeozoic time would result in a higher rate of heat production and consequently a higher heat flow compared with present crustal conditions. High heat flow could have been achieved by: (1) higher thermal gradients in sialic crust that would be compatible with the observed prevalent granulite—high amphibolite facies of Archaean gneiss terrain; and (2) a greater length of zones of upwelling mantle and accompanying eruption of mafic melts analogous to present-day ocean ridges (where considerable heat loss takes place), but perhaps on a more reticulate scale to allow a greater aggregate length. Rather than leading to separation of rigid lithospheric plates as at present, cycles of voluminous mafic eruptions seem to have led to loading and downsinking of relatively plastic crust resulting in the observed geometry of greenstone belts.

Many attempts have been made to compare the igneous rocks of greenstone belts directly to those found at both constructive and destructive present-day plate margins, in spite of notable dissimilarities. Perhaps the most convincing analogy to present-day igneous activity is plume-generated igneous rocks — see Condie (1975) for a synthesis integrating abundances, time sequence, and relative REE patterns of greenstone-belt igneous rocks.

12.5. IGNEOUS ACTIVITY IN PROTEROZOIC TIME

12.5.1. Introduction

Rocks of Proterozoic age record a much longer time span than do the preserved rocks of either the preceding Archaeozoic or succeeding Phanerozoic aeons. Despite this longer record of geological history Proterozoic igneous rocks are somewhat less distinctive than those preceding or succeeding them and they have in the main excited less interest although this situation is now rapidly changing. Because less is known and disseminated, one has considerable reservations about some of the generalizations that follow.

In megatectonic terms, on the one hand the Archaean is dominated by crustal reworking, the extrusion of mainly mafic volcanic sequences, and considerable vertical movements. On the other hand, we have now come to realize that the Phanerozoic rock record is dominated essentially by the lateral movement of large rigid lithospheric plates so beautifully documented by the present-day distribution of seismic foci, with igneous activity concentrated mainly around their margins (see Chapters 9 and 11), relatively few hotspots over inferred mantle plumes elsewhere (see Chapters 9 and 10), and possible control of some small-scale igneous activity along lithospheric fractures (see also Chapter 10).

The Archaean—Proterozoic boundary has been arbitrarily placed by some geologists at 2.4 Ga, the younger envelope of the many closely spaced ages of the "Kenoran orogeny" in the Canadian shield. However, the implicit concept of a widespread "terminating" orogeny is not tenable in view of the demonstrable great time span of formation of greenstone belts and interdigitating granitic activity. What seems to be a more precise conceptual basis for distinction is the termination of the typically Archaean style of greenstone belt—granite diapir tectonics and the initiation of essentially permanent stabilized continental crust, as shown, for example, by the age of earliest preserved undeformed sedimentary cover rocks and regional dyke swarms. This change in tectonic style, however, appears to be markedly diachronous. For example, the little-deformed sedimentary rocks of the Witwatersrand System in South Africa have yielded a K—Ar age of 2700 Ma and locally overlie the similarly undeformed volcanic rocks of the Dominion Reef Series that are possibly as old as 2800 Ma; this age range is significantly older than that of the age of formation of some typical greenstone belts in other parts of the world [see discussion of this problem by Read and Watson (1975, pp. 13, 178 and 179)].

The age of the inception of movement of the large plates so typical of Phanerozoic time is highly debatable — within a range of estimates of 1800 to 600 Ma! Although convincing ophiolites or high-pressure—low-temperature metamorphic belts have not been recorded from Proterozoic rocks, volcanic sequences comparable to SBZ rocks and possible plate boundary features are known from middle and upper Proterozoic rocks. Whereas we may have to live with a diachronous Archaean—Proterozoic boundary as tectonically defined and discussed above, the Proterozoic—Phanerozoic boundary is, of course, precisely dated by lowest Cambrian macroscopic hard-shelled fossils and begins ~570 Ma ago. It would be asking too much of coincidence for this age based on biological evolution to correspond exactly with a change in major tectonic style.

With some reservations therefore on lower and upper age limits, the distinctive and predominating tectonic style exhibited by Proterozoic rocks consists of wide metamorphic belts, some of them of considerable length. These belts in part rework and in part encircle older Archaean cratons, the assemblage remaining more or less relatively fixed though moving absolutely, as indicated by magnetic pole measurements. These Proterozoic metamorphic belts are apparently not to be construed therefore as representing the margins

of once relatively moving plates but rather as broad belts of intra-plate reactivation. Vast tracts of quartzo-feldspathic terrain were reworked and
radiometrically updated without significant expression of volcanism or
ocean-floor tectonics.

This kind of ensialic orogenesis apparently unrelated to plate tectonics has
been explained in terms of ductile spreading and contracting of the lower
crust (Wynne-Edwards, 1976), the so-called "millipede" model. What essentially is envisaged by Wynne-Edwards is an already thick but just subsolidus
Proterozoic sialic crust (as opposed to possibly thinner and certainly hotter
Archaean crust) moving absolutely over mantle that may still contain significant hotspots or hot lines. The result of these inferred thermal inhomogeneities in the Proterozoic mantle is, in general, not to produce sea-floor
spreading or mantle-derived volcanics working their way through brittle
lithosphere and crust (as in the Phanerozoic), but in the main to heat lower
crust so that the tectonic response of the latter in regions of any relatively
hot underlying mantle is ductile flow with the development of granitic
migmatite terrain below and high-temperature facies metamorphism above.
Locally, as in the case of anorthosite massifs, large volumes of mafic magmas
may be emplaced within sialic crust and cause local downwarping. Locally
too anatectic (?) acid magmas may penetrate to the epizone on an appreciable
scale. Stress may be taken up inhomogeneously by shearing rather than by a
more ductile response, and some mafic and ultramafic rocks may be emplaced
in shear belts.

These concepts do not necessarily exclude the development of plate-
tectonic like features in Proterozoic time. For example, new oceanic-type
crust has in fact developed in places by the rupture of continental crust and
substantial outpouring of mafic rocks. Locally also features reminiscent of
consuming plate boundaries are seen. Rocks characteristic of these environments were not, however, apparently developed on the widespread scale of
those in Phanerozoic time.

All in all, it seems that there is no simplistic model to encompass the
apparently multivaried response of Proterozoic crust to megatectonic forces.
As igneous rocks are conspicuous products of the varied inferred processes it
is appropriate to examine some examples singly, but the breadth of discussion
here cannot equal that provided, for example, by a multidisciplinary approach,
for example that of Windley (1977).

12.5.2. Early mafic intrusives

The Great Dyke of Rhodesia, dated at ~2500 Ma, is an outstanding
example of an undeformed early Proterozoic intrusion. It has a length, including satellite complexes, of 600 km, a width of ~5—8 km, and shows a
synclinal cross-section of layered cumulate rocks. The Great Dyke abruptly
transects typical Archaean greenstone belt and granite terrain, and younger

deformation is limited to a few tear faults. Examination of the layered sequence of the Great Dyke rocks, facilitated by the exposure presented by the synclinal cross-section and also by axial descent, reveals that the "Dyke" preserves parts of four contiguous lopoliths (Worst, 1958). The largest of these has revealed a remarkable *lateral* cryptic layering along the strike of major rhythmic units (C.J. Hughes, 1970a), themselves probably attributable to fresh inflows of magma. The greater part of the outcrop area of the Great Dyke consists of ultramafic olivine, chromite and orthopyroxene cumulates locally overlain by gabbroic cumulates. Several lines of evidence such as the overall composition of the preserved sequence, tholeiitic trend, compositions of cumulus olivine and orthopyroxene in lowermost rocks, abundance of chromium, and compositions of contemporaneous satellite dykes strongly indicate that the parental magma of the Great Dyke was akin to high-magnesium basalt (C.J. Hughes, 1977).

This 2500 Ma old crust was sufficiently rigid to undergo an apparent uniform stress pattern resulting in a discontinuous tensional break ~ 600 km in length, including the satellite dyke complexes. Calculations based on outcrop area, shape, compositions and geophysical restraints indicate that $\sim 3.5 \cdot 10^4$ km^3 of high-magnesium basalt was produced along this length, i.e. ~ 60 km^3 of magma per linear kilometre, within the geologically short period of the cooling time of the intrusion. This event, implying a high proportion of melt produced from the mantle did not lead to plate separation or development of a greenstone belt, two environments in which high-magnesium basalts are known to occur.

In Western Australia, a group of structurally analogous intrusions include the canoe-shaped Jimberlana norite body, 180 km long, maximum width 25 km, date tentatively 2420 Ma, and cutting metamorphic rocks dated at ~ 2670 Ma. The parental magma is inferred to be a high-magnesium basalt (Campbell and Borley, 1974). The subparallel and even larger Binneringie Dyke, 320 km long, maximum width 3.2 km, has vertical layering attributed to congelation cumulates (McCall and Peers, 1970). The presence of inverted pigeonite in marginal olivine-bronzite gabbros and olivine-bronzite quartz-gabbros indicates an evolved parental magma of tholeiitic affinity. It seems then in the Great Dyke suite we are witnessing a representative of a phenomenon that was not petro-chemically or structurally unique in late Archaeozoic —early Proterozoic time.

12.5.3. Proterozoic "orogenic" belts

Wide belts of crustal rocks, often containing abundant migmatite, and yielding a variety of metamorphic ages throughout Proterozoic time are known from all shield areas. Notable among them, because they have been more intensively studied, are rocks of the Churchill province (separating the Superior and Slave Archaean terrains) and the Grenville province of Canada, the Nagssugtoqidian belt in Greenland (lying north of the old Archaean terrain), the Laxfordian component of the Lewisian gneisses of Scotland, the Svecofennides and apparently penecontemporaneous Karelides (bordering older Archaean rocks on the north) which together underlie much of Finland and Sweden, the Limpopo belt of southern Africa (separating the Archaean rocks of Transvaal and Rhodesia), numerous belts in Central Africa (in part bordering older rocks), the Birrimides of West Africa, etc. (see Read and Watson, 1975).

These belts have several noteworthy features. By no means all pre-existing crust was reworked in Proterozoic time as shown by the preservation of Archaean shield fragments and of some early Proterozoic undeformed and unmetamorphosed sedimentary cover sequences; the belts do not show the strongly directional overthrust tectonic style typical of the Alpine and Himalayan chains, for example; volcanism of SBZ type is generally lacking; there are no ophiolites; blueschist metamorphism is absent; the belts are commonly very wide compared to Phanerozoic belts; in places older basement was extensively involved in the deformation and metamorphic recrystallization, notably in the Grenville province and as shown by the "mantled gneiss domes" of the Karelides (Eskola, 1949). This array of features led Wynne-Edwards to propose the millipede model outlined above.

Although these rocks fall primarily within the field of study of a metamorphic petrologist, several features of interest present themselves from the point of view of the igneous petrologist. One is the overall abundance of migmatites. Much of the classic work on these perplexing rocks (see Mehnert, 1968, for a definitive publication) was in fact accomplished by J.J. Sederholm and P.J. Holmquist on the Proterozoic rocks of Finland and Sweden, a tradition maintained by Scandinavian petrologists such as P. Eskola, V. Marmo and M. Härme.

12.5.4. Shear belts, some with associated mafic and ultramafic rocks

One feature of Proterozoic deformation is inhomogeneity both in metamorphic temperatures attained and in strain from place to place, the latter varying sometimes markedly over short distances within the same belt. Sutton and Watson (1951), for example, have described localized shear belts within terrain reworked by Laxfordian metamorphism that locally contains abundant migmatite apparently developed during the same tectonic episode. Perhaps a relative abundance of pre-existing granitic rocks was responsible for localising abundant migmatite; often too, interlayered sedimentary layers have suffered a great deal of strain. Wynne-Edwards (1976, pp. 934—941) has drawn attention to major "transformal shear zones" within the Grenville province and relates these features to differential movements between segments of crust creeping at slightly different rates.

Two prominent "fronts" border the Churchill province along its contact with the older Archaean rocks of the Slave province to the northeast and the Superior province to the southwest, known as the Thelon and Nelson fronts, respectively. Across both of these fronts there is a rapid transition from Archaean structural trends to younger Proterozoic trends paralleling the front. The Nelson front contains numerous pods of metamorphosed peridotite.

12.5.5. Anorthosites

Anorthosite massifs were apparently emplaced in mid-Proterozoic time into high amphibolite to granulite facies terrain. The greatest concentration of anorthosite is to be found in the Grenville province and adjacent areas to the north, notably Labrador. Known ages cluster fairly closely around 1400 Ma, and it is instructive therefore to examine an anorthosite body north of the Grenville front and thus unaffected by any younger metamorphism of Grenville age.

One such body is the Michikamau intrusion described by Emslie (1970). Situated in Labrador some 100 km north of the Grenville front, the Michikamau intrusion has an outcrop area of ~ 2000 km^2. A reconstruction allowing for tilting and faulting indicates that it is a massive funnel-shaped intrusion, depth at least 15 km. Within a thin envelope of chilled and presumably parental subalkalic basic magma composition there is a layered troctolitic cumulate sequence with structures comparable to those of other layered basic intrusions and containing mainly olivine and plagioclase as cumulate phases, grain sizes of 1—3 mm. Cryptic variation in the plagioclase, for example, ranges upward from An_{72} in lowest structural levels exposed to An_{50} in highest levels. Above this layered series there is a massive 3.5 km thick cap of crudely layered anorthosite formed by the *upward* accumulation of plagioclase crystals (the sole cumulus phase) some of them very large, commonly 1—3 cm, rarely up to 50 cm, presumably reflecting a long growth history while suspended in the magma before their eventual upward accumulation. The cryptic variation in this massive anorthosite layer ranges *downward* from compositions of An_{55} at the top to An_{44} at the bottom, thus with a curious overlap of cumulus plagioclase compositions in the apparently downward and upward accumulated sequences. Present in a "sandwich" layer between the prominently layered series below and the anorthosite above and also intruding the anorthosite are bodies of more or less homogeneous hedenbergite quartz-diorite and fayalite adamellite compositions, presumably representing samples of residual magmas resulting from strong fractionation.

In contrast to this easily comprehensible intrusion the bodies of anorthosite within the Grenville Province proper have proved much more difficult to interpret. Gravity data suggest that some of them may overlie large masses of basic rock, although some may have become detached from their basic roots during deformation. Mineralogically, these Grenville anorthosites show spectacular metamorphic recrystallization textures. Many show varying proportions of dark, strained, antiperthitic plagioclase primocrysts with curved twin lamellae within a matrix of pale, metamorphically recrystallized, granoblastic plagioclase showing an approach to straight-line boundaries and triple-point junctions; streaky patches of originally intercumulus dark mineral are also commonly recrystallized to granular aggregates of augite, hypersthene and opaque mineral. The deformation structures visible within the large plagioclases are thus a metamorphic phenomenon, not an igneous one. They gave credence at one time to a proposed origin for anorthosite by tectonic filter pressing of a crystallizing crystal mush rich in plagioclase (the more easily accessible anorthosite massifs within this whole area lie in fact within Grenville metamorphic domain). In this connection, however, it now appears probable that the anorthosites young southwards and that some may have been emplaced during a phase of Grenville metamorphism and under conditions approaching granulite facies. Experimental work has shown that under high P_{H_2O} the anorthite—diopside eutectic temperature is depressed and the eutectic composition is shifted markedly toward a more plagioclase-rich composition (see Fig. 6.19). Attempts to explain the overall plagioclase-rich composition of many exposed anorthosite massifs in terms of the crystallization of an unusually water-rich mafic magma fail conspicuously, however, to account for the marked absence of hydrous minerals either pyrogenetic or deuteric.

It seems therefore that the mechanism of upward accumulation of plagioclase on a massive scale within large mafic magma chambers of subalkaline composition as described by Emslie (1970) is probably the one that operated to produce the primary features of anorthosite massifs.

The ultimate origin of these remarkable subalkaline layered intrusions, the emplacement of which was apparently accompanied by significant downsagging of hot crust, is still controversial. Attempts to correlate them with plutonic equivalents of SBZ calc-alkaline volcanism are not convincing. Wynne-Edwards (1976, p.932) draws attention to an apparent approximate north—south alignment of anorthosite massifs in Labrador and the Grenville province and suggests quite simply that they mark the traces of mantle hotspots and accompanying basic eruption on a slowly moving Proterozoic crust that was too plastic to undergo brittle fracture like Phanerozoic lithosphere.

12.5.6. Distinctive acid eruptive rocks, notably Rapakivi granites

The world's classic occurrence of Rapakivi granites are centered around the Gulf of Finland where they form massifs up to 100 km across. They are mid-Proterozoic in age, dated at around 1700 Ma, post-orogenic and indeed anorogenic, and form typically epizonal, often drusy, intrusions of homophanous, porphyritic-textured granites, minor porphyry intrusions, and locally give rise to eruptive breccias (Savolahti, 1956). Many are fluorite-bearing and zircon is commonly a conspicuous accessory mineral. They are characterized by generally low to very low colour indices, silica contents in the range of 64—77 wt.%, and a high K_2O content, ~4.7—6.8 wt.%, considerably in excess of Na_2O, which is generally in the range 2.3—3.4 wt.%; they are thus greatly richer in K_2O than are ideal granite minimum compositions. In keeping with this distinctive feature of their chemistry, an orthoclase-rich alkali feldspar crystallized early, together with bipyramidal quartz phenocrysts, in most varieties. The most well-known textural feature of many (but not all) of the Rapakivi association rocks is a conspicuous mantling of a proportion of early, often ovoidal, pink orthoclase phenocrysts by white sodic plagioclase. The ovoidal shape is probably merely a reflection of the low entropy of fusion of K-rich sanidines. The conspicuous overgrowths have been attributed to progressive crystallization within high-level magma chambers resulting in a buildup of magmatic water in the remaining crystallizing magmatic liquid resulting in high-pressure ["magmatic overpressures" — see C.J. Hughes (1971)] so that the liquidus became depressed to the point of intersecting the solvus during the crystallization period. Thus two alkali-feldspars, one still orthoclase-rich in composition and the other relatively sodic, begin to crystallize in place of one, but both sets overgrow the conveniently existing early crystal nuclei.

Convincing analogies to this distinctive granitic association and questions of provenance remain obscure [see, however, Muehlberger et al. (1967), for a factual review of basement rocks in the continental interior of the U.S.A., including reference to three large volcanic-intrusive old provinces mainly of rhyolitic composition dated roughly 1200 Ma].

12.5.7. Massive mafic eruptions possibly constituting Proterozoic ocean crust

Dated at around 1100 Ma, the Keweenawan lavas of the Lake Superior region have been variously thought of as a young greenstone belt extruded on relatively more rigid and stable crust than Archaean, or as an early flood basalt association. Evidence of the tholeiitic affinity of the series and problems arising from variable degradation and metasomatism have been briefly outlined in Chapter 10.

A broad review of the Keweenawan Series (J.C. Green, 1977) shows that over a short period of ~20 Ma seven partly overlapping lava plateaus were erupted, predominantly from fissures, with individual dimensions of 2.5—7 km in thickness over areas of the order of 250 × 130 km. A rifting environment with possibly as much as 90 km spreading accompanied by subsidence resulted in the formation of a mafic crust, which seismic evidence suggests may locally attain a thickness of 55 km. Strong magnetic and gravity anomalies suggest that similar rocks and structures extend as far southwards as Kansas — the "mid-continent gravity high".

The timing of this massive mafic eruptive episode (including the apparently comagmatic Duluth gabbro, in the outcrop area of which Pigeon Point has given the mineral name pigeonite) overlaps with that of plutonism and deformation in the Grenville province to the east. In the Keweenawan lavas and associated eruptive rocks we may be witnessing therefore a (rare) example of formation of Proterozoic oceanic crust by crustal rupture, contrasting with a more general tectonic response of ensialic ductile spreading.

12.5.8. *Proterozoic belts with possible plate boundary features and associated igneous rocks*

J.M. Moore (1977) recorded several examples of Proterozoic volcanism in the Canadian shield and concluded that:

> "the operation of plate tectonic processes during this interval (Proterozoic), though unproven, is consistent with available data."

One of the clearest examples seems to be provided by the "Coronation syncline" situated on the northwestern flank of the Archaean Slave province. The sedimentary rocks of the geosyncline exhibit a miogeosynclinal carbonate facies and an eugeosynclinal, fine-grained clastic facies and are bounded further to the west by the Great Bear batholith. This is a substantial belt of epizonal plutons of granodioritic to granitic composition, $\sim 450 \times 125$ km in outcrop area, intruded into a thick pile of predominantly volcanic rocks (Hoffman and McGlynn, 1977, with reference to earlier work) that comprise in upward sequence: an apparently bimodal basalt—rhyolite assemblage; a cyclic sequence of andesites, dacites and rhyodacites, and some epiclastic rocks; and very thick, densely-welded ash flows of dacitic to rhyolitic compositions. Hoffman and McGlynn conclude that the batholith is subduction-related and intruded into terrain that began as an ensialic rift.

A comparable middle-Proterozoic sequence of mainly bimodal volcanics of calc-alkaline affinity including low-K tholeiites, abundant rhyolites, and some andesites, intruded by later elongate tonalite to granodiorite batholiths, has been described from northwest Queensland (I.H. Wilson, 1978) and is interpreted to have formed at a continental margin similar to that of the Andes.

The Labrador trough contains considerable amounts of mafic rocks and some ultramafic rocks leading some observers to make analogies with ophiolites. The belt extends in a northerly direction over a distance of ~ 1000 km, and is bordered on each side by Archaean gneisses, those on the eastern side remetamorphosed in Proterozoic time. The basin filling comprises two, possibly three, cycles of sedimentation, each comprising an upward sequence of orthoquartzite and limestone (with ironstone formation), passing upward into shales and flysch on the western flank and capped by the voluminous eruption of thick mafic flows, gabbro intrusions and ultramafic bodies on the eastern flank (Dimroth, 1970). Cross-sections indicate overturning towards the west, somewhat suggestive of the directional thrusting of Phanerozoic fold belts. However, a convincing analogy in sequence, structure, and scale to any Phanerozoic fold belt remains lacking.

After the excitement of the re-interpretation of contemporary and Phanerozoic igneous activity in terms of plate tectonics, it could be maintained that a greater challenge to igneous petrologists at the present day now lies in the interpretation of the igneous rocks of the Precambrian.

Chapter 13

PETROGENESIS OF IGNEOUS ROCKS

13.1. INTRODUCTION

In earlier decades the term petrogenesis applied to igneous rocks would in the main have been taken to refer to processes which we would now specify as differentiation. Bowen's (1928) *Evolution of the Igneous Rocks*, for example, refers mainly to differentiation, largely fractionation, processes. The identification of most of these processes of differentiation, reviewed in Chapter 7, can be firmly based on studies arising directly from the observation of field phenomena, and the processes themselves are now reasonably well appreciated and understood by petrologists.

Nowadays the term **petrogenesis** has taken on the connotation of the ultimate derivation of magma, in the case of mafic magmas primarily from the mantle. Advances in our understanding reflect a synthesis within the last decade of results from high-pressure experimental work, the implications of plate tectonics, geophysical interpretations, and increasingly abundant analytical data particularly on trace elements and isotopes.

The various distinguishable igneous rock series, reviewed in Chapters 8—12, internally differentiated though they may be, can be traced to significantly different parental magmas, themselves presumably originating from different primary magma types generated under different conditions in the mantle. The sites of generation of mafic magmas are not exposed, and thus a large element of fashionable inference, itself fairly well based on the recently developed experimental work at elevated pressures, is involved. For some intermediate and acid rocks generation may have occurred within lower crustal levels, but for these rocks too inference (and indeed much current debate) is involved.

In a recent historical context it is instructive to compare the contents and approach of the relatively short chapter on petrogenesis in F.J. Turner and Verhoogen (1960) with the more detailed treatment in Carmichael et al. (1974), particularly chapters 7 and 13, where an understanding of the "mantle—magma system" is seen as the crux of an understanding of igneous petrology. It would be prudent, however, to bear in mind their rider that (Carmichael et al., 1974, p.357):

> "the formation of basaltic magma in all its petrological diversity . . . is only very imperfectly understood"

and also, of course, that the evolution of many granitic rocks has apparently followed a much more intricate path than fractionation of mantle-derived magmas alone.

While the writer is agnostic concerning an ultimate precise understanding of mafic rock petrogenesis (each new lead seems to offer promise and then dissolve into complexity), some of the principles as presently conceived are presented in the following sections.

13.2. COMPOSITION OF UPPER MANTLE

Mantle rocks are those which occur between the outer boundary of the core at a depth of 2900 km and the Mohorovičić discontinuity (Moho) at the base of both continental and oceanic crust. Below the Moho the velocity of longitudinal seismic waves, v_p, jumps sharply to characteristic values of generally just over 8 km/s. Several lines of evidence concur in placing mafic magma generation within the upper mantle but below the base of the lithosphere. Thus geophysically defined and for the most part conveniently hidden from sight the mantle has been the victim of much overgeneralization. Actual samples that are available to us, however, constitute an alarmingly heterogeneous group:

(1) Massive units of harzburgite up to several kilometres thick in ophiolite complexes, associated in some with small bodies relatively richer in clinopyroxene.

(2) Eclogite nodules from kimberlites.

(3) Lherzolite nodules, usually garnetiferous, virtually restricted to kimberlites, associated with smaller amounts of nodules of dunite, harzburgite, and other combinations of the mineral constituents of lherzolite.

(4) Rare grospydite nodules from some kimberlites.

(5) Spinel lherzolite nodules characteristically found in many basanites and nephelinites and in some alkali basalts; the origin of this class of nodules is controversial, i.e. are they "average mantle" derived from some shallower depth than the lherzolite nodules of kimberlite, or are they (recrystallized) fractionation products of primary alkaline mafic magmas, or do they represent mantle depleted by an episode of partial melting that itself may or may not be related to the enclosing mafic magma?

Along with this tangible hand-specimen evidence there are several other indications of inhomogeneity:

(6) The v_p of uppermost lithospheric mantle, although characteristically just over 8 km/s, actually varies from below 8.0 to 8.4 km/s.

(7) A **low-velocity layer** (LVL), also known as **asthenosphere**, is present beneath lithosphere under oceanic crust and most continental crust. Its base is roughly at a depth of 250 km, and its top at a depth of 100—200 km under continental areas, 50—70 km in oceanic regions, and shallower still under ocean ridges. The observed subtly lower velocities and appreciable attenuation of S waves is consistent with the LVL containing ~1% partial melt phase (D.L. Anderson, 1970). Compositional restraints from seismic data are not

good for the LVL, but a composition of "pyrolite" (see p.450), mineralogically equivalent to a garnet- or spinel-lherzolite according to depth, is consistent with available data (D.L. Anderson, 1977, p.185); considerations based on MORB chemistry indicate that the LVL has been relatively depleted in more fusible components by one or more episodes of partial melting, at least under the oceans.

(8) Asthenospheric mantle with lower values of v_p down to 7.8 km/s underlies the Moho in places like Afar and Iceland, and is believed to signal the site of relatively hotter, rising mantle plumes which are inferred from the geochemistry of associated eruptive rocks to be significantly different in composition from average asthenosphere, i.e. richer in iron and incompatible elements.

(9) Evidence from mid-ocean ridge basalts (Erlank and Kable, 1976) reviewed in Chapter 9 indicates variable depletion within asthenospheric mantle, *not* relatable to plume activity.

(10) Inferred descending lithospheric slabs of observed high Q-values associated with Benioff zones are presumably transferring oceanic crust (transforming to eclogite below ~100 km depth) plus harzburgitic and other depleted mantle rocks to considerable depths within the mantle.

(11) There is a considerable variation in the proportion of eclogitic to lherzolitic nodules between different kimberlite occurrences. The precise significance of this is unclear, but high-density requirements from seismic data indicate that garnet (which is denser than magnesian olivine and pyroxenes) is probably quantitatively important in at least the lower part of the lithosphere (D.L. Anderson, 1977, p.182). Eclogite could originate either by the fractionation of mafic magmas at depth or by the recrystallization of subducted oceanic crustal material.

(12) Restraints imposed by the CaO/Al_2O_3 ratios of komatiites, believed to represent the products of high proportions of melt generated from mantle, leaving an olivine—orthopyroxene residue only, suggest a layered mantle (Cawthorn and Strong, 1975) with garnet content increasing in depth, at least in Archaean time.

(13) There is also the possibility of variation with time in the composition of mantle available for generation of mafic magmas. If one takes the present rate of eruption of basic rock to form oceanic crust, say at a slightly conservative round-figure estimate of 20 km³/a (see Chapter 9), and even disregarding the unknown but possibly significant amounts of mafic magma that may have crystallized below observable crustal levels within the mantle, and extends this rate back in time to 4.5 Ga (although the rate may well have been higher in the past), and assumes that to produce one part basaltic rock about four parts of mantle must have undergone an episode of partial melting, one can arrive at a figure of just over $300 \cdot 10^9$ km³ of mantle, reworked in this way and thus compositionally differentiated. This figure represents ~35% of the mantle.

In sum, although the mass of all crustal rock (ultimately derived from the mantle) amounts to no more than 0.42% of the mass of the Earth, or 0.62% of the mass of the mantle, it cannot by any means be assumed that immediately underlying mantle is, or has been, a constant, virtually infinite and unchanging source rock for igneous processes. Although therefore a consensus on an "average" upper-mantle composition would be a great convenience to igneous petrologists for model-making, such a concept appears to be singularly unfounded on fact.

From all the available evidence it would appear probable that the uppermost few kilometres of mantle is a refractory magnesium-rich harzburgite with very little Na, K and Ti, and strongly depleted in incompatible trace elements. Below this the predominant rock would be a lherzolite with variable proportions of eclogite. The lherzolite in addition to olivine, orthopyroxene and clinopyroxene would contain an aluminous phase — either spinel at relatively shallow depths or garnet at greater depths. At still greater depths and probably out of the range of genesis of all common magma there may be bodies of unknown extent of grospydite composition within lherzolitic mantle.

If one had to point, if not to an "average" mantle composition, at least to probable predominant mantle at the inferred depths of generation of most primordial mafic magmas, a garnet lherzolite composition would be the obvious choice. It has the required density to meet restrictions imposed by isostatic and seismic velocity data and has the capacity of generating a range of mafic melts on partial melting at appropriate pressures. From evidence of convergence of nodule compositions and consideration of the necessary generative role in magma production the work of P.G. Harris et al. (1967) indicated a narrow range of probable composition of undepleted mantle (Table 13.1). Mineralogically this inferred average mantle consists roughly of 65% of forsteritic olivine and just over 10% each of orthopyroxene, clinopyroxene and garnet. At shallower depths this composition could be represented by ~65% olivine, some 2% spinel, and the balance by aluminous ortho- and clinopyroxenes. Putting together analyses of garnet lherzolite nodules, the average analysis in Table 13.1 has been obtained (Carswell and Dawson, 1970). Note the strongly hypersthene-normative nature of this possibly best guess of the composition of average upper mantle relevant to igneous rock petrogenesis.

Some uncertainty attaches as to the possible existence in the upper mantle of small quantities of hydrous phases such as amphibole (potentially stable to ~80 km) and phlogopite (potentially stable to ~100 km). Although many nodules contain no hydrous minerals, and although some of the hydrous mineral components of nodules could have been produced by reaction with enclosing water-bearing mafic magmas during ascent to surface, a few nodules do contain apparent primary amphibole or phlogopite. A degree of uncertainty consequently attaches therefore to the possible tenor of H_2O and indeed of another major volatile constituent, CO_2, in the upper mantle, important components in considerations of petrogenesis.

TABLE 13.1

Possible approximations to upper-mantle compositions

	1	2	3
SiO_2	44.2	46.53	45.16
Al_2O_3	2.7	1.84	3.54
Fe_2O_3	1.1	—	0.46
FeO	7.3	6.70*	8.04
MnO	0.15	0.11	0.14
MgO	41.3	41.98	37.47
CaO	2.4	1.47	3.08
Na_2O	0.25	0.16	0.57
K_2O	0.015	0.15	0.13
TiO_2	0.1	0.33	0.71
P_2O_5	n.d.	0.02	0.06
Cr_2O_3	0.30	0.38	0.43
NiO	0.20	0.32	0.20
Total	100.015	99.99	99.99

1 = convergent composition indicated by analyses of ultramafic inclusions in kimberlites and mafic rocks and selected as being likely to represent average undepleted mantle (P.G. Harris et al., 1967, p.6366, table 2, analysis 3; passim for discussion).
2 = mean of nine analyses of garnet peridotite inclusions in South African kimberlite pipes (recast to 100 wt.% free of H_2O and CO_2) (Carswell and Dawson, 1970, p.170, table 3, analysis A).
3 = "pyrolite" (D.H. Green and Ringwood, 1967a, p.160, table 20).
n.d. = not determined.

*Total Fe as FeO.

Note also how, in a later paper (P.G. Harris et al., 1972), the compositional *inhomogeneity* of upper mantle was stressed and related to varying degrees of depletion by partial fusion and extraction of a liquid phase.

A more detailed discussion of these points is provided by Ringwood (1975) in his book *Composition and Petrology of the Earth's Mantle*.

13.3. EXPERIMENTAL WORK

13.3.1. Starting material for experimental work at high pressures

Judging from the success of classic experimental work at one atmosphere, mixtures of simple components should be expected to provide a good starting point. Indeed much experimental work has been done in investigating solidus and liquidus temperatures, and crystallizing phases for compositions in the

CMAS system, CaO—MgO—Al₂O₃—SiO₂, which of course includes the components forsterite, enstatite, diopside, pyrope and spinel, the magnesian end-members of all the mineral components of inferred mantle lherzolite.

To allow for the depressing effect on melting temperatures of other major oxides, notably iron and alkali oxides, other workers such as O'Hara (1968) chose to experiment as well on naturally occurring *garnet lherzolite nodules* from kimberlites.

D.H. Green and Ringwood (1967a) refer to a synthetic mixture of comparable composition, termed *"pyrolite"*, made up of three parts harzburgite and one part tholeiite (see Table 13.1). The rationale here was that partial melting of some mantle source rock presumably produced oceanic tholeiite and left behind a refractory residue equivalent to the harzburgitic alpine peridotite component of ophiolite. Recombining representative analyses of these two naturally occurring rock types in some reasonable, though admittedly arbitrary, proportion should give therefore an approximation to mantle starting material. The composition of pyrolite has been slightly modified twice since its original conception by Ringwood in 1966 so that references

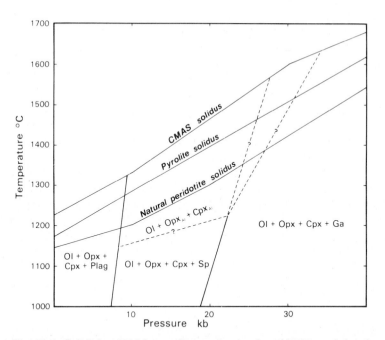

Fig. 13.1. Solidi for *CMAS*, pyrolite, and natural peridotite nodule compositions, and sub-solidus mineral assemblages appropriate to pyrolite and natural peridotite nodule compositions, at pressures up to 40 kbar (based on Jamieson, 1970, p.166, fig. 1; D.H. Green and Ringwood, 1967a, p.161, fig. 11). Note uncertainties about the possibility of Al₂O₃ being totally contained in aluminous pyroxenes at intermediate pressures and high temperatures, and about the position of the spinel—garnet boundary at high temperatures.

to pyrolites I, II and III are found in the literature (D.H. Green and Ringwood, 1967b).

The temperatures of beginning of dry melting of a natural peridotite, pyrolite, and appropriate *CMAS* composition at elevated pressures are included in Fig. 13.1.

13.3.2. Results of early experimental work

Naturally, one would like to know the compositions of liquids produced at various pressures by various proportions of partial melting; notwithstanding the above differences in starting material, results were in fact found to be similar.

An important consideration to bear in mind here is the difference between *fractional fusion* and *equilibrium fusion*, lucidly discussed by Presnall (1969). **Fractional fusion** (syn. fractional melting), wherein melt liquid is removed from the system more or less as soon as it is produced, will yield successively small quantities of the same eutectic composition until one of the solid phase components is all used up. Then, following some appreciable rise in temperature, melting will recommence at a eutectic temperature appropriate to the remaining phases, and so on. **Equilibrium fusion** (syn. equilibrium melting) on the other hand, where melt liquid remains in contact with and in equilibrium with the solid phases, will yield, following the production of some quantity of the same first eutectic liquid and exhaustion successively of solid phases, a slowly changing liquid composition initially along cotectic lines and eventually converging to the starting composition. In the general case, the compositional range of equilibrium melts during progressive degrees of partial melting is much narrower than that of successive fractional melts.

For example, let us assume a simple ternary system of three components, A, B and C, with three binary eutectics, E_1, E_2 and E_3, and one ternary eutectic, E_4, as in Fig. 13.2, and a starting composition, X. Fractional melting will yield batches of composition E_4 while the residual solid changes in composition from X to Y. Then, after an appreciable temperature rise, batches of liquid of composition E_1 will be produced while the residual solid changes in composition from Y to B. Finally, after a further appreciable rise in temperature, batches of liquid of composition B will be produced. Equilibrium melting, on the other hand, will yield some quantity of liquid of composition E_4 (given by the lever principle operating on line $Y-X-E_4$). On further melting the composition of the liquid will slowly change with concomitant slow rise in temperature and increasing proportion of melt along the cotectic line from E_4 to P, and finally along the line PX to X.

Now it is problematical at what proportion of melt to solid residue segregation and removal of the liquid fraction might occur. Estimates based on physical and compositional considerations range from a few per cent to several tens per cent of melt, i.e. generally fairly low. For this reason and the pragmatic one of experimental convenience, the first reported results per-

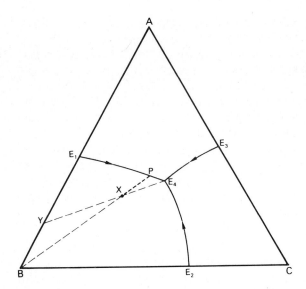

Fig. 13.2. Diagram of a simple ternary system with three binary eutectics and one ternary eutectic to illustrate the difference between fractional melting and equilibrium melting. See text for discussion.

tained to equilibrium melting. It should also be noted in this connection that where the proportion of melt is low enough that some amount of each of the four main original mantle phases (olivine, two pyroxenes, and an aluminous phase) remain, the melt composition is the same for fractional and equilibrium melting, i.e. a eutectic composition.

D.H. Green and Ringwood (1967a, pp. 164—167), working with pyrolite as starting material reported the following melt compositions:

(1) 0—5 kbar: for small degrees of partial melting, quartz-normative basalt
(2) 5—11 kbar: for 25% melt, high-alumina olivine tholeiite
(3) 11—22 kbar: for 20% melt, olivine alkali basalt; for 25% melt, olivine alkali basalt with only a little normative nepheline; for 30% melt, olivine tholeiite
(4) ~27 kbar: for 40% melt, picritic tholeiite with over 30% normative olivine

O'Hara (1968, pp. 79—82), working with natural lherzolitic peridotite (aluminous phase: plagioclase, spinel, or garnet according to depth) as starting material, and at lower temperatures in the solidus—liquidus temperature interval so that four phases remained, reported the following melt compositions:

(1) 5—8 kbar: high-alumina basalt
(2) 10—20 kbar: nepheline-normative basalt, becoming increasingly rich in normative olivine at higher pressures
(3) 20—30 kbar: hypersthene-normative basalt, becoming increasingly rich in normative olivine at higher pressures
(4) 30—40 kbar: "picritic" basalt, i.e. very rich in normative olivine

Minor discrepancies apart, it is clear that: (a) initial melts from anhydrous mantle can have critically undersaturated nepheline-normative compositions only within a limited intermediate pressure range; (b) that all other initial melts are tholeiitic, i.e. hypersthene-normative; (c) that melts from deeper levels will be olivine-rich in normative composition; and (d) that first melts can be quartz-normative only at pressures lower than 5 kbar (the pressure limit to the incongruent melting of magnesium-rich ferromagnesian pyroxenes). Note also that whereas initial melts resulting from a small degree of melting may have these distinctive compositions related to pressure in the first instance, increasing degrees of melting result in the incorporation of increasing amounts of orthopyroxene and olivine into the melt phase, necessarily resulting in the melt assuming an increasingly hypersthene-normative and olivine-rich composition.

13.4. THE MANTLE—MAGMA SYSTEM

13.4.1. Temperature regime in the mantle

Temperature gradients in upper mantle obviously vary from place to place. For example, below separating plates with their progressively thickening lithosphere away from the ridge axis there must be an upward component of movement of the underlying mantle thus bringing relatively hotter mantle to relatively shallow depths. Descending slabs of lithosphere below Benioff zones, on the other hand, must transport some relatively cold mantle to considerable mantle depths; according to whether the descending slab is just "slipping" along the Benioff zone or in part "falling" downwards, a complicated kinetic regime may exist in the wedge of mantle above the Benioff zone, possibly involving there too a component of upward transport of mantle. Below relatively stable areas of old continental crust however various xenoliths brought up by kimberlites (see Chapter 10) do afford an insight into thermal gradients in mantle, and it is instructive to draw up a diagram such as Fig. 13.3, showing the following data inserted on a $P-T$ grid:

(1) Liquidus temperature for "average" garnet lherzolite composition, increasing with depth at $\sim 10°C/kbar$.

(2) Dry solidus for the same composition.

(3) Wet solidus for the same composition (omitting minor complications in melting behaviour due to the possible presence of amphibole as a stable mantle phase at low pressures).

(4) Various observed upper crustal geothermal gradients.

(5) $P-T$ equilibration conditions of typical mantle xenoliths from kimberlite based on their clinopyroxene compositions, etc. (see, e.g., Meyer, 1977).

(6) The probable temperature at the depth of the inferred olivine—spinel inversion in the mantle (see discussion in Carmichael et al., 1974, pp. 344 and 345).

(7) Construction of an average subcontinental geotherm.

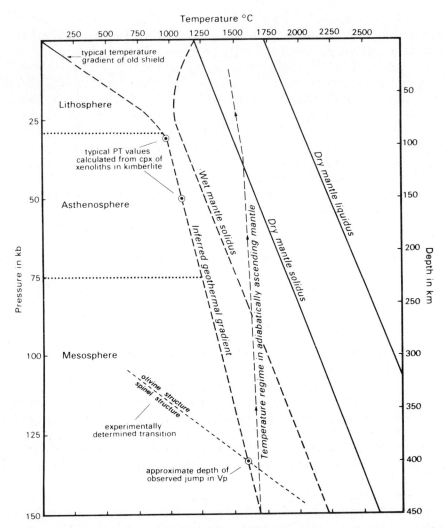

Fig. 13.3. Inferred subcontinental geothermal gradient and its relation to mantle solidus temperature and path of adiabatically ascending mantle (here arbitrarily assumed to commence ascent from a depth of 450 km). See text for discussion.

Note the high thermal gradients within rigid lithosphere contrasting with much lower gradients in sub-lithospheric mantle. It has been claimed that, given available estimates of mantle viscosity, and other factors such as density being equal, mantle convection would occur in the mesosphere to reduce temperature gradients in excess of the adiabatic gradient of ~1°C/kbar. Mantle temperatures apparently most nearly approach solidus temperatures (or might even exceed "wet" solidus temperatures) within a broad range below the lithosphere corresponding to the independently determined low-velocity layer. The actual amount of melting in the asthenosphere would, however, be heavily buffered by the small available amounts of water in probable mantle lithologies.

13.4.2. Partial melting in the mantle

Given the solidus for lherzolitic peridotite and probable distribution of temperature within the mantle, it is apparent therefore that partial melting on any significant scale cannot occur in a stable situation. The main trigger for melting would appear to be vertical upward movement of mantle material. Conduction of heat through mantle is a very slow process, and to a first approximation it may be ignored when considering the thermal regime of large bodies of moving mantle material. During conditions of pure adiabatic rise (or fall) no heat is conducted away from (or into) a moving parcel of mantle. Rising mantle will expand volumetrically very slightly in response to a lower confining pressure and this expansion is necessarily accompanied by a slight fall in temperature known as the adiabatic lapse rate (conversely, descending mantle would heat up slightly at the same adiabatic rate). A quantitative discussion of melting by adiabatic decompression is given by Cawthorn (1975) with particular reference to the simplified treatment by Carmichael et al. (1974, pp. 354—357).

Several possibilities of mafic magma generation within mantle, already strongly hinted at in Chapters 9—12, are thus:

(1) The "active" convective and near-adiabatic ascent of "blobs", "plumes", "lines", or "convention-cell limbs" of hot mantle material from depth that undergo a fall in temperature less rapidly than the mantle geothermal gradient and thus gradually climb towards solidus conditions during ascent. On intersecting the solidus any further climb towards liquidus temperatures would be buffered by the withdrawal of the necessary latent heat of fusion accompanying partial melting.

(2) Although the lithosphere is far too rigid to permit convection within it, where plates are separating (and thickening away from the ridges) there must be a compensatory upward component of movement in the LVL below the ridges (even apart from any "active" plume effects), to quite shallow depths, again facilitating partial melting. This possibility, presumably realized by MORB, is the "passive" case of the first possibility.

(3) Abnormally high amounts of water in mantle rocks, possibly released from a descending underlying lithospheric slab containing a proportion of "wet" oceanic crust, could lead to more extensive partial melting in mantle overlying a Benioff zone, at the depth of the LVL.

(4) A special case is the relatively rapid ascent of mantle material to isostatically compensate for the gratuitous excavation of enormous astrobleme craters [see discussion in Chapter 12 and reference to treatment by Hulme (1974) of this effect].

13.4.3. Upward ascent of magma

The proportions of melt generated by partial melting in different parts of a kinetic model of a rising mantle diapir spreading out laterally at asthenosphere depths have been calculated by Oxburgh and Turcotte (1968), and shown diagrammatically in Fig. 10.6. Although this model is based on one particular set of assumptions on temperature, dimensions, rate and direction of movement, etc., one thing is clear: there is not in this situation a localized volume where partial melting takes place but rather a broad region where different proportions of melt in equilibrium with mantle would be anticipated extending over a long time period. Presumably, there would be a tendency for melt to percolate upwards because of its lower density into the marginal upper parts of the region of melting where the inherent degree of melting is low because the mantle host is at, or near, its solidus temperatures. Indeed, at some problematical stage a different physical regime must supersede the generation of melt in equilibrium with mantle, and magma will segregate into discrete liquid bodies and rise. This is an interesting topic on which a synthesis of theoretical and experimental results allied with interpretation of the textures of some partially melted lherzolite nodules with a quenched melt phase should shed light in the future. The melt will continue its ascent to upper crustal levels in conduits within relatively cooler and brittle lithosphere and crust, at temperatures that are subliquidus with respect to the melt. In this situation one would expect the melt to behave much more as a closed system undergoing crystal fractionation.

An appreciation of the probable scale, complexity and time span of the whole process of magma generation in the mantle tends to dwarf the simplistic models in which petrologists are prone to indulge*. Note that one can distinguish, at least in theory, between a **first melt** generated by an approach to equilibrium partial melting in the mantle, and a **primary magma** referring to discrete segregated bodies of magma, necessarily following partial melting and any accompanying zone-refining processes during segregation.

Several of the anticipated complexities that may accompany the process of magma evolution at mantle depths include the following:

(1) *Zone refining*: this is the industrial process of purifying ingots of precious metal by passing them slowly over a heat source so that a zone of partial melt is generated and slowly passes along the ingot. Impurities of lower melting point are progressively concentrated in the small fraction of melt which can be removed at the end leaving behind a purer bar of the precious metal. By analogy with this it is conceived that during partial melting and movement of interstitial melt liquid within mantle a melt will eventually

---

*Consider the complexities and difficulties in interpretation presented by *accessible* exposures in migmatite terrain where presumably analogous partial melting (of sialic crust) has occurred!

be generated that could be greatly richer in incompatible elements than that which it would be possible to generate by any process of equilibrium melting with a fixed volume of mantle. The magnitude of this process is difficult to assess experimentally, and would be expected to be counterbalanced by the failure of the melt to equilibrate with more than the outer portion of crystals in a meshwork containing partial melt due to the extreme slowness of diffusion processes within crystals. Clearly, mantle grainsize could be a significant factor in determining the incompatible element contents of derived melts (see p.461). Nevertheless, some process akin to zone refining seems to be demanded, particularly by high contents of incompatible elements in the more alkaline mafic magmas (see discussion by P.G. Harris, 1957).

(2) *Wall-rock reaction*: this is a somewhat similar concept to the above, but is taken to refer specifically to any further interaction of magma and confining mantle in varying (but in all probability less) degrees of intimacy during the *post-segregation* stage.

(3) *Fractionation*: within the ascending magma crystallization will occur on cooling ideally where protected from wall-rock reaction in conduits by a congealed zone of crystallized selvedge against relatively cooler country rock. Note that *rapid* ascent of magma could result in a magma becoming superheated to above liquidus temperatures at lower pressures, if the temperature lowering by conductive heat loss failed to meet the difference between adiabatic and melting-point gradients. The nature of phases crystallizing in depth from a magma of specified composition isolated from its mantle progenitor and behaving as a closed system can be determined by experiment. The probable importance of fractionation under these conditions has been most strongly stressed by O'Hara (1965). Jamieson (1970) pointed out that this fractionation of ascending mafic magmas can be a complicated polythermic and polybaric sequential phenomenon. It could, for example, possibly occur at temperatures very low in the solidus—liquidus interval, accompanied by wall-rock reaction and equilibration with a four-phase mantle, in such a way as to keep the evolving magma composition close to equilibrium compositions for first-generated melts at the respective pressures. In this event it might be impossible to distinguish magma evolution from magma generation on the basis of major-element compositions alone. On the other hand, fractionation might occur only to a limited extent at temperatures at or near liquidus temperatures, and magma could theoretically ascend through a pressure regime that should (if re-equilibration with a four-phase mantle had ensued) have resulted in a different magma composition. The possibilities become so elastic that it may become difficult, some would say profitless (see discussion in Carmichael et al., 1974, p.628), to model them.

(4) *Assimilation*: in a sense the continued development of increasing proportions of melt during the stage of partial melting is an assimilation process, but this is still considered to lie in the realm of magma generation. Segregated primary magma rising through conduits in subsolidus country rock would to

some extent be isolated from possible assimilation processes by congealed margins. Furthermore, following the principles of N.L. Bowen (see Chapter 7) mafic magmas could not assimilate mantle rocks readily; the assimilation of sialic crustal rock by mafic magma remains, however, a real possibility. One point, made strongly by Fyfe (1978), is that failing a relatively rapid ascent to the surface (i.e. where not in a tensional regime presumably favoring the development of cracks), "ponding" of mafic magmas might occur at the mantle—continental crust interface, as mafic magma would ascend relatively quickly with respect to relatively dense mantle, but much less readily with respect to crustal material (unless a lithospheric hydrostatic head existed on a continuous magmatic "plumbing system" to drive the mafic magma relentlessly and rapidly to the surface). Over a long period of time therefore, considerable assimilation of sialic country rock might occur at lower crustal levels, a possibility alluded to in Chapter 7 and indeed strongly indicated by strontium-isotopic data and field evidence of SBZ plutonic complexes.

13.5. TRACE ELEMENTS IN MAFIC ROCKS

From the experimental work quoted in the previous sections, it is reasonably clear that small degrees of melting will produce melts in equilibrium with several mantle phases, and with compositions appropriate to specific pressures and temperatures that can be experimentally determined. For increasing degrees of melting, after augite and an aluminous phase have been dissolved, it is also clear that all melts derived from lherzolitic peridotite must become increasingly hypersthene- and olivine-normative, and increasingly magnesium-rich as progressively increasing amounts of magnesian hypersthene and olivine are taken up into the melt phase. Consideration of the possible effects of fractionation on primary magmas introduces further complexities into petrogenesis. As forthrightly set out by Gast (1968), the trace-element contents of mafic rocks vary far more widely than do their major-element contents (ideally determined from fresh, aphyric mafic flows) and these trace-element concentrations must have some significance in petrogenesis, and may throw some light on these complexities.

As examples of what might happen during petrogenesis and of the principles involved, let us consider the anticipated behaviour of two very different trace elements, Ni and Rb.

Experimental work has shown that in an olivine—mafic melt system in equilibrium, and indeed in a peridotite—mafic melt system, Ni is strongly partitioned into the solid phase. At low concentrations of Ni, assuming equilibrium melting and making some other simplifying assumptions, a partition coefficient defined as: (concentration of Ni in the liquid phase)/ (concentration of Ni in solid phase) would be constant. Assuming a Ni content of 3200 ppm in mantle and a partition coefficient of 1:12 for nickel between melt and olivine, it is a matter of easy arithmetic to construct the curves shown in Fig. 13.4A, depicting Ni concentrations in both liquid melt and remaining mantle, for varying proportions of melt under equilibrium conditions. This treatment ignores the effect of the several phases undergoing selective fusion (although, in fact, olivine is both a major component of mantle and a residual component of mantle undergoing progressive fusion, and is also the major repository of mantle nickel) and indeed other complexities (see, e.g., H. Sato, 1977; Mysen, 1978). In principle, however, it is clear that magma equilibrating with average mantle for an appreciable range of proportions of melt would have a somewhat uniform, lower Ni content. Once isolated from its mantle source, however, and undergoing early fractionation of olivine crystals (typical of nearly all mafic melts), the nickel content of such a melt would be very rapidly depleted. Applying the same melt/olivine partition coefficient, Fig. 13.4B shows the great change on the concentration of nickel in the melt originally containing say 360 ppm Ni effected by modest amounts of olivine fractionation. Ni contents are thus not an effective discriminant between mafic magmas produced by different degrees of partial melting of mantle (although Ni would be expected to be significantly high in melts such as komatiites produced by an inferred extraordinarily high degree of partial melting), but nickel content should be a very effective measure of early fractionation of olivine.

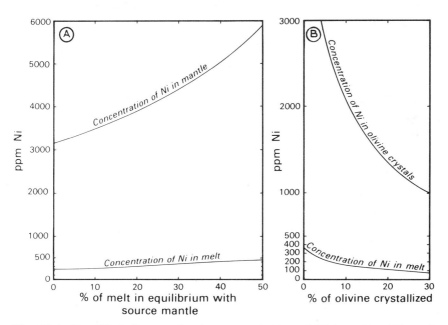

Fig. 13.4. Simplified diagram showing concentration of Ni in melts: (A) produced by increasing degrees of equilibrium partial melting of mantle; and (B) during fractionation of olivine. See text for discussion.

Rubidium, on the other hand, is an example of an *incompatible element*, i.e. one that does not enter readily into the lattices of major mantle mineral phases, and is therefore highly concentrated in initial melts. Assuming a mantle concentration of say 0.5 ppm Rb and a partition coefficient of 30, this time in favour of the liquid phase, the concentrations of Rb for varying proportions of melt are as shown in Fig. 13.5A, again assuming idealised theoretical equilibrium conditions, making some simplifying assumptions, and ignoring the probable great accentuating effect of zone-refining processes in nature. In contrast to the behaviour of Ni, there is a very large and distinctive difference indeed, other things being equal, between the Rb contents of melts produced by very low degrees of melting and those produced by higher degrees of melting. Assuming as before a modest amount of early olivine fractionation, and even assuming that no Rb at all enters into fractionating olivine, Fig. 13.5B shows that the tenor of Rb in the melt containing say 6 ppm Rb will not increase very much during this early olivine fractionation, or fractionation of any other mineral such as orthopyroxene into which Rb does not enter in any significant extent.

The concentration of Rb, and indeed other incompatible elements, thus provides a **trace-element signature,** capable of surviving a degree of later

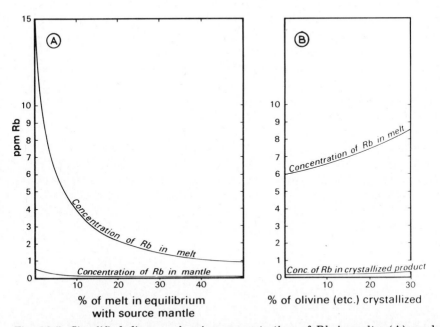

Fig. 13.5. Simplified diagram showing concentration of Rb in melts: (A) produced by increasing degrees of equilibrium partial melting of mantle; and (B) during fractionation of olivine and other phases towards which Rb is strongly incompatible. See text for discussion.

fractionation without major modification, and giving a strong indication, other things being equal, of the proportion of melting, possibly enhanced by zone-refining processes, that operated during the *production* of the primary magma.

Absolute or relative depletion in certain specific elements will signal the *fractionation* of phases, into which they are highly or even moderately partitioned, from primary and ensuing parental magmas, e.g., Ni by olivine, Eu by plagioclase, V by magnetite, Ti by kaersutitic amphibole, Na/K ratio by omphacite, HREE by garnet and clinopyroxene. The variation of amounts of these and indeed all other elements within members of an igneous rock series should therefore reveal a self-consistent pattern of the sequence and amount of fractionating solidus minerals, if fractionation is the means of differentiation.

There is yet another possibility in that a relative depletion in certain trace elements may signal not necessarily fractionation, but processes of magma generation and equilibration with a specific phase or phases. A prominent example of this is provided by the differing chondrite-normalized REE abundances of mafic rocks already alluded to in Chapter 9. Garnet, for example, readily accommodates the smaller (heavier) REE into its crystal lattices but not the lighter REE, so that generation of small quantities of magma at depths where garnet is the stable aluminous mantle phase results in melts strongly enriched in the LREE (see Gast, 1968, for a lucid discussion of this).

Much of the above discussion is unrealistically predicated on an assumption of equilibrium conditions. During partial melting, presumably along grain boundaries, there is probably not time for diffusion to occur so as to equilibrate the still solid crystal interiors with melt (indeed the study of mantle nodules reveals gross isotopic disequilibrium between different mantle phases, indicating that isotopic equilibration by diffusion processes between coexisting minerals in the mantle is not attained even over extremely long periods; see P.G. Harris et al., 1972; Hofmann and Hart, 1978). The ancitipated effects of partitioning will be dampened by this factor, but will be enhanced by zone-refining processes, in some complex manner.

In sum, trace-element concentrations in mafic melts should have a critical capability in modeling processes of magma generation and evolution. See particularly the discussion by Allègre and Minster (1978, with references), and methods of setting up models to test various hypotheses on trace-element behaviour.

13.6. STATUS OF PARENTAL MAGMA IN PETROGENETIC WORK

Within the different igneous rock series with mafic progenitors there is abundant evidence of differentiation, predominantly it seems by processes of fractionation. Various indices such as high M- and Mg-values, high MgO contents, high Ni, low Rb and Zr, etc., of fresh non-accumulative lava compositions point consistently to the parental magma of a series. In previous chapters the vital distinction has been drawn repeatedly between, on the one hand, variation *within* an igneous rock series and, on the other hand, such intrinsic differences as may exist *between* igneous rock series. By these deductive processes, with some allowance for the noise of geochemical data, we are led inexorably to the tangible, pragmatic concept of recognizable *parental magma types*. These, the end products as it were of our investigations at accessible crustal levels and the practical starting points of possible differential evolution at crustal levels, are now seen to be merely arbitrary points in a complex sequential history of evolution of mafic magmas, a history which the igneous petrologist would do well to bear in mind.

13.7. CONCLUSIONS

13.7.1. Tholeiites and alkali basalts

It has been estimated that tholeiites are two orders more abundant than alkali basalts, and that alkali basalts, in turn, are two orders more abundant than all the more alkalic mafic rocks lumped together. In practical terms, therefore, an understanding of the origins of tholeiite is an important first step in igneous petrogenesis, but the existence of significantly different parental mafic rocks cannot be ignored (as was the tendency in N.L. Bowen's great pioneering work).

From the now classic work of D.H. Green and Ringwood (1967a) and O'Hara (1968) referred to in the previous sections it is clear that, in the context of simple partial melting, alkali basalts can originate by low degrees of partial melting (wherein, without the operation of migration accompanied by zone-refining processes, segregation of magma into discrete batches may be difficult?) at pressures roughly between 10 and 20 kbar. Hypersthene-normative tholeiitic melts, however, will be generated at all other pressures, and with increasing degrees of melting at all pressures, and they will be increasingly olivine-rich in composition with greater proportions of melt and also at depths greater than those of possible alkali basalt generation. If these hypersthene- and olivine-normative melts undergo olivine fractionation, then only tholeiitic parental magmas can result. The absolute abundance of tholeiite and voluminous nature of most tholeiite eruptions seem to fit these conditions well.

The most abundant tholeiite is MORB. One recent detailed discussion (Malpas, 1978) of the generation of oceanic lithosphere calls for generation of a primary Mg-rich basaltic liquid by ~23% partial melting of spinel lherzolite at 18—22 kbar followed by early fractionation in the main of olivine.

The concept of significantly different mantle sources, in particular the probable effect of melting within plumes of relatively fertile mantle to produce OIT and associated parental magmas is, of course, the subject of ongoing debate (see discussion in Chapter 9, p.285).

The evolved nature of many continental flood basalts has been commented on in Chapter 10 (p.352). Most are tholeiitic with low M-values and many are quartz-normative. An origin by contamination with continental crust seems to be too facile and does not fit the span of available isotopic and trace-element data. Under anhydrous or nearly anhydrous conditions over-saturated tholeiites can only co-exist with peridotitic mantle at low pressures incommensurate with the lithospheric thickness below continental crust. Presumably flood basalts in particular have undergone a long history of fractionation before eruption to crustal levels. The enigma posed by the high $^{87}Sr/^{86}Sr$ ratios of many flood basalts remains unsolved.

In view of the association of relatively minor amounts of alkali basalt with earlier voluminous tholeiite in Hawaii, it was tempting to postulate derivation of alkali basalt from tholeiite in some way. Fractionation processes at low pressures involving phases of observed early crystallization cannot, however, cross the critical thermal divide between alkali basalt and tholeiite compositions (see pp. 184—186). Pyroxene fractionation of a tholeiitic liquid at depth could theoretically yield a critically undersaturated liquid of alkali basalt major-element chemistry, but this process along with more vaguely conceived processes of alkali transfer do not account for the distinctive trace-element signature of alkali basalts. Gast (1968) drew attention to this in his classic paper, and Schwarzer and J.J.W. Rogers (1974) have documented a world-wide alkali basalt association occurring in differing tectonic environments, many of them lacking associated tholeiite. Presumably, therefore, processes of generation involving lower degrees of partial melting than those connected with tholeiite lead specifically to alkali basalt compositions.

13.7.2. Komatiites and high-magnesium basalts

The basic petrogenetic consideration here is that by reason of the ratio of their magnesia to iron oxide contents these rocks seem to have originated by a high proportion of melting of mantle. *Equilibration* of magma with average mantle with residual olivine in the compositional range Fo_{90}—Fo_{92} will yield, following the work of Roeder and Emslie (1970), melts with Mg-values [i.e. $Mg^{2+}/(Mg^{2+} + Fe^{2+})$] in the range 73—77 or, more roughly, M-values [i.e. $Mg^{2+}/(Mg^{2+} + Fe^{2+} + Fe^{3+})$] of ~69—73. These values are not quite reached, for example, by the most MgO-rich MORB groundmass compositions and the most MgO-rich non-accumulative alkaline mafic eruptive rocks of the Hawaiian archipelago (see Tables 9.1 and 9.9). Komatiites, however, have M-values ranging from these values up to 86. Such liquids (which indeed crystallize early olivines as highly magnesian as Fo_{95}) would only equilibrate with solid phases significantly more magnesium-rich than those of inferred average mantle. Such liquids could be derived by higher degrees of equilibrium partial melting of mantle in which increasing proportions of melt generated result in *both* melt and residual mantle acquiring higher MgO/FeO ratios. This seems at first sight a mathematical paradox, but follows from the increasing *proportion* of melt generated that always contains the higher FeO/MgO ratio.

In keeping with this inferred mode of origin, TiO_2 contents of rocks of the komatiite family are in the general range of 0.2—0.6 wt.%, the lowest of all mafic magmas. Ti is, of course, an incompatible element with respect to mantle mineralogy and would preferentially be incorporated in relatively higher concentrations within first-formed melts. Given an increasing proportion of melt and the small finite original absolute abundance of Ti in mantle, the content of Ti will decrease systematically in concentration in liquids derived by progressively higher degrees of partial equilibrium melting (like Rb in the trace-element example discussed on p.460). Contents of other incompatible elements, for example, Zr, are relatively low in komatiites, although the low content of some incompatible elements may in part reflect metasomatism during degradation of these old, often metamorphosed rocks (Ti and Zr are relatively stable during degradation — see Chapter 14).

The puzzle perhaps is, despite indications that many "primary" tholeiitic magmas may have been relatively rich in MgO, why so many reach crustal levels with relatively low MgO levels.

A more complex origin for komatiites by a two-stage melting process is presented by Arndt (1977).

13.7.3. Basanites, nephelinites, and melilitites

Examples of basanite and nephelinite have been mentioned in connection with oceanic islands such as the Comores and Hawaii. Some melilite-bearing rocks are also found in Oahu. The most complete spectrum of these highly alkaline rocks is provided by the East Africa rift-valley system. A pattern of increasing content of alkalis in these rocks coupled with increasing amounts of incompatible trace elements suggests derivation by small proportions of melt equilibrating with mantle rocks under zone-refining conditions at intermediate pressures, but deep enough in the majority of cases to result in a distinctive pattern of enrichment in LREE presumably by interaction with a garnet solid phase. These rocks are typically explosive in their eruptive behaviour, and their magmas contain significant volatiles, in part H_2O and also in part CO_2, as shown by the common association of ancilliary amounts of carbonatite with the more alkaline members.

In connection with the petrogenesis of these highly alkaline rocks it has been shown (D.H. Green, 1970) that the presence of H_2O in the melt phase can exert a strong influence on the sequence of fractionating phases in the liquidus—solidus interval. In particular, the field of crystallization of pyroxenes, particularly orthopyroxene, is greatly enlarged relatively to that of olivine at intermediate and high pressures. Between 13 and 18 kbar, the fractionation of aluminous orthopyroxene and clinopyroxene from olivine basalt magma under these conditions (or conversely small degrees of partial melting of water-bearing pyrolite) could produce olivine nephelinite. At higher pressures, in the range of 18—27 kbar, olivine melilite nephelinite could similarly be produced. Pyroxenes, of course, have a higher silica content than both olivine and basic magmas, so that the fractionation of pyroxene or the persistence of pyroxenes as stable phases in equilibrium with melt should depress the silica content of the melt.

13.7.4. Kimberlites

Kimberlitic magmas are generated by small proportions of melting at great depths. At these pressures experimental studies indicate that a highly picritic (MgO-rich) first melt will be in equilibrium with four-phase mantle. Zone-refining processes have led to pronounced, although variable, enrichment in incompatible elements including radiogenic [87]Sr. Although comparable to komatiites in their range of MgO content, kimberlites thus differ sharply from komatiites in their chemistry and mode of origin. It is also now apparent that during the evolution of kimberlitic magmas large dissolved quantities of CO_2 and H_2O were involved and have in large measure escaped from the magma system before final crystallization so that rock analyses do not yield compositions that can be directly related to primary melt compositions in which CO_2 was a major component. For a discussion of kimberlites and their relationship to the mantle, see Dawson (1972).

13.7.5. Andesites and associated rocks

With the exception of back-arc spreading environments in areas of both oceanic and continental crust, in which basic rocks akin to MORB and acid rocks of probable crustal generation predominate respectively, andesites and associated rocks have been erupted in a non-tensional environment. Fissure eruptions of basalt, for example, are not generally seen in active andesitic terrain — an apparent exception, the famous Tarawera basalt fissure eruption of 1886 (see Cole, 1970) in the North Island of New Zealand was actually emplaced in a tensional back-arc regime characterized by voluminous rhyolite eruption and numerous normal faults. It seems therefore that, in general, parental SBZ magmas may penetrate the lithosphere slowly, that considerable fractionation may well occur en route to upper crustal levels, and that indeed ponding at the mantle—crust interface might be anticipated as suggested by Fyfe (1978).

Many features are consistent with andesites having evolved in a water-rich magma system: (1) a highly explosive eruptive behaviour compared to other subalkaline magmas; (2) the presence, at least in some andesites, of hydroxyl-bearing phenocryst phases (in extruded representatives these phenocrysts of hornblende and biotite are conspicuously rimmed by magnetite and pyroxene attributable to reaction with a lava dehydrated by vesiculation); (3) corrosion features and oscillatory zoning commonly seen in the plagioclase phenocrysts of andesites that can be most readily explained in terms of a significant (and changing) content of P_{H_2O} in high-level magma chambers; and (4) the equilibration temperatures of magnetite—ilmenite pairs (where present) which point to a lower temperature of actual crystallization than that which would be anticipated in a water-poor magma of otherwise comparable composition.

A high proportion of Fe_2O_3 to FeO even allowing for the effects of postconsolidation oxidation (Chayes, 1969), and the presence in many andesites of phenocrysts of basaltic hornblende and magnetite (the latter commonly overlooked in petrographic descriptions!) suggests, in addition to a significant water content, an elevated oxygen fugacity*.

The production of andesites is apparently highly specific to SBZ regions, although a few continental andesites, such as the Paleogene andesites of San Juan Mountains, Colorado, bear a more tenuous tie to any postulated Benioff zone. Most andesites, however, are indubitably erupted within remarkably narrow elongate belts marked by a sharp volcanic front on the side of the belt nearer the trench at a vertical distance above the Benioff zone of 100 km. They commonly show a degree of polarity (enrichment in K, incompatible elements, and LREE) transversely away from the trench. Most andesite provinces show a unimodal distribution of rock types with intermediate varieties greatly predominating; a frequently observed continuity to relatively more basic members suggests, however, a series derived from parental magmas of basaltic or at least basaltic andesite in composition.

Chemically, a distinctive trait is a lack of iron enrichment in intermediate members compared with tholeiites, the other important division of subalkaline rocks. The low-K tholeiites of island arcs (the island-arc tholeiite series) are intermediate in this respect in that they show a small degree of iron enrichment. Another feature of all andesitic rocks is a low TiO_2 content throughout.

One early speculation on the origin of andesites, based on their intermediate compositions and occurrence mainly at continental margins, held that they were formed by the assimilation of sialic crustal material by (unspecified, but presumably tholeiitic) basic magmas. This speculation now appears to be largely unfounded, at least as far as volcanic members are concerned. Some andesites are erupted in areas of wholly oceanic crust, for example, the Scotia arc, Tonga—Kermadec chain and western Aleutian Islands, thus precluding assimilation of continental crustal rocks. Many andesites also have $^{87}Sr/^{86}Sr$-values of ~0.7035, a value close to overall average mantle, and

---

*The fugacity of a volatile solute approximates to its vapour pressure and comes to equal it in very dilute solutions. For a precise definition and discussion see Krauskopf (1967, pp. 230 and 231).

not much greater than average values for MORB, which is believed to have been derived from depleted mantle. These isotopic data preclude assimilation of continental crustal material on any significant scale (some acid members of andesitic associations, however, have significantly higher $^{87}Sr/^{86}Sr$-values consistent with a crustal derivation). Furthermore unless any assimilation process were curiously selective, one is hard put to explain the observed concentrations of disparate elements such as Ni, Rb, REE, U, etc., on the basis of any mix of basalt and crustal material (see S.R. Taylor, 1969). Note, however, that many features of plutonic SBZ associations in areas of continental crust appear to demand a lengthy sequence of crustal assimilation processes.

Another early speculation was that andesites, dacites and rhyolites represent nothing more than the "normal" fractionation derivatives of tholeiitic magmas. However, in view of notable petrographic and chemical differences between the andesitic and tholeiitic kindreds, the latter well exemplified by the Skaergaard intrusion and the lavas of Thingmuli, this speculation cannot be sustained.

Two contrasted theories of origin of acid volcanic rocks in a SBZ environment, both predicated on a comparable data base for the voluminous rhyolites of the North Island of New Zealand, are those of Ewart and Stipp (1968) and of Lewis (1968a). Lewis drew attention to the apparent chemical continuum between these acid rocks and the spatially associated intermediate and basic rocks. Ewart and Stipp, on the other hand, concluded that the rhyolites originated independently of the andesites and basalts by remelting of crustal material (Mesozoic greywackes themselves of calc-alkaline parentage). It is interesting to compare the arguments and rationales of these two papers. Widely differing $^{87}Sr/^{86}Sr$ ratios of acid rocks from other SBZ provinces in areas of continental crust have been recorded. Low initial ratios, i.e. around 0.7035 and comparable with associated andesites, are consistent with fractionation [see, for example, Lowder and Carmichael (1970); see also Lanphere et al. (1980) for an account of Sr-isotopic geochemistry of the mid-Tertiary rhyolites of the Sierra Madre Occidental, Mexico, the largest contiguous rhyolite province in the world, where andesites, dacites and rhyolites all have initial $^{87}Sr/^{86}Sr$ ratios in the fairly narrow range of 0.7042—0.7050, showing no systematic variation with composition, and suggesting therefore an origin of the series by fractionation]. High initial ratios, even in excess of 0.720, however, strongly indicate crustal refusion (see Clemons and Long, 1971). These data strongly suggest that more than one process has been operative in the production of acid members of andesitic series to different degrees at different times and places. However, our main concern here is with the generation of andesites, without the additional complications apparently afforded by their spatially associated acid rocks.

Osborn (1962) suggested that, given a parental tholeiitic magma, any significantly higher content of water in the magma would lead to higher than average oxygen fugacities (by virtue of the equilibria set up by the partial dissociation of magmatic water into oxygen and hydrogen, and the partial preferential escape of the latter from the magma by diffusion processes). This, in turn, would increase the ferric/ferrous ratio in the magma and result in the early crystallization and possible fractionation of magnetite. This would result in a lack of iron enrichment in intermediate members of andesite series and would also lead to the production of greater amounts of over-saturated residual liquids (as magnetite contains no SiO_2 at all), two observed chemical features of the oregenic andesite association. S.R. Taylor (1969), however, pointed out that early magnetite fractionation must necessarily deplete a basic melt in V and Sc, a depletion that is not found in andesites. Magnetite is, of course, often a phenocryst phase in andesites, but only in some of the intermediate members of andesite suites, and need not therefore have necessarily played any role in controlling a liquid line of descent in the basic to intermediate range of composition.

A plausible method of generating the distinctive orogenic andesite associa-tion, again predicated on relatively high H_2O contents of primordial mafic magmas, is by fractionation involving a component of amphibole (Cawthorn and O'Hara, 1976). High P_{H_2O} leads to a marked increase in the stability field of amphibole relative to that of plagioclase. Analyses of amphibole crystal-lized experimentally from mafic melts under conditions of high water content but not necessarily approaching P_{total} show them to be kaersutitic (i.e. with contents of TiO_2 up to roughly 8 wt.%) and to have SiO_2 contents of roughly 40 wt.%. Fractionation in nature of comparable amphibole could therefore produce the marked depletion in TiO_2 characteristic of all SBZ volcanic rocks and could also lead to the production of a silica-rich melt fraction. In addition, the experimentally produced amphiboles have Fe/Mg ratios such that their fractionation would curtail iron enrichment in the magma series. Not only is amphibole fractionation attractive in theory, and indicated by experiment, but also nodules containing amphiboles of kaersutitic compo-sitions are found in some andesitic rocks. The fractionation sequence olivine—augite—hornblende—magnetite is, of course, well established for Alaskan-pipe-type intrusive complexes, considered to be probable andesite feeder pipes (see Chapter 11). The observed abundance of plagioclase phenocrysts in intermediate SBZ volcanic rocks reflects later crystallization from evolved melts at relatively low pressures where amphibole is not stable.

Given the apparently strict correlation of andesites with Benioff zones, further speculation has centered on the ultimate origin of primary magmas of the andesitic associations (fractionated as they may be by the time they arrive at upper crustal levels). S.R. Taylor (1969) after reviewing the trace-element chemistry of andesitic rocks, proposed a "two-stage" origin for andesites, viz. by: (1) partial melting of mantle to form oceanic crust at spreading ridges; and (2) subsequent partial melting of this oceanic crust where it forms a part of the subducting lithospheric slab below Benioff zones to form parental andesitic magma.

T.H. Green and Ringwood (1968a), in a series of runs on "dry" melts of calc-alkaline rocks ranging from tholeiitic to rhyolitic in composition at 30-kbar pressure, showed that the melt with the lowest liquidus temperature was andesitic in composition, lower than basalt, and more surprisingly, lower than rhyolite. Phases crystallizing at or near the liquidus comprised garnet*, omphacitic clinopyroxene and quartz. On the basis of this work it is thus possible to envisage the production under dry conditions of andesite as an anatectic product of subducted mafic oceanic crust metamorphosed to eclogite at appropriate depths.

There is an obvious objection, however, to this simple method of generating andesitic magmas in that there does seem to be a compositional continuum in the full spectrum of SBZ series from intermediate and acid members back to more basic members. The basic members of the island-arc tholeiites in particular bear a close resemblance to MORB basalts. Given the tectonic setting of MORB basalt only mantle processes can be invoked in its genesis. With this in mind it is unwise therefore to eschew processes of partial melting of mantle in considering andesite petrogenesis. It is, of course, not essential to have a parental basic member present and exposed in substantial proportions to substantiate a series derived from it. What then might be the *special* conditions for generating the distinctive SBZ volcanic series assuming mantle involvement?

In experiments on the partial melting of basalt conducted at $P_{H_2O} = 0.6$ P_{total} at the rather low pressures of 2, 5, and 8 kbar, respectively, Holloway and Burnham (1972) showed that:

ol + cpx ± mag	will coexist with a "high alumina basalt" melt, SiO_2 content 50—53%
ol + cpx + mag + amph	will coexist with an "andesite" melt, SiO_2 content 53—63%
ol + cpx + mag + amph + plag	will coexist with a "dacite" melt, SiO_2 content >63%

They noted, as would be expected, the improbably high proportion of melting necessary to generate the more basic members of this sequence from *basic* rock. Whereas generation of andesitic melts by partial melting of basalt under these conditions remains a technical possibility, experimental data such as these are, of course, also compatible with the fractionation crystallization of a mafic assemblage including amphibole from primary basaltic melts to produce an andesitic suite.

Current fractionation models for low-K SBZ series begin with an H_2O-rich basic magma. This could be produced from mantle overlying the Benioff

---

*Some dacitic rocks contain phenocrysts of garnet (Oliver, 1961), probably crystallized at lower pressures (T.H. Green and Ringwood, 1968b) than those of the experiments quoted above.

zone and melting there could be facilitated by an access of water driven off from the descending lithospheric slab that carries with it an upper layer of somewhat hydrous oceanic crust as presciently suggested by Coats (1962). Given probable temperature regimes this essentially basic oceanic crustal material would be expected to recrystallize to eclogite (via an intermediate amphibolite stage) at vertical depths of ~ 80—100 km, the latter depth, of course, corresponding with the observed depth to Benioff zone below the volcanic front. In this model, therefore, the contribution of the downgoing oceanic crustal material to SBZ volcanic activity is water rather than anatexis. Note that, under hydrous conditions, experimental work (Kushiro, 1972a) has shown that the pressure limit to the incongruent melting of enstatite is extended from ~ 5 kbar (under "dry" conditions) to 25 kbar; a slightly over-saturated basic magma could therefore evolve within the mantle under these conditions.

Indicative of the complexity and uncertainty surrounding andesite petro-genesis are syntheses of the petrological evolution of island-arc systems presented by Ringwood (1974, 1977), with reference to the controls exercised by trace-element data including those of REE distribution. Ringwood envisages partial melting of mantle immediately above Benioff zones at ~ 100 km depth to produce primordial tholeiitic magmas that fractionate, predominantly by olivine crystallization, to yield the IAT series. For the more "mature" phase of island-arc development, comprising the more typically calc-alkaline series, Ringwood proposes sequential eclogite and amphibole fractionation from primary melts that may have had a complex history of generation by partial melting of subducted eclogitic ocean crustal material and subsequent reaction with mantle.

13.7.6. Granitic rocks

It is perhaps ironic that the genesis of granitic rocks, the commonest igneous rocks of continental crust, is still shrouded in uncertainty and debate, although trace-element, especially REE, studies are providing some constraints (see, e.g., Hanson, 1978). There would seem to be no less than four different ways in which granitic rocks could originate, thus affording an avenue to resolving rigid attitudes associated with the "granite controversy" of a generation ago.

(1) Advanced *fractionation* of parental mafic magmas could result in the production of necessarily small quantities of residual felsic liquids. Igneous rocks formed by the crystallization of such liquids would be expected to have low colour indices and high Fe/Mg ratios (reflecting the fractionation of mafic phases) and possibly a peralkaline chemistry (the "plagioclase effect" reflecting plagioclase fractionation) except in felsic rocks derived from those parental tholeiite magmas which were low in alkalis to start with. In addition, the overall compositions should approach those of the two ternary minima,

i.e. either a quartz-rich ideal granite composition or an undersaturated phonolite composition. Such rocks would typically be anticipated in a "volcanic association" as they originated from liquid magmas. Initial $^{87}Sr/^{86}Sr$-values should be low, deviating from parental magma values only to the extent of assimilation of crustal material. Probable examples include the acid intrusions associated with the upper parts of the Skaergaard and Bushveld intrusions (tholeiitic parental magma), the peralkaline granites of Afar (transalkaline parental magma), and the nepheline—syenite ring dykes of Mount Kenya (alkali basalt parental magma).

(2) *Anatexis* of continental crustal material might occur where large volumes of mafic magma had been emplaced in the crust within a short period of time, or in other areas of exceptionally high thermal gradients in the crust. Resultant melts should have minimum compositions, approaching ideal granite compositions the more felsic the crustal material being remelted. Rocks crystallized from such melts would also occur in a volcanic association but would not have trace-element contents indicative of prolonged fractionation from a mafic progenitor. Initial $^{87}Sr/^{86}Sr$ ratios should be variable and high, equalling those of the rocks being remelted. Probable examples include *some* of the Tertiary granites of the Hebrides, Scotland, (Moorbath and J.D. Bell, 1965) associated with large central volcanoes, the intrusive and extrusive acid rocks of the Taupo zone, New Zealand, and younger acid rocks (mainly extruded as ignimbrites) of the Basin and Range province.

(3) Under conditions of tectonic crustal thickening and regional metamorphism, *migmatite complexes* (typically developed from the gneissic basement below a thick, regionally metamorphosed clastic wedge) may intrude diapirically to mesozonal (and epizonal?) levels, while undergoing the complex, long time-span processes of "granitization". These comprise the mineralogical, chemical, physical and kinetic changes discussed on p. 75, and result in the eventual emplacement of increasingly homogeneous granitic rocks of increasingly felsic compositions at increasingly high crustal levels. In most, broad compositional inhomogeneities would persist, cognate inclusions may be abundant, and an igneous foliation would reflect emplacement not as a liquid but as a crystal mush. As well as intrinsic inhomogeneities some recognizably composite complexes might span a compositional range from diorite through quartz-diorite and granodiorite to adamellite. Only exceptionally, where evolutionary processes were greatly enhanced, would there be significant amounts of more felsic rocks, which would more typically occur only as small pockets of pegmatite and other differentiates. Mafic rocks would be lacking except as remnants derived from the pre-existing metamorphic component of the migmatite. Relative to the fractionates and anatectic melts, quartz contents would remain low and colour indices high (indeed most deep-seated granitic plutonic rocks have lower quartz contents and higher colour indices than superficially comparable rocks of the volcanic association). A peraluminous condition would be common. Initial $^{87}Sr/^{86}Sr$

ratios would be high, equalling those of the rocks being granitized. In sum, these are the S-type granites of Chappell and A.J.R. White (1974). Examples include the Coast Range batholith (Hutchison, 1970), and more problematically the foliated leucogranites of Manaslu, Nepal, (Hamet and Allègre, 1976), the latter reflecting perhaps an unusually high component of anatexis.

(4) *Assimilation* of crustal material by mantle-derived magmas on a massive scale could yield progressively more felsic products. This process has apparently operated in Andean-type non-tensional SBZ environments accompanied by "ponding" of rising magmas at the mantle—crust interface (Fyfe, 1978). A typical range of resultant products would comprise cumulitic gabbros, hornblende gabbros, diorites, quartz-diorites, granodiorites and adamellites, with the more felsic members predominating at relatively shallow erosion depths, and intermediate members predominating overall. Initial $^{87}Sr/^{86}Sr$ ratios would be low (~ 0.704) in mafic members, progressively increasing in more felsic members to values typically in the range 0.706—0.708. These are the I-type granites of Chappell and White. Examples include the Coastal batholith of Peru and the Southern California and Sierra Nevada batholiths, to which reference has been made in Chapters 7 and 11.

Chapter 14

DEGRADATION OF IGNEOUS ROCKS

14.1. INTRODUCTION

A majority of igneous rocks in the geological record have suffered meta-morphism, and degradation of a low-grade type is particularly evident in many older volcanic rocks. Just as a study of recent fresh volcanic rocks is a vital link in understanding the processes of the new global tectonics, so also should a study of older volcanic rocks in the geological record contribute to our understanding of the past pattern of tectonic regimes and crustal evolu-tion. The proper study of degraded rocks, however, has until recently tended to remain a grey area between the preferred fields of study of most igneous and metamorphic petrologists, and one on which many texts are reticent. Certain authors, and quite possibly teachers, have been willing, it seems, to send students out into a real world of abundant degraded igneous rocks armed only with a cognizance of the less common fresh ones. Certain basic concepts are, however, very easy to grasp.

Igneous rocks, more particularly volcanic rocks, have mineralogies that originated at, or allowing for supercooling and quench effects, not far below liquidus temperatures under surface pressure conditions. Considered on the metamorphic facies principle these temperatures represent by far the highest possible attainable of all crustal rocks. Not surprisingly, therefore, many of the mineral assemblages and mineral solid-solution compositions that volcanic rocks contain are not in equilibrium at ambient surface and upper-crustal temperatures.

The very existence therefore of "igneous" mineralogies, textures and mineral compositions, namely those resulting from the crystallization of a silicate melt, reflects a losing battle fought by equilibrium considerations against kinetic considerations, resulting from quick cooling. The relatively more slowly cooled plutonic rocks do, of course, show such mineralogical adjustments to falling temperature as exsolution, recrystallization, and given sufficient available H_2O, CO_2, etc., the formation of varying amounts of deuteric minerals that are in fact stable at lower temperatures. With the exception of some granitic rocks (many of which never crystallized from melt in the first place) however, an origin by magmatic crystallization is apparent from the mineralogies and textures of fresh igneous rocks.

However far re-equilibration may have gone during the cooling of the igneous body, further change becomes inhibited by extremely low velocities of reaction. Chemical reactions approximately double in speed for a temperature rise of 10°C; therefore they proceed a thousand times faster (or slower) for a temperature rise (or fall) of 100°C; a million times for 200°C; a billion times for 300°C; a trillion times for 400°C; a quadrillion for 500°C, and so on. In other words, any mineralogical re-equilibration that a volcanic rock, for example, does not undergo during initial cooling between liquidus temperatures and air temperature, a time measurable in a few days, weeks, or months, should not be expected to be achieved during all the aeons of geologic time.

This happy state of affairs for the igneous petrographer, however, is not to be. Igneous rocks cannot escape burial to various depths and attaining there temperatures appropriate to the geothermal gradients at the various times of their burial history; circulating groundwaters may ensure the accession of amounts of H_2O and CO_2 adequate for the formation of new low-temperature minerals. With respect to this low-temperature re-equilibration it is important to be aware that (C.J. Hughes, 1972b):

> "It is not necessarily a matter of attaining the temperature of around 400° at which several prograde endothermic reactions in sedimentary rocks occur and define the lower temperature boundary of the greenschist facies. In this respect, the experimental illustration by Eskola (1937) of 'the spilite reaction' whereby a pure albite was produced from anorthite in the presence of water, soda, silica, and CO_2 at temperatures between 264° and 331° should be noted. It is thus merely a matter of attaining a temperature at which the velocity of chemical reaction in an already unstable mineralogical assemblage becomes appreciable."

In the ensuing discussion we will restrict ourselves mainly to the re-equilibration of volcanic rocks in greenschist and sub-greenschist facies conditions, as it is volcanic rocks that have proved to be particularly useful in distinguishing between different series of igneous rocks and relating them to global tectonic regimes.

14.2. SPILITE AND KERATOPHYRE

We could refer to volcanic rocks that prove on thin-section examination to have a mineralogy that has obviously re-equilibrated under low-temperature conditions quite simply as "degraded". However, one often wants to answer the question — degraded what? For reasons that will be apparent the answer to this question is commonly not immediately ascertainable. The terms spilite and keratophyre are therefore being increasingly employed in a broad context to denote degraded volcanic rocks, and without any connotation of magmatic affinity. Recent attempts at definitions are as follows (Cann, 1969, p.1):

> Spilites are "rocks of basaltic texture in which greenschist mineralogy is completely or almost completely developed".

and (C.J. Hughes, 1975, p.425):

> "Keratophyre is a retrograded fine-grained igneous rock of intermediate to acid composition, showing little or no penetrative deformation, in which a pyrogenetic mineralogy is mimetically replaced by an assemblage appropriate to greenschist or lower facies mineralogy."

In respect of the definition of Cann we should note that mineralogical re-equilibration may well have occurred at temperatures below the formal greenschist facies range as shown by the common occurrence in spilites of typical zeolite-facies minerals such as prehnite and pumpellyite. The term propylite could perhaps be included in this list of degraded igneous rocks and has been defined as follows (Gary et al., 1972):

> "a hydrothermally altered andesite resembling a greenstone and containing calcite, chlorite, serpentine, quartz, pyrite, and iron ore."

Propylites, however, include a high proportion of rocks that have undergone syn-volcanic alteration or alteration related to intrusions, thus contrasting with the more widespread burial metamorphism that has been responsible for the degradation of most spilites and keratophyres.

The history of the terms *spilite* and *keratophyre* has been a checkered one. The existence of primary spilite was claimed for many years. For example, Amstutz (1968) categorically stated that:

> "if the fabric relations are considered, it is evident that many spilites and keratophyres must be primary, because no secondary diffusion process is able to mimetically reproduce a primary process."

Evidence for the secondary origin for spilitic rocks has, however, been conclusively detailed by Vallance (1960, 1969). The present position has been reasonably stated by Carmichael et al. (1974, p.560) who conclude that the final synthesis of Vallance leaves scarcely any room for further doubt. It is interesting to contrast this succinct summary by Carmichael et al. with the long and uneasy treatment of the subject of spilite and keratophyre in the earlier book by F.J. Turner and Verhoogen (1960, pp. 258—272). One reason for the confusion surrounding the origin of these rocks has undoubtedly been the largely mimetic textures of the spilite—keratophyre suite (C.J. Hughes, 1972b, p.516):

> "At temperatures significantly below 400° there may not be sufficient energy to nucleate and grow some or all of the new phases in the rock, and the mineralogical changes that occur in the production of spilite and keratophyre have perforce to be in the main replacement processes. The absence of shear stress (Cann, 1969, p.17) is probably another important factor inhibiting nucleation and favouring mimetic textures. The difficulty of Amstutz (1968) regarding the fabrics of these rocks is thus resolved, but it is worthy of note that one of the most deceiving features of these rocks *is* the presence of both primary structures and mimicked primary textures."

The usage of the term keratophyre has also been in question but in a slightly different context. As a result of classical petrographical study, a set of anchimetamorphic terms specifically for degraded pre-Tertiary igneous rocks was used on the continent of Europe, particularly in Germany — i.e. the recognition of degraded igneous rocks is not new! Unfortunately, British and American petrographers were reluctant to accept the anchimetamorphic nomenclature. The definition of anchimetamorphism in the American Geological Institute *Glossary of Geology* (Gary et al., 1972) for example, reads as follows:

> "A term introduced by Harrassowitz (1927) and still used by German authors to indicate changes in mineral content of rocks under temperature and pressure conditions prevailing in the region between the Earth's surface and the zone of true metamorphism, i.e. approximately in the zones of weathering and ground-water circulation. The term was never accepted by Anglo-Saxon authors."

In a recent exchange (Lehmann, 1974, 1977; C.J. Hughes, 1975, 1978), it is clear that the anchimetamorphic term keratophyre, at least for some petrographers, had also acquired a narrow compositional connotation, although it is fair to say that current usage follows a broader definition as outlined above.

14.3. METASOMATISM IN DEGRADED VOLCANIC ROCKS

For purposes of assigning magmatic affinity to degraded volcanic rocks, one might legitimately ask why bother about degradation, as it is in the main chemical parameters rather than mineralogical ones that characterize the different igneous rock series. Analyses of fine-grained rocks, whatever may be their reconstituted mineralogy, should therefore indicate their magmatic affinity. This, however, is to ignore the possibility of metasomatism accompanying degradation.

Metasomatism in this context has been a surprisingly touchy subject. One subjective reason for this might be that with the recent upsurge in availability of analytical methods and facilities, many petrologists have as never before been able to amass a great deal of analytical data. Someone who may question the very significance of these arrays of precise figures is conceived of as hostile. More objectively, a legitimate question to be asked of anybody who proposes metasomatism is to enquire how he might know what the original composition was (if indeed it were different).

Consider however the chemistry involved in the degradation of a basalt, essentially labradorite and augite (plus opaque mineral), to common spilite, composed of albite and chlorite (plus leucoxene). The gains and losses in major-element content involved in the two major mimetic substitutions can be indicated, at least to a crude first approximation, as follows:

	Si	Al	Mg, Fe	Ca	Na	K
Augite to chlorite	−	+	0	−	0	0
Labradorite to albite	+	−	0	−	+	−

Consider first potassium: plagioclase of volcanic rocks contains up to several per cent of orthoclase molecule in crystalline solution, the albite of spilite virtually nil. In spilite, adularia may be found pseudomorphing some of the plagioclase although commonly it is absent. Sericite may occur within saussuritized plagioclase; commonly it too is lacking from the limpid albite of spilitic rocks and does not nucleate elsewhere in the rock. In view of the necessary pervasive influx of H_2O through the rock to convert augite to chlorite (H_2O content around 12 wt.%) and the apparent lack of potassium-bearing minerals in common spilite, it is quite conceivable, therefore, that potassium could have been leached out of a spilitized rock by metasomatizing solutions, particularly as potassium carbonate has an extremely high solubility in water. Consider calcium: there would seem to be a gross loss in calcium consequent on spilitization. Outcrops of degraded basalts commonly reveal, however, veinlets, veins, and vesicle fillings rich in calcium minerals such as calcite, epidote and prehnite, together with more generalized patches relatively rich in these minerals — the different domains so well documented by the work of R.E. Smith (1968). Also, on the scale of a thin-section, veinlets, small amygdales, and sporadic developments of these minerals may be visible. Often, however, they are lacking from a sample of hand-specimen size, and it would appear therefore that, at least at the level of size of a sample collected for analysis, calcium may have been lost during degradation. There would also appear to have been an indisputable nett gain in sodium as the secondary albite of spilite must account for considerably more sodium than the more calcic plagioclase that it mimetically replaces. By the same token, in the (rare) recorded instances of K-feldspar-rich spilites (poeneites) a marked metasomatic addition of potassium must have taken place. Among trace elements, Li shows a marked increase in abundance from ~12 ppm in many fresh basaltic rocks to an average of 75 ppm in spilites (see discussion in D.M. Shaw et al., 1977).

A metasomatism affecting spilite specimens of hand-specimen dimensions is thus not an idle or forced speculation. Indeed in degraded pillow lavas, for example, there is an obvious compositional differentiation between the cores of pillow lavas, their chlorite-rich rims, and intervening hyaloclastic material, and any scattered calcium-rich domains. The tendency undoubtedly is to collect from the average-looking (but in fact sodium-enriched) pillow core material and hope for the best. In degraded acid rocks there is commonly a lack of such obvious differentiation as that between pillow core and rim, and degraded acid rocks may appear deceptively uniform in outcrop. The chemistry of samples of such rocks, however, commonly reveals alkali contents that clearly do not coincide with those of any fresh igneous rocks (C.J. Hughes, 1972b).

Leaving aside here the questions of the timing and the *scale* of metasomatism, i.e. between pillow core, rim, and intervening hyaloclastic material in spilitic pillow lavas (Vallance, 1965; Cann, 1969), or between different "domains" in a degraded basaltic lava pile (R.E. Smith, 1968), or between different flows and parts thereof and associated fragmental rocks in sequences of degraded rhyolites (Battey, 1955), or within a yet more open system with possible bulk gains or losses of some components over large volumes of rock, it is commonly apparent that, on the scale of a sampled and analysed hand-specimen metasomatism has undeniably occurred [see the alkali plots in C.J. Hughes (1972b) and accompanying rationale]. Moreover, the metasomatism has in many rocks patently affected the content of such major elements as alkalis and lime that are often critical to definitions of magmatic affinity. In dealing with degraded rocks, therefore, such a widely used variation diagram as a total alkalis vs. silica plot becomes unreliable (many degraded tholeiites plot in the field of alkali basalt). So also may be calculations of alkali-lime index (Peacock, 1931), the suite index of Rittmann (1962), and the alkalinity ratio of J.B. Wright (1969a). Also suspect are diagrams combining variation in alkalis with variation in Mg/Fe ratio in some form or other such as the familiar *AFM* triangle, etc.

In sum, whereas igneous rock series are recognized and defined, when dealing with recent or young fresh volcanic rocks, on the basis of: (1) tectonic setting; (2) petrography; and (3) analytical data, in older and degraded volcanic rocks the situation is a great deal more complex because, in general, item (1) is not known (indeed, one often would like to know it!), item (2) is prone to severe alteration, and item (3) likely is affected by allochemical alteration to an unknown extent.

14.4. DETERMINATION OF MAGMATIC AFFINITY OF DEGRADED VOLCANIC ROCKS

Clearly we have a problem; how are we going to tackle it? Surprisingly, even some quite recent papers have chosen to *deny* the problem, and faced with the mineralogy of degraded rocks, some writers go to extraordinary lengths to maintain that the albite and chlorite of spilites are primary crystallization products of "spilitic" mafic magmas. A more fashionable approach is to *ignore* the problem; i.e. whatever degree of degradation there may be in the rocks in question is acknowledged, as is also the possibility of metasomatism, but then appears a disclaimer, often barely noticeable among an imposing array of geochemical data, declaring that the effects of metasomatism may safely be dismissed. However, in recent years a growing number of researchers have shown great concern in coming to grips with the problem of magmatic affinity of degraded volcanic rocks and various lines of attack have emerged as follows.

14.4.1. Application of conventional criteria to the analytical data

One approach is to analyse numerous samples for major and trace elements, to discard those which appear to be markedly anomalous (due presumably t relatively pronounced alteration and metasomatism), and to see what trends may emerge from the resultant raw data. This approach assumes that, at least in some samples, sufficient of the pristine chemistry may show through a veil of later allochemical alteration processes to characterize the rocks.

A successful application of this approach was the delineation of an island-arc tholeiite series and an overlying calc-alkaline andesitic suite within a 5 km thick, largely submarine, Ordovician sequence in Notre Dame Bay, Newfoundland, now degraded into low greenschist facies. From an initial collection of 300 samples showing relatively little signs of alteration in outcrop, 60 were analysed for the trace elements Rb, Sr, Zr, Ba, Ni, Cr, Cu, Zn and Co, and of this group 29 were selected and analysed for major-element content. Rocks collected from the lower and upper parts of the succession fall into two distinct groups though with some scatter; the upper andesitic rocks are clearly higher in Al_2O_3, K_2O, and associated large-cation trace-elements (see Kean and Strong, 1975, pp. 103—112, for discussion of data).

A study of Proterozoic Keweenawan basic lavas (Jolly and R.E. Smith, 1972) revealed compositional gradations leading back from clearly altered and metasomatised samples through partly degraded rocks to rocks that were apparently the least altered and metasomatised, and thus probably most closely approximating the original lava compositions.

14.4.2. Petrography

Any assignation of magmatic affinity based on analyses of degraded volcanic rocks must be in harmony with what can be seen or deduced of the original petrography. The retention of petrography as an essential working tool has been very clearly emphasized by Moorhouse (1970) in the introduction to *A Comparative Atlas of Textures of Archaean and Younger Volcanic Rocks*:

> "Undeformed Archean volcanic rocks frequently preserve their textures with remarkable fidelity and can therefore provide approximate clues to their original unaltered composition . . . Despite the lack of deformation, chemical changes have undoubtedly taken place in many of these volcanic rocks and so it is difficult if not impossible to classify the lava flow in terms reflecting its original composition."

Fortunately, the largely mimetic style of degradation commonly permits identification of phenocrysts, and the abundance and order of appearance of phenocryst phases in rocks of an igneous rock series is of course an informative observation. For example, the typical olivine—plagioclase phenocryst assemblage of oceanic tholeiites, more rarely with augite in addition, contrasts with the more frequent appearance of augite as a cotectic phase along with olivine and plagioclase in ocean-island tholeiites. The phenocryst sequence olivine—augite—plagioclase is more typical of alkali basalts. An abundance of plagioclase phenocrysts, often with sieve textures, together with augite ± olivine ± orthopyroxene is characteristic of relatively basic members of many andesitic series, and so on.

Groundmass textures are more prone to be obliterated as a result of degradation. Nevertheless the presence of groundmass olivine, characteristic of alkalic series in general, can sometimes be discerned; distinctive basalt groundmass textures may survive degradation, as may the various distinctive andesitic textures of plagioclase microlites in a fine-grained acid mesostasis.

The relative abundance of basic, intermediate and acid rocks in a degraded suite may itself be a guide to magmatic affinity (see, e.g., P.E. Baker, 1968).

The association of pillow breccias and associated hyalotuff material with a pillow lava sequence suggests a relatively shallow depth of formation (see Chapter 2) and negates a MORB environment.

14.4.3. Plutonic equivalents of extrusive rocks

Plutonic rocks are massive in nature, coarser in grain size, and do not contain any highly reactive glass; they are not vesicular and are not intimately interbedded with waterlogged porous sediments. As a consequence, relative to volcanic rocks, plutonic rocks tend to resist the mineralogical effects of mild degradation processes. Their chemistry likewise may remain relatively unaltered and a study of plutonic members of an igneous association might therefore provide evidence of magmatic affinity. However, there are some inherent difficulties in working with plutonic rocks. Among basic plutonic rocks, the deduction of magmatic type from a cumulate sequence is notoriously difficult (see, e.g., Chapter 11, Section 11.3.2, on Alaskan pipes). The abundant granitic rocks emplaced in a SBZ environment may have had complicated histories of assimilation and autometasomatism superimposed on fractionation (see Chapters 7 and 11), and thus may not have compositions equivalent to the liquid line of descent as seen in the compositions of erupted lavas.

14.4.4. Immobile trace elements

Because of metasomatism associated with degradation, many of the parameters that are used in assigning magmatic affinite to fresh rocks are thus of questionable value for degraded volcanic rocks. Along with Na, K and Ca, it would seem highly likely that the amounts of chemically similar dispersed trace elements such as Rb, Cs, Sr and Ba in groups I and II of the periodic table would also be prone to metasomatic change and thus dangerous or useless for discrimination purposes. However, it could be that Fe, Ti and other transition trace elements such as Ni, Cr and V might be relatively resistant to solution and transportation in the various hydrothermal solutions presumably involved in degradation processes; Zr, if largely locked up in chemically stable zircon, might also persist in unchanged concentrations during degradation. And there might well be more relatively *"immobile"* elements.

Clearly, to be of any value in determining the magmatic affinity of degraded rocks, selected trace elements, as well as being immobile to a high degree during degradation, should also be of value as chemical discriminants between the fresh rocks of various igneous rock series. They should also be reasonably easy to determine analytically. The 1970's has witnessed an important field of investigation with these three criteria in mind.

The establishing of certain elements as immobile is an empirical process. In this regard the warning of Vallance (1974, pp. 94 and 95) should be heeded from the outset: elements that may prove to be relatively immobile during a certain set of conditions of degradation (including pressure, temperature, P_{H_2O}, P_{CO_2}, pH of solution and salinity) may unfortunately prove to be mobile under other conditions. As an illustration of this are the differences summarized by Pearce (1975, p.49) between the chemical changes accompanying weathering of ocean-floor basalt on the one hand and during greenschist facies metamorphism on the other.

Among relatively early workers in the field Herrmann and Wedepohl (1970) recognized that spilitization of Variscan basaltic rocks of Devonian—Carboniferous age in northwest Germany was a metasomatic process in a partially open system. This involved a loss of Si and Ca, and a gain of H_2O, CO_2 and Na; absolute abundances of Y and lanthanide REE, however, remained probably unchanged. Cann (1970) showed that abundances of Y, Ti, Zr and Nb in some ocean-floor basalts seemed to be unaffected even by severe secondary alteration processes whereas Rb and K, and to a lesser extent Sr were mobile.

Work by Pearce and Cann (1971) concerned the Ti, Zr and Y contents of five groups of fresh mafic volcanic rocks: ocean-floor basalts, Hawaiian tholeiites, alkali olivine basalt from Flores, Azores, island-arc tholeiites from Japan, and andesites and basaltic andesites from New Zealand. Data plots and discriminant analysis of the data showed good characterization of these five differing magma types. Degraded mafic volcanic rocks from ophiolite sequences were shown to have contents of these elements comparable with fresh ocean-floor basalts.

Pearce and Cann (1973) later extended this work in several ways:

(1) A further classificatory subdivision of fresh mafic volcanic rocks, as shown in Fig. 14.1, a necessary prelude to assigning magmatic affinity and a useful synopsis of different parental mafic magma types as then perceived. Note, however, the additional complexities that have since come to light, for example, with respect to subtle variations in SBZ series rocks and ocean-floor basalts and their inferred derivation from variable or more fertile mantle sources (see Chapters 9 and 11).

(2) The consideration of other trace-element indicators: Nb was included in addition to Ti, Zr and Y; Sr was also considered because it is such a useful discriminant in fresh rocks and appears to be fairly stable in uncarbonated, slightly degraded rocks although a counter-indication to its employment is

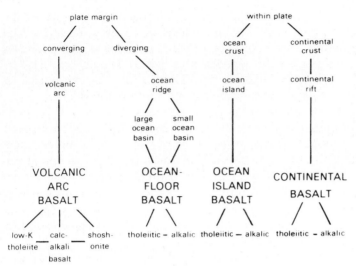

Fig. 14.1. A classificatory scheme for basic volcanic rocks based on their tectonic setting, proposed by Pearce and Cann (1973, p.291, fig. 1).

albitisation of plagioclase (actually commonly virtually complete in many spilites). Other elements experimented with were Ba, Cr and Ni: Ba, however, was found to be particularly mobile during degradation, much more so than Sr; Cr and Ni remain relatively stable during degradation, Cr more so than Ni, but the contents of both fluctuate greatly even in fresh mafic rocks because they are very sensitive to even small amounts of early fractionation of olivine and pyroxene (and chrome-spinel).

(3) The collection of a considerable number of sources of representative analyses (Pearce and Cann, 1973, table 1) of mafic rocks (defined as having a combined (CaO + MgO) content between 12 and 20 wt.% (note that Pearce and Cann restrict their analysis to mafic rocks).

(4) The development of a flow sheet (Pearce and Cann, 1973, fig. 6) for the determination of magmatic affinity of degraded mafic volcanic rocks using several data plots. Two, which have been widely quoted and used, are reproduced here as Figs. 14.2 and 14.3. Note also the use of the Y/Nb ratio for an indication of alkalinity: a low Y/Nb ratio correlates with alkali basalts, and a high Y/Nb ratio with tholeiites.

Later work by Pearce (1975) showed a good discrimination between ocean-floor basalts and island-arc tholeiites, using a Ti—Cr plot (Fig. 14.4).

Phosphorus, although showing significant mobility under certain conditions of degradation, was included with Zr, Ti, Y and Nb by Floyd and Winchester (1975) who first demonstrated a good discrimination between fresh tholeiite and alkali basalts, then between altered representatives (Winchester and Floyd, 1976), using several plots, viz. TiO_2—Zr, TiO_2—Y/Nb, P_2O_5—Zr, TiO_2—Zr/P_2O_5 and Nb/Y—Zr/P_2O_5.

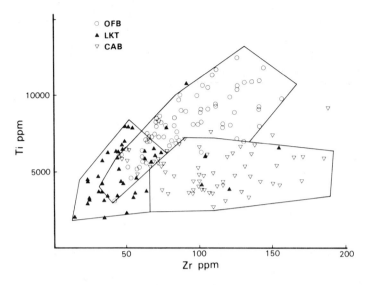

Fig. 14.2. Discrimination diagram, using Ti and Zr. Ocean-floor basalts (OFB) plot in fields *D* and *B*, low-potassium tholeiites (LKT) in fields *A* and *B*, and calc-alkali basalts (CAB) in fields *C* and *B* (after Pearce and Cann, 1973, p.295, fig. 2).

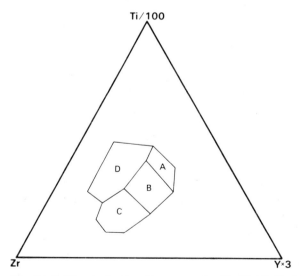

Fig. 14.3. Discrimination diagram, using Zr, Ti (divided by 100), and Y (multiplied by 3). "Within-plate" basalts (i.e. oceanic-island and continental basalts) plot in field *D*, ocean-floor basalts in field *B*, low-potassium tholeiites (island-arc tholeiites) in fields *A* and *B*, and calc-alkali basalts in fields *C* and *B* (after Pearce and Cann, 1973, p.295, fig. 3).

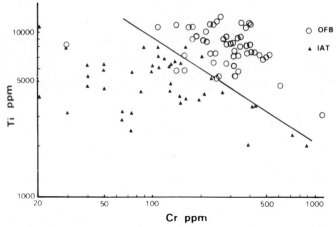

Fig. 14.4. Discrimination diagram, using Ti and Cr to separate ocean-floor basalts (OFB) from island-arc tholeiites (IAT) (after Pearce, 1975, p.48, fig. 5).

Following Herrmann and Wedepohl (1970), Herrmann et al. (1974) presented REE data from spilitic rocks from the Mid-Atlantic Ridge, the Variscan geosyncline, and several localities in Switzerland, that indicated that the REE contents of these degraded basalts apparently remained essentially unchanged during low-grade metamorphism up to 400°C. D.A. Wood et al. (1976), however, studied element mobility during various degrees of zeolite facies metamorphism in basaltic members of the lava pile of eastern Iceland, well documented by earlier work as to age, zeolite zones, and depth of burial. They confirmed that whereas Ti, Zr, Y, Nb, Ta and P were apparently unaffected by metasomatic transport under these conditions, the LREE, however, (along with K, Rb and Sr) had undergone significant mobilization in all sampled basalts except the most shallowly buried (less than 600 m). Basalt samples from depths in the lava pile deeper than this revealed a significant enrichment in LREE. Wood et al. drew attention to this mobility in view of the heavy dependence on Sr-isotopic data and LREE data such as La/Sm ratios in recent models of possible plume origins for certain fresh basaltic rocks (see Chapter 9, Section 9.3, dealing with oceanic-island tholeiite). Floyd (1977) similarly discussed LREE enrichment and apparently differing patterns of REE behaviour in modern and ancient spilites. Hellman et al. (1979), in a study of hydrous burial metamorphism of volcanic rocks from three localities coupled with an evaluation of published work, conclude that the mobility of REE under these conditions is highly variable and can be described by: "1. gross REE and selective LREE enrichment; 2. REE movement around a primary mean; 3. gross REE depletion; 4. selective REE mobility." The authors caution against uncritical application of raw REE analytical data to petrogenetic models.

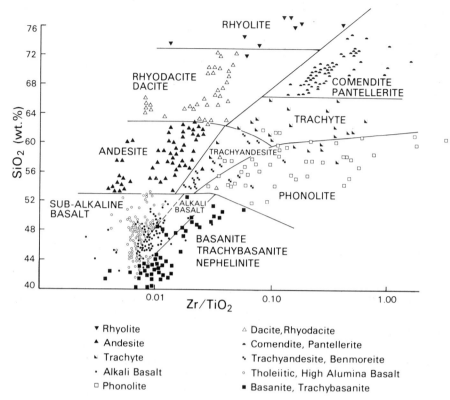

Fig. 14.5. Discrimination diagram, using Zr, Ti, and SiO$_2$ contents for common volcanic rock series (after Winchester and Floyd, 1977, p.333, fig. 2).

Winchester and Floyd (1977) have given a useful compilation of concentrations of trace elements believed to be immobile in several igneous rock series of differing alkalinity. They have added Ce, Ga and Sc to the now "traditional" list of Zr, Ti, Y and Nb, and furthermore have included intermediate and acid members of series in addition to the data for basic rocks covered in the now classic work of J.A. Pearce, J.R. Cann, and others. Fig. 14.5, a relatively simple Zr/TiO$_2$—SiO$_2$ plot, is indicative of the spread obtained between different series.

This and other diagrams presented by Winchester and Floyd (1977) coupled with those of Pearce and Cann (1973) and Pearce (1975) appear to have practical promise for the trace-element discrimination of degraded volcanic rocks at the time of writing.

14.4.5. Relict augite phenocrysts

The chemical compositions of clinopyroxene phenocrysts and groundmass crystals are typical of the magma type from which they crystallized (Kushiro,

1960; Le Bas, 1962; Coombs, 1963). Generally speaking, the more alkalic the parental mafic magma the richer in Ti, Al and normative nepheline its clinopyroxenes will be. In many degraded basaltic rocks with augite phenocrysts it is observed that the augite phenocrysts commonly appear fresh whereas the matrix pyroxene may be completely replaced by chlorite. Smaller grain size, permitting easier access to intergranular metasomatising fluids, coupled with possibly metastable quench compositions (e.g., sub-calcic augites in some classes of tholeiite) may likely have contributed to this relatively pronounced degradation of the groundmass pyroxene. The apparent freshness of the augite phenocrysts could be illusory: after all, deceptively limpid albite may pseudomorph primary plagioclase in the same rock. Nevertheless, Vallance (1969) suggested the possibility that the compositions of relict clinopyroxene in spilitic rocks may provide a pointer to their original magmatic affinity.

A successful application of this method using probe analysis of the clinopyroxenes of a variably spilitised Deccan flow (Vallance, 1974) showed that even some of the groundmass pyroxenes in this flow retained a sub-calcic iron-rich composition in keeping with their known tholeiitic affinity. More recently, it has been demonstrated that although clinopyroxenes may show slight primary differences in composition (notably Ti and Al contents) related to cooling rates, metastable clinopyroxene compositions can survive an episode of low-temperature—high-pressure metamorphism (Mével and Velde, 1976). The chemistry of relict porphyritic clinopyroxene (and hornblende) in otherwise completely degraded rocks of andesitic petrography was successfully utilized by Barron (1976), who included useful references to papers dealing with the chemistry of indisputably *fresh* clinopyroxenes from various igneous rock series.

Another useful compilation of pyroxene compositions from known magma types classified into ocean-floor basalts, volcanic-arc basalts, within-plate tholeiites, and within-plate alkali basalts is given by Nisbet and Pearce (1977). The conclusion is reached that the classification into magmatic affinity by analysis of even a single "unknown" clinopyroxene would have a 70% chance of success (although discrimination of some classes of basalt is relatively easier than others). Nisbet and Pearce (1977, p.159) conclude that:

"Pyroxenes from within-plate alkalic basalts are easiest to recognize; pyroxenes from volcanic-arc basalts and within-plate tholeiites can be distinguished from each other, but not in many cases from pyroxenes belonging to ocean-floor basalts."

Note, however, possible problems in interpretation in that clinopyroxene compositions may be related to variables other than magmatic major-element composition. Barberi et al. (1971) found that high P_{H_2O} and high P_{O_2} favour the early crystallization of clinopyroxene relative to plagioclase in basalt melts, and vice versa. Thus, in an alkali basalt melt crystallizing, for example, under unusually low P_{H_2O} and P_{O_2}, early crystallization of plagioclase could reduce the content of Al in the remaining melt and the subsequently crystal-lizing augite might have a composition more resembling that of tholeiitic augites.

14.5. DISCUSSION

The question of the recognition of magmatic affinity of older degraded igneous rocks is a most topical one. It is witnessing a considerable prolifera-tion of published papers accompanied by a great increase in perception of complications, in these features very typical of a new scientific field. Surely there is much more to come. Factors contributing to the complexity include the following.

First, there are the inherent questions of subtle mantle inhomogeneities at the present day and their reflection in derived mafic magmas, and the inevitable "noise" of geochemical data even on fresh rocks. These contem-porary sources of variation have to be integrated with a consideration of possible overall variation of mantle composition, megatectonic processes, and derived igneous rock series with time.

Secondly, although some of the recently published work deals with differ-ences of content of immobile elements between different *igneous rock series*, much of it deals with the immobile-element content of their *mafic members*. In this regard, it has been demonstrated several times in preceding chapters how *within* the class of rocks accepted as basaltic, for example, considerable fractionation can have occurred. This early fractionation can have a very great effect on the contents of those elements such as Cr and Ni which are strongly partitioned into early fractionating phases, and also a significant effect on the contents of residual elements such as Ti and Zr that are incom-patible with the early fractionating minerals, without much accompanying change, for example, in the SiO_2 content. See, for example, an approximately two-fold variation of Ti contents of MORB groundmass compositions attri-butable to fractionation processes (Fig. 9.1). For more precise comparative purposes, we should be tying immobile-element content to comparable fractionation stages as measured by such conventional and reasonably unaffected parameters as M-values or Zr content.

Thirdly, and possibly most significant of all, we should heed the warning of Vallance referred to above and reckon with different degrees of element mobility with different types and conditions of alteration. See, for example, the plea for more rigorous testing made by R.E. Smith and S.E. Smith (1976) and their identification of several contrasted alteration lithologies. Condie et al. (1977), for example, documented the differing chemical changes accom-panying the degradation of tholeiite flows of the Barberton greenstone belt that have been either significantly epidotized or carbonatized reflecting differences in P_{CO_2} from place to place during degradation. An interesting summary of the effects of initial submarine alteration of basaltic pillow lavas is provided by Scott and Hajash (1976), who refer to the numerous ambiguities and uncertainties that still remain and conclude by quoting Mark Twain's delightful comment,

> "The researches of many commentators have already thrown much darkness on this subject, and it is probable that, if they continue, we shall soon know nothing at all about it."

EPILOGUE

*The true test of a university's success is not the discoveries
and inventions which its professors make or the books they write,
but the sort of people its students turn out to be.*

Clarence Tracy

REFERENCES

A.A.G.S. (Association of African Geological Surveys), 1963. *Geological map of Africa; scale 1:5,000,000.* Assoc. Afr. Geol. Surv., Paris.

Abbott, M.J., 1969. Petrology of the Nandewar volcano, N.S.W., Australia. *Contrib. Mineral. Petrol.,* 20: 115—134.

Allègre, C.J. and Condomines, M., 1976. Fine chronology of volcanic processes using ^{238}U—^{230}Th systematics. *Earth Planet. Sci. Lett.,* 28: 395—406.

Allègre, C.J. and Minster, J.F., 1978. Quantitative models of trace element behaviour in magmatic processes. *Earth Planet. Sci. Lett.,* 38: 1—25.

Amstutz, G.C., 1968. Les laves spilitiques et leurs gîtes minéraux. *Geol. Rundsch.,* 57: 936—954.

Anderson, A.T., 1976. Magma mixing: petrological process and volcanological tool. *J. Volcanol. Geotherm. Res.,* 1: 3—33.

Anderson, D.L., 1970. Petrology of the mantle. *Spec. Pap. Mineral. Soc. Am.,* 3: 85—93.

Anderson, D.L., 1975. Chemical plumes in the mantle. *Bull. Geol. Soc. Am.,* 86: 1593—1600.

Anderson, D.L., 1977. Composition of the mantle and core. *Ann. Rev. Earth Planet. Sci.,* 5: 179—202.

Anderson, E.M., 1951. *The Dynamics of Faulting and Dyke Formation.* Oliver and Boyd, London, 206 pp.

Andrews, A.J. and Fyfe, W.S., 1976. Metamorphism and massive sulphide generation in oceanic crust. *Geosci. Can.,* 3: 84—94.

Appleton, J.D., 1972. Petrogenesis of potassium-rich lavas from the Roccamonfina Volcano, Roman region, Italy. *J. Petrol.,* 13: 425—456.

Arndt, N.T., 1977. Ultrabasic magmas and high-degree melting of the mantle. *Contrib. Mineral. Petrol.,* 64: 205—221.

Arndt, N.T. and Brooks, C., 1978. Iron-rich basaltic komatiites in the early Precambrian Vermilion district — Discussion. *Can. J. Earth Sci.,* 15: 856—857.

Atkins, F.B., 1969. Pyroxenes of the Bushveld Intrusion, South Africa. *J. Petrol.,* 10: 222—249.

Badham, J.P.N. and Morton, R.D., 1976. Magnetite—apatite intrusions and calc-alkaline magmatism, Camsell River, N.W.T. *Can. J. Earth Sci.,* 13: 348—354.

Bailey, D.K., 1975. Experimental petrology relating to oversaturated peralkaline volcanics: a review. *Bull. Volcanol.,* 38: 637—652.

Bailey, D.K., 1977. Lithosphere control of continental rift magmatism. *J. Geol. Soc. London,* 133: 103—106.

Bailey, D.K. and Schairer, J.F., 1964. Feldspar—liquid equilibria in peralkaline liquids — the orthoclase effect. *Am. J. Sci.,* 262: 1198—1206.

Bailey, E.B., 1960. *Geology of Ben Nevis and Glen Coe and the surrounding country.* Mem. Geol. Surv. Scotl., Her Majesty's Stationery Office, Edinburgh, 307 pp.

Bailey, E.B., Clough, C.T., Wright, W.B., Richey, J.E. and Wilson, G.V., 1924. *Tertiary and post-Tertiary geology of Mull, Loch Aline, and Oban.* Mem. Geol. Surv. Scotl., Her Majesty's Stationery Office, Edinburgh, 445 pp.

Bailey, R.A., Dalrymple, G.B. and Lanphere, M.A., 1976. Volcanism, structure, and geochronology of Long Valley Caldera, Mono County, California. *J. Geophys. Res.,* 81: 725—744.

Baker, B.H. and Wohlenberg, J., 1971. Structure and evolution of the Kenya Rift Valley. *Nature (London)*, **229**: 538—542.

Baker, B.H., Goles, G.G., Leeman, W.P. and Lindstrom, M.M., 1977. Geochemistry and petrogenesis of a basalt—benmoreite—trachyte suite from the southern part of the Gregory Rift, Kenya. *Contrib. Mineral. Petrol.*, **64**: 303—332.

Baker, I., 1969. Petrology of the volcanic rocks of Saint Helena Island, South Atlantic. *Bull. Geol. Soc. Am.*, **80**: 1283—1310.

Baker, P.E., 1968. Comparative volcanology and petrology of the Atlantic island-arcs. *Bull. Volcanol.*, **32**: 189—206.

Baker, P.E., 1973. Islands of the South Atlantic. *In*: A.E.M. Nairn and F.G. Stehli (Editors), *The Ocean Basins and Margins, Vol. 1, The South Atlantic*. Plenum, New York, N.Y., pp. 493—553.

Baker, P.E., Gass, I.G., Harris, P.G. and Le Maitre, R.W., 1964. The volcanological report of the Royal Society expedition to Tristan da Cunha, 1962. *Philos. Trans. R. Soc. London, Ser. A*, **256**: 439—575.

Baldwin, R.B., 1949. *The Face of the Moon*. University of Chicago Press, Chicago, Ill., 273 pp.

Baldwin, R.B., 1963. *The Measure of the Moon*. University of Chicago Press, Chicago, Ill., 488 pp.

Baldwin, R.B., 1965. *A Fundamental Survey of the Moon*. McGraw-Hill, New York, N.Y., 149 pp.

Balk, R., 1937. Structural behaviour of igneous rocks. *Mem. Geol. Soc. Am.*, **5**, 177 pp.

Barberi, F., Bizouard, H. and Varet, J., 1971. Nature of the clinopyroxene and iron enrichment in alkalic and transitional basaltic magmas. *Contrib. Mineral. Petrol.*, **33**: 93—107.

Barberi, F., Tazieff, H. and Varet, J., 1972. Volcanism in the Afar depression: its tectonic and magmatic significance. *Tectonophysics*, **15**: 19—29.

Barberi, F., Ferrara, G., Santacroce, R., Treuil, M. and Varet, J., 1975. A transitional basalt—pantellerite sequence of fractional crystallization, the Boina Centre (Afar Rift, Ethiopia). *J. Petrol.*, **16**: 22—56.

Barker, P.F., 1972. A spreading centre in the east Scotia Sea. *Earth Planet. Sci. Lett.*, **15**: 123—132.

Barker, P.F. and Griffiths, D.H., 1972. The evolution of the Scotia Ridge and Scotia Sea. *Philos. Trans. R. Soc. London, Ser. A*, **271**: 151—183.

Barr, S.M. and Chase, R.L., 1974. Geology of the northern end of Juan de Fuca Ridge and sea-floor spreading. *Can. J. Earth Sci.*, **11**: 1384—1406.

Barrière, M., 1976. Flowage differentiation: limitation of the "Bagnold Effect" to the narrow intrusions. *Contrib. Mineral. Petrol.*, **55**: 139—145.

Barron, B.J., 1976. Recognition of the original volcanic suite in altered mafic volcanic rocks at Sofala, New South Wales. *Am. J. Sci.*, **276**: 604—636.

Barth, T.F.W., 1952. *Theoretical Petrology*. Wiley, New York, N.Y., 387 pp.

Barth, T.F.W., 1959. Principles of classification and norm calculations of metamorphic rocks. *J. Geol.*, **67**: 135—152.

Bateman, P.C., 1961. Granitic formations in the east-central Sierra Nevada near Bishop, California. *Bull. Geol. Soc. Am.*, **72**: 1521—1538.

Bateman, P.C., 1977. IUGS classification of granitic rocks: a critique — Comment and Reply. *Geology*, **5**: 252—254.

Bateman, P.C. and Dodge, F.C.W., 1970. Variations of major chemical constituents across the central Sierra Nevada batholith. *Bull. Geol. Soc. Am.*, **81**: 409—420.

Battey, M.H., 1955. Alkali metasomatism and the petrology of some keratophyres. *Geol. Mag.*, **92**: 104—126.

Beeson, M.H. and Jackson, E.D., 1970. Origin of the garnet pyroxenite xenoliths at Salt Lake crater, Oahu. *Spec. Pap., Mineral. Soc. Am.*, **3**: 95—112.

Behrendt, J.C., Drewry, D.J., Jankowski, E. and Grim, M.S., 1980. Aeromagnetic and radio echo ice-sounding measurements show much greater area of the Dufek Intrusion, Antarctica. *Science*, 209: 1014—1017.

Bell, J.D., 1976. The Tertiary intrusive complex on the Isle of Skye. *Proc. Geol. Assoc.*, 87: 247—271.

Bell, K. and Powell, J.L., 1969. Strontium isotopic studies of alkalic rocks: the potassium-rich lavas of the Birunga and Toro—Ankole regions, East and central-Equatorial Africa. *J. Petrol.*, 10: 536—572.

Bhattacharji, S., 1967. Mechanics of flow differentiation in ultramafic and mafic sills. *J. Geol.*, 75: 101—112.

Bickle, M.J., Martin, A. and Nisbet, E.G., 1975. Basaltic and peridotitic komatiites and stromatolites above a basal unconformity in the Belingwe greenstone belt, Rhodesia. *Earth Planet. Sci. Lett.*, 27: 155—162.

Billings, M.P., 1945. Mechanics of igneous intrusion in New Hampshire. *Am. J. Sci.*, 243A: 40—68.

Björnsson, A., Saemundsson, K., Einarsson, P., Tryggvason, E. and Grönvold, K., 1977. Current rifting episode in north Iceland. *Nature (London)*, 266: 318—322.

Black, L.P., Gale, N.H., Moorbath, S., Pankhurst, R.J. and McGregor, V.R., 1971. Isotopic dating of very early Precambrian amphibolite facies gneisses from the Godthaab district, west Greenland. *Earth Planet. Sci. Lett.*, 12: 245—259.

Black, R. and Girod, M., 1970. Late Palaeozoic to Recent igneous activity in West Africa and its relationship to basement structure. In: T.N. Clifford and I.G. Gass (Editors), *African Magmatism and Tectonics*, Oliver and Boyd, Edinburgh, pp. 185—210.

Blander, M., Planner, H.N., Keil, K., Nelson, L.S. and Richardson, N.L., 1976. The origin of chondrules: experimental investigation of metastable liquids in the system $Mg_2SiO_4-SiO_2$. *Geochim. Cosmochim. Acta*, 40: 889—896.

Bonatti, E., Harrison, C.G.A., Fisher, D.E., Honnorez, J., Schilling, J.-G., Stipp, J.J. and Zentilli, M., 1977. Eastern volcanic chain (Southeast Pacific): a mantle hot line. *J. Geophys. Res.*, 82: 2457—2478.

Bond, A. and Sparks, R.S.J., 1976. The Minoan eruption of Santorini, Greece. *J. Geol. Soc. London*, 132: 1—16.

Borley, G.D., 1967. Potash-rich volcanic rocks from southern Spain. *Mineral. Mag.*, 36: 364—379.

Bostock, H.H., 1965. Clearwater complex, New Quebec. *Can. Geol. Surv. Pap.*, 64-45, 17 pp.

Bottinga, Y. and Weill, D.F., 1972. The viscosity of magmatic silicate liquids: a model for calculation. *Am. J. Sci.*, 272: 438—475.

Bowden, P. and Turner, D.C., 1974. Peralkaline and associated ring-complexes in the Nigeria—Niger province, West Africa. In: H. Sørensen (Editor), *The Alkaline Rocks*. Wiley, London, pp. 330—351.

Bowden, P., van Breeman, O., Hutchison, J. and Turner, D.C., 1976. Palaeozoic and Mesozoic age trends for some ring complexes in Niger and Nigeria. *Nature (London)*, 259: 297—299.

Bowen, N.L., 1913. The melting phenomena of the plagioclase feldspars. *Am. J. Sci.*, 4th Ser., 35: 577—599.

Bowen, N.L., 1914. The ternary system diopside—forsterite—silica. *Am. J. Sci., 4th Ser.*, 38: 207—264.

Bowen, N.L., 1915. The crystallization of haplobasaltic, haplodioritic, and related magmas. *Am. J. Sci., 4th Ser.*, 40: 161—185.

Bowen, N.L., 1928. *The Evolution of the Igneous Rocks*. Princeton University Press, Princeton, N.J., 332 pp. (Note also facsimile edition, 1956, with new introduction by J.F. Schairer and a complete bibliography of the writings of Bowen, Dover Publications, New York, N.Y.)

Bowen, N.L., 1945. Phase equilibria bearing on the origin and differentiation of alkaline rocks. *Am. J. Sci.*, (Daly Volume), 243A: 75—89.

Bowen, N.L. and Andersen, O., 1914. The binary system MgO—SiO$_2$. *Am. J. Sci.*, 4th Ser., 37: 487—500.

Bowen, N.L. and Schairer, J.F., 1935. The system MgO—FeO—SiO$_2$. *Am. J. Sci.*, 5th Ser., 29: 151—217.

Bowes, D.R. and Wright, A.E., 1967. The explosion-breccia pipes near Kentallen, Scotland, and their geological setting. *Trans. R. Soc. Edinburgh*, 67: 109—143.

Boyd, F.R., 1961. Welded tuffs and flows in the rhyolite plateau of Yellowstone Park. *Bull. Geol. Soc. Am.*, 72: 387—426.

Boyd, F.R., 1973. A pyroxene geotherm. *Geochim. Cosmochim. Acta*, 37: 2533—2546.

Bradley, J., 1965. Intrusion of major dolerite sills. *Trans. R. Soc. N.Z.*, 3: 27—55.

Bridgwater, D., Keto, L., McGregor, V.R. and Myers, J.S., 1976. Archean gneiss complex of Greenland. In: A. Escher and W.S. Watt (Editors), *Geology of Greenland*, Geological Survey of Greenland, Godthaab, pp. 18—75.

Bridgwater, D., Collerson, K.D. and Myers, J.S., 1978. The development of the Archaean gneiss complex of the North Atlantic region. In: D.H. Tarling (Editor), *Evolution of the Earth's Crust*. Academic Press, London, pp. 19—69.

Briggs, N.D., 1976a. Welding and crystallization zonation in Whakamaru ignimbrite, central North Island, New Zealand. *N.Z. J. Geol. Geophys.*, 19: 189—212.

Briggs, N.D., 1976b. Recognition and correlation of subdivisions within the Whakamaru ignimbrite, central North Island, New Zealand. *N.Z. J. Geol. Geophys.*, 19: 463—501.

Brock, P.W.G., 1974. The sheeted dike layer of the Betts Cove Ophiolite Complex does not represent spreading. *Can. J. Earth Sci.*, 11: 208—210.

Brooks, C., James, D.E. and Hart, S.R., 1976. Ancient lithosphere: its role in young continental volcanism. *Science*, 193: 1086—1094.

Brooks, C.K., 1976. The Fe$_2$O$_3$/FeO ratio of basalt analyses: an appeal for a standardized procedure. *Bull. Geol. Soc. Denm.*, 25: 117—120.

Brooks, C.K., Nielsen, T.F.D. and Petersen, T.S., 1976. The Blosseville Coast basalts of east Greenland: their occurrence, composition, and temporal variations. *Contrib. Mineral. Petrol.*, 58: 279—292.

Brown, G.M., 1956. The layered ultrabasic rocks of Rhum, Inner Hebrides. *Philos. Trans. R. Soc. London, Ser. B*, 240: 1—53.

Brown, G.M., 1963. Melting relations of Tertiary granitic rocks in Skye and Rhum. *Mineral. Mag.*, 30: 533—562.

Brown, G.M., 1977. Two major igneous events in the evolution of the moon. *Philos. Trans. R. Soc. London, Ser. A*, 286: 439—451.

Brown, G.M. and Vincent, E.A., 1963. Pyroxenes from the late stages of fractionation of the Skaergaard intrusion, east Greenland. *J. Petrol.*, 4: 175—197.

Brown, G.M., Vincent, E.A. and Brown, P.E., 1957. Pyroxenes from the early and middle stages of fractionation of the Skaergaard intrusion, east Greenland. *Mineral. Mag.*, 31: 511—543.

Brown, G.M., Holland, J.G., Sigurdsson, H., Tomblin, J.F. and Arculus, R.J., 1977. Geochemistry of the Lesser Antilles volcanic island arc. *Geochim. Cosmochim. Acta*, 41: 785—801.

Brown, P.E., 1967. Major element composition of the Loch Coire migmatite complex, Sutherland, Scotland. *Contrib. Mineral. Petrol.*, 14: 1—26.

Bryan, W.B., 1972. Morphology of quench crystals in submarine basalts. *J. Geophys. Res.*, 77: 5812—5819.

Bryan, W.B. and Moore, J.G., 1977. Compositional variations of young basalts in the Mid-Atlantic rift valley near lat. 36°49'N. *Bull. Geol. Soc. Am.*, 88: 556—570.

Bucher, W.H., 1963. Cryptoexplosion structures caused from without or from within the earth? ("astroblemes" or "geoblemes"?). *Am. J. Sci.*, 261: 597—649.

Buddington, A.F., 1959. Granite emplacement with special reference to North America. *Bull. Geol. Soc. Am.*, **70**: 671—747.

Buddington, A.F. and Lindsley, D.H., 1964. Iron—titanium oxide minerals and synthetic equivalents. *J. Petrol.*, **5**: 310—357.

Buist, D.S., 1959. The composite sill of Rudh' an Eireannaich, Skye. *Geol. Mag.*, **96**: 247—252.

Bullard, F.M., 1962. *Volcanoes: in history, in theory, in eruption.* University of Texas Press, Austin, Texas, 441 pp.

Burke, K. and Dewey, J.F., 1973. Plume-generated triple junctions: key indicators in applying plate tectonics to old rocks. *J. Geol.*, **81**: 406—433.

Burnham, C.W., 1967. Hydrothermal fluids at the magmatic stage. In: H.L. Barnes (Editor), *Geochemistry of Hydrothermal Ore Deposits.* Holt, Rinehart and Winston, New York, N.Y., pp. 34—76.

Burns, R.G., 1970. *Mineralogical Applications of Crystal Field Theory.* Cambridge, London, 224 pp.

Burns, R.G. and Fyfe, W.S., 1964. Site of preference energy and selective uptake of transition-metal ions from a magma. *Science*, **144**: 1001—1003.

Bussell, M.A., Pitcher, W.S. and Wilson, P.A., 1976. Ring complexes of the Peruvian coastal batholith: a long-standing subvolcanic regime. *Can. J. Earth Sci.*, **13**: 1020—1030.

Cameron, E.N. and Desborough, G.A. 1964. Origin of certain magnetite-bearing pegmatites in the eastern part of the Bushveld Complex, South Africa. *Econ. Geol.*, **59**: 197—225.

Campbell, I.H. and Borley, G.D., 1974. The geochemistry of pyroxenes from the Lower Layered Series of the Jimberlana Intrusion, Western Australia. *Contrib. Mineral. Petrol.*, **47**: 281—297.

Cann, J.R., 1969. Spilites from the Carlsberg Ridge, Indian Ocean. *J. Petrol.*, **10**: 1—19.

Cann, J.R., 1970. Rb, Sr, Y, Zr and Nb in some ocean floor basaltic rocks. *Earth Planet. Sci. Lett.*, **10**: 7—11.

Cann, J.R., 1971. Major element variations in ocean-floor basalts. *Philos. Trans. R. Soc. London, Ser. A*, **268**: 495—505.

Carey, S.N. and Sigurdsson, H., 1978. Deep-sea evidence for distribution of tephra from the mixed magma eruption of the Soufrière on St. Vincent, 1902: ash turbidites and air fall. *Geology*, **6**: 271—274.

Carlisle, D., 1963. Pillow breccias and their aquagene tuffs, Quadra Island, British Columbia. *J. Geol.*, **71**: 48—71.

Carmichael, I.S.E., 1962. Pantelleritic liquids and their phenocrysts. *Mineral. Mag.*, **33**: 86—113.

Carmichael, I.S.E., 1963. The crystallization of feldspar in volcanic acid liquids. *Q.J. Geol. Soc. London*, **119**: 95—113.

Carmichael, I.S.E., 1964. The petrology of Thingmuli, a Tertiary volcano in eastern Iceland. *J. Petrol.*, **5**: 435—460.

Carmichael, I.S.E., 1967a. The mineralogy of Thingmuli, a Tertiary volcano in eastern Iceland. *Am. Mineral.*, **52**: 1815—1841.

Carmichael, I.S.E., 1967b. The mineralogy and petrology of the volcanic rocks from the Leucite Hills, Wyoming. *Contrib. Mineral. Petrol.*, **15**: 24—66.

Carmichael, I.S.E., Nicholls, J. and Smith, A.L., 1970. Silica activity in igneous rocks. *Am. Mineral.*, **55**: 246—263.

Carmichael, I.S.E., Turner, F.J. and Verhoogen, J., 1974. *Igneous Petrology*, McGraw-Hill, New York, N.Y., 739 pp.

Carr, J.M., 1954. Zoned plagioclases in layered gabbros of the Skaergaard Intrusion, east Greenland. *Mineral. Mag.*, **30**: 367—375.

Carswell, D.A. and Dawson, J.B., 1970. Garnet peridotite xenoliths in South African kimberlite pipes and their petrogenesis. *Contrib. Mineral. Petrol.*, **25**: 163—184.

Cawthorn, R.G., 1975. Degrees of melting in mantle diapirs and the origin of ultrabasic liquids. *Earth Planet. Sci. Lett.*, **27**: 113—120.

Cawthorn, R.G. and O'Hara, M.J., 1976. Amphibole fractionation in calc-alkaline magma genesis. *Am. J. Sci.*, **276**: 309—329.

Cawthorn, R.G. and Strong, D.F., 1975. The petrogenesis of komatiites and related rocks as evidence for a layered upper mantle. *Earth Planet. Sci. Lett.*, **21**: 369—375.

Cawthorn, R.G., Curran, E.B. and Arculus, R.J., 1973. A petrogenetic model for the origin of the calc-alkaline suite of Grenada, Lesser Antilles. *J. Petrology*, **14**: 327—337.

Chapin, C.E. and Elston, W.E. (Editors), 1979. Ash-Flow Tuffs. *Spec. Pap. Geol. Soc. Am.*, **180**, 211 pp.

Chappell, B.W. and White, A.J.R., 1974. Two contrasting granite types. *Pac. Geol.*, **8**: 173—174.

Chase, R.L., 1977. J. Tuzo Wilson knolls: Canadian hotspot. *Nature (London)*, **266**: 344—346.

Chayes, F., 1969. The chemical composition of Cenozoic andesite. In: A.R. McBirney (Editor), *Proceedings of the Andesite Conference*. Bull. Oreg. Dep. Miner. Ind., **65**: 1—11.

Christiansen, R.L. and Lipman, P.W., 1972. Cenozoic volcanism and plate-tectonic evolution of the western United States, II. Late Cenozoic. *Philos. Trans. R. Soc. London, Ser. A*, **271**: 249—284.

Church, W.R. and Riccio, L., 1974. The sheeted dike layer of the Betts Cove Ophiolite Complex does not represent spreading — Discussion. *Can. J. Earth Sci.*, **11**: 1499—1502.

Clark, S.P. (Editor), 1966. Handbook of physical constants. *Mem. Geol. Soc. Am.*, **97**, 587 pp.

Clarke, D.B., 1970. Tertiary basalts of Baffin Bay: possible primary magma from the mantle. *Contrib. Mineral. Petrol.*, **25**: 203—224.

Clarke, D.B., McKenzie, C.B., Muecke, G.K. and Richardson, S.W., 1976. Magmatic andalusite from the South Mountain Batholith, Nova Scotia. *Contrib. Mineral. Petrol.*, **56**: 279—287.

Clemons, R.E. and Long, L.E., 1971. Petrologic and Rb—Sr isotopic study of the Chiquimula pluton, southeastern Guatemala. *Bull. Geol. Soc. Am.*, **82**: 2729—2740.

Clifford, P.M., 1968. Flood basalts, dike swarms, and sub-crustal flow. *Can. J. Earth Sci.*, **5**: 93—96.

Coats, R.R., 1962. Magma type and crustal structure in the Aleutian arc. *Monogr. Am. Geophys. Union*, **6**: 92—109.

Cobbing, E.J. and Pitcher, W.S., 1972. The Coastal Batholith of central Peru. *J. Geol. Soc. London*, **128**: 421—460.

Cobbing, E.J., Pitcher, W.S. and Taylor, W.P., 1977. Segments and super-units in the coastal batholith of Peru. *J. Geol.*, **85**: 625—631.

Cole, J.W., 1970. Petrography of the rhyolite lavas of Tarawera volcanic complex. *N.Z. J. Geol. Geophys.*, **13**: 903—924.

Coleman, R.G., 1977. *Ophiolites: Ancient Oceanic Lithosphere?* Springer, Berlin, 229 pp.

Coleman, R.G. and Peterman, Z.E., 1975. Oceanic plagiogranite. *J. Geophys. Res.*, **80**: 1099—1108.

Colgate, S.A. and Sigurgeirsson, T., 1973. Dynamic mixing of water and lava. *Nature (London)*, **244**: 552—555.

Collerson, K.D. and Fryer, B.J., 1978. The role of fluids in the formation and subsequent development of early continental crust. *Contrib. Mineral. Petrol.*, **67**: 151—167.

Compston, W., McDougall, I. and Heier, K.S., 1968. Geochemical comparison of the Mesozoic basaltic rocks of Antarctica, South Africa, South America, and Tasmania. *Geochim. Cosmochim. Acta*, **32**: 129—149.

Condie, K.C., 1975. Mantle-plume model for the origin of Archaean greenstone belts based on trace element distributions. *Nature (London)*, 258: 413—414.

Condie, K.C., Viljoen, M.J. and Kable, E.J.D., 1977. Effects of alteration on element distributions in Archaean tholeiites from the Barberton greenstone belt, South Africa. *Contrib. Mineral. Petrol.*, 64: 75—89.

Cook, E.F., 1965. Stratigraphy of Tertiary volcanic rocks in eastern Nevada. *Nev. Bur. Mines, Rep.*, 11, 61 pp.

Cook, H.E., 1968. Ignimbrite flows, plugs, and dikes in the southern part of the Hot Creek Range, Nye County, Nevada. *Mem. Geol. Soc. Am.*, 116: 107—152.

Coombs, D.S., 1963. Trends and affinities of basaltic magmas and pyroxenes as illustrated on the diopside—olivine—silica diagram. *Spec. Pap. Mineral. Soc. Am.*, 1: 227—250.

Cox, K.G., 1970. Tectonics and vulcanism of the Karroo Period and their bearing on the postulated fragmentation of Gondwanaland. In: T.N. Clifford and I.G. Gass (Editors), *African Magmatism and Tectonics*. Oliver and Boyd, Edinburgh, pp. 211—235.

Cox, K.G., 1972. The Karroo volcanic cycle. *J. Geol. Soc. London*, 128: 311—336.

Cox, K.G., 1978a. Kimberlite pipes. *Sci. Am.*, April 1978, pp. 120—130.

Cox, K.G., 1978b. Komatiites and other high-magnesia lavas: some problems. *Philos. Trans. R. Soc. London., Ser. A*, 288: 599—609.

Cox, K.G. and Bell, J.D., 1972. A crystal fractionation model for the basaltic rocks of the New Georgia Group, British Solomon Islands. *Contrib. Mineral. Petrol.*, 37: 1—13.

Cox, K.G., MacDonald, R. and Hornung, G., 1967. Geochemical and petrographic provinces in the Karroo basalts of southern Africa. *Am. Mineral.*, 52: 1451—1474.

Cox, K.G., Gass, I.G. and Mallick, D.I.J., 1970. The peralkaline volcanic suite of Aden and Little Aden, south Arabia. *J. Petrol.*, 11: 433—461.

Cox, K.G., Bell, J.D. and Pankhurst, R.J., 1979. *The Interpretation of Igneous Rocks*. Allen and Unwin, London, 450 pp.

Crandell, D.R. and Mullineaux, D.R., 1967. Volcanic hazards at Mount Rainier, Washington. *U.S. Geol. Surv. Bull.*, 1238, 26 pp.

Crandell, D.R. and Mullineaux, D.R., 1975. Technique and rationale of volcanic-hazards appraisals in the Cascade Range, northwestern United States. *Environ. Geol.*, 1: 23—32.

Cross, W., Iddings, J.P., Pirsson, L.V. and Washington, H.S., 1902. A quantitative chemico-mineralogical classification and nomenclature of igneous rocks. *J. Geol.*, 10: 555—690.

Crowder, D.F., McKee, E.H., Ross, D.C. and Krauskopf, K.B., 1973. Granitic rocks of the White Mountains area, California—Nevada: age and regional significance. *Bull. Geol. Soc. Am.*, 84: 285—296.

Curtis, G.H., 1968. The stratigraphy of the ejecta of the 1912 eruption of Mount Katmai and Novarupta, Alaska. *Mem. Geol. Soc. Am.*, 116: 153—210.

Daly, R.A., 1933. *Igneous Rocks and the Depths of the Earth*, Hafner, New York, N.Y., 598 pp. (reprinted, 1968).

Davies, H.L., 1971. Peridotite—gabbro—basalt complex in eastern Papua, an overthrust plate of oceanic mantle and crust. *Bull. Aust. Bur. Miner. Resour.*, 128, 48 pp.

Davis, D.W., Gray, J., Cumming, G.L. and Baadsgaard, H., 1977. Determination of the ^{87}Rb decay constant. *Geochim. Cosmochim. Acta*, 41: 1745—1749.

Dawson, J.B., 1966. Oldoinyo Lengai — an active volcano with sodium carbonatite lava flows. In: O.F. Tuttle and J. Gittins (Editors), *Carbonatites*. Wiley, New York, N.Y., pp. 155—168.

Dawson, J.B., 1967. Geochemistry and origin of kimberlite. In: P.J. Wyllie (Editor), *Ultramafic and Related Rocks*. Wiley, New York, N.Y., pp. 269—278.

Dawson, J.B., 1972. Kimberlites and their relation to the mantle. *Philos. Trans. R. Soc. London, Ser. A*, 271: 297—311.

Dawson, J.B., 1977. Sub-cratonic crust and upper mantle models based on xenolith suites in kimberlite and nephelinitic diatremes. *J. Geol. Soc. London*, 134: 173—184.

Dawson, J.B. and Hawthorne, J.B., 1973. Magmatic sedimentation and carbonatitic differentiation in kimberlite sills at Benfontein, South Africa. *J. Geol. Soc. London*, 129: 61—85.

De, A., 1974. Silicate liquid immiscibility in the Deccan Traps and its petrogenetic significance. *Bull. Geol. Soc. Am.*, 85: 471—474.

de Albuquerque, C.A.R., 1977. Geochemistry of the tonalitic and granitic rocks of the Nova Scotia southern plutons. *Geochim. Cosmichim. Acta*, 41: 1—13.

Deer, W.A., Howie, R.A. and Zussman, J., 1966. *An Introduction to the Rock-forming Minerals*. Longmans and Green, London, 528 pp.

Dence, M.R., 1972. Meteorite impact craters and the structure of the Sudbury Basin. *Spec. Pap. Geol. Assoc. Can.*, 10: 7—18.

Dewey, H. and Flett, J.S., 1911. On some British pillow-lavas and the rocks associated with them. *Geol. Mag.*, 8: 202—209; 241—248.

Dickey, Jr., J.S., Frey, F.A., Hart, S.R. and Watson, E.B., 1977. Geochemistry and petrology of dredged basalts from the Bouvet triple junction, South Atlantic. *Geochim. Cosmochim. Acta*, 41: 1105—1118.

Dietz, R.S., 1963. Cryptoexplosion structures — A discussion. *Am. J. Sci.*, 261: 650—664.

Dimroth, E., 1970. Evolution of the Labrador geosyncline. *Bull. Geol. Soc. Am.*, 81: 2717—2742.

Dimroth, E., Cousineau, P., Leduc, M. and Sanschagrin, Y., 1978. Structure and organization of Archean subaqueous basalt flows, Rouyn—Noranda area, Quebec, Canada. *Can. J. Earth Sci.*, 15: 902—918.

Doell, R.R., Dalrymple, G.B., Smith, R.L. and Bailey, R.A., 1968. Paleomagnetism, potassium—argon ages, and geology of rhyolites and associated rocks of the Valles Caldera, New Mexico. *Mem. Geol. Soc. Am.*, 116: 211—248.

Drever, H.I., 1953. A note on the field relations of the Shiant Isles picrite. *Geol. Mag.*, 90: 159—161.

Drewes, H., Fraser, G.D., Snyder, G.L. and Barnett, H.F., 1961. Geology of Unalaska Island and adjacent insular shelf, Aleutian Islands, Alaska. *Bull. U.S. Geol. Surv.*, 1028-S: 583—676.

D.S.I.R. (Department of Scientific and Industrial Research), 1972. *Geological map of New Zealand*. N.Z. Dep. Sci. Ind. Res., Wellington, scale 1: 1,000,000.

Duncan, A.M., 1978. The trachybasaltic volcanics of the Adrano area, Mount Etna, Sicily. *Geol. Mag.*, 115: 273—285.

Dunham, A.C., 1970. The emplacement of the Tertiary igneous complex of Rhum. In: G. Newall and N. Rast (Editors), *Mechanism of Igneous Intrusion*. Gallery Press, Liverpool, pp. 23—32.

du Toit, A.L., 1937. *Our Wandering Continents*, Oliver and Boyd, Edinburgh, 366 pp.

Edgar, A.D. and Condliffe, E., 1978. Derivation of K-rich ultramafic magmas from a peridotitic mantle source. *Nature (London)*, 275: 639—640.

Edmonds, E.A., McKeown, M.C. and Williams, M., 1975. *British Regional Geology: South West England*. Her Majesty's Stationery Office, London, 4th ed., 136 pp.

Eggler, D.H., 1972. Water-saturated and undersaturated melting relations in a Parícutin andesite and an estimate of water content in the natural magma. *Contrib. Mineral. Petrol.*, 34: 261—271.

Ehlers, E.G., 1972. *The Interpretation of Geological Phase Diagrams*. Freeman, San Francisco, Calif., 280 pp.

Eichelberger, J.C., 1975. Origin of andesite and dacite: evidence of mixing at Glass Mountain in California and at other circum-Pacific volcanoes. *Bull. Geol. Soc. Am.*, 86: 1381—1391.

Ekren, E.B. and Byers, F.M., 1976. Ash-flow fissure vent in west-central Nevada. *Geology*, 4: 247—251.

Elsdon, R., 1969. The structure and intrusive mechanism of the Kap Edvard Holm layered gabbro complex, east Greenland. *Geol. Mag.*, **106**: 46—56.

Emeleus, C.H., 1963. Structural and petrographic observations on layered granites from southern Greenland. *Spec. Pap. mineral. Soc. Am.*, **1**: 22—29.

Emeleus, C.H. and Upton, B.G.J., 1976. The Gardar period in southern Greenland. In: A. Escher and W.S. Watt (Editors), *Geology of Greenland*. Grønl. Geol. Unders., pp. 152—181.

Emmons, R.C., 1940. The contribution of differential pressure to magmatic differentiation. *Am. J. Sci.*, **238**: 1—21.

Emslie, R.F., 1970. The geology of the Michikamau Intrusion, Labrador. *Pap. Geol. Surv. Can.*, **68-57**, 85 pp.

Engel, A.E.J., Engel, C.G. and Havens, R.G., 1965. Chemical characteristics of oceanic basalts and the upper mantle. *Bull. Geol. Soc. Am.*, **76**: 719—734.

Erlank, A.J. and Kable, E.J.D., 1976. The significance of incompatible elements in mid-Atlantic ridge basalts from 45°N with particular reference to Zr/Nb. *Contrib. Mineral. Petrol.*, **54**: 281—291.

Ernst, W.G., 1971. Metamorphic zonations on presumably subducted lithospheric plates from Japan, California, and the Alps. *Contrib. Mineral. Petrol.*, **34**: 43—59.

Ernst, W.G., 1976. *Petrologic Phase Equilibria*. Freeman, San Francisco, Calif., 333 pp.

Eskola, P., 1937. An experimental illustration of the spilite reaction. *Bull. Comm. Géol. Finl.*, **119**: 61—68.

Eskola, P., 1949. The problem of mantled gneiss domes. *Q.J. Geol. Soc. London*, **104**: 461—476.

Ewart, A., 1965a. Petrology of the andesites. In: B.N. Thompson, L.O. Kermode and A. Ewart (Editors), *New Zealand Volcanology — Central Volcanic Region*. N.Z. Dep. Sci. Ind. Res., Info. Ser., **50**: 86—93.

Ewart, A., 1965b. Supplementary petrographic notes on the Matahina ignimbrite. In: B.N. Thompson, L.O. Kermode and A. Ewart (Editors), *New Zealand Volcanology — Central Volcanic Region*. N.Z. Dep. Sci. Ind. Res., Info. Ser., **50**: 129—132.

Ewart, A., 1976. Mineralogy and chemistry of modern orogenic lavas — some statistics and implications. *Earth Planet. Sci. Lett.*, **31**: 417—432.

Ewart, A. and Healy, J., 1965. Te Whaiti ignimbrites at Murupara. In: B.N. Thompson, L.O. Kermode and A. Ewart (Editors), *New Zealand Volcanology — Central Volcanic Region*. N.Z. Dep. Sci. Ind. Res., Info. Ser., **50**: 121—125.

Ewart, A. and Stipp, J.J., 1968. Petrogenesis of the volcanic rocks of the central North Island, New Zealand, as indicated by a study of $^{87}Sr/^{86}Sr$ ratios, and Sr, Rb, K, U, and Th abundances. *Geochim. Cosmochim. Acta*, **32**: 699—736.

Ewart, A., Taylor, S.R. and Capp, A.C., 1968. Geochemistry of the pantellerites of Mayor Island, New Zealand. *Contrib. Mineral. Petrol.*, **17**: 116—140.

Exley, C.S. and Stone, M., 1964. The granitic rocks of south-west England. In: K.F.G. Hosking and G.J. Shrimpton (Editors), *Present Views on Some Aspects of the Geology of Cornwall and Devon*. R. Geol. Soc. Cornwall, 150th Anniv. Vol., **131** 131—184.

Fairbairn, H.W., 1953. Precision and accuracy of chemical analysis of silicate rocks. *Geochim. Cosmochim. Acta*, **4**: 143—156.

Farrand M.G., 1960. The distribution of some elements across the contacts of four xenoliths. *Geol. Mag.*, **97**: 488—493.

Faure, G., 1977. *Principles of Isotope Geology*. Wiley, New York, N.Y., 464 pp.

Faure, G. and Powell, J.L., 1972. *Strontium Isotope Geology*. Springer, New York, N.Y., 188 pp.

Fenner, C.N., 1948. Incandescent tuff flows in southern Peru. *Bull. Geol. Soc. Am.*, **59**: 879—893.

Ferguson, A.K. and Cundari, A., 1975. Petrological aspects and evolution of the leucite bearing lavas from Bufumbira, south west Uganda. *Contrib. Mineral. Petrol.*, 50: 25—46.

Ferguson, J., 1964. Geology of the Ilimaussaq alkaline intrusion, south Greenland. *Medd. Grϕnl.*, 172(4), 82 pp.

Ferguson, J. and Currie, K.L., 1971. Evidence of liquid immiscibility in alkaline ultrabasic dykes at Callander Bay, Ontario. *J. Petrol.*, 12: 561—585.

Ferguson, J. and Pulvertaft, T.C.R., 1963. Contrasted styles of igneous layering in the Gardar province of south Greenland. *Spec. Pap. Mineral. Soc. Am.*, 1: 10—21.

Fisher, R.V., 1964. Maximum size, median diameter, and sorting of tephra. *J. Geophys. Res.*, 69: 341—355.

Fisher, R.V., 1966. Rocks composed of volcanic fragments and their classification. *Earth-Sci. Rev.*, 1: 287—298.

Fisher, R.V. and Waters, A.C., 1970. Base surge bed forms in maar volcanoes. *Am. J. Sci.*, 268: 157—180.

Fiske, R.S., 1963. Subaqueous pyroclastic flows in the Ohanapecosh Formation, Washington. *Bull. Geol. Soc. Am.*, 74: 391—406.

Fiske, R.S., Hopson, C.A. and Waters, A.C., 1963. Geology of Mount Rainier National Park, Washington. *Prof. Pap. U.S. Geol. Surv.*, 444, 93 pp.

Fitton, J.G. and Hughes, D.J., 1977. Petrochemistry of the volcanic rocks of the island of Principe, Gulf of Guinea. *Contrib. Mineral. Petrol.*, 64: 257—272.

Flower, M.F.J., Robinson, P.T., Schmincke, H.-U. and Ohnmacht, W., 1977. Magma fractionation systems beneath the Mid-Atlantic Ridge at 36—37°N. *Contrib. Mineral. Petrol.*, 64: 167—195.

Floyd, P.A., 1977. Rare earth element mobility and geochemical characterization of spilitic rocks. *Nature (London)*, 269: 134—137.

Floyd, P.A. and Winchester, J.A., 1975. Magma type and tectonic setting discrimination using immobile elements. *Earth Planet. Sci. Lett.*, 27: 211—218.

Foland, K.A. and Faul, H., 1977. Ages of the White Mountain intrusives — New Hampshire, Vermont, and Maine. *Amer. J. Sci.*, 277: 888—904.

Fournier, R.B., 1968. A granitic dike swarm at its source. *Mem. Geol. Soc. Am.*, 116: 249—274.

Francis, E.H., 1970. Bedding in Scottish (Fifeshire) tuff-pipes and its relevance to maars and calderas. *Bull. Volcanol.*, 34: 697—712.

Francis, P., 1976. *Volcanoes*, Penguin Books, Harmondsworth, 368 pp.

Frantsesson, E.V., 1970. *The Petrology of the Kimberlites.* Aust. Natl. Univ., Canberra, A.C.T., Dep. Geol. Publ., No. 150, 195 pp. (English edition; translated from the Russian by D.A. Brown).

Freeberg, J.H., 1966. Terrestrial impact structures — a bibliography. *Bull. U.S. Geol. Surv.*, 1220, 91 pp.

Freeberg, J.H., 1969. Terrestrial impact structures — a bibliography, 1965—68. *Bull. U.S. Geol. Surv.*, 1320, 39 pp.

French, B.M., 1967. Sudbury structure, Ontario: some petrographic evidence for origin by meteorite impact. *Science*, 156: 1094—1098.

French, B.M., 1968. Shock metamorphism as a geological process. In: B.M. French and N.M. Short (Editors), *Shock Metamorphism of Natural Materials.* Mono Book Corporation, Baltimore, Md., pp. 1—17.

Frye, K., 1974. *Modern Mineralogy.* Prentice-Hall, Englewood Cliffs, N.J., 325 pp.

Fudali, R.F. and Melson, W.G., 1971. Ejecta velocities, magma chamber pressure and kinetic energy associated with the 1968 eruption of Arenal volcano. *Bull. Volcanol.*, 35: 383—401.

Fyfe, W.S., 1978. The evolution of the earth's crust: modern plate tectonics to ancient hot spot tectonics? *Chem. Geol.*, 23: 89—114.

Gale, G.H., 1973. Paleozoic basaltic komatiite and ocean-floor type basalts from northeastern Newfoundland. *Earth Planet. Sci. Lett.*, 18: 22—28.

Garson, M.S., 1966. Carbonatites in Malawi. In: O.F. Tuttle and J. Gittins (Editors), *Carbonatites*. Wiley, New York, N.Y., pp. 33—71.

Gary, M., McAfee, R. and Wolf, C.L., 1972. *Glossary of Geology*. American Geological Institute, Washington, D.C., 805 pp.

Gass, I.G., 1970. Tectonic and magmatic evolution of the Afro-Arabian dome. In: T.N. Clifford and I.G. Gass (Editors), *African Magmatism and Tectonics*. Oliver and Boyd, Edinburgh, pp. 285—300.

Gass, I.G., 1980. The Troodos Massif: its role in the unravelling of the ophiolite problem and its significance in the understanding of constructive plate margin processes. In: A. Panayiotou (Editor), *Ophiolites, Proceedings of the International Ophiolite Symposium, Cyprus, 1979*. Geological Survey Department, Nicosia, pp. 23—35.

Gass, I.G., Neary, C.R., Plant, J., Robertson, A.H.F., Simonian, K.O., Smewing, J.D., Spooner, E.T.C. and Wilson, R.A.M., 1975. Comments on "The Troodos ophiolite complex was probably formed in an island arc", by A. Miyashiro; and subsequent correspondence, by A. Hynes and A. Miyashiro. *Earth Planet. Sci. Lett.*, 25: 236—238.

Gass, I.G., Chapman, D.S., Pollack, H.N. and Thorpe, R.S., 1978. Geological and geophysical parameters of mid-plate volcanism. *Philos. Trans. R. Soc. London, Ser. A*, 288: 581—597.

Gast, P.W., 1968. Trace element fractionation and the origin of tholeiitic and alkaline magma types. *Geochim. Cosmochim Acta*, 32: 1057—1086.

Ghose, N.C., 1976. Composition and origin of Deccan basalts. *Lithos*, 9: 65—73.

Gibb, F.G.F., 1968. Flow differentiation in the xenolithic ultrabasic dykes of the Cuillins and the Strathaird Peninsula, Isle of Skye, Scotland. *J. Petrol.*, 9: 411—443.

Gibson, I.L., 1975. A review of the geology, petrology and geochemistry of the volcano Fantale. *Bull. Volcanol.*, 38: 791—802.

Gibson, I.L. and Walker, G.P.L., 1963. Some composite rhyolite/basalt lavas and related composite dykes in eastern Iceland. *Proc. Geol. Assoc. London*, 74: 301—318.

Gilbert, C.M., 1938. Welded tuff in eastern California. *Bull. Geol. Soc. Am.*, 49: 1829—1862.

Gill, J.B., 1970. Geochemistry of Viti Levu, Fiji, and its evolution as an island arc. *Contrib. Mineral. Petrol.*, 27: 179—203.

Gill, J.B., 1976. Composition and age of Lau Basin and Ridge volcanic rocks: implications for evolution of an interarc basin and remnant arc. *Bull. Geol. Soc. Am.*, 87: 1384—1395.

Gilluly, J., 1963. The tectonic evolution of the western United States. *Q.J. Geol. Soc. London*, 119: 133—174.

Gittins, J., 1966. Summaries and bibliographies of carbonatite complexes. In: O.F. Tuttle and J. Gittins (Editors), *Carbonatites*. Wiley, New York, N.Y., pp. 417—541.

Gold, D.P., 1967. Alkaline ultrabasic rocks in the Montreal area, Quebec. In: P.J. Wyllie (Editor), *Ultramafic and Related Rocks*. Wiley, New York, N.Y., pp. 288—302.

Goldschmidt, V.M., 1954. *Geochemistry*. Clarendon, Oxford, 730 pp. (edited by A. Muir).

Goldsmith, J.R., 1976. Scapolites, granulites, and volatiles in the lower crust. *Bull. Geol. Soc. Am.*, 87: 161—168.

Goles, G.G., 1976. Some constraints on the origin of phonolites from the Gregory Rift, Kenya, and inferences concerning basaltic magmas in the rift system. *Lithos*, 9: 1—8.

Goodwin, A.M. and Ridler, R.H., 1970. The Abitibi orogenic belt. *Pap. Geol. Surv. Can.*, 70-40, 1—30.

Goranson, R., 1936. Silicate—water systems: the solubility of water in albite-melt. *Trans. Am. Geophys. Union*, 17: 257—259.

Gow, A.J., 1968. Petrographic and petrochemical studies of Mt. Egmont andesites. *N.Z. J. Geol. Geophys.*, 11: 166—190.

Grapes, R.H., 1975. Petrology of the Blue Mountain complex, Marlborough, New Zealand. *J. Petrol.*, 16: 371—428.

Green, D.H., 1970. A review of experimental evidence on the origin of basaltic and nephelinitic magmas. *Phys. Earth Planet. Inter.*, 3: 221—235.

Green, D.H., 1972. Archaean greenstone belts may include terrestrial equivalents of lunar maria? *Earth Planet. Sci. Lett.*, 15: 263—270.

Green, D.H. and Ringwood, A.E., 1967a. The genesis of basaltic magmas. *Contrib. Mineral. Petrol.*, 15: 103—190.

Green, D.H. and Ringwood, A.E., 1967b. The stability fields of aluminous pyroxene peridotite and garnet peridotite and their relevance in upper mantle structure. *Earth Planet. Sci. Lett.*, 3: 151—160.

Green, D.H., Edgar, A.D., Beasley, P., Kiss, E. and Ware, N.G., 1974. Upper mantle source for some hawaiites, mugearites, and benmoreites. *Contrib. Mineral. Petrol.*, 48: 33—43.

Green, H.W. and Gueguen, Y., 1974. Origin of kimberlite pipes by diapiric upwelling in the upper mantle. *Nature (London)*, 249: 617—620.

Green, J.C., 1977. Keweenawan plateau volcanism in the Lake Superior region. *Spec. Pap. Geol. Assoc. Can.*, 16: 407—422.

Green, T.H. and Ringwood, A.E., 1968a. Genesis of the calc-alkaline igneous rock suite. *Contrib. Mineral. Petrol.*, 18: 105—162.

Green, T.H. and Ringwood, A.E., 1968b. Origin of garnet phenocrysts in calc-alkaline rocks. *Contrib. Miner. Petrol.*, 18: 163—174.

Greene, H.G., Dalrymple, G.B. and Clague, D.A., 1978. Evidence for northward movement of the Emperor seamounts. *Geology*, 6: 70—74.

Grieg, J.W., 1927. Immiscibility in silicate melts. *Am. J. Sci.*, 5th Ser., 13: 1—44.

Grieg, J.W. and Barth, T.F.W., 1938. The system $Na_2O \cdot Al_2O_3 \cdot 2SiO_2$ (nepheline, carnegieite)—$Na_2O \cdot Al_2O_3 \cdot 6SiO_2$ (albite). *Am. J. Sci.*, 5th Ser., 35A: 94—112.

Griffin, W.L. and Heier, K.S., 1973. Petrological implications of some corona structures. *Lithos*, 6: 315—335.

Grindley, G.W., 1965. Tongariro National Park; stratigraphy and structure. In: B.N. Thompson, L.O. Kermode and A. Ewart (Editors), *New Zealand Volcanology: central volcanic region.* N.Z. Dep. Sci. Ind. Res., Info. Ser., 50: 79—86.

Grossman, L., 1975. The most primitive objects in the solar system. *Sci. Am.*, 232(2): 30—38.

Gunn, B.M., 1966. Modal and element variation in Antarctic tholeiites. *Geochim. Cosmochim. Acta*, 30: 881—920.

Guy-Bray, J.V. (Editor), 1972. New developments in Sudbury geology. *Spec. Pap. Geol. Assoc. Can.*, 10, 124 pp.

Haggerty, S.E., 1978. The Allende meteorite: evidence for a new cosmothermometer based on Ti^{3+}/Ti^{4+}. *Nature (London)*, 276: 221—225.

Haller, J., 1971. *Geology of the East Greenland Caledonides*, Wiley, New York, N.Y., 413 pp.

Hamet, J. and Allègre, C.J., 1976. Rb—Sr systematics in granite from central Nepal (Manaslu): significance of the Oligocene age and high $^{87}Sr/^{86}Sr$ ratio in Himalayan orogeny. *Geology*, 4: 470—472.

Hamilton, D.L., Burnham, C.W. and Osborn, E.F., 1964. The solubility of water and effects of oxygen fugacity and water content on crystallization in mafic magmas. *J. Petrol.*, 5: 21—39.

Hamilton, E.I., 1963. The isotopic composition of strontium in the Skaergaard Intrusion, east Greenland. *J. Petrol.*, 4: 383—391.

Hanson, G.N., 1978. The application of trace elements to the petrogenesis of igneous rocks of granitic composition. *Earth Planet. Sci. Lett.*, 38: 26—43.

Hanson, G.N., 1980. Rare earth elements in petrogenetic studies of igneous systems. *Annu. Rev. Earth Planet. Sci.*, 8: 371—406.

Harker, A., 1904. The Tertiary igneous rocks of Skye. *Mem. Geol. Surv. Scotl., Her Majesty's Stationery Office, Glasgow.*

Harker, A., 1909. *The Natural History of Igneous Rocks.* Macmillan, New York, N.Y., 384 pp.

Harker, A., 1954. *Petrology for Students.* Cambridge University Press, London, 8th ed., 283 pp. (revised by C.E. Tilley, S.R. Nockolds and M. Black).

Härme, M., 1965. On the potassium migmatites of southern Finland. *Bull. Comm. Géol. Finl.,* 219, 43 pp.

Härme, M., 1966. Experimental anatexis and genesis of migmatite — A reply. *Contrib. Mineral. Petrol.,* 12: 13—14.

Harris, P.G., 1957. Zone refining and the origin of potassic basalts. *Geochim. Cosmochim. Acta,* 12: 195—208.

Harris, P.G., 1969. Basalt type and African Rift Valley tectonism. *Tectonophysics,* 8: 427—436.

Harris, P.G., Reay, A. and White, I.G., 1967. Chemical composition of the upper mantle. *J. Geophys. Res.,* 72: 6359—6369.

Harris, P.G., Kennedy, W.Q. and Scarfe, C.M., 1970. Volcanism versus plutonism — the effect of chemical composition. In: G. Newall and N. Rast (Editors), *Mechanism of Igneous Intrusion.* Gallery Press, Liverpool, pp. 187—200.

Harris, P.G., Hutchison R. and Paul, D.K., 1972. Plutonic xenoliths and their relation to the upper mantle. *Philos. Trans. R. Soc. London, Ser. A,* 271: 313—323.

Harrassowitz, H., 1927. Anchimetamorphose, das Gebiet zwischen Oberflächen- und Tiefenum-wandlung der Erdrinde. *Oberhess. Ges. Natur- Heilkd. Giessen, Naturwiss. Abt., Ber.,* 12: 9—15.

Harry, W.T. and Emeleus, C.H., 1960. Mineral layering in some granite intrusions of S.W. Greenland. *Rep. 21st Int. Geol. Congr.,* Part 14, pp. 172—181.

Hatch, F.H., Wells, A.K. and Wells, M.K., 1972. *Petrology of the Igneous Rocks.* Thomas Murby, London, 13th ed., 551 pp.

Hatherton, T. and Dickinson, W.R., 1969. The relationship between andesitic volcanism and seismicity in Indonesia, the Lesser Antilles, and other island arcs. *J. Geophys. Res.,* 74: 5301—5310.

Hawkes, D.D., 1966. Differentiation of the Tumatumari—Kopinang dolerite intrusion, British Guiana. *Bull. Geol. Soc. Am.,* 77: 1131—1158.

Hawkins, D.W., 1976. *Emplacement, petrology, and geochemistry of ultrabasic to basic intrusives at Aillik Bay, Labrador.* M.Sc. Thesis, Memorial University, St. John's, Nfld., 236 pp.

Hawkins, J.W., 1976. Petrology and geochemistry of basaltic rocks of the Lau Basin. *Earth Planet. Sci. Lett.,* 28: 283—297.

Hawthorne, J.B., 1975. Model of a kimberlite pipe. *Phys. Chem. Earth,* 9: 1—15.

Healy, J., 1962. Structure and volcanism in the Taupo Volcanic Zone, New Zealand. *Am. Geophys. Union, Geophys. Monogr.,* 6: 151—157.

Heier, K.S. and Compston, W. 1969. Rb—Sr isotopic studies of the plutonic rocks of the Oslo region. *Lithos,* 2: 133—145.

Heinrich, E.W., 1965. *Microscopic Identification of Minerals.* McGraw-Hill, New York, N.Y., 414 pp.

Hekinian, R., Moore, J.G. and Bryan, W.B. 1976. Volcanic rocks and processes of the Mid-Atlantic Ridge Rift Valley near 36°49'N. *Contrib. Mineral. Petrol.,* 58: 83—110.

Hellman, P.L., Smith, R.E. and Henderson, P., 1979. The mobility of the rare earth elements: evidence and implications from selected terrains affected by burial meta-morphism. *Contrib. Mineral. Petrol.,* 71: 23—44.

Henderson, P. and Gijbels, R., 1976. Trace element indicators of the genesis of the Rhum layered intrusion, Inner Hebrides. *Scott. J. Geol.,* 12: 325—333.

Henley, R.W. and McNabb, A., 1978. Magmatic vapor plumes and ground-water interaction in porphyry copper emplacement. *Econ. Geol.*, **73**: 1—20.

Herrmann, A.G. and Wedepohl, K.H., 1970. Untersuchungen an spilitischen Gesteinen der variskischen Geosynkline in Nordwestdeutschland. *Contrib. Mineral. Petrol.*, **29**: 255—274.

Herrmann, A.G., Potts, M.J. and Knake, D., 1974. Geochemistry of the rare earth elements in spilites from the oceanic and continental crust. *Contrib. Mineral. Petrol.*, **44**: 1—16.

Hess, H.H., 1960. Stillwater igneous complex, Montana. *Mem. Geol. Soc. Am.*, **80**, 230 pp.

Higazy, R.A., 1954. Trace elements of volcanic ultrabasic potassic rocks of southwestern Uganda and adjoining part of the Belgian Congo. *Bull. Geol. Soc. Am.*, **65**: 39—70.

Hills, E.S., 1959. Cauldron subsidences, granitic rocks, and crustal fracturing in S.E. Australia. *Geol. Rundsch.*, **47**: 543—561.

Hills, E.S., 1963. *Elements of Structural Geology.* Wiley, New York, N.Y., 483 pp.

Hoffman, P.F. and McGlynn, J.C., 1977. Great Bear Batholith: a volcano—plutonic depression. *Spec. Pap. Geol. Assoc. Can.*, **16**: 169—192.

Hofmann, A.W. and Hart, S.R., 1978. An assessment of local and regional isotopic equilibrium in the mantle. *Earth Planet. Sci. Lett.*, **38**: 44—62.

Holland, J.G. and Brown, G.M., 1972. Hebridean tholeiitic magmas: a geochemical study of the Ardnamurchan cone sheets. *Contrib. Mineral. Petrol.*, **37**: 139—160.

Holloway, J.R. and Burnham, C.W., 1972. Melting relations of basalt with equilibrium water pressure less than total pressure. *J. Petrol.*, **13**: 1—29.

Holmes, A., 1920. *The Nomenclature of Petrology.* Allen and Unwin, London, 284 pp. (facsimile edition, 1972, Hafner, New York, N.Y.).

Holmes, A., 1930. *Petrographic Methods and Calculations.* Van Nostrand, New York, N.Y., 516 pp.

Holmes, A., 1965. *Principles of Physical Geology.* Nelson, London, 1288 pp.

Holmes, A. and Harwood, H.F., 1932. Petrology of the volcanic fields east and southeast of Ruwenzori, Uganda. *Q.J. Geol. Soc. London*, **88**: 370—442.

Honnorez, J. and Kirst, P., 1975. Submarine basaltic volcanism: morphometric parameters for discriminating hyaloclastites from hyalotuffs. *Bull. Volcanol.*, **39**: 441—465.

Hughes, C.J., 1960. The Southern Mountains igneous complex, Isle of Rhum. *Q.J. Geol. Soc. London*, **116**: 111—138.

Hughes, C.J., 1970a. Major rhythmic layering in ultramafic rocks of the Great Dyke of Rhodesia with particular reference to the Sebakwe area. *Spec. Publ., Geol. Soc. S. Afr.*, **1**: 594—609.

Hughes, C.J., 1970b. The significance of biotite selvedges in migmatites. *Geol. Mag.*, **107**: 21—24.

Hughes, C.J., 1971. Anatomy of a granophyre intrusion. *Lithos*, **4**: 403—415.

Hughes, C.J., 1972a. Note on the variability of granophyric texture. *Bull. Geol. Soc. Am.*, **83**: 2419—2421.

Hughes, C.J., 1972b. Spilites, keratophyres, and the igneous spectrum. *Geol. Mag.*, **109**: 513—527.

Hughes, C.J., 1975. Keratophyre defined. *Neues Jahrb. Mineral., Monatsh.*, pp. 425—430.

Hughes, C.J., 1976. Volcanogenic cherts in the late Precambrian Conception Group, Avalon Peninsula, Newfoundland. *Can. J. Earth Sci.*, **13**: 512—519.

Hughes, C.J., 1977. Parental magma of the Great Dyke of Rhodesia — voluminous late Archaean high magnesium basalt. *Trans. Geol. Soc. S. Afr.*, **79**: 179—182.

Hughes, C.J., 1978. Keratophyre defined — Further discussion in light of reply by E. Lehmann. *Neues Jahrb. Mineral., Monatsh.*, pp. 308—310.

Hughes, C.J. and Brückner, W.D., 1971. Late Precambrian rocks of eastern Avalon Peninsula, Newfoundland — a volcanic island complex. *Can. J. Earth Sci.*, **8**: 899—915.

Hughes, C.J. and Farrant, J.R., 1963. The geology of ¼° field sheet No. 26, Cape Coast and Saltpond districts, with particular reference to the occurrence of berylliferous pegmatites. *Bull. Geol. Surv. Ghana*, **29**, 64 pp.

Hughes, C.J. and Hussey, E.M., 1976. M and Mg values in igneous rocks; proposed usage and a comment on currently employed Fe_2O_3 corrections. Geochim. Cosmochim. Acta, 40: 485—486.

Hulme, G., 1974. Generation of magma at lunar impact crater sites. Nature (London), 252: 556—558.

Hurlbut, C.S. and Griggs, D.T., 1939. Igneous rocks of the Highwood Mountains, Montana, Part 1. The laccoliths. Bull. Geol. Soc. Am., 50: 1043—1112.

Hurley, P.M., Bateman, P.C., Fairbairn, H.W. and Pinson, W.H., 1965. Investigation of initial $^{87}Sr/^{86}Sr$ ratios in the Sierra Nevada plutonic province. Bull. Geol. Soc. Am., 76: 165—174.

Hutchison, W.W., 1970. Metamorphic framework and plutonic styles in the Prince Rupert region of the central Coast Mountains, British Columbia. Can. J. Earth Sci., 7: 376—405.

Hyndman, D.W., 1972. Petrology of Igneous and Metamorphic Rocks. McGraw-Hill, New York, N.Y., 533 pp.

Irvine, T.N., 1974. Petrology of the Duke Island ultramafic complex, southeastern Alaska. Mem. Geol. Soc. Am., 138: 240 pp.

Irvine, T.N. and Baragar, W.R.A., 1971. A guide to the chemical classification of the common volcanic rocks. Can. J. Earth Sci., 8: 523—548.

Irvine, T.N. and Smith, C.H., 1967. The ultramafic rocks of the Muskox Intrusion, Northwest Territories, Canada. In: P.J. Wyllie (Editor), Ultramafic and Related Rocks. Wiley, New York, N.Y., pp. 38—49.

Irving, A.J. and Green, D.H., 1976. Geochemistry and petrogenesis of the Newer Basalts of Victoria and South Australia. J. Geol. Soc. Aust., 23: 45—66.

Ishizaka, K. and Yanagi, T., 1977. K, Rb, and Sr abundances and Sr isotopic composition of the Tanzawa granitic and associated gabbroic rocks, Japan: low-potash island arc plutonic complex. Earth Planet. Sci. Lett., 33: 345—352.

Jackson, E.D., 1961. Primary textures and mineral associations in the ultramafic zone of the Stillwater Complex, Montana. Prof. Pap., U.S. Geol. Surv., 358, 106 pp.

Jacobson, R.R.E., MacLoed, W.N. and Black, R., 1958. Ring-complexes in the Younger Granite province of northern Nigeria. Mem. Geol. Soc. London, 1, 71 pp.

Jaeger, J.C., 1968. Cooling and solidification of igneous rocks. In: H.H. Hess and A. Poldervaart (Editors), Basalts, Vol. 2., Wiley, New York, N.Y., pp. 503—536.

Jahns, R.H., 1955. The study of pegmatites. Econ. Geol., 50: 1025—1130.

Jahns, R.H. and Burnham, C.W., 1969. Experimental studies of pegmatite genesis, I. A model for the derivation and crystallization of granitic pegmatites. Econ. Geol., 64: 843—864.

Jakeš, P. and White, A.J.R., 1971. Composition of island arcs and continental growth. Earth Planet. Sci. Lett., 12: 224—230.

Jakeš, P. and White, A.J.R., 1972. Major and trace element abundances in volcanic rocks of orogenic areas. Bull. Geol. Soc. Am., 83: 29—40.

Jakobsson, S.P., 1972. Chemistry and distribution pattern of Recent basaltic rocks in Iceland. Lithos, 5: 365—386.

James, D.E., Brooks, C. and Cuyubamba, A., 1976. Andean Cenozoic volcanism: Magma genesis in the light of strontium isotopic composition and trace-element geochemistry. Bull. Geol. Soc. Am., 87: 592—600.

Jamieson, B.G., 1970. Differentiation of ascending basic magma. In: G. Newall and N. Rast (Editors), Mechanism of Igneous Intrusion. Gallery Press, Liverpool, pp. 165—176.

Jamieson, B.G. and Clarke, D.B., 1970. Potassium and associated elements in tholeiitic basalts. J. Petrol., 11: 183—204.

Janse, A.J.A., 1969. Gross Brukkaros, a probable carbonatite volcano in the Nama Plateau of South West Africa. Bull. Geol. Soc. Am., 80: 573—586.

Janse, A.J.A., 1971. Monticellite bearing porphyritic peridotite from Gross Brukkaros, South West Africa. *Trans. Geol. Soc. S. Afr.*, **74**: 45—55.

Johanssen, A., 1931—1939. *A Descriptive Petrography of the Igneous Rocks*, Vols. 1—4. University of Chicago Press, Chicago, Ill.

Johnson, R.W., Mackenzie, D.E. and Smith, I.E.M., 1978. Volcanic rock associations at convergent plate boundaries: reappraisal of the concept using case histories from Papua New Guinea. *Bull. Geol. Soc. Am.*, **89**: 96—106.

Johnston, R., 1953. The olivines of the Garbh Eilean Sill, Shiant Isles. *Geol. Mag.*, **90**: 161—171.

Johnstone, G.S., 1973. *British Regional Geology: The Grampian Highlands.* Her Majesty's Stationery Office, Edinburgh, 3rd ed., with amendments, 107 pp.

Jolly, W.T. and Smith, R.E., 1972. Degradation and metamorphic differentiation of the Keweenawan tholeiitic lavas of northern Michigan, U.S.A. *J. Petrol.* **13**: 273—309.

Jones, J.G., 1970. Interglacial volcanoes of the Laugarvatn region, southwest Iceland, II. *J. Geol.*, **78**: 127—140.

Jones, O.T. and Pugh, W.J., 1948. The form and distribution of dolerite masses in the Builth—Llandrindod inlier, Radnorshire. *Q.J. Geol. Soc. London*, **104**: 71—98.

Joplin, G.A., 1968. The shoshonite association: a review. *J. Geol. Soc. Aust.*, **15**: 275—294.

Joplin, G.A., 1971. *A Petrography of Australian Igneous Rocks.* Angus and Robertson, Sydney, N.S.W., 3rd ed., 153 pp.

Karig, D.E., 1970. Ridges and basins of the Tonga—Kermadec island arc system. *J. Geophys. Res.*, **75**: 239—254.

Karig, D.E., 1971. Origin and development of marginal basins in the western Pacific. *J. Geophys. Res.*, **76**: 2542—2561.

Karner, F.R., 1968. Compositional variation in the Tunk Lake granite pluton, southeastern Maine. *Bull. Geol. Soc. Am.*, **79**: 193—222.

Katsui, Y., 1961. Petrochemistry of the Quaternary volcanic rocks of Hokkaido and surrounding areas. *J. Fac. Sci., Hokkaido Univ., Ser. IV, Geol. Mineral.*, **11**: 1—58.

Katsui, Y., 1963. Evolution and magmatic history of some Krakatoan calderas in Hokkaido, Japan. *J. Fac. Sci., Hokkaido Univ., Ser. IV, Geol. Mineral.*, **11**: 631—650.

Kay, R.W. and Gast, P.W., 1973. The rare earth content and origin of alkali-rich basalts. *J. Geol.*, **81**: 653—682.

Kay, R.W., Hubbard, N.J. and Gast, P.W., 1970. Chemical characteristics and origin of oceanic ridge volcanic rocks. *J. Geophys. Res.*, **75**: 1585—1613.

Kean, B.F. and Strong, D.F., 1975. Geochemical evolution of an Ordovician island arc of the central Newfoundland Appalachians. *Am. J. Sci.*, **275**: 97—118.

Kennedy, G.C., 1955. Some aspects of the role of water in rock melts. *Spec. Pap., Geol. Soc. Am.*, **62**: 489—503.

Kennedy, W.Q., 1933. Trends of differentiation in basaltic magmas. *Am. J. Sci.*, **25**: 239—256.

Kennedy, W.Q. and Anderson, E.M., 1938. Crustal layers and the origin of magmas. *Bull. Volcanol.*, **3**: 24—41.

Kerr, P.F., 1959. *Optical Mineralogy.* McGraw-Hill, New York, N.Y., 442 pp.

Kidd, R.G.W. and Cann, J.R., 1974. Chilling statistics indicate an ocean-floor spreading origin for the Troodos Complex, Cyprus. *Earth Planet. Sci. Lett.*, **24**: 151—155.

King, B.C., 1965. The nature and origin of migmatites: metasomatism or anatexis. In: W.S. Pitcher and G.W. Flinn (Editors), *Control of Metamorphism.* Oliver and Boyd, Edinburgh, pp. 219—234.

King, B.C., 1970. Volcanicity and rift tectonics in East Africa. In: T.N. Clifford and I.G. Gass (Editors), *African Magmatism and Tectonics.* Oliver and Boyd, Edinburgh, pp. 263—283.

King, B.C. and Chapman, G.R., 1972. Volcanism of the Kenya rift valley. *Philos. Trans. R. Soc. London, Ser. A*, **271**: 185—208.

Kirkpatrick, R.J., 1975. Crystal growth from the melt: a review. *Am. Mineral.*, **60**: 798—814.

Kistler, R.W. and Peterman, Z.E., 1973. Variations in Sr, Rb, K, Na, and initial $^{87}Sr/^{86}Sr$ in Mesozoic granitic rocks and intruded wall rocks in central California. *Bull. Geol. Soc. Am.*, **84**: 3489—3512.

Kistler, R.W., Evernden, J.F. and Shaw, H.R., 1971. Sierra Nevada plutonic cycle, Part 1. Origin of composite granite batholiths. *Bull. Geol. Soc. Am.*, **82**: 853—868.

Korringa, M.K., 1973. Linear vent area of the Soldier Meadow Tuff, an ash-flow sheet in northwestern Nevada. *Bull. Geol. Soc. Am.*, **84**: 3849—3866.

Krauskopf, K.B., 1967. *Introduction to Geochemistry*. McGraw-Hill, New York, N.Y., 721 pp.

Krauskopf, K.B., 1968. A tale of ten plutons. *Bull. Geol. Soc. Am.*, **79**: 1—18.

Krishnamurthy, P. and Cox, K.G., 1977. Picrite basalts and related lavas from the Deccan Traps of western India. *Contrib. Mineral. Petrol.*, **62**: 53—75.

Kristjansson, L. (Editor), 1974. *Geodynamics of Iceland and the North Atlantic area — Proceedings of the NATO Advanced Study Institute held in Reykjavik, Iceland, 1—7 July, 1974.* Reidel, Dordrecht, 323 pp.

Kuno, H., 1959. Origin of Cenozoic petrographic provinces of Japan and surrounding areas. *Bull. Volcanol.*, **20**: 37—76.

Kuno, H., 1960. High-alumina basalt. *J. Petrol.*, **1**: 121—145.

Kuno, H., 1966. Lateral variation of basalt magma type across continental margins and island arcs. *Bull. Volcanol.*, **29**: 195—222.

Kuno, H., 1969. Andesite in time and space. In: A.R. McBirney (Editor), *Proceedings of the Andesite Conference.* Oreg. Dep. Geol. Miner. Ind. Bull., **65**: 13—20.

Kuno, H., Ishikawa, T., Katsui, Y., Yagi, K., Yamasaki, M. and Taneda, S., 1964. Sorting of pumice and lithic fragments as a key to eruptive and emplacement mechanism. *Jpn. J. Geol. Geogr.*, **35**: 223—238.

Kushiro, I., 1960. Si—Al relation in clinopyroxenes from igneous rocks. *Am. J. Sci.*, **258**: 548—554.

Kushiro, I., 1972a. Effect of water on the composition of magmas formed at high pressures. *J. Petrol.*, **13**: 311—334.

Kushiro, I., 1972b. Determination of liquidus relations in synthetic silicate systems with electron probe analysis: the system forsterite—diopside—silica at 1 atmosphere. *Am. Mineral.*, **57**: 1260—1271.

Kushiro, I., Yoder, H.S. and Mysen, B.O., 1976. Viscosity of basaltic and andesitic liquids at high pressures. *Annu. Rep. Dir. Geophys. Lab., Carnegie Inst., Washington*, **75**: 614—618.

Lambert, I.B. and Sato, T., 1974. The Kuroko and associated ore deposits of Japan: a review of their features and metallogenesis. *Econ. Geol.*, **69**: 1215—1236.

Lanphere, M.A., Cameron, K.L. and Cameron, M., 1980. Sr isotopic geochemistry of voluminous rhyolitic ignimbrites and related rocks, Batopilas area, western Mexico. *Nature (London)*, **286**: 594—597.

Larsen, E.S., 1938. Some new variation diagrams for groups of igneous rocks. *J. Geol.*, **46**: 505—520.

Larsen, E.S., 1940. Petrographic province of central Montana. *Bull. Geol. Soc. Am.*, **51**: 887—948.

Larsen, E.S., 1948. Batholith of southern California. *Mem. Geol. Soc. Am.*, **29**, 182 pp.

Larsen, O. and Sørensen, H., 1960. Principles of classification and norm calculations of metamorphic rocks: a discussion. *J. Geol.*, **68**: 681—683.

Leake, B.L., 1978. Nomenclature of amphiboles. *Can. Mineral.*, **16**: 501—520.

Le Bas, M.J., 1962. The role of aluminum in igneous clinopyroxenes with relation to their parentage. *Am. J. Sci.*, **260**: 267—288.

Le Bas, M.J., 1977. *Carbonatite-Nephelinite Volcanism — An African Case History.* Wiley, New York, N.Y., 347 pp.

Leeman, W.P. and Vitaliano, C.J., 1976. Petrology of McKinney basalt, Snake River Plain, Idaho. *Bull. Geol. Soc. Am.*, **87**: 1777—1792.

Lehmann, E., 1974. Keratophyres, so-called keratophyres and allied rocks, especially spilites. *Neues Jahrb., Mineral. Abh.*, **122**: 268—290.

Lehmann, E., 1977. Reply to the note "Keratophyre defined", of C.J. Hughes. *Neues Jahrb Mineral., Monatsh.*, pp. 193—198.

Le Maitre, R.W., 1976. Some problems of the projection of chemical data into mineralogical classifications. *Contrib. Mineral. Petrol.*, **56**: 181—189.

Le Pichon, X., 1968. Sea-floor spreading and continental drift. *J. Geophys. Res.*, **73**: 3661—3697.

Leveson, D.L., 1973. Origin of comb layering and orbicular structure, Sierra Nevada batholith, California — Discussion. *Bull. Geol. Soc. Am.*, **84**: 4005—4006.

Lewis, J.F., 1968a. Trace elements, variation in alkalis, and the ratio $^{87}Sr/^{86}Sr$ in selected rocks from the Taupo volcanic zone. *N.Z. J. Geol. Geophys.*, **11**: 608—629.

Lewis, J.F., 1968b. Tauhara volcano, Taupo zone, Part II. Mineralogy and petrology. *N.Z. J. Geol. Geophys.*, **11**: 651—684.

Lipman, P.W., 1965. Chemical comparison of glassy and crystalline volcanic rocks. *Bull. U.S. Geol. Surv.*, **1201-D**: 1—24.

Lobjoit, W.M., 1969. The granitization of metabasites in the Wiawso—Bibiani area, western Ghana. *J. Min. Geol.*, 4: 39—52.

Loewinson-Lessing, F.Y., 1954. *A Historical Survey of Petrology.* Oliver and Boyd, Edinburgh, 112 pp. (English edition; translated from the Russian by S.I. Tomkeieff).

Lofgren, G., 1974. An experimental study of plagioclase crystal morphology: isothermal crystallization. *Am. J. Sci.*, **274**: 243—273.

Lorenz, V., 1975. Formation of phreatomagmatic maar-diatreme volcanoes and its relevance to kimberlite diatremes. *Phys. Chem. Earth*, 9: 17—27.

Lowder, G.G. and Carmichael, I.S.E., 1970. The volcanoes and caldera of Talasea, New Britain: geology and petrology. *Bull. Geol. Soc. Am.*, **81**: 17—38.

Lowell, J.D. and Guilbert, J.M., 1970. Lateral and vertical alteration—mineralization zoning in porphyry ore deposits. *Econ. Geol.*, **65**: 373—408.

Lozej, G.P. and Beales, F.W., 1975. The unmetamorphosed sedimentary fill of the Brent meteorite crater, south-eastern Ontario. *Can. J. Earth Sci.*, **12**: 606—628.

Lyons, P.C., 1976. IUGS classification of igneous rocks: a critique. *Geology*, 4: 425—426.

Lyons, P.C., 1977. IUGS classification of granitic rocks: a critique — Comment and reply. *Geology*, 5: 254—255.

Macdonald, G.A., 1949. Petrography of the island of Hawaii. *Prof. Pap., U.S. Geol. Surv.*, 214-D, 96 pp.

Macdonald, G.A., 1955. Catalogue of the active volcanoes of the world including solfatara fields, Part 3. Hawaiian Islands, Int. Assoc. Volcanol., Naples, 37 pp.

Macdonald, G.A., 1968. Composition and origin of Hawaiian lavas. *Mem. Geol. Soc. Am.*, **116**: 477—522.

Macdonald, G.A., 1972. *Volcanoes*, Prentice-Hall, Englewood Cliffs, N.J., 510 pp.

Macdonald, G.A. and Katsura, T., 1964. Chemical composition of Hawaiian lavas. *J. Petrol.*, 5: 82—133.

Macdonald, R., 1969. The petrology of alkaline dykes from the Tugtutoq area, south Greenland. *Bull. Geol. Soc. Denm.*, **19**: 257—282.

Malpas, J.G., 1977. Petrology and tectonic significance of Newfoundland ophiolites, with examples from the Bay of Islands. In: R.G. Coleman and W.P. Irwin (Editors), *North American Ophiolites.* Bull. Oreg. Dep. Geol. Miner. Ind., **95**: 13—23.

Malpas, J.G., 1978. Magma generation in the upper mantle, field evidence from ophiolite suites, and application to the generation of oceanic lithosphere. *Philos. Trans. R. Soc. London, Ser. A*, **288**: 527—546.

Malpas, J.G. and Strong, D.F., 1975. A comparison of chrome-spinels in ophiolites and mantle diapirs of Newfoundland. *Geochim. Cosmochim. Acta*, 39: 1045—1060.

Manson, V., 1967. Geochemistry of basaltic rocks: major elements. In: H.H. Hess and A. Poldervaart (Editors), *Basalts*, Vol. 1. Wiley, New York, N.Y., pp. 215—269.

Manton, W.I., 1968. The origin of associated basic and acid rocks in the Lebombo—Nuanetsi igneous province, southern Africa, as implied by strontium isotopes. *J. Petrol.*, 9: 23—39.

Marmo, V., 1967. On the granite problem. *Earth-Sci. Rev.*, 3: 7—29.

Marsh, B.D. and Carmichael, I.S.E., 1974. Benioff zone magmatism. *J. Geophys. Res.*, 79: 1196—1206.

Marshall, P., 1935. Acid rocks of the Taupo—Rotorua district. *Trans. R. Soc. N.Z.*, 64 (Part 3): 323—366.

Mason, B., 1966. *Principles of Geochemistry*. Wiley, New York, N.Y., 3rd ed., 329 pp.

Mathews, W.H., 1947. "Tuyas", flat-topped volcanoes in northern British Columbia. *Am. J. Sci.*, 245: 560—570.

Mathews, W.H., Thorarinsson, S. and Church, N.B., 1964. Gravitative settling of olivine in pillows of an Icelandic basalt. *Am. J. Sci.*, 262: 1036—1040.

Maxwell, J.A., 1968. *Rock and Mineral Analysis*. Wiley, New York, N.Y., 584 pp.

McBirney, A.R., 1969. Andesitic and rhyolitic volcanism of orogenic belts. In: P.J. Hart (Editor), *The Earth's Crust and Upper Mantle*. Monogr. Am. Geophys. Union, 13: 501—507.

McBirney, A.R., 1974. Factors governing the intensity of explosive andesitic eruptions. *Bull. Volcanol.*, 37: 443—453.

McBirney, A.R., 1975. Differentiation of the Skaergaard Intrusion. *Nature (London)*, 253: 691—694.

McBirney, A.R. and Noyes, R.M., 1979. Crystallization and layering of the Skaergaard Intrusion. *J. Petrol.*, 20: 487—554.

McCall, G.J.H. and Peers, R., 1970. Geology of the Binneringie Dyke, Western Australia. *Geol. Rundsch.*, 60: 1174—1263.

McDougall, I., 1962. Differentiation of the Tasmanian dolerites: Red Hill dolerite—granophyre association. *Bull. Geol. Soc. Am.*, 73: 279—316.

McDougall, I., 1976. Geochemistry and origin of the Columbia River Group, Oregon and Washington. *Bull. Geol. Soc. Am.*, 87: 777—792.

McGetchin, T.R. and Ullrich, G.W., 1973. Xenoliths in maars and diatremes with inferences for the Moon, Mars, and Venus. *J. Geophys. Res.*, 78: 1833—1853.

McIntyre, D.B., 1968. Impact metamorphism at Clearwater Lake, Quebec. In: B.M. French and N.M. Short (Editors), *Shock Metamorphism of Natural Materials*. Mono Book Corporation, Baltimore, Md., pp. 363—366.

McKee, E.H. and Silberman, M.L., 1970. Geochronology of Tertiary igneous rocks in central Nevada. *Bull. Geol. Soc. Am.*, 81: 2317—2328.

McKenzie, C.B. and Clarke, D.B., 1975. Petrology of the South Mountain Batholith, Nova Scotia. *Can. J. Earth Sci.*, 12: 1209—1218.

McQuillin, R., Bacon, M. and Binns, P.E., 1975. The Blackstones Tertiary igneous complex. *Scott. J. Geol.*, 11: 179—192.

Mehnert, K.R., 1968. *Migmatites*. Elsevier, Amsterdam, 405 pp.

Mével, C. and Velde, D., 1976. Clinopyroxenes in Mesozoic pillow lavas from the French Alps: influence of cooling rate on compositional trends. *Earth Planet. Sci. Lett.*, 32: 158—164.

Meyer, H.O.A., 1977. Mineralogy of the upper mantle: a review of the minerals in mantle xenoliths from kimberlite. *Earth Sci. Rev.*, 13: 251—281.

Middlemost, E.A.K., 1971. Classification and origin of the igneous rocks. *Lithos*, 4: 105—130.

Misch, P., 1968. Plagioclase compositions and non-anatectic origin of migmatitic gneisses in northern Cascade Mountains of Washington State. *Contrib. Mineral. Petrol.*, 17: 1—70.

Moberly, R., 1972. Origin of lithosphere behind island arcs, with reference to the western Pacific. *Mem. Geol. Soc. Am.*, **132**: 35—55.

Mohr, P.A., 1971. Ethiopian rift and plateaus; some volcanic petrochemical differences. *J. Geophys. Res.*, **76**: 1967—1984.

Moorbath, S. and Bell, J.D., 1965. Strontium isotope abundance studies and rubidium—strontium age determinations on Tertiary igneous rocks from the Isle of Skye, north-west Scotland. *J. Petrol.*, **6**, 37—66.

Moorbath, S., Sigurdsson, H., and Goodwin, R. 1968. K—Ar ages of the oldest exposed rocks in Iceland. *Earth Planet. Sci. Lett.*, **4**: 197—205.

Moorbath, S., O'Nions, R.K., Pankhurst, R.J., Gale, N.H. and McGregor, V.R., 1972. Further rubidium—strontium age determinations on the very early Precambrian rocks of the Godthaab district, west Greenland. *Nature (London), Phys. Sci.*, **240**: 78—82.

Moore, J.G., 1965. Petrology of deep-sea basalt near Hawaii. *Am. J. Sci.*, **263**: 40—52.

Moore, J.G., 1967. Base surge in recent volcanic eruptions. *Bull. Volcanol.*, **30**: 337—363.

Moore, J.G., 1970. Water content of basalt erupted on the ocean floor. *Contrib. Mineral. Petrol.*, **28**: 272—279.

Moore, J.G. and Evans, B.W., 1967. The role of olivine in the crystallization of the prehistoric Makaopuhi tholeiitic lava lake, Hawaii. *Contrib. Mineral. Petrol.*, **15**: 202—223.

Moore, J.G. and Lockwood, J.P., 1973. Origin of comb layering and orbicular structure, Sierra Nevada batholith, California — Reply. *Bull. Geol. Soc. Am.*, **84**: 4007—4010.

Moore, J.M., 1977. Orogenic volcanism in the Proterozoic of Canada. *Spec. Pap. Geol. Assoc. Can.*, **16**: 127—148.

Moorhouse, W.W., 1959. *The Study of Rocks in Thin Section.* Harper and Row, New York, N.Y., 514 pp.

Moorhouse, W.W., 1970. A comparative atlas of textures of Archaean and younger volcanic rocks. *Spec. Pap., Geol. Assoc. Can.*, **8** (edited by J.J. Fawcett).

Morgan, W.J., 1972. Plate motions and deep mantle convection. *Mem. Geol. Soc. Am.*, **132**: 7—22.

Morse, S.A., 1970. Alkali feldspars with water at 5 kb pressure. *J. Petrol.*, **11**: 221—251.

Muehlberger, W.R., Denison, R.E. and Lidiak, E.G., 1967. Basement rocks in continental interior of United States. *Bull. Am. Assoc. Petrol. Geol.*, **51**: 2351—2380.

Muir, I.D. and Tilley, C.E., 1961. Mugearites and their place in alkali igneous rock series. *J. Geol.*, **69**: 186—203.

Mullan, H.S. and Bussell, M.A., 1977. The basic rock series in batholithic associations. *Geol. Mag.*, **114**: 265—280.

Murata, K.J. and Richter, D.H., 1966. Chemistry of the lavas of the 1959—60 eruption of Kilauea volcano, Hawaii. *Prof. Pap. U.S. Geol. Surv.*, **537-A**, 26 pp.

Murray, C.G., 1972. Zoned ultramafic complexes of the Alaskan type: feeder pipes of andesitic volcanoes. *Mem. Geol. Soc. Am.*, **132**: 313—335.

Myers, J.S., 1975. Cauldron subsidence and fluidization: mechanisms of intrusion of the Coastal Batholith of Peru into its own volcanic ejecta. *Bull. Geol. Soc. Am.*, **86**: 1209—1220.

Mysen, B.O., 1978. Experimental determination of nickel partition coefficients between liquid, pargasite, and garnet peridotite minerals and concentration limits of behaviour according to Henry's Law at high pressure and temperature. *Am. J. Sci.*, **278**: 217—243.

Mysen, B.O. and Boettcher, A.L., 1975. Melting of a hydrous mantle, I. Phase relations of natural peridotite at high pressures and temperatures with controlled activities of water, carbon dioxide, and hydrogen. *J. Petrol.*, **16**: 520—548.

Naldrett, A.J., 1964. Ultrabasic rocks of the Porcupine district and related nickel deposits. Ph.D. Thesis, Queen's University, Kingston, Ont. (unpublished).

Naldrett, A.J. and Kullerud, G., 1967. A study of the Strathcona Mine and its bearing on the origin of the nickel—copper ores of the Sudbury district, Ontario. *J. Petrol.*, 8: 453—531.

Naldrett, A.J. and Mason, G.D., 1968. Contrasting Archean ultramafic igneous bodies in Dundonald and Clergue Townships, Ontario. *Can. J. Earth Sci.*, 5: 111—143.

Naldrett, A.J., Bray, J.G., Gasparrini, E.L., Podolsky, T. and Rucklidge, J.C., 1970. Phase layering and cryptic variation in the Sudbury nickel irruptive. *Spec. Publ., Geol. Soc. S. Afr.*, 1: 532—546.

Nash, W.P. and Wilkinson, J.F.G., 1970. Shonkin Sag Laccolith, Montana, I. Mafic minerals and estimates of temperature, pressure, oxygen fugacity and silica activity. *Contrib. Mineral. Petrol.*, 25: 241—269.

Needham, H.D., Choukroune, P., Cheminée, J.L., Le Pichon, X., Franchetau, J. and Tapponnier, P., 1976. The accreting plate boundary: Ardoukôba Rift (northeast Africa), and the oceanic rift valley. *Earth Planet. Sci. Lett.*, 28: 439—453.

Newall, G. and Rast, N. (Editors), 1970. *Mechanism of Igneous Intrusion*. Gallery Press, Liverpool, 380 pp.

Nicholls, J. and Carmichael, I.S.E., 1969. A commentary on the absarokite—shoshonite—banakite series of Wyoming, U.S.A. *Schweiz. Mineral. Petrogr. Mitt.*, 49: 47—64.

Niem, A.R., 1977. Mississippian pyroclastic flow and ash-fall deposits in the deep-marine Ouachita flysch basin, Oklahoma and Arkansas. *Bull. Geol. Soc. Am.*, 88: 49—61.

Niggli, P., 1952. The chemistry of the Keweenawan lavas. *Am. J. Sci., Bowen Volume*, 250A: 381—412.

Nisbet, E.G. and Pearce, J.A., 1977. Clinopyroxene composition in mafic lavas from different tectonic settings. *Contrib. Mineral. Petrol.*, 63: 149—160.

Nixon, P.H. (Editor), 1973. *Lesotho Kimberlites*. Lesotho National Development Corporation, Maseru, 350 pp.

Noble, D.C., 1972. Some observations on the Cenozoic volcano-tectonic evolution of the Great Basin, western United States. *Earth Planet. Sci. Lett.*, 17: 142—150.

Noble, D.C., Sargent, K.A., Mehnert, H.H., Ekren, E.B. and Byers, F.A., 1968. Silent Canyon volcanic center, Nye County, Nevada. *Mem. Geol. Soc. Am.*, 110: 65—75.

Nockolds, S.R. and Allen, R., 1953. The geochemistry of some igneous rock series, Part I. *Geochim. Cosmochim. Acta*, 4: 105—142.

Nockolds, S.R., O'B. Knox, R.W. and Chinner, G.A., 1978. *Petrology for students*. Cambridge University Press, London, 435 pp. (rev. edition of Harker, 1954).

Norton, D. and Knight, J., 1977. Transport phenomena in hydrothermal systems: cooling plutons. *Am. J. Sci.*, 277: 937—981.

Oftedahl, C., 1960. Permian rocks and structures of the Oslo region. In: O. Holtedahl (Editor), *Geology of Norway*. Nor. Geol. Unders., 208: 298—343.

O'Hara, M.J., 1965. Primary magmas and the origin of basalts. *Scott. J. Geol.*, 1: 19—40.

O'Hara, M.J., 1968. The bearing of phase equilibria studies in synthetic and natural systems on the origin and evolution of basic and ultrabasic rocks. *Earth-Sci. Rev.*, 4: 69—133.

O'Hara, M.J., 1973. Non-primary magmas and dubious mantle plume beneath Iceland. *Nature (London)*, 243: 507—508.

O'Hara, M.J., 1975. Is there an Icelandic mantle plume? *Nature (London)*, 253: 708—710.

O'Hara, M.J., 1977. Geochemical evolution during fractional crystallization of a periodically refilled magma chamber. *Nature (London)*, 266: 503—507.

Oliver, R.L., 1961. The Borrowdale volcanic and associated rocks of the Scafell area, English Lake District. *Q.J. Geol. Soc. London*, 117: 377—417.

Ollier, C., 1969. *Volcanoes*. Massachusetts Institute of Technology Press, Cambridge, Mass., 177 pp.

Osborn, E.F., 1962. Reaction series for subalkaline igneous rocks based on different oxygen pressure conditions. *Am. Mineral.*, 47: 211—226.

Osborn, E.F. and Tait, D.B., 1952. The system diopside—forsterite—anorthite. *Am. J. Sci.*, *Bowen Vol.*, **250A**: 413—433.

Oxburgh, E.R. and Turcotte, D.L., 1968. Mid-ocean ridges and geotherm distribution during mantle convection. *J. Geophys. Res.*, **73**: 2643—2661.

Page, R.W. and McDougall, I., 1972. Ages of mineralization of gold and porphyry copper deposits in the New Guinea Highlands. *Econ. Geol.*, **67**: 1034—1048.

Panayiotou, A. (Editor), 1980. *Ophiolites — Proceedings of the International Ophiolite Symposium, Cyprus, 1979*. Geological Survey Department, Nicosia, 781 pp.

Pankhurst, R.J., 1977. Open system crystal fractionation and incompatible element variation in basalts. *Nature (London)*, **268**: 36—38.

Park, C.F., 1961. A magnetite "flow" in northern Chile. *Econ. Geol.*, **56**: 431—441.

Parsons, W.H., 1968. *Structures and origin of volcanic rocks: Montana—Wyoming—Idaho*. Guideb. Summer Field Course, Wayne State Univ., Detroit, Mich., 74 pp.

Parsons, W.H., 1969. Criteria for the recognition of volcanic breccias: review. *Mem. Geol. Soc. Am.*, **115**: 263—304.

Paul, D.K., Potts, P.J., Rex, D.C. and Beckinsale, R.D., 1977. Geochemical and petrogenetic study of the Girnar igneous complex, Deccan volcanic province, India. *Bull. Geol. Soc. Am.*, **88**: 227—234.

Pauling, L., 1960. *The Nature of the Chemical Bond*. Cornell University Press, Ithaca, N.Y., 3rd ed., 644 pp.

Peacock, M.A., 1931. Classification of igneous rock series. *J. Geol.*, **39**: 54—67.

Pearce, J.A., 1975. Basalt geochemistry used to investigate past tectonic environments on Cyprus. *Tectonophysics*, **25**: 41—67.

Pearce, J.A. and Cann, J.R., 1971. Ophiolite origin investigated by discriminant analysis using Ti, Zr, and Y. *Earth Planet. Sci. Lett.*, **12**: 339—349.

Pearce, J.A. and Cann, J.R., 1973. Tectonic setting of basic volcanic rocks determined using trace element analyses. *Earth Planet. Sci. Lett.*, **19**: 290—300.

Pearce, J.A. and Flower, M.F.J., 1977. The relative importance of petrogenetic variables in magma genesis at accreting plate margins: a preliminary investigation. *J. Geol. Soc. London*, **134**: 103—127.

Peckover, R.S., Buchanan, D.J. and Ashby, D.E.T.F., 1973. Fuel—coolant interactions in submarine volcanism. *Nature (London)*, **245**: 307—308.

Peterman, Z.E. and Hedge, C.E., 1971. Related strontium isotopic and chemical variations in oceanic basalts. *Bull. Geol. Soc. Am.*, **82**: 493—500.

Petersen, J.S., 1978. Structure of the larvikite—lardalite complex, Oslo region, Norway, and its evolution. *Geol. Rundsch.*, **67**: 330—342.

Philpotts, A.R., 1967. Origin of certain iron—titanium oxide and apatite rocks. *Econ. Geol.*, **62**: 303—315.

Philpotts, A.R., 1969. In: G. Pouliot (Editor), *Guidebook to the Geology of the Monteregian Hills*. Mineral. Assoc. Can., Montreal, Que., plate 1, 169 pp.

Philpotts, A.R., 1976. Silicate liquid immiscibility: its probable extent and petrogenetic significance. *Am. J. Sci.*, **276**: 1147—1177.

Philpotts, A.R., 1977. Archean variolites — quenched immiscible liquids: — Discussion. *Can. J. Earth Sci.*, **14**: 139—144.

Pichler, H. and Schiering, W., 1977. The Thera eruption and late Minoan-IB destructions on Crete. *Nature (London)*, **267**: 819—822.

Pichler, H. and Zeil, W., 1969. Andesites of the Chilean Andes. In: A.R. McBirney (Editor), *Proceedings of the Andesite Conference*. Oreg. Dep. Geol. Miner. Ind. Bull, **65**: 165—174.

Pike, J.E.N. and Schwarzman, E.C., 1977. Classification of textures in ultramafic xenoliths. *J. Geol.*, **85**: 49—61.

Pinkerton, H. and Sparks, R.S.J., 1978. Field measurements of the rheology of lava. *Nature (London)*, **276**: 383—384.

Pirsson, L.V., 1905. Petrography and geology of the igneous rocks of the Highwood Mountains, Montana. *Bull. U.S. Geol. Surv.*, 237.

Pitcher, W.S., 1974. The Mesozoic and Cenozoic batholiths of Peru. *Pac. Geol.*, 8: 51—62.

Pitcher, W.S., 1978. The anatomy of a batholith. *J. Geol. Soc. London*, 135: 157—182.

Pitcher, W.S. and Berger, A.R., 1972. *The Geology of Donegal: A Study of Granite Emplacement and Unroofing.* Wiley, New York, N.Y., 435 pp.

Pitcher, W.S. and Read, H.H., 1959. The main Donegal Granite. *Q.J. Geol. Soc. London*, 114: 259—305.

Poldervaart, A., 1964. Chemical definition of alkali basalts and tholeiites. *Bull. Geol. Soc. Am.*, 75: 229—232.

Poldervaart, A. and Hess, H.H., 1951. Pyroxenes in the crystallization of basaltic magma. *J. Geol.*, 59: 472—489.

Poldervaart, A. and Parker, A.B., 1964. The crystallization index as a parameter of igneous differentiation in binary variation diagrams. *Am. J. Sci.*, 262: 281—289.

Powell, J.L. and Bell, K., 1970. Strontium isotopic studies of alkalic rocks: localities from Australia, Spain, and the western United States. *Contrib. Mineral. Petrol.*, 27: 1—10.

Presnall, D.C., 1969. The geometrical analysis of partial fusion. *Am. J. Sci.*, 267: 1178—1194.

Presnall, D.C. and Bateman, P.C., 1973. Fusion relations in the system $NaAlSi_3O_8$—$CaAl_2Si_2O_8$—$KAlSi_3O_8$—SiO_2—H_2O and generation of granitic magmas in the Sierra Nevada batholith. *Bull. Geol. Soc. Am.*, 84: 3181—3202.

Prider, R.T., 1960. The leucite lamproites of the Fitzroy Basin, Western Australia. *J. Geol. Soc. Aust.*, 6: 71—118.

Prinz, M., 1967. Geochemistry of basaltic rocks: trace elements. In: H.H. Hess and A. Poldervaart (Editors), *Basalts.* Wiley, New York, N.Y., pp. 271—323.

Proffett, J.M., 1977. Cenozoic geology of the Yerington district, Nevada, and implications for the nature and origin of Basin and Range faulting. *Bull. Geol. Soc. Am.*, 88: 247—266.

Propach, G., 1976. Models of filter differentiation. *Lithos*, 9: 203—209.

Puchelt, H. and Emmermann, R., 1977. REE characteristics of ocean floor basalts from the MAR 37°N (Leg 37 DSDP). *Contrib. Mineral. Petrol.*, 62: 43—52.

Pyke, D.R., Naldrett, A.J. and Eckstrand, O.R., 1973. Archaean ultramafic flows in Munro Township, Ontario. *Bull. Geol. Soc. Am.*, 84: 955—978.

Ramberg, H., 1956. Pegmatites in west Greenland. *Bull. Geol. Soc. Am.*, 67: 185—214.

Rankin, A.H. and Le Bas, M.J., 1974. Liquid immiscibility between silicate and carbonate melts in naturally occurring ijolite magma. *Nature (London)*, 250: 206—209.

Read, H.H., 1957. *The Granite Controversy.* Murby, London, 430 pp.

Read, H.H. and Watson, J., 1975. *Introduction to Geology, Volume 2, Earth History, Part I. Early Stages of Earth History*, Macmillan, London, 221 pp.

Reid, A.M., Donaldson, C.H., Dawson, J.B., Brown, R.W. and Ridley, W.I., 1975. The Igwisi Hills extrusive "kimberlites". *Phys. Chem. Earth*, 9: 199—218.

Reynolds, D.L., 1954. Fluidization as a geological process, and its bearing on the problem of intrusive granites. *Am. J. Sci.*, 252: 577—613.

Reynolds, J.H., 1960. The age of the elements in the solar system. *Sci. Am.*, 203(5): 171—182.

Rhodes, R.C., 1975. New evidence for impact origin of the Bushveld Complex, South Africa. *Geology*, 3: 549—554.

Richey, J.E., 1928. The structural relations of the Mourne granites. *Q.J. Geol. Soc. London*, 83: 653—688.

Richey, J.E., 1932. Tertiary ring-structures in Britain. *Trans. Geol. Soc. Glasgow*, 19: 42—140.

Richey, J.E., MacGregor, A.G. and Anderson, F.W., 1961. *British Regional Geology: The Tertiary Volcanic Districts.* Her Majesty's Stationery Office, Edinburgh, 3rd ed., 120 pp.

Richter, F.M., 1973. Convection and the large-scale circulation of the mantle. *J. Geophys. Res.*, **78**: 8735—8745.

Ridley, W.I., 1970. The petrology of the Las Canadas volcanoes, Tenerife, Canary Islands. *Contrib. Mineral. Petrol.*, **26**: 124—160.

Ringwood, A.E., 1955a. The principles governing trace element distribution during magmatic crystallization, Part I. The influence of electronegativity. *Geochim. Cosmochim. Acta*, **7**: 189—202.

Ringwood, A.E., 1955b. The principles governing trace-element behaviour during magmatic crystallization, Part II. The role of complex formation. *Geochim. Cosmochim. Acta*, **7**: 242—254.

Ringwood, A.E., 1974. The petrological evolution of island arc systems. *J. Geol. Soc. London*, **130**: 183—204.

Ringwood, A.E., 1975. *Composition and Petrology of the Earth's Mantle*. McGraw-Hill, New York, N.Y., 618 pp.

Ringwood, A.E., 1977. Petrogenesis in island arc systems. In: M. Talwani and W.C. Pitman III (Editors), *Island Arcs, Deep Sea Trenches, and Back-Arc Basins*. Am. Geophys. Union, Maurice Ewing Ser., **1**: 311—324.

Rittmann, A., 1962. *Volcanoes and Their Activity*. Wiley, New York, N.Y., 305 pp.

Robertson, F., 1962. Crystallization sequence of minerals leading to formation of ore deposits in quartz monzonitic rocks in the northwestern part of the Boulder Batholith, Montana. *Bull. Geol. Soc. Am.*, **73**: 1257—1276.

Robinson, G.D., Klepper, M.R. and Obradovich, J.D., 1968. Overlapping plutonism, volcanism, and tectonism in the Boulder Batholith region, western Montana. *Mem. Geol. Soc. Am.*, **116**: 557—576.

Robson, G.R. and Tomblin, J.F., 1966. *Catalogue of active volcanoes of the world including solfatara fields, Part 20, West Indies*. International Association of Volcanology, Naples, 56 pp.

Roedder, E., 1951. Low temperature liquid immiscibility in the system K_2O—FeO—Al_2O_3—SiO_2. *Am. Mineral.*, **36**: 282—286.

Roedder, E., 1978. Silicate liquid immiscibility in magmas and in the system K_2O—FeO—Al_2O_3—SiO_2: an example of serendipity. *Geochim. Cosmochim. Acta*, **42**: 1597—1617.

Roedder, E. and Weiblen, P.W., 1970. Silicate liquid immiscibility in lunar magmas, evidenced by melt inclusions in lunar rocks. *Science*, **167**: 641—644.

Roeder, P.L. and Emslie, R.F., 1970. Olivine—liquid equilibrium. *Contrib. Mineral. Petrol.*, **29**: 275—289.

Rogers, N.W., 1977. Granulite xenoliths from Lesotho kimberlites and the lower continental crust. *Nature (London)*, **270**: 681—684.

Rogers, N.W. and Gibson, I.L., 1977. The petrology and geochemistry of the Creag Dubh composite sill, Whiting Bay, Arran, Scotland. *Geol. Mag.*, **114**: 1—8.

Roobol, M.J., 1976. Post-eruptive mechanical sorting of pyroclastic material — an example from Jamaica. *Geol. Mag.*, **113**: 429—440.

Ross, C.S. and Smith, R.L., 1961. Ash flow tuffs: their origin, geological relations, and identification. *Prof. Pap., U.S. Geol. Surv.*, **366**, 81 pp.

Sabine, P.A., 1978. Progress on the nomenclature of volcanic rocks, carbonatites, melilite-rocks and lamprophyres. *Geol. Mag.*, **115**: 463—466.

Saggerson, E.P. and Williams, L.A.J., 1964. Ngurumanite from southern Kenya and its bearing on the origin of rocks in the northern Tanganyika alkaline district. *J. Petrol.*, **5**: 40—81.

Sahama, T.G., 1962. Petrology of Mt. Nyiragongo: a review. *Trans. Edinburgh Geol. Soc.*, **19**: 1—28.

Sato, H., 1977. Nickel content of basaltic magmas: identification of primary magmas and a measure of the degree of olivine fractionation. *Lithos*, **10**: 113—120.

Sato, T., 1974. Distribution and geological setting of the kuroko deposits. *Spec. Iss., Min. Geol.*, 6: 1—9.

Savolahti, A., 1956. The Ahvenisto massif in Finland. *Bull. Comm. Géol. Finl.*, 174: 83—96.

Scarfe, C.M., 1977. Viscosity of a pantellerite melt at one atmosphere. *Can. Mineral.*, 15: 185—189.

Schairer, J.F., 1950. The alkali feldspar join in the system $NaAlSiO_4$—$KAlSiO_4$—SiO_2. *J. Geol.*, 58: 512—517.

Schiener, E.J., 1970. Sedimentology and petrography of three tuff horizons in the Caradocian sequence of the Bala area, North Wales. *Geol. J.*, 7: 25—46.

Schilling, J.-G., 1973a. Afar mantle plume: rare earth evidence. *Nature (London)*, 242: 2—5.

Schilling, J.-G., 1973b. Iceland mantle plume: geochemical study of Reykjanes Ridge. *Nature (London)*, 242: 565—571.

Schilling, J.-G., 1973c. Iceland mantle plume. *Nature (London)*, 246: 141—143.

Schilling, J.-G., 1975. Azores mantle blob: rare-earth evidence. *Earth Planet. Sci. Lett.*, 25: 103—115.

Schilling, J.-G. and Noe-Nygaard, A., 1974. Faeroe—Iceland plume: rare-earth evidence. *Earth Planet. Sci. Lett.*, 24: 1—14.

Scholz, C.H., Barazangi, M. and Sbar, M.L., 1971. Late Cenozoic evolution of the Great Basin, western United States, as an ensialic interarc basin. *Bull. Geol. Soc. Am.*, 82: 2979—2990.

Schwarzer, R.R. and Rogers, J.J.W., 1974. A worldwide comparison of alkali olivine basalts and their differentiation trends. *Earth Planet. Sci. Lett.*, 23: 286—296.

Scott, R.B. and Hajash, A., 1976. Initial submarine alteration of basaltic pillow lavas: a microprobe study. *Am. J. Sci.*, 276: 480—501.

Searle, E.J., 1960. Petrochemistry of the Auckland basalts. *N.Z. J. Geol. Geophys.*, 3: 23—40.

Searle, E.J., 1964. Volcanic risk in the Auckland metropolitan district. *N.Z. J. Geol. Geophys.*, 7: 94—100.

Self, S., 1976. The recent volcanology of Terceira, Azores. *J. Geol. Soc. London*, 132: 645—666.

Self, S. and Gunn, B.M., 1976. Petrology, volume and age relations of alkaline and saturated peralkaline volcanics from Terceira, Azores. *Contrib. Mineral. Petrol.*, 54: 293—313.

Seward, D., 1974. Age of New Zealand Pleistocene substages by fission-track dating of glass shards from tephra horizons. *Earth Planet. Sci. Lett.*, 24: 242—248.

Shand, S.J., 1950. *Eruptive Rocks*. Thomas Murby, 4th ed., London, 488 pp.

Shannon, R.D. and Prewitt, C.T., 1969. Effective ionic radii in oxides and fluorides. *Acta Crystallogr.*, B25: 925—946.

Shaw, D.M., Vatin-Pérignon, N. and Muysson, J.R., 1977. Lithium in spilites. *Geochim. Cosmochim. Acta*, 41: 1601—1607.

Shaw, H.R., 1965. Comments on viscosity, crystal settling, and convection in granitic magmas. *Am. J. Sci.*, 263: 120—152.

Shaw, H.R., Wright, T.L., Peck, D.L. and Okamura, R., 1968. The viscosity of basaltic magma: an analysis of field measurements in Makaopuhi lava lake, Hawaii. *Am. J. Sci.*, 266: 225—264.

Shea, J.H., 1968. A tale of ten plutons — Discussion. *Bull. Geol. Soc. Am.*, 79: 1243.

Sigurdsson, H., 1977. Generation of Icelandic rhyolites by melting of plagiogranites in the oceanic layer. *Nature (London)*, 269: 25—27.

Sigvaldason, G.E., 1968. Structure and products of subaquatic volcanoes in Iceland. *Contrib. Mineral. Petrol.*, 18: 1—16.

Sigvaldason, G.E., Steinthorsson, S., Oskarsson, N. and Imsland, P., 1974. Compositional variation in recent Icelandic tholeiites and the Kverkfjöll hot spot. *Nature (London)*, 251: 579—582.

Sillitoe, R.H., 1972. A plate tectonic model for the origin of porphyry copper deposits. *Econ. Geol.*, 67: 184—197.

Sillitoe, R.H., 1974. Tectonic segmentation of the Andes: implications for magmatism and metallogeny. *Nature (London)*, 250: 542—545.

Silvestri, S.C., 1965. Proposal for a genetic classification of hyaloclastites. *Bull. Volcanol.*, 25: 315—321.

Simons, F.S., 1963. Composite dike of andesite and rhyolite at Klondyke, Arizona. *Bull. Geol. Soc. Am.*, 74: 1049—1056.

Smith, F.G., 1963. *Physical Geochemistry*. Addison—Wesley, Reading, Mass., 624 pp.

Smith, I.E., 1972. High-potassium intrusives from southeastern Papua. *Contrib. Miner. Petrol.*, 34: 167—176.

Smith, P.J., 1978. Age of Midway revised. *Nature (London)*, 272, 12.

Smith, R.E., 1968. Redistribution of major elements in the alteration of some basic lavas during burial metamorphism. *J. Petrol.*, 9: 191—219.

Smith, R.E. and Smith, S.E., 1976. Comments on the use of Ti, Zr, Y, Sr, K, P, and Nb in classification of basaltic magmas. *Earth Planet. Sci. Lett.*, 32: 114—120.

Smith, R.L., 1960. Ash flows. *Bull. Geol. Soc. Am.*, 71: 795—842.

Smith, R.L. and Bailey, R.A., 1968. Resurgent cauldrons. *Mem. Geol. Soc. Am.*, 116: 613—662.

Sobolev, N.V., 1972. *Deep-seated Inclusions in Kimberlites and the Problem of the Composition of the Earth's Mantle*. Aust. Nat. Univ., Canberra, A.C.T., Dep. Geol., Publ. 210, 38 pp. (English edition; translated from the Russian by D.A. Brown).

Sørenson, H., 1958. The Ilimaussaq batholith: a review and discussion. *Medd. Grønl.*, 162(3), 48 pp.

Sparks, R.S.J. and Walker, G.P.L., 1973. The ground surge deposit: a third type of pyroclastic rock. *Nature (London), Phys. Sci.*, 241: 62—64.

Sparks, R.S.J. and Walker, G.P.L., 1977. The significance of vitric-enriched air-fall ashes associated with crystal-enriched ignimbrites. *J. Volcanol. Geotherm. Res.*, 2: 329—341.

Sparks, R.S.J., Self, S. and Walker, G.P.L., 1973. Products of ignimbrite eruptions. *Geology*, 1: 115—118.

Sparks, R.S.J., Pinkerton, H. and Macdonald, R., 1977a. The transport of xenoliths in magmas. *Earth Planet. Sci. Lett.*, 35: 234—238.

Sparks, R.S.J., Sigurdsson, H. and Wilson, L., 1977b. Magma mixing: a mechanism for triggering acid explosive eruptions. *Nature (London)*, 267: 315—318.

Steven, T.A. and Lipman, P.W., 1976. Calderas of the San Juan volcanic field, southwestern Colorado. *Prof. Pap., U.S. Geol. Surv.*, 958, 35 pp.

Stewart, F.H., 1965. Tertiary igneous activity. In: G.Y. Craig (Editor), *The Geology of Scotland*. Oliver and Boyd, London, pp. 417—465.

Streckeisen, A.L., 1967. Classification and nomenclature of igneous rocks — Final report of an enquiry. *Neues Jahrb. Mineral. Abh.*, 107: 144—214.

Streckeisen, A.L. (Chairman), 1973. Plutonic rocks: classification and nomenclature recommended by the IUGS subcommission on the systematics of igneous rocks. *Geotimes*, 18(10), Oct. 1973, pp. 26—30.

Strong, D.F., 1972. The petrology of the lavas of Grande Comore. *J. Petrol.*, 13: 181—217.

Strong, D.F. and Harris, A., 1974. The petrology of Mesozoic alkaline intrusives of central Newfoundland. *Can. J. Earth Sci.*, 11: 1208—1219.

Strong, D.F. and Malpas, J.G., 1975. The sheeted dike layer of the Betts Cove Ophiolite Complex does not represent spreading: further discussion. *Can. J. Earth Sci.*, 12: 894—896.

Sukheswala, R.N. and Poldervaart, A., 1958. Deccan basalts of the Bombay area, India. *Bull. Geol. Soc. Am.*, 69: 1475—1494.

Sun, S.-S. and Nesbitt, R.W., 1978. Geochemical regularities and genetic significance of ophiolitic basalts. *Geology*, 6: 689—693.

Sutton, J. and Watson, J., 1951. The pre-Torridonian metamorphic history of the Loch Torridon and Scourie areas in the north-west Highlands, and its bearing on the chronological classification of the Lewisian. *Q.J. Geol. Soc. London*, 106: 241—307.

Sutulov, A., 1974. *Copper Porphyries*. University of Utah Printing Services, Salt Lake City, Utah, 200 pp.

Swanson, D.A., Wright, T.L. and Helz, R.T., 1975. Linear vent systems and estimated rates of magma production and eruption for the Yakima Basalt on the Columbia Plateau. *Am. J. Sci.*, 275: 877—905.

Tarling, D.H. (Editor), 1978. *Evolution of the Earth's Crust*. Academic Press, London, 443 pp.

Tatsumoto, M., Knight, R.J. and Allègre, C.J., 1973. Time differences in the formation of meteorites as determined from the ratio of lead-207 to lead-206. *Science*, 180: 1279—1283.

Taylor, F.C. and Dence, M.R., 1969. A probable meteorite origin for Mistastin Lake, Labrador. *Can. J. Earth Sci.*, 6: 39—45.

Taylor, H.P., 1967. The zoned ultramafic complexes of southeastern Alaska. In: P.J. Wyllie (Editor), *Ultramafic and Related Rocks*. Wiley, New York, N.Y., pp. 97—121.

Taylor, S.R., 1969. Trace element chemistry of andesites and associated calc-alkaline rocks. In: A.R. McBirney (Editor), *Proceedings of the Andesite Conference*. Bull. Oreg. Dep. Geol. Miner. Ind., 65: 43—64.

Taylor, S.R., 1975. *Lunar Science: A Post-Apollo View*. Pergamon, New York, N.Y., 372 pp.

Taylor, S.R. and Gorton, M.P., 1977. Geochemical application of spark source mass spectrography, III. Element sensitivity, precision and accuracy. *Geochim. Cosmochim. Acta*, 41: 1375—1380.

Tazieff, H., 1970. The Afar triangle. *Sci. Am.*, 222(2): 32—40.

Tazieff, H., Varet, J., Barberi, F. and Giglia, G., 1972. Tectonic significance of the Afar (or Danakil) Depression. *Nature (London)*, 235: 144—147.

Thompson R.N., 1975. Primary basalts and magma genesis, II. Snake River Plain, Idaho, U.S.A. *Contrib. Mineral. Petrol.*, 52: 213—232.

Thompson, R.N., Esson, J. and Dunham, A.C., 1972. Major element chemical variation in Eocene lavas of the Isle of Skye, Scotland. *J. Petrol.*, 13: 219—253.

Thompson, T.B. and Giles, D.L., 1974. Orbicular rocks of the Sandia Mountains, New Mexico. *Bull. Geol. Soc. Am.*, 85: 911—916.

Thorarinsson, S., 1968. Some problems of volcanism in Iceland. *Geol. Rundsch.*, 57: 1—20.

Thorarinsson, S., Einarsson, T. and Kjartansson, G., 1960. On the geology and geomorphology of Iceland. *Int. Geogr. Congr., Norden, 1960, Excurs. Guide E.I.1*, pp. 135—169.

Thornbury, W.D., 1965. *Regional Geomorphology of the United States*. Wiley, New York, N.Y., 609 pp.

Thornton, C.P. and Tuttle, O.F., 1960. Chemistry of igneous rocks, 1. Differentiation index. *Am. J. Sci.*, 258: 664—684.

Tilley, C.E., 1950. Some aspects of magmatic evolution. *Q.J. Geol. Soc. London*, 106: 37—61.

Tilley, C.E., 1960. Differentiation of Hawaiian basalts — Some variants in lava suites of dated Kilauean eruptions. *J. Petrol.*, 1: 47—55.

Turcotte, D.L. and Oxburgh, E.R., 1978. Intra-plate volcanism. *Philos. Trans. R. Soc. London, Ser. A*, 288: 561—579.

Turner, F.J. and Verhoogen, J., 1960. *Igneous and Metamorphic Petrology*. McGraw-Hill, New York, N.Y., 694 pp.

Tuttle, O.F. and Bowen, N.G., 1958. Origin of granite in the light of experimental studies in the system $NaAlSi_3O_8—KAlSi_3O_8—SiO_2—H_2O$. *Mem. Geol. Soc. Am.*, 74, 153 pp.

Tuttle, O.F. and Gittins, J. (Editors), 1966. *Carbonatites*. Wiley, New York, N.Y., 591 pp.

Tyrrell, G.W., 1926. *The Principles of Petrology*. Methuen, London, 349 pp.

Upton, B.G.J., 1960. The alkaline igneous complex of Kûngnât Fjeld, south Greenland. *Medd. Grønl.*, 123(4), 145 pp.

Upton, B.G.J., 1974. The alkaline province of southwest Greenland. In: H. Sørensen (Editor), *The Alkaline Rocks*. Wiley, London, pp. 221—238.

Ussing, N.V., 1912. Geology of the country around Julianehaab, Greenland. *Medd. Grønl.*, 38, 376 pp.

Vallance, T.G., 1960. Concerning spilites. *Proc. Linn. Soc. N.S.W.*, 85: 8—52.

Vallance, T.G., 1965. On the chemistry of pillow lavas and the origin of spilites. *Mineral. Mag.*, 34: 471—481.

Vallance, T.G., 1969. Spilites again: some consequences of the degradation of basalts. *Proc. Linn. Soc. N.S.W.*, 94: 8—51.

Vallance, T.G., 1974. Spilitic degradation of a tholeiitic basalt. *J. Petrol.*, 15: 79—96.

van Bemmelen, R.W., 1949. *The Geology of Indonesia*, Vol. 1A. Government Printing Office, The Hague, 732 pp.

Veevers, J.J. and Cotterill, D., 1976. Western margin of Australia: a Mesozoic analog of the East African rift system. *Geology*, 4: 713—717.

Viljoen, M.J. and Viljoen, R.P., 1969. An introduction to the geology of the Barberton granite—greenstone terrain. *Spec. Publ., Geol. Soc. S. Africa*, 2: 9—304 (plus 9 following papers and a combined bibliography).

Vincent, E.A. and Phillips, R., 1954. Iron—titanium oxide minerals in layered gabbros of the Skaergaard intrusion, East Greenland, Part I. Chemistry and ore-microscopy. *Geochim. Cosmochim Acta*, 6: 1—26.

von Eckermann, H., 1966. Progress of research on the Alnö carbonatite. In: O.F. Tuttle and J. Gittins (Editors), *Carbonatites*. Wiley, New York, N.Y., pp. 3—31.

von Engelhardt, W. and Stöffler, D., 1974. Ries meteorite crater, Germany. *Fortschr. Mineral.*, 52: 103—122 (in three consecutive parts).

von Gruenewaldt, G., 1968. The Rooiberg felsite north of Middelburg and its relation to the layered sequence of the Bushveld Complex. *Trans. Geol. Soc. S. Afr.*, 71: 153—172.

von Platen, H., 1965. Experimental anatexis and genesis of migmatites. In: W.S. Pitcher and G.W. Flinn (Editors), *Controls of Metamorphism*. Oliver and Boyd, Edinburgh, pp. 203—218.

Wachendorf, H., 1973. The rhyolitic lava flows of the Lebombos (SE-Africa). *Bull. Volcanol.*, 37: 515—529.

Wade, A. and Prider, R.T., 1940. The leucite-bearing rocks of the West Kimberley area, Western Australia. *Q.J. Geol. Soc. London*, 96: 39—97.

Wadsworth, W.J., 1961. The layered ultrabasic rocks of south-west Rhum, Inner Hebrides. *Philos. Trans. R. Soc. London, Ser. B*, 244: 21—64.

Wager, L.R., 1960. The major element variation of the layered series of the Skaergaard intrusion and a re-estimation of the average composition of the hidden layered series and of the successive residual magmas. *J. Petrol.*, 1: 364—398.

Wager, L.R. and Brown, G.M., 1968. *Layered Igneous Rocks*. Oliver and Boyd, London, 588 pp.

Wager, L.R. and Deer, W.A., 1939. Geological investigations in East Greenland, Part 3. The petrology of the Skaergaard Intrusion, Kangerdlugssuaq, east Greenland. *Medd. Grønl.*, 105(4): 1—352.

Wager, L.R. and Mitchell, R.L. 1951. The distribution of trace elements during strong fractionation of basic magmas — a further study of the Skaergaard Intrusion, east Greenland. *Geochim. Cosmochim. Acta*, 1: 129—208.

Wager, L.R., Brown, G.M. and Wadsworth, W.J., 1960. Types of igneous cumulates. *J. Petrol.*, 1: 73—85.

Wager, L.R., Vincent, E.A., Brown, G.M. and Bell, J.D., 1965. Marscoite and related rocks of the Western Red Hills Complex, Isle of Skye. *Philos. Trans. R. Soc. London, Ser. A,* **257**: 273—307.

Walker, F., 1940. The differentiation of the Palisade diabase, New Jersey. *Bull. Geol. Soc. Am.,* **51**: 1059—1106.

Walker, F. and Poldervaart, A., 1949. Karroo dolerites of the Union of South Africa. *Bull. Geol. Soc. Am.,* **60**: 591—706.

Walker, G.P.L., 1958. Geology of the Reydarfjördur area, eastern Iceland. *Q.J. Geol. Soc. London,* **114**: 367—393.

Walker, G.P.L., 1963. The Breiddalur central volcano, eastern Iceland. *Q.J. Geol. Soc. London,* **119**: 29—63.

Walker, G.P.L., 1965. Some aspects of Quaternary volcanism in Iceland. *Trans. Leicester Lit. Philos. Soc.,* **59**: 25—40.

Walker, G.P.L., 1971. Viscosity control of the composition of ocean floor volcanics. *Philos. Trans. R. Soc. London, Ser. A,* **268**: 727—729.

Walker, K.R., 1969. A mineralogical, petrological, and geochemical investigation of the Palisades Sill, New Jersey. *Mem. Geol. Soc. Am.,* **115**: 175—187.

Wasson, J.T., 1974. *Meteorites.* Springer, Berlin, 316 pp.

Watanabe, K. and Katsui, Y., 1976. Pseudo-pillow lavas in the Aso caldera, Kyushu, Japan. *J. Jpn. Assoc. Mineral. Petrol. Econ. Geol.,* **71**: 44—49.

Waters, A.C., 1961. Stratigraphic and lithologic variations in the Columbia River basalt. *Am. J. Sci.,* **259**: 583—611.

Watkinson, D.H. and Wyllie, P.J., 1971. Experimental study of the composition join $NaAlSiO_4$—$CaCO_3$—H_2O and the genesis of alkalic rock—carbonatite complexes. *J. Petrol.,* **12**: 357—378.

Watson, E.B., 1976. Glass inclusions as samples of early magmatic liquid: determinative method and application to a South Atlantic basalt. *J. Volcanol. Geotherm. Res.,* **1**: 73—84.

Weaver, S.D., Sceal, J.S.C. and Gibson, I.L., 1972. Trace-element data relevant to the origin of trachytic and pantelleritic lavas in the East African rift system. *Contrib. Mineral. Petrol.,* **36**: 181—194.

Wells, M.K., 1953. The structure and petrology of the hypersthene-gabbro intrusion, Ardnamurchan, Argyllshire. *Q.J. Geol. Soc. London,* **109**: 367—397.

White, A.J.R., 1966. Genesis of migmatites from the Palmer region of South Australia. *Chem. Geol.,* **1**: 165—200.

White, R.W., 1966. Ultramafic inclusions in basaltic rocks from Hawaii. *Contrib. Mineral. Petrol.,* **12**: 245—314.

White, W.M. and Bryan, W.B., 1977. Sr-isotope, K, Rb, Cs, Sr, Ba, and rare-earth geochemistry of basalts from the FAMOUS area. *Bull. Geol. Soc. Am.,* **88**: 571—576.

White, W.M., Schilling, J.-G. and Hart, S.R., 1976. Evidence for the Azores mantle plume from strontium isotope geochemistry of the central North Atlantic. *Nature (London),* **263**: 659—663.

Whitney, J.A., 1975. Vapor generation in a quartz monzonite magma: a synthetic model with application to porphyry copper deposits. *Econ. Geol.,* **70**: 346—358.

Whittaker, E.J.W. and Muntus, R., 1970. Ionic radii for use in geochemistry. *Geochim. Cosmochim. Acta,* **34**: 945—956.

Williams, D.A.C., 1972. Archaean ultramafic, mafic and associated rocks, Mt. Monger, Western Australia. *J. Geol. Soc. Aust.,* **19**: 163—188.

Williams, D.L. and von Herzen, R.P., 1974. Heat loss from the earth: new estimate. *Geology,* **2**: 327—328.

Williams, Hank and Malpas, J.G., 1972. Sheeted dikes and brecciated dike rocks within transported igneous complexes, Bay of Islands, western Newfoundland. *Can. J. Earth Sci.,* **9**: 1216—1229.

Williams, Howel, 1942. *The geology of Crater Lake National Park Oregon, with a reconnaissance of the Cascade Range southward to Mount Shasta.* Carnegie Inst. Washington Publ., 540, 162 pp.

Williams, Howel, Turner, F.J. and Gilbert, C.M., 1954. *Petrography.* Freeman, San Francisco, Calif., 406 pp.

Williams, L.A.J., 1969. Volcanic associations in the Gregory Rift Valley, East Africa. *Nature (London),* 224: 61—64.

Williams, L.A.J., 1972. The Kenya Rift volcanics: a note on volumes and chemical composition. *Tectonophysics,* 15: 83—96.

Willmott, W.F., 1972. Late Cainozoic shoshonitic lavas from Torres Strait, Papua and Queensland. *Nature (London), Phys. Sci.,* 235: 33—35.

Wilson, I.H., 1978. Volcanism on a Proterozoic continental margin in northwestern Queensland. *Precambrian Res.,* 7: 205—235.

Wilson, J.T., 1973. Mantle plumes and plate motions. *Tectonophysics,* 19: 149—164.

Wilson, L., 1972. Explosive volcanic eruptions, II. The atmospheric trajectories of pyroclasts. *Geophys. J.R. Astron. Soc.,* 30: 381—392.

Winchester, J.A. and Floyd, P.A., 1976. Geochemical magma type discrimination: application to altered and metamorphosed basic igneous rocks. *Earth Planet. Sci. Lett.,* 28: 459—469.

Winchester, J.A. and Floyd, P.A., 1977. Geochemical discrimination of different magma series and their differentiation products using immobile elements. *Chem. Geol.,* 20: 325—343.

Windley, B.F. (Editor), 1976. *The Early History of the Earth.* Wiley, New York, N.Y., 619 pp.

Windley, B.F., 1977. *The Evolving Continents.* Wiley, Chichester, 385 pp.

Winkler, H.E.F., 1947. Kristalgrösse und Abkühlung. *Heidelberger Mineral. Petrogr. Beitr.,* 1: 86—104.

Wise, W.S., 1969. Geology and petrology of the Mt. Hood area: a study of High Cascade volcanism. *Bull. Geol. Soc. Am.,* 80: 969—1006.

Wood, B.J., 1974. The solubility of alumina in orthopyroxene coexisting with garnet. *Contrib. Mineral. Petrol.,* 46: 1—15.

Wood, B.J. and Banno, S., 1973. Garnet—orthopyroxene and orthopyroxene—clinopyroxene relationships in simple and complex systems. *Contrib. Mineral. Petrol.,* 42: 109—124.

Wood, D.A., Gibson I.L. and Thompson, R.N., 1976. Elemental mobility during zeolite facies metamorphism of the Tertiary basalts of eastern Iceland. *Contrib. Mineral. Petrol.,* 55: 241—254.

Worst, B.G., 1958. The differentiation and structure of the Great Dyke of Southern Rhodesia. *Trans. Geol. Soc. S. Afr.,* 61: 283—354.

Wright, J.B., 1969a. A simple alkalinity ratio and its application to questions of non-orogenic granite genesis. *Geol. Mag.,* 106: 370—384.

Wright, J.B., 1969b. Olivine nodules and related inclusions in trachyte from the Jos Plateau, Nigeria. *Mineral. Mag.,* 37: 370—374.

Wright, J.V. and Walker, G.P.L., 1977. The ignimbrite source problem: significance of a co-ignimbrite lag-fall deposit. *Geology,* 5: 729—732.

Wright, T.L., 1971. Chemistry of Kilauea and Mauna Loa lava in space and time. *Prof. Pap., U.S. Geol. Surv.,* 735, 40 pp.

Wright, T.L., Grolier, M.J. and Swanson, D.A., 1973. Chemical variation related to the stratigraphy of the Columbia River basalt. *Bull. Geol. Soc. Am.,* 84: 371—386.

Wyllie, P.J. (Editor), 1967. *Ultramafic and Related Rocks.* Wiley, New York, N.Y., 464 pp.

Wyllie, P.J. and Huang, W.L., 1976. Carbonation and melting reactions in the system $CaO—MgO—SiO_2—CO_2$ at mantle pressures with geophysical and petrological applications. *Contrib. Mineral. Petrol.,* 54: 79—107.

Wyllie, P.J. and Tuttle, O.F., 1960. The system CaO—CO_2—H_2O and the origin of carbonatites. *J. Petrol.*, 1: 1—46.

Wynne-Edwards, H.R., 1976. Proterozoic ensialic orogenesis: the millipede model of ductile plate tectonics. *Am. J. Sci.*, **276**: 927—953.

Yagi, K., 1953. Recent activity of Usu Volcano, Japan, with special reference to the formation of Syowa Sinzan. *Trans. Am. Geophys. Union*, **34**: 449—456.

Yoder, H.S., 1965. Diopside—anorthite—water at five and 10 kilobars and its bearing on explosive volcanism. *Annu. Rep. Dir. Geophys. Lab., Carnegie Inst., Washington*, **64**: 82—89.

Yoder, H.S. and Tilley, C.E., 1962. Origin of basaltic magmas: an experimental study of natural and synthetic rock systems. *J. Petrol.*, **3**: 342—532.

Zartman, R.E., 1977. Geochronology of some alkalic rock provinces in eastern and central United States. *Annu. Rev. Earth Planet. Sci.*, **5**: 257—286.

Zies, E.G., 1946. Temperature measurements at Parícutin Volcano. *Trans. Am. Geophys. Union*, **27**: 178—180.

NAME INDEX

Abbott, M.J., 93
Allègre, C.J., 19, 44, 203, 461, 472, 515
Allen, R., 262
Amstutz, G.C., 475
Andersen, O., 182
Anderson, A.T., 210
Anderson, D.L., 317, 318, 446, 447
Anderson, E.M., 57, 58, 69, 72
Anderson, F.W., 511
Andrews, A.J., 68
Appleton, J.D., 323, 327
Arculus, R.J., 492, 494
Arndt, N.T., 431, 464
Atkins, F.B., 188

Baadsgaard, H., 495
Badham, J.P.N., 207
Bacon, M., 507
Bailey, D.K., 204, 264, 364
Bailey, E.B., 46, 70, 71, 247, 255
Bailey, R.A., 41, 47, 248, 496
Baker, B.H., 330, 331, 332, 334, 335
Baker, I., 301, 303, 304
Baker, P.E., 39, 307, 308, 311, 371, 372,
     373, 480
Baldwin, R.B., 51, 417
Balk, R., 84
Banno, S., 317
Baragar, W.R.A., 97, 260, 385
Barazangi, M., 513
Barberi, F., 267, 341, 486, 515
Barker, P.F., 371, 372
Barnett, H.F., 496
Barr, S.M., 295
Barrière, M., 225
Barth, T.F.W., 96, 101, 356
Bateman, P.C., 118, 218, 219, 503
Battey, M.H., 478
Beales, F.W., 420
Beasley, P., 500
Beckinsale, 510
Beeson, M.H., 310
Behrendt, J.C., 64
Bell, J.D., 138, 214, 270, 323, 325, 366,

471, 495, 517
Bell, K., 324
Berger, A.R., 133
Bhattacharji, S., 60
Bickle, M.J., 429
Billings, M.P., 73
Binns, P.E., 507
Bizouard, H., 490
Björnsson, A., 30
Black, L.P., 423
Black, R., 73, 360, 503
Blander, M., 164
Boettcher, A.L., 318
Bonatti, E., 295
Bond, A., 48
Borley, G.D., 18, 323, 327, 437
Bostock, H.H., 420
Bottinga, Y., 152, 153
Bowden, P., 359, 360
Bowen, N.L., 74, 93, 139, 146, 150, 154,
     163, 167, 169, 178, 192, 193, 204, 205,
     211, 224, 231, 252, 301, 319, 445
Bowes, D.R., 54
Boyd, F.R., 151, 317
Bradley, J., 61
Bray, J.G., 509
Bridgwater, D., 423
Briggs, N.D., 43, 393
Brock, P.W.G., 68, 279
Brögger, W.C., 223, 319, 356
Brongniart, A., 64
Brooks, C., 355, 431, 503
Brooks, C.K., 272, 298, 346, 352, 354
Brown, G.M., 63, 85, 132, 133, 187, 188,
     206, 214, 231, 232, 234, 238, 239, 240,
     366, 384, 400, 402, 403, 417, 516, 517
Brown, P.E., 77, 492
Brown, R.W., 511
Brückner, W.D., 38
Bryan, W.B., 160, 276, 501
Bucher, W.H., 421
Buddington, A.F., 76, 148
Buist, 209
Bullard, F.M., 29, 147

LOCALITY INDEX

GENERAL INDEX

Aa lava, 31, 147
Absarokite, 377, 386
Accessory mineral, 100, 136
Accidental inclusions, 133, 215
   in kimberlite, 315
Accumulative rocks, 109
   in feldsparphyric mafic rocks, 270
   in Hawaii tholeiite series, 290
   in Kartala alkali basalt series, 267
   in "porphyritic central type" of Mull, 255
   in Thingmuli tholeiite series, 246
Accuracy of analysis, 92
Achondrite, achondritic meteorite, 415
Acicular crystal, 134
Acid igneous rocks, 100
Acmite, 99
   see also aegirine
Actinides, 5
Adamellite, 116, 117, 218
Adcumulate, 235
Adiabatic gradient in mantle, 454
Adularia, 477
Aegirine, 22
   in peralkaline rocks, 102
Aegirine-augite, 21
Aenigmatite (syn. cossyrite), 22
   in Mayor Island pantellerite, 395
AFM variation diagram, 264, 266, 278, 374, 384, 478
Age relations of intrusive rocks, 86, 126
Agglomerate, 54
Agmatite, 132
Agpaitic index, 119
Air-fall material, 54
Åkermanite, 20
Alaskite, 119
Albite, 23
   in complex pegmatite, 223
   in perthite, 26, 145—146
   in spilite, 474, 476
Albite rims, 144, 196
Alkali basalt,
   definition, 106—107

classification, 108
occurrence, 295—296
evolution of name, 254—256
petrographic contrast to tholeiite, 297—298
compositional distinction from tholeiite, 256, 297, 299
in basalt tetrahedron, 186
petrogenesis, 463
Alkali basalt series, 255—259, 295—307
   in Afro-Arabian dome, 337—341
   in Auckland, 394
   in Basin and Range province, 399
   in East African Rift Valley, 331—333
   in Gardar province, 367
   in Grande Comore, 305
   in Grenada, 400
   in Hawaii, 299—300
   in Hebrides, 366
   in Oslo region, 356
   in St. Helena, 303, 304
   in SBZ terrain, 377, 384, 385
   in Terceira, 301—302
Alkali content of mafic rocks, 106
Alkali feldspar, 23, 103
   see also Albite, Anorthoclase, Micro-
         cline, Perthite, Sanidine
Alkali feldspar granite, alkali feldspar syenite, 117
Alkali-lime index, 260, 478
Alkali olivine basalt, 109, 255
Alkali rocks (definition of Holmes, 1920), 251
Alkali transfer, 203, 301, 303
Alkalic series, 260
Alkalic-calcic series, 260, 383
Alkaline basic rocks, 256
Alkalinity ratio (of J.B. Wright, 1969), 260, 478
Allanite, 20
   in Shaler pluton, 228
Allotriomorphic texture, 136
Alluvial facies, 38
Almandine garnet, 20